ISBN 978-1-332-58645-5
PIBN 10002996

1 MONTH OF
FREE
READING

at

www.ForgottenBooks.com

By purchasing this book you are eligible for one month membership to ForgottenBooks.com, giving you unlimited access to our entire collection of over 1,000,000 titles via our web site and mobile apps.

To claim your free month visit: www.forgottenbooks.com/free2996

The Navajo Church, New Mexico; wind sculpture of cross-bedded sandstone

AN INTRODUCTION TO GEOLOGY

SECOND EDITION REVISED THROUGHOUT

BY

WILLIAM B. SCOTT, Ph.D., LL.D.

BLAIR PROFESSOR OF GEOLOGY AND PALÆONTOLOGY
IN PRINCETON UNIVERSITY

" There rolls the deep where grew the tree.
O earth what changes hast thou seen !
There where the long street roars, hath been
The stillness of the central sea.

" The hills are shadows, and they flow
From form to form and nothing stands;
They melt like mists, the solid lands,
Like clouds they shape themselves and go."

TENNYSON.

With numerous illustrations from drawings by BRUCE HORSFALL
and from many new photographs.

New York
THE MACMILLAN COMPANY
1911

TO

A. A. P. S.

𝔗𝔥𝔦𝔰 𝔅𝔬𝔬𝔨 𝔦𝔰 𝔇𝔢𝔡𝔦𝔠𝔞𝔱𝔢𝔡

IN GRATEFUL RECOGNITION OF AN EVER READY

AND INSPIRING SYMPATHY

FROM THE PREFACE TO THE FIRST EDITION

This book had its origin in the attempt to write an introductory work, dealing principally with American Geology, upon the lines of Sir Archibald Geikie's excellent little " Class-Book." In spite of vigorous efforts at compression, it has expanded to its present size, though the difference from the " Class-Book," in this respect, lies not so much in the quantity of matter as in the larger size of the type and illustrations.

The book is intended to serve as an introduction to the science of Geology, both for students who desire to pursue the subject exhaustively, and also for the much larger class of those who wish merely to obtain an outline of the methods and principal results of the science. To the future specialist it will be of advantage to go over the whole ground in an elementary course, so that he may appreciate the relative significance of the various parts, and their bearing upon one another. This accomplished, he may pursue his chosen branch much more intelligently than if he were to confine his attention exclusively to that branch from the beginning of his studies.

Students, and only too often their instructors, are apt to prefer a text-book upon which they can lean with implicit confidence, and which never leaves them in doubt upon any subject, but is always ready to pronounce a definite and final opinion. They dislike being called upon to weigh evidence and balance probabilities, and to suspend judgment when the testimony is insufficient to justify a decision. This is a habit of mind which should be discouraged ; for it deludes the learner into the belief that he knows the subject when he has only acquired some one's opinions and dogmas, and renders further progress exceedingly difficult to him. In no science are there more open questions than in Geology, in none are changes of view more frequent, and in none, consequently, is it more important to emphasize the distinction between fact and inference, between observation and hypothesis. An open-minded hospitality for new facts is essential to intellectual advance.

*   *   *   *   *.   *   ..

In preparing this book, I have of course availed myself of material wherever it was to be found, but I wish to acknowledge my special obligations to the text-books of Dana, Le Conte, Geikie, Green, Prestwich, Credner, Kayser, Neumayr, Koken, de Lapparent, and Jukes-Brown. From the last-named writer is taken the arrangement of the Dynamical Agencies, which experience in the class-room has led me to consider as the best.

PRINCETON, N.J.,
   Jan. 15, 1897.

PREFACE TO THE SECOND EDITION

THE ten years that have passed since the first publication of this book have been years fruitful of results in geological knowledge. Some departments of the subject have been fairly revolutionized and in all there has been great progress, so that any text-book ten years old is necessarily left far behind in the general advance. Revision, indeed rewriting, had become imperative to incorporate the most important and significant parts of the newer results, as well as to remove as many of the defects as I might be able to do. The increase in size is an extremely regrettable feature, but I have not seen my way to avoid it, for it is largely due to the much greater number of illustrations, and these were needed in the interests of clearness.

While many minor changes have been made, the general plan of the book remains the same, for experience has convinced me that the somewhat rigidly conventional arrangement of topics, which has sufficiently evident drawbacks, is of actual assistance to the beginner. It avoids confusing him by any premature attempt to point out the infinitely ramifying relations of every fact of nature. One of the keenest pleasures of intellectual growth is the continual discovery of these unsuspected relations, but for the beginner the simpler and more obvious line of reasoning is the more profitable.

The labour of revision has been greatly lightened by those admirable store-houses of geological learning: the second edition of Kayser's "Lehrbuch," the fourth edition of Geikie's "Text-Book," and the "Geology" of Chamberlin and Salisbury. To all of these my obligations are great, especially for the bibliographies which they contain and which have rendered the collection of the newer technical literature of monographs and papers a much less onerous task than it could otherwise have been.

In this new edition I have introduced a very considerable number of brief quotations, at the request of some of those who have employed the book as a convenient work of reference and who desire to know the authority upon which the more novel or less familiar statements have been made.

It gives me great pleasure to express my thanks to the many friends who have assisted me in my undertaking. To Mr. C. W. Hayes and Mr. Bailey Willis, of the United States Geological Survey, I am under particular obligations for the kindness which enabled me to profit by the magnificent collection of photographs which the Survey has gathered. It so happened that, but for this kindness, obstacles of a temporary nature would have prevented my enjoyment of this privilege. Mr. Willis was also kind enough to give highly valued assistance and counsel in many other directions.

Professor W. H. Hobbs sent me proofs and manuscript of unpublished books and papers on seismological subjects, a service which it is difficult to describe adequately. Professor Bumpus, director of the American Museum of Natural History, New York, Professor Osborn, Professor R. B. Young of Johannesburg, and Professor R. W. Brock of Kingston, Ontario, have all been most liberal in supplying me with photographs and other means of illustration. My colleagues in the Geological Department of Princeton University have rendered assistance that was literally invaluable; Professor C. H. Smyth, Jr., has read the proofs and has made very many useful and timely suggestions, and Dr. W. J. Sinclair took many photographs especially for the book and has given me the benefit of his experience in using it. Greatest of all are my obligations to Mr. Gilbert van Ingen, to whom the book owes much of whatever good it may possess; he made a large number of the photographs, prepared the maps, selected the invertebrate fossils for the plates, and supervised the admirable drawings upon which Mr. Horsfall has expended such pains and skill, and gave much useful assistance in the stratigraphical part.

As to the figures in the plates, a word of explanation is required. A few only are original; the great majority are taken from monographs by well-known writers, but almost all have been so modified by restoration or otherwise that it did not seem proper to put the responsibility upon the original authority.

During the past ten years I have received many letters containing criticisms of the book and suggestions for its improvement in one or other particular. So far as lay in my power, I have endeavoured to profit by these criticisms and suggestions, and I wish to thank those who have taken the trouble to write them for my benefit.

Finally, I venture to express the hope that the new edition may find a place of usefulness in a crowded field, notwithstanding the defects of which I am very well aware but have not been able to remedy in the time at my disposal.

PRINCETON, N.J.,
  Oct. 10, 1907.

CONTENTS

xiii

CHAPTER V

PAGE

CHAPTER VI

CHAPTER VII

CHAPTER VIII

CHAPTER IX

PART II

STRUCTURAL OR TECTONIC GEOLOGY

CHAPTER X

CHAPTER XI

CHAPTER XII

CHAPTER XIII

CHAPTER XIV

CHAPTER XV

CHAPTER XVI

CHAPTER XVII

PART III

GEOMORPHOLOGY

CHAPTER XVIII

CHAPTER XIX

CHAPTER XX

CHAPTER XXI

CHAPTER XXII

CHAPTER XXIII

PART IV

HISTORICAL GEOLOGY

CHAPTER XXIV

CHAPTER XXV

CONTENTS

LIST OF ILLUSTRATIONS

xvii

CONTENTS

LIST OF ILLUSTRATIONS

xvii

PLATES

INTRODUCTION

Geology is the study of the structure, history, and development of the earth and its inhabitants, as revealed in the rocks.

From this definition it is apparent that the central problem in geology is the deciphering of the earth's history, and that the historical standpoint is dominant throughout. For this purpose it is necessary to apply the results and principles of all those sciences which can aid us in interpreting the record contained in the rocks. Astronomy, physics, chemistry, mineralogy, physical geography, botany, and zoölogy are all needed in the task, and geology as a true science did not become possible until the other sciences were sufficiently advanced to afford a solid foundation for it.

The history of the earth involves vast periods of time, to be measured only in millions of years,—no one can say how many,—so that all our familiar conceptions of " ancient " and " modern," derived from the history of our own race, must be greatly changed before they can be applied to geological time. In reaching its present condition, the earth has passed through many stages of change in its geographical, climatic, and biological relations, most of these stages leaving behind them records which are preserved in the successive layers of rock.

In order to read the record contained in the rocks, it is first of all necessary to learn the language in which it is written. This can be done only through an intimate acquaintance with all the methods in which rocks are made, and with the changes which the rocks undergo. This, in turn, implies a knowledge of all those processes which are now at work in modifying and changing the globe internally and on the surface. Just because our knowledge of these methods and processes is often incomplete and vague, do we so frequently find the geological record ambiguous, open to several interpretations, or even quite unintelligible. Again, many changes go on under conditions which render direct observation impossible, either because they are confined to the deep interior of the earth and are thus beyond our reach, or because their operation is so slow that a lifetime is all too short for their detection. In such cases we must deduce the invisible cause from the visible effect, but it is often extremely difficult, or even impracticable, from many possible causes, to select the real and rightful one. Hence come the wide differences of opinion which the interpretation so often calls forth.

As a living and growing science, geology is subject to continual change, a change which is by no means a simple advance from one point to another, but an unending revision of opinions, a perpetual tearing down and rebuilding.

To many intelligent people this continual modification of scientific opinion, which is a necessary consequence of advancing knowledge, is a source of annoyance. This attitude of mind comes from a failure to discriminate between fact on the one hand, and inference, or hypothesis, on the other. Accurately observed facts may be added to, but they remain trustworthy; the changeable element is the inference which is drawn from the facts. These inferences are of very different degrees of certainty. Some such deductions which were made centuries ago remain unshaken to-day, while others of far more recent date have proved illusory. Thus, when we find a rock composed of cemented sand-grains, arranged in regular beds or layers, and full of marine shells, we infer that it was formed under the sea, and further that the land where the rock is now found was once covered by the sea. Such inferences are practically certain, because they explain all the known facts and are in conflict with none. On the other hand, the hypotheses of Cuvier and others as to the character of the earth's development, and the manner in which the successive assemblages of animals and plants were called into being, have been long abandoned.

In the process of reasoning from the known to the unknown,

the inferences become the more uncertain, the farther we recede from demonstrable facts. Hypotheses are assumptions which we make to explain and coördinate large numbers of facts, and so long as their true nature is understood, they are useful, indeed indispensable, means of reaching the truth. The objection is that they are too often taught as though they were established beyond dispute. A true hypothesis will prove to be in harmony with newly discovered facts, which will take their place under it simply and naturally. A false hypothesis, on the other hand, may be in accordance with all the facts known at the time when it was proposed, but the progress of discovery will bring to light facts which are inconsistent with the hypothesis, until it is plainly seen to be inadequate and misleading. Yet even a false hypothesis may serve a useful purpose, for it puts before us a definite problem, instead of a mere catalogue of uncorrelated observations. The pathway of every science is strewn with wrecks of hypotheses which have been used, worn out, and thrown aside. In all our thinking and reasoning the distinction between hypothesis and fact must be steadily held in view.

Geology is a unit and, though for the purpose of orderly treatment, it is necessary to divide the subject into various provinces, it should be clearly understood that these provinces are rather the various aspects and phases of the same science than actual divisions. Every part of the subject is so intimately related to every other part, that any possible arrangement involves the more or less violent separation of things that belong together and requires much anticipation and repetition. The past is meaningless unless we understand the present, and a full understanding of the present can only be gained through a knowledge of the past, yet it is obvions that we cannot deal with both past and present simultaneously. Although it is an undoubted evil, some classification is necessary, if we would avoid losing ourselves in bewildering labyrinths of detail.

The departments into which geology is usually divided are as follows: —

1. **Dynamical Geology**, or the study of the forces which are now at work in modifying the surface of the earth, and of the chemical and mechanical changes which they effect. This is the key by which we may interpret past changes.

2. **Structural Geology**, or the study of the materials of which the earth is composed and of the manner in which they are arranged; together with such explanations of the modes in which the arrangement was produced as may be inferred from the structure.

3. **Geomorphology** (also called Physiographical Geology, or Physiography) is an examination of the topographical features of the earth and of the manner in which they were produced. Primarily, this subject is a province of physical geography, but it is a valuable adjunct to geology.

The three foregoing divisions together constitute a larger division, which is called *Physical Geology*, and which is contrasted with —

4. **Historical Geology.** — This is the study of the earth's history, the changes of level between land and sea, of topography, of climate, and of the successive groups of animals and plants which have lived upon the globe. The historical is the dominant standpoint in geology, the main problem of which is to interpret the records of the earth's history. The other departments are the means to this great end.

While the geologist needs the help of almost all the other physical and natural sciences, he has his peculiar province in the *rocks* which make up the accessible crust of the earth. These rocks are aggregates of a comparatively few common minerals, called, for that reason, the *rock-forming* minerals. A study of the processes now going on shows that rocks are formed in various ways and, in accordance with these modes of formation, they may be grouped in three great classes: I. *Igneous Rocks*, or those which have solidified by cooling from a state of fusion and are therefore not divided into layers or beds, are either glassy or crystalline, and are composed of complex minerals. The igneous rocks have forced their way upward from the earth's interior, thus penetrating the overlying rocks in various ways. A familiar example of this group is lava. II. *Sedimentary*, or *Stratified Rocks*, those which were accumulated under water or on land in a series of successive beds, or *strata*, from material derived from the disintegration of older rocks, and are generally fragmentary, or non-crystalline, and composed of simpler minerals than those which make up the igneous rocks. Speaking broadly, the beds of the sedimentary rocks were originally laid down in a horizontal position, and hence when they are found to be tilted, inclined, or folded, it follows that they have been disturbed from their original attitudes. III. *Metamorphic Rocks*, or those igneous or sedimentary rocks which have been more or less profoundly reconstructed in place, often with the generation of entirely new minerals.

In the accessible part cf the earth's crust, rocks of all kinds (other than loose materials, such as sand) are divided into pieces, by vertical and horizontal partings, which are called *joints* (see p. 369). In addition, the rocks are divided into still larger masses, or *blocks*, by a profounder system of fissures, and planes of dislocation, or *faults* (see p. 353). The blocks are of all sizes, up to thousands of square miles and down to areas of a few square feet, and thus the surface of the earth has been well compared to a vast mosaic of rock-pieces.

The crust of our planet is called the *lithosphere*, a shell of rocks of unknown thickness. Within the lithosphere is the great mass of the earth, or *centrosphere*, concerning which we know only that it is highly heated, of great density, and under enormous pressure. The surface of the globe is very irregular and covered with elevations and depressions. The deeper depressions are filled with water and constitute the ocean basins which in area bear to the land the proportion of 2.54 : 1, and this incomplete envelope of water is the *hydrosphere*. If the surface of the earth were smooth, the ocean would cover it entirely to a depth of nearly two miles. Finally, the *atmosphere* is a gaseous envelope, which encloses the earth completely.

CHAPTER A

THE ROCK-FORMING MINERALS

Of the simple undecomposable substances which chemists call elements, and of which rather more than seventy have been discovered, only sixteen enter at all largely into the composition of the earth's crust, so far as this is accessible to observation. It is estimated that 98 % of the crust is made up of the following eight elements, arranged in the order of abundance, with the percentages as calculated by F. W. Clarke.

Oxygen	47.07	Calcium	3.44
Silicon	28.06	Magnesium	2.40
Aluminium	7.90	Sodium	2.43
Iron	4.43	Potassium	2.45

The remaining eight elements, titanium, carbon, sulphur, hydrogen, chlorine, phosphorus, manganese, and barium, are far less abundant, but still of considerable importance.

Only two of these elements, carbon and sulphur, are found in a more or less impure state as minerals or rock masses; the others occur as compounds, formed by the union of two or more of them.

A *mineral* is a natural, inorganic substance, which has a homogeneous structure, definite chemical composition and physical properties, and usually a definite crystal form.

Crystals are solids of more or less regular and symmetrical shape, bounded, usually, by plane surfaces. The number of known crystal forms is very great, and yet they may be all grouped in six systems, which are characterized by the relations of their axes. The *axes* of a crystal are imaginary lines, which connect the centres of opposite faces, or opposite edges, or opposite solid angles, and which intersect one another at a point in the interior of the crystal.

The Systems of Crystal Forms have received many names, the following being those which are most generally used in this country: —

I. **Isometric System** (monometric, cubical, regular). — In this system the three axes are of equal length and intersect one another at right angles.

II. **Tetragonal System** (dimetric, pyramidal). — The axes intersect at right angles, but while the lateral axes are of equal length, the vertical axis is longer or shorter than the laterals.

III. **Hexagonal System.** — Here four axes are employed, three equal lateral axes intersecting at angles of 60°, and a vertical axis, which is perpendicular to and longer or shorter than the laterals.

IV. **Orthorhombic System** (rhombic, trimetric). — The three axes intersect at right angles and are all of different lengths.

V. **Monoclinic System** (monosymmetric, oblique). — All three axes are of different lengths; the two laterals are at right angles to each other, while the third is oblique to one of the former.

VI. **Triclinic System** (anorthic, asymmetric). — Three axes of unequal lengths and oblique to one another.

It is important to bear in mind the relations which the forms sustain toward one another. For example, a regular octahedron may be derived from a cube by evenly paring off the eight solid angles, until the planes thus produced intersect one another, the centres of the faces of the cube becoming the apices of the solid angles of the octahedron. Conversely, a cube may be formed from an octahedron by symmetrically truncating the angles, until the planes thus formed intersect. By slicing away the twelve edges of a cube or an octahedron a dodecahedron will result. These crystal forms are, therefore, so related as to be all derivable one from another, and the relations of their axes remain unchanged; all three forms may be assumed by the same mineral, and they thus properly belong in the same system. Similar relations may be observed between the crystal forms of the other systems.

It might be supposed that the crystal systems and the relations of their imaginary axes were merely mathematical devices to reach a convenient classification of forms. Such a conclusion would, however, be a very erroneous one. Crystalline form is an expression of molecular structure, and the physical properties of minerals are closely related to their mathematical figure. It is clear that these physical properties are not inherent in the molecules of the mineral, but are conditioned by the way in which the molecules are built up into the crystal. Amorphous substances refract light equally in all directions, and are thus called *isotropic;* but when an amorphous substance crystallizes, it assumes the qualities proper to its crystal form. Thus water is isotropic, while the hexagonal crystals of ice are singly refractive in only one direction, doubly refractive in all others. The same substance may, under different circumstances, crystallize in different systems, and will then display the properties appropriate to each system.

Not only the refractive powers of a crystal, but also its mode of expansion when heated, and its conductivity of electricity and heat are controlled by the molecular structure which determines its shape.

The crystals of the isometric system, which have their three axes of equal length, are singly refractive in all directions, expand equally when heated, and conduct heat and electricity equally in all directions. Those of the tetragonal and hexagonal systems, which have one axis longer or shorter than the others, are doubly refractive along the lateral axes, expand equally when heated, and show equal conductivity along these axes. Along the principal axis they are singly refractive, expand to a different degree when heated, and display a different conductivity along this axis than along the others. In the orthorhombic, monoclinic, and triclinic systems, which have all the axes of unequal lengths, the crystals are singly refractive in two directions; they expand unequally and .conduct differently along all their axes.

The optical properties of minerals are of great value in the study of rocks, and by the aid of the polarizing microscope very minute crystals may be identified.

Cleavage (see p. 11) is still another physical property, the dependence of which upon crystal form is very clear.

Most inorganic substances which are solid under any circumstances are capable of assuming a crystal form, so that solidification and crystallization are usually identical. For the formation of large and regular crystals, it is necessary that the process be gradual and that space be given for the individual crystals to grow. Usually crystallization begins at many points simultaneously, and the crystals crowd upon one another, resulting in a mass of more or less irregular crystalline grains. The same substance which, when very rapidly solidified, forms an amorphous glass, will give rise to distinct crystals, if slowly solidified.

Crystallization requires that the molecules be free to move upon each other, and thus to arrange themselves in a definite fashion. It may take place either by the deposition of a solid from solution, by cooling from a state of fusion, or by solidification from the condition of vapour. In all cases the size and regularity of the crystals depend upon the time and space allowed for their growth. In a manner not yet understood, amorphous solids may be converted into crystalline aggregates. This has been observed in the case of certain glassy volcanic rocks, which, though amorphous when first solidified, have gradually become crystalline, without losing their solidity, and a similar change has been observed in certain artificial glasses. This process is called *devitrification.*

The actual steps of crystallization may be observed by slowly evaporating a solution of some crystalline salt under the microscope. The first visible step in the process is the appearance of innumerable dark points in the fluid, which rapidly grow, until their spherical shape is made apparent. The globules then begin to move about rapidly and arrange themselves in straight lines, like strings of beads, and next suddenly coalesce into straight rods. The rods arrange themselves into layers, and thus build up the crystals so rapidly, that it is hardly possible to follow the steps of change. In certain glassy rocks, which solidified too quickly to allow crystallization to take place, the incipient stages of crystals, in the form of globules and hair-like rods, may be detected with the microscope.

Forms and Combinations. — A *form* is an assemblage of faces, all of which have similar relations to the axes. Two or more forms occurring as a single crystal constitute a *combination.* Only forms belonging to the same system can occur in combination, but, even with this limitation, the variety and complexity of crystals are very great. Certain forms occur which may be regarded as developed from other forms by the suppression of one-half or three-quarters of the faces of the latter.

Irregularities of growth (distortion) are very common, some faces of a form being larger than others, while certain faces may even be obliterated; but however great the variation, the angle at which corresponding faces meet invariably remains constant for each mineral.

Massive and imperfectly crystallized minerals may consist of grains, fibres, or thin layers (*laminæ*).

Hardness. — The hardness of minerals is a useful means of identifying them. For this purpose they are referred to a scale of hardness, ranging from such soft substances as may be readily scratched with the finger-nail, to the hardest known substance, diamond. The degree of hardness is expressed by the numerical place of the mineral in the scale, and intermediate grades are indicated by fractions. Thus a mineral which is scratched by quartz, and scratches orthoclase with equal ease, has a hardness of 6.5. The scale is as follows: —

1. Talc.	6. Orthoclase.
2. Gypsum.	7. Quartz.
3. Calcite.	8. Topaz.
4. Fluorite.	9. Sapphire.
5. Apatite.	10. Diamond.

Cleavage. —Many minerals split more or less readily in certain fixed directions, while in other directions they break irregularly. This property is called *cleavage.* Cleavage is uniform in different crystals of the same mineral, and is parallel to actual or possible crystal faces.

Pseudomorphs occur when one mineral assumes the crystal form proper to another. This may take place either by the addition or the removal of certain constituents, or some constituents may be removed and others substituted for them. The entire substance of a mineral may be removed and its place taken, molecule by molecule, by another, retaining the form, sometimes even the cleavage, of the first. The study of pseudomorphs is often of the greatest service, as throwing light upon the history of the rock in which they occur.

Compound crystals are formed by the joining of simple crystals. When two half-crystals are united along a plane in such a way that their faces and axes do not correspond, they are said to be twinned. When the twinning is repeated along numerous parallel planes, the crystal is a *polysynthetic twin.* Two crystals united at the ends to form a right angle are called *geniculate*, while two geniculate crystals may be so combined as to form a cross, and then are said to be *cruciform.*

Rock-forming Minerals. —The number of known minerals is large and constantly increasing, but only a few enter in any important way into the constitution of the earth's crust. We now proceed to a consideration of these constituent minerals, which are called rock-forming minerals, because the rocks are aggregations of them. It must be emphasized that the student can gain no real knowledge of minerals or rocks by merely reading

about them; it is necessary that he should familiarize himself with actual specimens.

A. MINERALS COMPOSED OF SILICA

Next to oxygen, silicon is by far the most abundant constituent of the earth's crust, though never occurring alone. It is united with oxygen to form silica (SiO_2) or enters into the formation of more complex compounds.

1. **Quartz** (SiO_2) is anhydrous silica in a crystalline state and is one of the most abundant of minerals. It belongs in the hexagonal system, and crystallizes in hexagonal prisms capped by six-sided pyramids, or in double six-sided pyramids, or in modifications of these forms. It is insoluble in acids, except hydrofluoric, and only very slowly soluble in boiling caustic alkalies.

Quartz has no cleavage and is very hard ($H = 7$), scratching glass readily, while it cannot be scratched with a knife; the specific gravity (sp. gr.) is 2.6.

When pure and symmetrically crystallized, quartz is transparent, colourless, and lustrous (rock crystal), but it more commonly is found in dull masses. Many different colours are produced by minute quantities of foreign substances in the crystals.

2. **Chalcedony** occurs in spheroidal or stalactitic masses, composed of more or less concentric shells. The chemical composition and behaviour of this mineral are the same as in quartz, but the specific gravity is somewhat lower, and the optical properties are different. Chalcedony has a waxy appearance, and is translucent or semi-opaque, and of various pale colours.

3. **Flint** and **Chert** are mixtures of hydrated and anhydrous silica. They occur in amorphous masses of neutral or dark colours, and are opaque, or somewhat translucent in thin pieces.

B. MINERALS COMPOSED OF SILICATES

There are several silicic acids, which form a very extensive series of compounds with various metallic bases. As rock-forming minerals the silicates are of the first importance.

I. The Felspar Group

The felspars are essentially silicates of alumina (Al_2O_3) together with potash, soda, or lime. ·*Orthoclase* and *microcline* are potash felspars (K_2O, Al_2O_3, 6 SiO_2); *albite* is a soda felspar (Na_2O, Al_2O_3, 6 SiO_2); and *anorthite*, a lime felspar (CaO, Al_2O_3, 2 SiO_2). From the combination of these two series are formed: the lime-soda series, *oligoclase*, *andesine*, and *labradorite*, the plagioclases, and the potash-soda series, *anorthoclase*.

The felspars crystallize in either the monoclinic or triclinic systems, but the forms of the crystals are very much alike. With few exceptions, these minerals are of pale colours and, except when decomposing, are very hard.

1. *Monoclinic Felspars*

Orthoclase is a potash felspar (K_2O, Al_2O_3, 6 SiO_2=K,Al, Si_3O_8), though soda may replace part of the potash. Hardness =6, sp. gr. =2.54–2.57. Orthoclase crystallizes in oblique rhombic prisms and is very generally twinned; there are two sets of cleavage planes, which intersect at a right angle and have thus given its name to the mineral. Orthoclase is usually dull and turbid, which is due to the presence of various alteration products, and even thin sections under the microscope are commonly hazy. *Sanidine* is a glassy, transparent variety of orthoclase, which is found in lavas of late geological date. Its clearness is due to the absence of the decomposition products, which render ordinary orthoclase turbid.

2. *Triclinic Felspars*

The minerals of this series are grouped together under the comprehensive term of *Plagioclase*, because of the difficulty of distinguishing them from each other under the microscope; they are very generally characterized by polysynthetic twinning, which makes fine parallel lines on the basal cleavage planes. Chemically, they are isomorphous mixtures of albite and anorthite. The following table (from Lévy and Lacroix) gives the composition of the various members of this series, representing the soda-felspar constituent, or *albite*, by Ab, and the lime-felspar constituent, or *anorthite*, by An: —

NAME	COMPOSITION	SPECIFIC GRAVITY
Albite	Ab	2.62
Oligoclase	$Ab_{10}An_3$	2.65
Andesine	Ab_2An_1	2.67
Labradorite . . .	Ab_2An_3	2.70
Anorthite	An	2.75

It will be observed that the specific gravity increases with the lime constituent, and the fusibility diminishes in the same proportion. Anorthite is decomposed by hydrochloric acid, labradorite is slightly attacked by it, while the other members of the series are not affected.

Anorthoclase is a triclinic potash-soda felspar (Ab_2Or_1), but is less common than the plagioclases as a constituent of rocks.

Microcline has the composition of orthoclase and plays a similar rôle in rocks, but crystallizes in the triclinic instead of the monoclinic system.

II. The Felspathoid Group

These minerals are very closely allied to the felspars in chemical composition and geological relations, but differ from them in crystal form and physical properties. They have a much more restricted distribution than the felspars, but have, nevertheless, an important bearing upon the classification of certain groups of rocks in which they occur.

1. **Nepheline** is a silicate of potash, soda, and alumina $(Na, K)_2 O(Al_2O_3, 2 SiO_2)$. It crystallizes in transparent and colourless six-sided prisms, of the hexagonal system. $H = 5.5–6$; sp. gr. $= 2.6$. It is an important constituent of certain lavas.

2. **Leucite** is composed as follows: $K_2O, Al_2O_3, 4 SiO_2$, with some of the potash replaced by soda. It crystallizes in twenty-

four-sided figures (trapezohedrons), which belong to the tetragonal system, but can be distinguished from the isometric only by very careful measurement. H$= 5.5-5.6$; sp. gr.$= 2.44-2.56$.

III. THE MICA GROUP

These minerals have a complex chemical composition, and are so variable that it is difficult to give formulæ for them; they are silicates of alumina, together with potash, lithia, magnesia, iron, or manganese. When crystallized, the micas all form six-sided prisms, which, though of hexagonal habit, are in reality monoclinic. All varieties have a remarkably perfect cleavage, and split into thin, elastic, and flexible leaves, by which they may be readily recognized. They are quite soft, and most of them may be scratched with the finger-nail.

1. **Muscovite** may be selected as the most important and wide-spread of the numerous alkaline micas, with the general formula, K_2O, $3 Al_2O_3$, $6 SiO_2$, $2 H_2O$. It is a lustrous, silvery-white mineral, usually transparent and colourless in thin leaves; it has a specific gravity of $2.76-3.1$, and a hardness of $2.1-3$.

Sericite is a silvery or pale green form of muscovite, which is an alteration product and often is derived from a felspar.

2. **Biotite** is the most important and widely disseminated of the numerous dark-coloured, ferromagnesian micas. This mineral is black or dark green in mass, and smoky even in thin leaves; chemically it is a silicate of potash, alumina, iron, and magnesia. In hardness and specific gravity it differs little from muscovite.

IV. THE AMPHIBOLE AND PYROXENE GROUPS

These two groups contain parallel series of minerals of similar chemical composition, but differing in their crystallization and physical properties. In composition they range from silicates of magnesia to those of lime and lime-alumina, while iron is present in most of them. They belong to the orthorhombic and monoclinic systems, and can be distinguished by their cleavage. The pyroxenes have a prismatic cleavage of nearly 90°, while in the amphiboles the angles are 124° 30′ and 55° 30′. The orthorhombic amphiboles are rare and unimportant as rock-forming minerals, but the pyroxenes of this form are widely distributed, though less so than the monoclinic.

a. *Orthorhombic Pyroxenes* are silicates of magnesia and iron $(Mg, Fe)O, SiO_2$.

1. **Enstatite** has less than 5 % of FeO.

2. **Bronzite** has 5–14 % of FeO.

3. **Hypersthene** has more than 14 % of FeO.

The colour becomes darker and the optical properties change with the increase in the percentage of iron.

b. *Monoclinic Pyroxenes.*

1. **Augite.** — This very abundant and important mineral is a silicate of lime, magnesia, iron, and alumina $(Ca, Mg, Fe)O$, $(Al, Fe)_2O_3, 4 SiO_2$. Sp. gr. = 3.3–3.5; H = 5–6. It crystallizes in oblique rhombic prisms, and in colour is green to black and opaque.

2. **Diallage** is a variety of augite, usually of a green colour, which is distinguished by its laminated structure, with lustrous faces.

c. *Monoclinic Amphiboles.*

1. **Hornblende,** like augite, which it closely resembles in chemical composition, is among the most important of rock-forming minerals. In colour it is usually green, brown, or black, and it crystallizes in modified oblique rhombic prisms. Sp. gr. = 2.9–3.5; H. = 5–6.

2. **Tremolite** is a silicate of magnesia and lime $(Ca, Mg)O$, SiO_2. This mineral is pale green or white and occurs in laminæ or long, blade-like crystals.

3. **Actinolite** resembles tremolite in composition, with the addition of iron $(Ca, Mg, Fe)O, SiO_2$. Colour, green; sp. gr. = 3–3.2; usually occurs in long and thin crystals. A fibrous variety

of tremolite or actinolite, in which the fibres are often like flexible threads and may be woven into cloth, is called *asbestus*.

V. The Olivine Group

Olivine is the only mineral of this group of sufficient importance to require mention; it is a silicate of magnesia and iron, $2(Mg, Fe)O\ SiO_2$, though the percentage of iron varies greatly. Sp. gr.$=3.2-3.5$; $H=6.5-7$. Olivine crystallizes in the orthorhombic system, and occurs in prisms, flat tables, or irregular grains. The colour varies from olive-green to yellow, or it may be colourless, and usually the irregular grains look like fragments of bottle glass.

VI. The Epidote Group

Epidote is a silicate of alumina, with iron and lime, the different varieties being named according to the preponderance of one or other base. Iron-epidote (*Pistazite*) forms monoclinic crystals, with a specific gravity of $3.2-3.5$, and a hardness of $6-7$. Lime-epidote (*Zoisite*), which has little or no iron, is orthorhombic.

VII. The Garnet Group

The **Garnets** are highly complex silicates of alumina, iron, lime, magnesia, chromium, and manganese, though in most cases only two or three of the bases are present in any considerable quantity, and the different varieties have received names in accordance with the predominating bases. They usually crystallize as dodecahedrons, or twelve-sided figures. Sp. gr.$=3.4-4.3$; $H=6.5-7.5$. Clear and brilliantly coloured garnets are considerably used in jewellery.

The commonest variety (*Almandine*) is a silicate of alumina and iron, and is usually red.

C. OTHER SILICATES, CHIEFLY DECOMPOSITION PRODUCTS

Many of the complex silicates, when long exposed to the action of the weather and of percolating waters, become more or less

c

profoundly changed chemically, a change which is known as *alteration* and forms an early stage of decay. One of the commonest of these changes is hydration, or the taking up of water into chemical union, and this may be accompanied by the loss of soluble ingredients, or the replacement of some constituents by others.

I. ZEOLITES

In this group are included a large number of minerals, which are hydrated silicates of alumina, potash, soda, lime, etc. They all contain water and hence boil and effervesce when heated before the blowpipe. All these minerals are products of decomposition and do not occur as original constituents of rocks.

II. TALC AND CHLORITE GROUPS

1. **Chlorite.** — Under this name are grouped a number of closely allied minerals, which are hydrated silicates of alumina, magnesia, and iron. They are soft minerals, with a hardness of 1–1.5 and a specific gravity of 2.6–2.96, and are usually of a green colour. They crystallize in the monoclinic system, with a pseudo-hexagonal symmetry. These minerals are laminated and split readily into thin leaves, as do the micas, from which they may be distinguished by the fact that the leaves are not elastic.

The chlorites result from the decomposition of hornblende, augite, or the magnesian micas.

2. **Talc** is a hydrated silicate of magnesia, $3\,MgO$, $4\,SiO_2$, H_2O; the water varies in amount to as much as 7 per cent. Sp. gr. $= 2.56$–2.8; $H = 1$. It is of a white or pale green colour, with a pearly lustre and a greasy, soapy feeling to the touch. Talc is rarely found crystallized; the crystals have a false hexagonal symmetry, and it is doubtful whether they should be referred to the orthorhombic or monoclinic systems. Usually it occurs in flakes or foliated masses, which split into thin, non-elastic leaves. Talc results from the alteration of magnesian minerals.

3. **Steati**te, or Soapstone, has the same composition as talc, but is not foliated, and may be much harder, as much as 2.5.

4. **Serpentine** is a hydrated silicate of magnesia and iron: $3(Mg, Fe)O, 2 SiO_2, 2 H_2O$. It does not crystallize, but is rather common in pseudomorphs. Sp. gr. $= 2.5-2.65$; $H = 2.5-4$. Its proper colour is green, but it is usually mottled with red or yellow by iron stains. Serpentine is generally formed from the decay of olivine, less commonly from augite, or hornblende.

Kaolinite is the hydrated silicate of alumina, Al_2O_3, $2 SiO_2$, $2 H_2O$. It is usually soft and plastic, but orthorhombic crystals of pseudo-hexagonal symmetry may be sometimes detected with the microscope. Kaolinite arises from the decomposition of the felspars and especially of orthoclase.

Glauconite is a hydrated silicate of iron and potassium, with small quantities of alumina, lime, magnesia, and soda. It is of a green colour, soft and friable.

D. CALCAREOUS MINERALS

1. **Calcite**, carbonate of lime, $CaCO_3$. Sp. gr. $= 2.72$; $H = 3$. This mineral crystallizes in the hexagonal system, in a great variety of forms; rhombohedrons and scalenohedrons are common; hexagonal prisms and pyramids less so. Cleavage is very perfect, parallel to the faces of a rhombohedron, and the mineral breaks up into rhombohedrons when struck a sharp blow. Calcite is rapidly attacked, even by cold and weak acids, CO_2 escaping with effervescence. When pure, as in Iceland spar, the mineral is colourless, very transparent, and lustrous, and displays the phenomenon of double refraction strongly; but more commonly it is cloudy or white, or stained red or yellow by iron. It is soluble in water holding CO_2, affording calcium bicarbonate which is found in nearly all natural waters. It is widely diffused among the rocks, and in a state of varying purity forms great masses of limestone.

2. **Aragonite** ($CaCO_3$) is somewhat harder and heavier than

calcite, with a specific gravity of 2.93 and a hardness of 3.5-4, and crystallizes in compound prismatic forms which belong to the orthorhombic system. It has not the marked cleavage of calcite and is less stable as a rule; when heated it is converted into calcite and falls into tiny rhombohedrons of that mineral.

3. **Dolomite** is a carbonate of lime and magnesia $(Ca, Mg) CO_3$; it resembles calcite in appearance, and crystallizes in rhombohedrons which often have curved faces. Sp. gr. = 2.8-2.9; H = 3.5-4. Dolomite may be readily distinguished from calcite by the fact that cold acids affect it but little.

4. **Gypsum**, hydrated sulphate of lime, $CaSO_4$, $2 H_2O$. Sp. gr. = 2.31-2.33; H = 1.5-2. It crystallizes in right rhomboidal prisms, belonging to the monoclinic system, and cleaves into thin, non-elastic leaves. When pure, gypsum is transparent and colourless, but is often stained by iron. This mineral occurs largely in granular masses, from which plaster of Paris is made by calcining the gypsum and so driving off the water of crystallization. *Alabaster* is a gypsum of especially fine grain, mottled in pale colours, or white. *Selenite* is a transparent variety.

5. **Anhydrite**, $CaSO_4$, is sulphate of lime without water; it is harder and heavier than gypsum (Sp. gr. = 2.9-2.98; H = 3-3.5), and crystallizes in a different system, the orthorhombic. The crystals have three sets of cleavage planes, which intersect each other at right angles.

6. **Apatite** is a phosphate and chloride or fluoride of calcium, $3(Ca_3P_2O_8)$, $2(Ca, Cl, F)$. Sp. gr. = 2.92-3.25; H = 5. It crystallizes in hexagonal prisms, terminated by hexagonal pyramids, and also occurs in masses. It is sometimes transparent and colourless, but more commonly opaque brown or green. Apatite is soluble in acids, and in water containing carbon dioxide, or ammonia; and gives rise to a valuable plant food.

7. **Fluorite**, fluoride of calcium, CaF_2. Sp. gr. = 3.01-3.25; H = 4. Crystallizes in the isometric system, usually in cubes, and has a perfect octahedral cleavage. When pure, fluorite is either clear and colourless, or blue, green, yellow, or brown.

E. IRON MINERALS

1. **Hæmatite**, or Specular Iron, is ferric oxide, Fe_2O_3. Sp. gr. = 4.5–5.3; H = 6.5. Crystallizes in rhombohedrons, or more commonly, in nodular masses. The colour is black, steel-grey, or red, and always is red when the mineral is finely powdered. Hæmatite frequently contains earthy and other impurities and is one of the most important ores of iron.

2. **Limonite**, or Brown Hæmatite, is hydrated ferric oxide (2 Fe_2O_3, 3 H_2O) containing more than 14% of water. It is softer than hæmatite and of a yellow or brown colour. Sp. gr. =3.6–4; H = 5–5.5.

3. **Magnetite** is the black oxide of iron, Fe_3O_4 (or FeO, Fe_2O_3). Sp. gr. = 4.9–5.2; H = 5.5–6.5. Crystallizes in the isometric system, usually in octahedrons, sometimes in dodecahedrons. This mineral is strongly magnetic and is black in colour, with a bluish-black metallic lustre, when viewed in reflected light. Magnetite is widely diffused in certain classes of rocks, and also occurs in veins and beds, which form an important source of supply of the metal.

4. **Ilmenite** is an oxide of iron and titanium $(Ti, Fe)_2O_3$. Sp. gr. = 4.5–5.2; H = 5–6. When crystallized, this mineral is rhombohedral, but is generally massive.

5. **Siderite** is ferrous carbonate, $FeCO_3$. Sp. gr. = 3.7–3.9; H = 3.5–4.5. Crystallizes in rhombohedrons, the faces of the crystals frequently much curved, and often the crystals are very much flattened. When fresh, the mineral is grey or brown. It is but slightly acted on by cold acids; hot acids dissolve it with effervescence. Mixed with clay, siderite forms clay iron-stone, a valuable ore.

6. **Pyrite**, or Iron Pyrites, bisulphide of iron, FeS_2. Sp. gr. = 4.9–5.2; H = 6–6.5. Crystallizes in the isometric system, usually in cubes, sometimes in dodecahedrons, and has a very characteristic brassy lustre and colour, to which it owes the popular name of "fools' gold." It is very hard, cannot be scratched

with a knife, and strikes fire, like flint, when struck with steel.
The mineral is soluble in nitric acid: it is widely disseminated
in the rocks.

7. **Marcasite**, or White Iron Pyrites, has the same composition
as pyrite, but crystallizes in the orthorhombic system, in modified
prisms, but more commonly occurs in nodular masses, with a radial
structure. It has the same hardness as pyrite, but is not quite
so heavy. Sp. gr. = 4.68–4.85. In colour it is paler than pyrite,
with a tendency to grey, green, or even black. It decomposes
very readily, and after a few months' exposure, even to dry air,
often crumbles to a whitish powder.

The iron minerals are seldom largely represented in any given
rock, with the exception of the ore beds; but iron is one of the
most widely diffused of substances, few rocks being altogether free
from it, and its various compounds play a very important rôle as
colouring-matter in the rocks.

PART I

DYNAMICAL GEOLOGY

WE have already seen that the chief task of geology is to construct a history of the earth, to determine how and in what order the rocks were formed, through what changes they have passed, and how they reached their present position. The logical order of treatment might seem to require that we should first learn what the rocks are, of what they are composed, and how they are arranged, before attempting to explain these facts. In such a study, however, we should meet with so much that would be quite unintelligible, that a more convenient way will be to begin with a study of the agencies which are at work upon and within the earth, and which tend to modify it in one or other particular. In other words, we must employ the present order of things as a key by means of which to decipher the hieroglyphics of the past, and proceed from what may be directly observed to past changes which can only be inferred.

We might assume that the present was so radically different from the far-distant past, that the one could throw no light upon the other. Such an assumption, however, would be most illogical, for there is nothing to support it. There is no reason to imagine that physical and chemical laws are different now from what they have always been, and the more we study the earth, the more clearly we perceive that its history is a continuous whole, determined by factors of the same sort as are now continuing to modify it. Some geologists have assumed that these agencies have always acted with just the same intensity as they do to-day; but this assumption is neither necessary, nor in itself probable. There is, on the contrary, much reason to believe that while certain forces act with greater efficiency at the present time than they did in the past, others act with less.

An attentive examination of the changes which are now in progress on the surface of the earth, will show us that nothing terrestrial is quite stable or unchangeable, but that there is a slow, ceaseless circulation of matter taking place on the surface and within the crust of the globe. Matter, chemistry teaches us, is indestructible, and, disregarding the relatively insignificant amount of material which reaches us from outer space in the form of meteorites, the sum total of matter composing the globe remains constant. But while practically nothing is added to or taken away from the materials which make up the earth's crust, ceaseless cycles of change continually alter the position, physical relations, and chemical combinations of those materials. This circulation of matter may be aptly compared to the changes which take place in the body of a living animal, only, of course, they are of a different kind and are effected at an infinitely slower rate. In the animal body, so long as life lasts, old tissues break down into simpler compounds and are ejected, while new tissues are built up out of fresh material. So, on the earth rock-masses decay, their particles are swept away, accumulate in a new place, perhaps far distant from their source, and are consolidated into new rocks, which in their turn are attacked and yield materials for further combinations. The study of the physical and chemical changes in the bodies of animals and plants constitutes the science of physiology, and by analogy we may call dynamical geology the physiology of the earth's crust. Analogies, however, must not be pushed too far, or they land us in absurdity. One essential difference between the earth and a living organism suggests itself at once; namely, that the former is self-contained, and neither ejects old material nor receives new, but employs the same matter over and over again in ever-varying combinations. The animal or plant, on the contrary, continually takes in new material from without, in the shape of food, and ejects the waste resulting from the breaking down of tissue.

Although the earth needs no fresh supplies of matter, its dynamical operations are, to a very large extent, maintained by energy from without; namely, from the sun. The circulation of the winds and waters, the changes of temperature, and the activities of living beings, all depend upon the sun's energy, and were that withdrawn, only such changes as are brought about by the earth's internal heat could continue in operation.

The study of dynamical agencies, subterranean and surface, necessarily gathers together an enormous mass of detail. But we need concern ourselves with only so much of this as throws light upon the earth's history, so that the sciences of dynamical geology and physical geography, though having much in common, are not coextensive. In order to make clear the operations of the forces which tend to modify the surface of the earth, it is necessary that we should classify and arrange them, so that they may be treated in a more or less logical order. However, in making such a classification, it is impossible to avoid entirely a certain arbitrariness of arrangement, since we must consider separately agencies that act together. Natural phenomena are not due to single causes, but to combinations and series of causes, and yet, to make them intelligible, they must be treated singly or in simple groups, else we shall be confronted by a chaos of uncorrelated facts. The career of a raindrop, from its first condensation to its entrance into the sea through the mouth of some river, is a continuous one, yet rain and rivers are distinct geological agencies and do different kinds of work. Again, the very important way in which the various dynamical agents modify, check, or augment one another, must not be overlooked in a systematic arrangement of these agents.

Some of the agencies that we shall consider may seem, at first sight, to be very trivial in their effects, but it must be remembered that they appear so only because of the short time during which we observe them. For enormously long periods of time they have been steadily at work, and their cumulative effects must not be left out of account in estimating the forces which have made the earth what we find it.

Much as may be learned by the study of the operation of the forces which are still at work in modifying the earth, this method of study is yet insufficient to solve all geological problems. Many of the changes which have indisputably taken place are such as no man has ever observed, because they are brought about so slowly or so deep down within the crust that no direct observation is possible, and we can only infer the mode of procedure by examining the result. No human eye has ever witnessed the birth of a mountain range, or has seen the beds of solid rock folded and crumpled like so many sheets of paper, or observed the processes by which a rock is changed in all its essential characteristics; "metamorphosed," as it is technically called. All such problems must be discussed in connection with structural geology.

The dynamical agencies may, primarily, be divided into two classes: I, *the Subterranean Agencies*, which act, or at least originate, at considerable depths within the earth; and II, *the Surface or Superficial Agencies*, whose action takes place at or near the surface of the earth. The former are due to the inherent energy of the earth, and their seat is primarily subterranean, though their effects are very frequently apparent at the surface. These agencies are also called *igneous* (from *ignis*, fire), which is a misnomer; but the term is nevertheless in common use. The surface agents are those which are derived from the energy of the sun.

The logical order of treatment of these subjects is to begin with the subterranean agencies, because the most ancient rocks of the earth's crust were doubtless formed by these forces, and the circulation of matter upon and through the crust started originally from *igneous rocks*.

The subterranean agents originate, primarily at least, either below or deep within the lithosphere, while the surface agents originate at the surface and penetrate to varying depths. Thus it happens that at certain levels and along certain lines the two classes of agents have a common place of activity and coöperate to produce effects which neither could produce alone. We may regard the lithosphere as composed of a series of concentric shells, in each of which the conditions of pressure and temperature differ from those of the other shells and, in consequence, the characteristic chemical and mechanical processes are different in each. Depth is thus a controlling factor of great importance in the operation of the dynamical agencies. " Under any given set of conditions, minerals tend to form which remain permanent under those conditions " (Van Hise), but when new conditions arise a readjustment begins and new combinations are formed.

The outermost of the concentric lithosphere-shells (the thickness of which cannot be definitely stated, but may be as much as 20,000–25,000 feet) is characterized by a tendency of the minerals to change from more complex to simpler compounds. At greater depths, this tendency is reversed and the change is from simpler to more complex compounds. Changes of this class constitute the process of *metamorphism*, to which the metamorphic rocks owe their origin, but, locally, metamorphism may occur at much higher levels and even very near the surface. It may be doubted whether any chemical recombinations occur at very great depths.

The boundaries of the concentric shells are not always definite, one shell gradually passing into another, nor are they always fixed, but may fluctuate from time to time within certain limits.

SECTION I

SUBTERRANEAN OR IGNEOUS AGENCIES

CHAPTER I

. DIASTROPHISM. EARTHQUAKES

THE subterranean agencies are those which are due to the earth's own inherent energy and arise deep within the earth's interior, though they are often displayed at the surface in a most striking manner. No problems of geology are more difficult and obscure than those connected with the internal constitution of the earth, and satisfactory explanations of the subterranean processes have not yet been devised.

These agencies fall naturally into two great groups: I, *Diastrophism*, or the movements of the earth's crust; and II, *Vulcanism*, or the phenomena of volcanoes, geysers, thermal springs, etc., while a third set of phenomena, *Earthquakes*, is intimately associated with each of the others, but, on account of its great importance, will require separate treatment. It is extremely probable that all of these so-called igneous processes are but different manifestations of the same forces, in ways that we cannot yet clearly understand, but, until the physical constitution of the earth's interior shall have been determined, such unity of origin cannot be definitely proved.

Diastrophism is the general term for differential movements of the lithosphere, whether upward, downward, or horizontal, and whether slow and imperceptible, or sudden and violent. These movements are of different kinds and may be classified as follows: I, *Orogenic* (Greek *Oros*, a mountain), the upheaval of long and relatively narrow belts of land by compression and crumpling of the rocks. As will be shown in a later chapter, the orogenic processes take place at considerable depths below the surface and hence cannot be directly observed; they are included here merely for the sake of completeness, for the study of them belongs properly to structural geology. II, *Epeirogenic* (Greek *Epeiros*, a continent), the broad uplift or depression of areas of the land or of the sea-bottom, in which the strata are not folded or crumpled, but may be tilted or may retain their original horizontal attitude. Movements of this class may be distinguished as (1) *Warping*, or *Bradyseism* (Greek *Bradys*, slow, and *Seismos*, earthquake), which is a broad gentle curving of the surface upward, *upwarp*, or downward, *downwarp;* (2) direct upheaval or depression, with fracturing and dislocation of the rocks, which may be accompanied by a tilting of the strata. Diastrophic movements of this class are almost, if not quite, invariably associated with earthquakes and can be most conveniently studied in connection with the latter.

Warpings or bradyseisms manifest themselves most clearly as changes of relative level between land and sea, because even slight changes of that character are often easily detected, while in the interior of the continents they can be demonstrated only under exceptionally favourable circumstances.

CHANGES OF LEVEL

Permanent changes of level frequently accompany earthquakes, but these are sudden and appear to be nearly always the result of dislocation or faulting. By change of level, in the general sense, is meant the gradual elevation or subsidence of land, with reference to the sea, over considerable areas. Such movements are very slow and hence are apt to escape observation, so that there is much dispute as to the facts and still more as to their interpretation.

The change may be in the land or in the sea; any important and permanent change in the bed of the sea must affect its surface, but then such changes must be widespread. On the other hand,

movements of the land may be either locally restricted or of great
extent. The absolute direction of the movements we have no
means of determining; that is, whether at a given point the earth's
radius is shortened or lengthened. The movement may be
always downward, but at different rates in adjoining areas, or
may be sometimes in one direction and sometimes in the other.
In view of these uncertainties, it has been proposed to avoid the
use of the terms, " elevation and depression of land," and to
substitute for them " negative and positive displacements of the
coast-line," respectively. For the sake of convenience, it will be
best to retain the older and more current terms, without insisting
that in all cases the land moves rather than the sea.

It is by no means true that all displacements of the coast-line
are diastrophic in origin; other processes that produce similar
results must be carefully distinguished from actual changes of
level. Thus, in many places the sea is advancing upon the land
by cutting back its shore, and areas which once were covered with
farms and villages are now permanently under water; but this is
not due to any sinking of the land. Another process which
simulates depression is the settling of loose masses of sediment,
which sometimes allows the sea to cover a flat coast. In other
places the sea is building up the coast by depositing sand upon
it, extending it seaward, and rivers build their deltas out into
the sea, but such changes are not diastrophic.

Along coast-lines the evidences of elevation are more obvious
than those of depression, because the traces of marine action are
always present on land which has recently risen from the sea,
while a submerged land-surface is soon changed and buried out
of sight.

Evidences of Elevation. — On certain coasts long inhabited by
civilized man, ancient structures like quays and bridges, which
were built in the water, may now be found high above it. Such
changes have been noted in the Mediterranean lands, especially
in southern Italy and the island of Crete. The so-called *Serapeum* at Pozzuoli, near Naples, is a famous and much discussed

example of repeated oscillations upward and downward. This
structure was built in Roman times and probably began to sink
while still in use, as appears from the two ancient pavements, one
above the other. Three large monolithic columns of marble,
about 40 feet high, are still standing erect, and on each of them
is a belt about 10 feet above the ground and 9 feet wide, honey-
combed by the boring mollusc, *Lithodomus*, which still abounds
in the neighbouring bay, and many of the shells were actually

FIG. 1. — Columns of the "Serapeum "; Pozzuoli, Italy

found in the columns. Evidently, the building was once submerged to a depth of nearly 20 feet, and when under water, the
columns were attacked and perforated by the mollusc. Just
when the reëlevation began is not definitely known, but there is
some documentary evidence to show that it was in progress in
the early years of the sixteenth century and was probably completed in 1538, when a volcanic eruption in the neighbourhood
resulted in the formation of Monte Nuovo (see p. 66). For

nearly a century past a slow movement of subsidence has been going on.

Rocks and cliffs long exposed to the action of the surf are worn and marked in a characteristic fashion and cut into terraces, and when found above the level at which the sea can now reach them, are evidences of upheaval at that point. Such well-defined sea-marks high above the present sea-level are common in the high latitudes of the northern hemisphere and, in many places, the change is still in progress. The Scandinavian peninsula shows slow changes of level, which constitute an upwarp; the south coast of Sweden is stationary or sinking slightly, but elsewhere the movement is upward and increases in amount towards the interior, the successive terraces rising toward the heads of the numerous deep fjords which indent the coast. These facts have been strongly disputed, but have recently been emphatically reaffirmed by the Swedish geologist De Geers, who shows that the isobasic curves, connecting points of equal elevation. form ellipses, the long axis of which coincides with the water-shed between Sweden and Norway. Along this line, the elevation is at a maximum, reaching nearly 1000 feet, and diminishing from the axis toward the periphery. Such a result can be explained only by an elevation of the land, not by a withdrawal of the sea, which could not have changed the level of the terraces.

" Raised beaches," filled with the remains of marine animals, are a decisive proof of a rise of the land, or a fall in the sea, and evidence of a similar kind is given by raised coral reefs. Such raised beaches now far above the sea occur in Scandinavia, Great Britain, the West Indies, the west coast of South America, the Red Sea, and elsewhere. The eastern coast of North America shows marks of relatively late elevation, increasing in amount northward. At the mouth of the Connecticut, the highest beach is 40 to 50 feet above sea-level, at Boston it is 75 to 100 feet, on the coast of Maine it is 200, and on that of Labrador 500 feet. On the eastern shore of Hudson's Bay the marine terraces and beaches extend up to 700 feet above sea-level.

Still another kind of evidence of recent elevation may often be gained from the form and character of the coast itself, as will be explained in Part III.

In the geological period (Pleistocene) immediately preceding the recent one, in which we are living, several immense lakes existed in the interior of North America, some around the basins of the present Great Lakes, others in Utah and Nevada. The ancient shore-lines of these vanished lakes may still be seen, for the most part, in admirable preservation; when first formed by the action of the waters, these beaches must have been level, but accurate surveys show that they are no longer so, but have undergone extensive warpings.

Wherever rocks of marine origin occur on land, they prove the elevation of the area where they are found. The great importance of the process is shown by the fact that the larger part of all the continents is composed of rocks which were laid down in the sea and are of all geological dates.

Evidences of Depression. — As ancient structures on long-inhabited coasts sometimes show elevation, they likewise sometimes show depression. On the north coast of Egypt ancient rock-cut tombs are now visible beneath the waters of the Mediterranean. The testimony of old buildings shows that the eastern end of the island of Crete is sinking, while the west and south coasts are rising. The Roman mole at Pozzuoli has sunk, as is shown by the mooring-rings for ships, now permanently below sea-level. South of Stockholm, in Sweden, the remains of an ancient hut were found in place, 65 feet below the surface, buried in marine deposits which contain shells of the same species now living in the Baltic. On the west coast of Greenland the sinking is so rapid as to have attracted the attention of the natives.

Buried forests found below sea-level indicate subsidence. Such forests occur in the delta of the Mississippi, on the shores of Chesapeake Bay and at many points on the sea-coast of the southern and middle Atlantic States, notably in New Jersey, where the coast is sinking at a rate estimated at 2 feet per century. Submerged

D

forests are also found on the coast of Holland and along the whole north coast of Germany, both on the North and Baltic Seas.

A river channel invaded and covered by the sea is still another proof of depression, because a river flowing into the sea cannot excavate the sea-bottom below the level of its mouth. Very many such instances are known, of which it will suffice to mention two or three. The ancient channel of the Hudson has been traced by soundings out to the edge of the continental platform, more than 100 miles southeast of Sandy Hook. In the same manner the channel of the St. Lawrence may be followed out through the Straits of Belle Isle, and that of the Congo extends out 70 miles from the west coast of Africa, with a depth of nearly 1000 fathoms.

The apparently contradictory evidence in the case of the St. Lawrence channel, which indicates depression, and that of the Labrador coast, which is rising, is not so in reality, for the movements are successive, not simultaneous.

Coral reefs often give proof of depression, for, as most of the reef-building corals cannot live in water more than 20 fathoms deep, a greater thickness of the reef than this indicates a slow sinking, at a rate not exceeding the upward growth of the coral. Borings made in the South Pacific island of Funafuti show that that reef exceeds 1100 feet in thickness, and must therefore have been gradually depressed. Another obvious proof of subsidence is a great thickness of shallow water deposits; for, if the sea-bottom did not sink, the shallow water would soon be filled up and the coast-line advanced. The study of the materials now accumulating on the ocean-floor enables us to determine the depth of water in which ancient deposits were formed, and applying this knowledge, we learn (to give only one example) that from the Hudson River southward, the coastal plain of the Atlantic States is covered by very thick, shallow-water, marine beds, as is revealed by the numerous artesian wells which have been driven through them. The fact that these beds are now part of a land-surface indicates, of course, that they have been elevated subsequently to their formation.

Finally, the form and topography of a coast may betray its recent subsidence, as will be more fully explained in Part III.

As regards the oscillations of level which are now going on, it is not definitely known whether they proceed continuously at a uniform rate, or spasmodically with intervals of complete rest. In the case of a succession of marine terraces, one above another, the movement cannot have been uniform, or else a continuous slope would have been produced; each terrace and beach indicates a pause, during which the waves cut the rocky shelf, or accumulated the beach, while the steep slope between two successive terraces points to a relatively rapid rise.

The following table, which exhibits the data gathered chiefly by Kayser, will be serviceable as showing the extent and character of the diastrophic movements which, it is inferred, are still, or have recently been in progress along the principal coast-lines of the world. In this table no account is taken of the movements which have here and there been detected in the interior of the different continents, such as those already mentioned for North America, and others which have been observed in northern Switzerland.

North America

RISING	SINKING
East coast of Greenland.	West coast of Greenland.
East coast down to 45° N. lat.	East coast from 45° N. lat. to end
Gulf of Mexico and Antilles.	of Florida.
Pacific coast.	East coast of Central America.

South America

Pacific coast, except that of Peru.	Coast of Peru.
Atlantic coast from mouth of La Plata to 20° S. lat.	Atlantic coast, except Uruguay and South Brazil.

Asia

Entire north coast and east coast to 30° N. lat.	East coast of southern China and Tonkin.
South coast and Malay Archipelago.	Laccadive and Maldive Islands.
Asia Minor.	

Australasia

Australia, south coast, and Tasmania.	Australia, northeast coast.
East coast of New Zealand.	West coast of New Zealand.
Pacific coast of New Guinea.	South coast of New Guinea.
Solomon, New Hebrides, Samoan, Sandwich Islands, and many others.	Caroline, Marshall, Gilbert, Tonga, Society Islands, etc.

Africa

Coast of Red Sea; east coast, west coast up to Gulf of Guinea.	Atlantic coast of Morocco, east coast of Tripoli.
	North coast of Egypt: Gulf of Guinea.

Europe

Peloponnesus, Sicily, Sardinia, Ligurian coast, Balearic Islands, south coast of Spain; west coast of France, Ireland and Scotland; Scandinavia.	England; north coast of France, the Netherlands and Germany.

From this table it is apparent that few coasts are stationary, but that almost all are, or have lately been, in movement, and further that upheaval greatly preponderates over subsidence. Still another significant result of these observations is that areas of opposite movement may be in close juxtaposition, as on the two sides of the Baltic, the east and west coasts of Greenland, the eastern coast of North America and Asia, and many other regions. In such cases the movement must be in the land rather than in the sea.

EARTHQUAKES

An earthquake is caused by a series of elastic waves due to a sudden shock in the earth's interior; the visible phenomena at the surface are produced by the outcropping of these waves and by the movements of the soft and inelastic soil, which is set in motion by the outcropping waves. While the elastic waves, which in mode of transmission resemble those of sound, are very regular in hard and homogeneous rocks, the actual movements of a given particle at the surface are highly irregular and confused, as is well shown in the wire model, Fig. 2. This model, constructed from the records of seismographs, gives in magnified form the movements of an earth-particle from the 20th to the 40th second of the shock, the numbers indicating the position of the particle at each successive second. The *seismographs* referred to are recording instruments, commonly horizontal pendulums, which register on paper strips the various components, horizontal and vertical, of the movements. So delicate are these

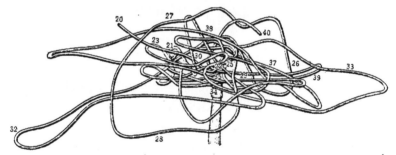

FIG. 2. — Magnified model, showing the movements of a surface particle of earth from the 20th to the 40th second of a shock. (Omori)

instruments, that they register even those violent shocks which originate at the very antipodes of the observatory where the instrument is installed.

The study of seismographic records has brought to light many highly significant facts, among others that minute and insensible *tremors* of the earth are almost incessant, but some, at least, of these tremors are due to atmospheric changes and it is not known how large a proportion of them are of subterranean origin. The term " earthquake " is usually restricted to those movements of the ground which can be felt, though the distinction is a somewhat arbitrary one. Another very important result of the seismographic observations is that when a very distant earthquake is registered, three series of waves are indicated, viz., the 1st and 2d phases of the preliminary tremors, and the larger waves of the main shock. Those first to arrive, called the preliminary tremors, are believed to be transmitted through the mass of the earth along the chord of the arc included between the point of origin and the point of observation. The preliminary tremors include two of the three series of waves, known as phases. The heterogeneous mass of rocks which forms the outermost crust of the earth does not

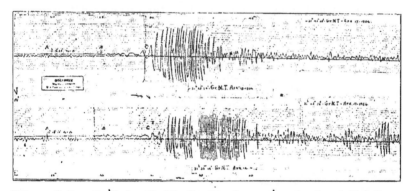

FIG. 3.—Seismographic record of the San Francisco earthquake of 1906, U. S. Coast Survey observatory, Cheltenham, Md. A, Preliminary tremors, 1st phase; B, Preliminary tremors, 2d phase; C, Main shock. The upper record shows the north-south component, and the lower gives the east-west component. (Bauer)

permit the transmission of any simple form of wave-motion, and it is only at a distance of about 10° of arc of the earth's surface (about 700 miles) that the three different kinds of waves begin to appear upon the instrumental records. The preliminary tremors, which pass through the subcrustal region of the earth and travel at a higher rate of speed than the waves which follow the surface, appear, as already mentioned, in two phases. The waves of the first phase are believed to be the normal, or compressional waves, and those of the second phase to be the transverse or distortional waves, the two known kinds of wave motion which can be transmitted through a homogeneous solid. The waves of the third series are longer and slower (*i.e.* of greater amplitude and period) and constitute the " main shock "; they are believed to follow the curvature of the earth's surface.

FIG. 4. — Earthquake regions of the Eastern Hemisphere. (de Montessus de Ballore)

Distribution of Earthquakes. — Sensible earthquakes are very numerous, not less than 30,000 is the estimated number per annum; of course, the great majority of these are very light. While any part of the earth's surface may be visited by earthquakes, there is a very great difference between different regions in regard to their *seismicity, i.e.* the frequency and violence of the shocks which affect them. The main seismic regions, when platted upon a map, are found to be arranged in two great-circle belts, one of which encloses the Pacific Ocean and the other girdles the whole

FIG. 5. — Earthquake regions of the Western Hemisphere. (de Montessus de Ballore)

earth. The latter includes the Mediterranean region, the Azores, the bed of the Atlantic westward from the Azores to the West Indies, those islands themselves, Central America, Hawaii, Japan, China, India, Afghanistan, Persia, and Asia Minor.

It must not be supposed that these belts are uninterrupted

zones of seismic activity; they are rather seismic tracts separated by other tracts of low seismicity. For example, the eastern Aleutian Islands and the Alaskan coast form a region of frequent and sometimes very violent quakes, while the coast between Alaska and California is not often shaken. California is an earthquake region, as is also southern Mexico, and Central America has a very high degree of seismicity, but there is a long interval before the earthquake region of Ecuador is reached. Though the belts are thus discontinuous, it is nevertheless significant that the separate seismic tracts are arranged in belts.

The regions most subject to earthquakes are those which have the steepest slopes and are associated with the great lines of corrugation of the earth's surface. A sea-bottom steeply descending from the shore is apt to be unstable, especially if high mountains arise near the coast, while a low-lying coastal plain and adjoining gently sloping sea-bottom are usually stable.

Beside the main seismic regions above enumerated, there are many others where the shocks, though not infrequent, are seldom violent. Examples of such regions are New England, Switzerland, Austria, and South Germany.

Although earthquakes are commonly perceptible upon the land, the most frequent seats of disturbance are in the bed of the sea. These submarine quakes occur at all depths of water, and their frequency and violence are independent of the distance from volcanoes. In the sea there are regions quite free from quakes and others of a high degree of seismicity, but quakes also occur in an isolated and scattered manner.

In the Atlantic there are two remarkable seismic belts, one, already mentioned as part of the great earthquake zone, extending westward from the mouth of the Tagus in Portugal, the other nearly equatorial and reaching from the north shore of the Gulf of Guinea toward Brazil. In this second belt the sea-floor has precipitous slopes.

Submarine cables are frequently interrupted at the same points. Thus, the cable from the Lipari Islands to Sicily has been broken

five times at the same point; on October 4, 1884, three parallel
cables, about 10 miles apart, were simultaneously broken at the
base of the steep continental slope, 330 miles east of St. John, N.B.
Many similar instances might be given.

Classification of Earthquakes. — Earthquakes may be classified
in several ways, according to the purpose in view. With regard
to the manner of production, they may be grouped into *volcanic*
and *tectonic* quakes, which will be explained-when we take up the
causes of these phenomena. For our present purpose, which is
chiefly descriptive, it will be most convenient to divide earthquakes into: (1) *Macroseismic*, or large earthquakes which
" disturb continental areas and frequently disturb the world as
a whole " (Milne); and (2) *Microseismic*, or local earthquakes
which, as a rule, affect areas of only a few miles' radius, rarely as
much as 100 or 200 miles. While the macroseismic quakes are
due to a disturbance both of the earth's crust and of the homogeneous interior, those of the microseismic class " appear to be
the shiverings within the crust " (Milne).

It should be observed that these terms have been frequently
employed in senses very different from those here used, which
are taken from Professor Milne. That there should be no very
distinct line of demarcation between the two classes, is not surprising, for they represent different degrees of intensity or violence
in similar phenomena.

Phenomena of Earthquakes. — The phenomena of earthquakes
differ greatly in accordance with the number, duration, and
intensity of the shocks, and with the distance of the place of observation from that of the origin of the disturbance. One of the
greatest of modern earthquakes is that of northern India of 1897,
which is well summed up in the official report.

" On the afternoon of June 12, 1897, there burst upon the western
portion of Assam an earthquake which, for violence and extent,
has not been surpassed by any of which we have historic record.
Lasting about two and one-half minutes, it had not ceased at
Shillong before an area of 150,000 square miles had been laid in

ruins, all means of communication interrupted, the hills rent and cast down in landslips, and the plains fissured and riddled with

FIG. 6. — Earthquake fissure in limestone, Arizona. (U. S. G. S.)

vents, from which sand and water poured out in most astounding quantities; and ten minutes had not elapsed from the time when

Shillong was laid in ruins before about one and three-quarter million square miles had felt a shock, which was everywhere recognized as one quite out of the common." (R. D. Oldham.)

A great earthquake usually begins suddenly and without warning. A rumbling sound, quickly becoming a loud roar, accompanies or slightly precedes the movement of the ground, which is at first a trembling, then a shaking; and finally a rapid swaying, wriggling motion, describing a figure 8, which is extremely destructive and overthrows the buildings affected, and even in the open country it is impossible to keep one's feet. The surface of the ground has been repeatedly observed to rise in low, very swiftly moving waves, somewhat like those on the surface of water, upon the crests of which the soil opens in cracks, closing again in the wave-troughs. When the earth-waves traverse a forested region, the trees sway violently from side to side, like a field of ripe grain in the breeze. In the details of movement earthquakes differ greatly from one another; sudden and extremely violent vertical shocks may come from below, or the surface may writhe and twist in every direction, instead of rolling in waves; there may be only a single shock, or many successive ones.

Violent earthquakes, which affect extensive areas, are almost always followed by a succession of *after-shocks*, which may continue for weeks, months, or even years. These may be very violent, though never equalling the primary shock in this respect, but gradually die away, until the region once more comes to rest.

In the sea the elastic waves producing shock soon die away in the water. Observations made on the several ships affected by the same quake frequently show a lineal arrangement of the disturbances. A special manifestation of earthquakes in the bed of the sea is the *great sea-wave* (sometimes erroneously called the tidal wave), which is a gravity wave produced by disturbances of the sea-floor or by a submarine volcanic eruption. The great sea-wave, though not strikingly displayed in the open sea, piles up on the coast into enormous breakers, which often are more terribly destructive than the earth-waves themselves.

Effects of Earthquakes. — Strictly speaking, the geological effects of earthquakes are of less importance than is usually supposed. The violent shaking of the surface often brings about great land-slips in mountain regions, which precipitate enormous masses of earth and rock from the heights down into the valleys. A striking example of this was given by the earthquakes of northwestern Greece in 1870, in which the rockslides were on a gigantic scale. The falling masses may temporarily or permanently block the valleys, converting their streams into lakes.

On the other hand, the diastrophic forces which produce the earthquake often have other effects of the greatest importance. In all of the more violent quakes cracks and fissures of the ground are formed, which may close again or remain open, and may show a lineal, curved, zigzag, or radiating arrangement. Through these fissures great quantities of water and sand are often forced up from below and form little sand craters, or water-filled funnels on the surface. Frequently the fissures assume the character of *faults*, or dislocations, one side being raised, the other depressed, so that long *scarps*, or low cliffs, are left standing. A long list of such faults formed in modern earthquakes might be given, though the limitations of space forbid the mention of more than a few.

In 1811–1812, near New Madrid, Mo., hundreds of faults resulted from the violent earthquakes which shook that region, and a depressed area, 70 × 30 miles in extent, known as the " sunk country," was formed. The earthquake of Owen's Valley, Cal., in 1872, was accompanied by the formation of a fault 40 miles in length and with a vertical displacement, or *throw*, of 5–20 feet, along the eastern base of the Sierra Nevada. In the Sonora earthquake of 1887, in Arizona and Mexico, a zigzag fault, 35 miles long and with a maximum throw of 20 feet, was produced. The Sierra Teras, in Mexico, appears to have been raised in this movement, for a second fault, with opposite inclination, was formed on the eastern side of that range. The Japanese earthquake of 1891 was accompanied by a fault of 40 miles in length, with throws exceeding 33 feet in height.

Mention has already been made of the great Indian, or Assam, earthquake of 1897; it was accompanied by several large faults, from which lesser ones branch out. As a result of the earthquake of 1899, the region of Yakutat Bay, Alaska, was much disturbed, with an elevation of the coast at one point of 47 feet. "The change of level was differential, indicating a complex system of faulting on a large scale; and shattering of the rocks proves

FIG. 7.— Fault-scarp in the Neo Valley, Japan, earthquake of 1891.   The road shows a slight horizontal as well as vertical displacement.   (Milne)

much differential movement on a smaller scale." (Tarr and Martin.) In the San Francisco earthquake of 1906 two long lines of parallel faults were formed, with varying throw up to 20 feet.

Earthquake displacements may be horizontal as well as vertical, though the former are less obvious and have not therefore been so frequently observed. Lateral displacements amounting, in some localities, to as much as 20 feet, in a direction parallel with the fault-scarps, were a very marked feature of the San Francisco earthquake. Horizontal displacements were also observed after the Indian earthquake of 1897, the Sumatran of 1892, the Japanese of 1891, the Owen's Valley of 1872, and others.

The most remarkable of modern faults are those which have been detected by soundings in the floor of the eastern Mediterranean. The earthquake of October, 1873, off the western coast of Greece resulted in a scarp on the sea-floor with a depth of 2000 feet, where

FIG. 8. — Fence broken and shifted horizontally 15 feet, San Francisco earthquake of 1906

formerly the depth had been 1400 feet. In 1878 the cable to the island of Crete was broken in two places by a violent earthquake and the sea-floor had become so irregular that, in relaying the cable, it was found necessary to make a long detour. The cable from Zante to Crete was broken by an earthquake in August, 1886, and at the break soundings revealed an increased depth of 1300 feet. Some of the Mediterranean scarps are 3000–5000 feet high, but the time of formation of these enormous dislocations is **not** definitely known.

FIG. 9. — Horizontal shifting of the ground, San Francisco earthquake, 1906. Before the earthquake the road was straight. (Photograph by Sinclair)

A very common result of earthquakes is a change in the circulation of underground waters. Wells and springs go dry, while other springs are formed in new places, or old ones may be increased in volume. The changes in the form of the land surface produce corresponding changes in surface drainage; rivers are diverted into new channels or dammed into lakes, while streams intersected by fault-scarps form new cascades. Many new lakes resulted from the Indian earthquake of 1897.

The general results obtained from the study of the diastrophic movements which accompany earthquakes are thus summed up by Professor Hobbs: —

" 1. Appreciable surface dislocations appear to be formed only at the time of macroseisms, and the throws upon these planes stand in some relation to the magnitude of the disturbance.

" 2. The evident dislocations produced are generally of two orders of magnitude, those of the higher order being generally very limited in number, while those of the lower order are often quite numerous.

" 3. Earthquake dislocations are normal faults with hades approaching the vertical." That is to say, the *hade*, or inclination, of the fault-plane slopes downward toward the depressed side, as though the latter had merely slipped down the inclined plane.

" 4. The crustal movements indicated at the surface at the time of earthquakes appear to be due to an adjustment in position of individual blocks."

The formation of normal faults and open fissures during an earthquake shows that the stresses to which the outer portion of the crust yields are tensional, or stretching. Compression may also occur locally, as appears from the upward or lateral bending of railway rails, so frequent a phenomenon of great disturbances, but such compression is frequently due to the slipping of deep masses of soil and to be compensated by stretching at other points. The measurements of the railway lines after the great Indian earthquake of 1897 proved that whenever the rails had been bent at one point, they had been dragged apart by an equivalent amount at another. This refers to the movements of soil; horizontal faulting, on the other hand, implies a true compression.

E

Causes of Earthquakes. — In respect to their mode of causation, earthquakes are usually divided into two classes, *volcanic* and *tectonic,* though it is often impossible to determine to which of the two classes a given earthquake should be referred. The volcanic earthquakes, which are closely associated in time and space with volcanic eruptions, are due to steam explosions and to the struggles of the rising lava within the earth to escape. In their typical manifestation, volcanic quakes have a definite centre of origin, which is in or near a volcano, and are rarely felt over any very extensive area of country. The earthquake of 1883 in the island of Ischia was of terrible violence and completely destroyed the town of Casamicciola, with great loss of life, yet the shock was hardly perceptible at Naples, a distance of twenty-two miles. The great eruption of Mauna Loa in the Hawaiian Islands in 1868 was preceded for six days by earthquakes of gradually increasing force, until they became frightfully destructive. When the lava burst out from the volcano, the earthquakes rapidly died away. Extreme as was the violence of these shocks, they were almost confined to the southern side of the island; elsewhere they did little damage, and at a distance of 150 miles were barely sensible. The earthquakes of Central America and those at the base of Ætna and Vesuvius and other volcanoes are of this class.

Tectonic earthquakes are believed to be due to stresses in the interior of the earth which, when suddenly yielded to by the rocks, cause the jar and shock which generates the earthquake. Macroseisms are probably caused by the formation of new and great fractures and microseisms by settling along existing lines of fracture. Tectonic earthquakes are linear, the maximum destructiveness being along the line of fracture and rapidly diminishing transversely to this line; owing to the deeper position of their origins and the greater masses of rock involved in the movement, their effects are far more widely spread than those of the volcanic class. Another distinction from the latter is the succession of after-shocks which follow the great tectonic quakes and which last longest away from the primary line of fracture. These are due to the gradual readjustment of the mosaic of blocks, which, as we have already seen, makes up the outermost part of the earth's crust.

Just how the internal stresses above referred to are generated, is by no means clear. The explanation usually accepted is that the earth is slowly contracting on account of the loss of heat, and that the crust, which must follow the shrinking interior, is being crowded into a smaller space, with resultant ruptures and shocks. However, this contractional hypothesis is altogether rejected by several high authorities, and no very satisfactory substitute for it has been proposed.

On the other hand, it is contended by some geologists that all earthquakes are essentially volcanic in origin. These observers call attention to the " marked synchronism or close following of the major disturbances, whether volcanic or seismic, at distantly removed points of the earth's surface, at certain periods.' (Heil' prin.) It is undoubtedly true that such an association is indicated by many facts, but much remains to be learned before the full significance of these facts can be determined.

CHAPTER II

VOLCANOES

A VOLCANO is usually a conical mountain or hill, with an opening, or crater, through which various solid, molten, or gaseous materials are ejected. The essential part of the volcano is the opening, or vent, and some volcanoes consist of almost nothing else. The mountain, when present, is secondary and is formed by the materials which the volcano itself has piled up; it is thus the effect and in no sense the cause of the phenomena.

Present Distribution of Volcanoes. — The geographical distribution of volcanic vents has greatly changed at different periods of the earth's history. While there are some large land areas, like most of the Mississippi Valley, which appear never to have been visited by volcanic activity, such areas are comparatively few in number. In most regions we find distinct traces of such action, though it may have died out ages ago and though at present no active vent may occur for very great distances in any direction. Such is the case with the valley of the Connecticut and northern New Jersey, Ireland, Great Britain, and very many other countries.

We cannot definitely determine the number of vents which are at present in activity in various regions of the earth, because a volcano may remain dormant for centuries, and then break out again. Almost all tradition of the volcanic nature of Vesuvius had died away among the inhabitants of Italy, until the dreadful eruption of the year 79 A.D. showed that it had only been slumbering. Many volcanic regions, such as the western part of North and South America, and the East Indian islands, have been known to civilized man for only a few centuries, and in such regions the distinction between dormant and extinct vents cannot always be made.

Furthermore, the number of vents is constantly changing, new openings forming, and old ones closing up, while some that had escaped observation are not infrequently discovered. Another distinction which is often arbitrary, is that between independent volcanoes and mere subsidiary vents connected with larger ones. Several submarine volcanoes have been observed, but it is altogether probable that many more exist which have escaped detection. Making these allowances, the number of volcanoes now active may be estimated at about 328, of which rather more than one-third are situated in the continents, and the remainder on islands.

The active volcanoes are not scattered haphazard over the surface of the globe, but are arranged in belts or lines, which bear a definite relation to the great topographical features of the earth as well as to the seismic belts described in the preceding chapter. Two of these belts together encircle the Pacific Ocean; one on the west coast of the Americas runs from Alaska to Cape Horn, the other, a very long and sinuous band, running from Kamschatka through the islands parallel to the east coast of Asia, the East Indian and south Pacific islands, to the Antarctic circle, where it joins the American band.

As in the case of the seismic belts, the volcanic bands are not continuous, but consist of a series of volcanic tracts separated by others which have no volcanoes. The coincidence of the volcanic and earthquake belts which encircle the Pacific Ocean is very close and striking.

A third band occupies a ridge in the eastern bed of the Atlantic, from Iceland to beyond St. Helena, from which arise numerous volcanic islands and submarine vents. Included in this Atlantic band are Jan Mayen, Iceland, the Azores, Canary and Cape Verde Islands, Ascension, St. Helena, and, in the far south (38° S. lat.), Tristan d'Acunha. South of Iceland there are no known volcanoes for a great distance, until the Azores are reached, and on the east coast of the Americas are none at all.

A subsidiary belt, parallel to the Atlantic band, includes the volcanoes of east Africa, the Mascarene and Comores Islands, together with the extinct vents of Madagascar in the south and those of Armenia, Syria and Arabia in the north.

Other subsidiary volcanic belts are scattered along the great equatorial seismic zone mentioned in Chapter I. Thus, the volcanoes of southern Mexico and Central America have a general east-west arrangement and are in line with those of the West Indies. On another portion of the same zone are placed the Mediterranean vents. At the crossing of the equatorial and west Pacific zones are the volcanoes of the Philippines and Japan, and those of the former continue westward through Java, Sumatra, the Nicobar and Andaman Islands, to Burmah.

A very striking fact is the nearness of most active volcanoes to the sea; by far the greater number of vents are upon islands, and those of the continents are, with a few exceptions, not far from the coasts. Some of the volcanoes in Mexico and Ecuador are 150 miles from the ocean, and Kirunga in east Africa is more than 600 miles from the sea. Another relation which should be noted, is that between the volcanic bands and the mountain-chains, the bands running parallel to or coinciding with the mountains, as in the great volcanoes of the Andes. Not all coast-lines or all mountain chains have volcanoes associated with them, but where the mountains are near the seashore, volcanoes are usually, though not invariably, found. The seat of volcanic activity is frequently shifted, as we have learned, and it has been observed that this activity tends to die out of the older rocks and to make its appearance in those of a later date.

The relations of volcanoes to lines of fracture and faulting are much disputed. As we shall see, lava may force its way to the surface independently of such lines; nevertheless, " the great majority of recent and earlier eruptions are connected with fissures and zones of fracture in the earth's crust." (Kayser.)

Volcanic Eruptions. — The phenomena displayed by different volcanoes, or even by the same volcano at different times, vary greatly. It often seems difficult to believe that similar forces are involved, and that the divergences are due merely to different circumstances attending the outbreak. A careful comparison, however, of the varying phenomena brings to light a fundamental likeness in them all. Some vents, like Stromboli in the Mediterranean, are in an almost continual state of eruption of a quiet kind; others, like Vesuvius, have long periods of dormancy, broken by eruptions of terrible violence. In a general way, it may be said that the longer the period of quiet, the more violent and long-continued will the subsequent eruption be, while weak eruptions and those of short duration recur at brief intervals.

As one extreme of the various forms of volcanic activity should be regarded the *explosive type*, in which little or no lava is produced and, in some cases, even the finely shattered fragments of lava, the so-called volcanic ash, are absent. An instructive example of this kind is afforded by the eruption of the Japanese volcano Shirane in 1882, which consisted of a single tremendous explosion of steam, hurling a vast column of rock into the air, but without the emission of lava or ashes. Another eruption of the same kind was that of Bandai San, also a Japanese volcano, in 1888; a single terrific steam explosion blew away the greater part of the mountain, which was more than 2000 feet high, likewise without the formation of ash or lava.

The first recorded eruption of Vesuvius, which occurred in 79 A.D. and is described in two letters written to Tacitus by the younger Pliny, was of the explosive type, but was much more prolonged than those of the Japanese vents above-mentioned and was accompanied by the production of immense volumes of ashes and larger fragments. In this frightful paroxysm little or no molten lava was ejected, and so enormous was the quantity of ashes that at Misenum, across the bay of Naples, the sun was darkened, as Pliny reports, " not as on a moonless cloudy night, but as when the light is extinguished in a closed room . . . In order not to be covered by the falling ashes and crushed by their weight, it was often necessary to rise and shake them off." Herculaneum was overwhélmed with floods of ashes mixed with water, while Pompeii was completely buried in dry ashes and small fragments.

The explosive type of eruption is exhibited in its extreme form by several of the East Indian volcanoes, and preëminently by Krakatoa, the eruption of which in 1883 was the most frightful ever recorded. This volcanic island, situated in the Strait of Sunda, was little known, except that it had been in eruption in 1680. As

FIG. 10. — Pompeii, showing depth of volcanic accumulations. (Photograph by McAllister)

the island was uninhabited, the earliest stages of the outburst were not observed, but on May 20 a great cloud of steam was seen over the vent. The catastrophe occurred in August, when, besides the fearful devastation caused by the disturbances of the sea on the coasts of Sumatra and Java, the island itself was almost annihilated. Hardly one-third of its original surface was left above water, and where formerly was land are now depths of 100 to 150 fathoms of water. The force of the explosion produced waves in the atmosphere which were propagated around the whole earth, and the first one was observed in Berlin ten hours after the explosion. The ejected materials were all fragmentary and of an incredible volume; ashes were distributed over an area of 300,000 square miles, the greater part falling within a radius of eight miles around the island; stretches of water that had had an average depth of 117 feet were so filled up as to be no longer navigable. Enormous masses of pumice floated upon the sea and stopped navigation except for the most powerful steamers. Even more remark-

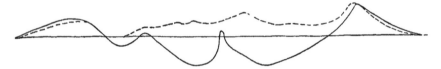

FIG. 11. — Profiles of Krakatoa. The full curved line is the present condition, the dotted line the condition before the eruption of 1883, while the horizontal line is that of sea-level. (Judd)

able is the fact that the finest dust, which was hurled into the upper atmosphere, remained suspended in the air for many months and was gradually diffused over the world. The wonderful, flaming-red sunsets which characterized the autumn and winter of 1883–1884, have been very generally ascribed to the refractive effects of the impalpably fine Krakatoa dust.

These tremendous explosions, even when they do not tear out one whole side of the mountain as in the case of Krakatoa, may blow off the top and thus leave a great crater ring many miles in circumference, within which subsequent eruptions may build up a new cone. When the volcanic activity dies out, the ring may be filled with water, forming a circular lake. Crater rings may also be formed in another way, illustrated by Crater Lake in Oregon. The glacial markings on the outer side of the mountain prove that the latter must once have been much higher than at present. On the other hand, the surrounding country displays no such quantity of fragments as would have been formed, had the top been blown off by an explosion. In this case, the upper portion of the cone was probably engulfed in the crater and perhaps remelted.

The year 1902 was made memorable by a series of excessively violent explosive eruptions, in some instances accompanied by frightful destruction and loss of life, in Central America and the

FIG. 12. — Crater Lake, Oregon. The Small island is a cone of eruption, built up after the formation of the crater ring

Lesser Antilles. In Nicaragua there was an unimportant eruption of Masaya (June 25), and Isalco in Salvador, after a pause of more than a year, began erupting on May 10, but this eruption was not of the explosive type and produced streams of lava. Far more violent was the outbreak of Santa Maria in Guatemala, a volcano which had been regarded as extinct. The eruptions began October

24 and were repeated, with diminishing energy, for more than a year, and were accompanied by an incredible quantity of ashes, which covered several hundred thousand square miles and, for a long distance around the vent, destroyed a great amount of property.

FIG. 13. — Gorge 200 feet deep filled by ash from La Soufrière, St. Vincent, eruption of 1902. (E. O. Hovey, courtesy of the American Museum of Natural History)

The islands of St. Vincent and Martinique in the Lesser Antilles were devastated by a series of fearful and nearly simultaneous eruptions, which in certain important respects differ from those of any other known volcanoes. The volcano of St. Vincent, known as La Soufrière (the last violent eruption of which had been in 1812), began to show signs of activity in February, 1901, by a

succession of earthquakes, which were repeated, with longer or
shorter intervals, until April, 1902, in the latter part of which they
increased in number and violence. The actual outbreak began
on May 6, 1902, in a series of tremendous steam explosions; May 7

FIG. 14. — Spine of Mt. Pelée.   (Photograph by Heilprin)

the eruption became continuous and on the same day occurred the
dreadful, descending " hot blast," a cloud of superheated steam
and other gases, mingled with red-hot particles of ash, which rushed
down the mountain and destroyed 1400 human lives. The erup-

tions, which were repeated at varying intervals and with different degrees of violence for considerably more than a year, were characterized by the absence of lava and by the vast quantity of finely divided ash ejected by the explosions.

The eruptions of Mont Pelée in Martinique were actually less violent, but far more destructive to life than those of St. Vincent. The previous outbreaks of Mont Pelée within historic times had been those of 1792 and 1851, both of which occurred with the same suddenness as the awful catastrophe of 1902. In the latter year slight earthquakes were noted on April 23, and on the 25th a heavy cloud of " smoke " appeared over the volcano. On May 2 the ejections of ash became heavier and more frequent, increasing until the 8th, when a descending cloud of hot vapours and glowing ash swept with terrible velocity down the ravine of the Rivière Blanche upon the city of St. Pierre, which, together with its 30,000 inhabitants, was instantly annihilated. The velocity of the air set in motion by the descending cloud was sufficiently great to hurl from its pedestal the great iron statue of Notre Dame de la Garde, weighing several tons, to a distance of more than 40 feet.

Mont Pelée had a long succession of subsequent eruptions of varying violence, especially on May 19, 20, and 25, June 6, July 9 and 13, August 25, 28, and 30, September 3, 1902; January 25, March 26, September 12 and 16, 1903; the last hardly less violent than the first terrible outbreak of May 8, 1902.

It is the descending clouds which lend such an exceptional character to the eruptions of St. Vincent and Martinique, but Mont Pelée also displayed certain other peculiar phenomena. While no lava streams were produced, very stiff and viscous lava appeared at the summit, filling up the old crater and forming a steep cone, through which protruded a lofty obelisk or spine, which, thrust up from below, grew irregularly in height, as it continually lost material by scaling off of the top and sides; eventually it fell altogether.

In all eruptions of the explosive kind, a few typical examples of which are described above, the active agency is obviously

exploding masses of intensely heated and compressed steam, and
all such eruptions are accompanied by gigantic steam-clouds,
which, condensing in the atmosphere, fall in rains of torrential
volume and violence. The hot water thus produced mingles with
the volcanic ash in the air and on the ground, forming streams of
hot mud, which are often more destructive than the lava flows
themselves. When cold, the mud sets into quite a firm rock,
called *tuff*.

FIG. 15. — Crater-floor of Kilauea, showing the lava lake, Hale-mau-mau. (Photograph by Libbey)

The opposite extreme of volcanic activity from the explosive
type is to be found in the volcanoes of the Sandwich Islands,
such as Mauna Loa and Kilauea. Here the eruptions are usually
not heralded by earthquakes; the lava is remarkably fluid and
simply wells up over the sides of the crater, pouring down the sides
of the mountain in streams which flow for many miles. More
commonly the walls of the crater are unable to withstand the enor-

mous pressure of the lava column, and the molten mass breaks through at some level below the crater, rising through the fissure in giant fountains, sometimes 1000 feet high. Even in the ordinary activity of Kilauea jets of 30 and 40 feet in height are thrown up. Hardly any ashes or other fragmental products are formed; and though the clouds of steam, the invariable accompaniments of volcanic outbursts, are present, yet the quantity of steam is relatively less than in those volcanoes in which explosions occur.

Between such extremes as the Hawaiian volcanoes on the one hand, and the explosive East Indian type (Krakatoa), on the other, we may find every intermediate gradation. The comparatively gentle operations of Stromboli, one of the Lipari Islands, northwest of Sicily, give an opportunity to observe directly the essential phenomena of a volcanic eruption. Though occasionally breaking out with violence, Stromboli has been in a state of almost continuous activity for more than 2000 years, and is, for long periods, in such exact equilibrium, that barometric changes have a marked effect upon its activity and the Mediterranean sailors make use of it as a weather signal.

The crater-floor is formed by hardened lava, the cracks in which glow at night from the heat of the molten mass below, and which is perforated by various openings. From some of these steam is given out, from others molten lava wells up occasionally. In openings of a third class the lava may be seen rising and sinking, until a great bubble forms on its surface and bursts with a loud roar, scattering the hardened lava scum about the crater in fragments of various sizes, some very fine, others coarse. The bubble is of steam, and when set free, the steam globule rises to join the cloud which always overhangs the mountain. The bursting of the bubble is followed by a rush of steam through the mass of the lava, the pressure is relieved, and the lava column sinks down out of sight, until the steam pressure again accumulates and the performance is repeated.

Evidently, one active agent in these phenomena is imprisoned steam in its struggles to escape. Different as are the manifesta-

tions at other volcanoes, steam is an important cause of the eruption in all cases, though the conditions under which it acts vary
widely. Little or no combustion is involved, and that not as a
cause, but as an effect of the activity.

In the modern eruptions of Vesuvius essentially the same phe-

FIG. 16. — Crater of Vesuvius in moderate eruption

nomena may be observed, but on a far grander and more terrible
scale. Earthquakes usually announce the coming eruption, increasing in force until the outbreak occurs. Terrific explosions
blow out fragments of all sizes, from great blocks to the finest and
most impalpable dust. The finer fragments arise chiefly from the
scattering of the partly hardened lava by the force of the explosion, but in part also from the crashing together of the blocks as

they rise and fall through the air. Inconceivable quantities of steam are given off with a loud roar, which is awe-inspiring in its great and steady volume. The condensation of such masses of vapour produces torrents of rain, which, mingling with the "ashes" and dust, gives rise to streams of hot mud that flow for long distances. Great floods of molten rock, or lava, issue from the crater, or burst their way through the walls of the cone, and pour down the mountain side, until they gradually stiffen by cooling.

During historic times Vesuvius has had long periods of dormancy and the violence of the subsequent outbreak has been, in a general way, proportionate to the length of the dormant period, though one of the most notable eruptions, that of 1906, occurred after quite a short period of rest. With the exception of a moderate outbreak in 1500, the mountain was quiet between 1139 and 1631; one of the three most violent recorded eruptions took place in the latter year.

Radically different as the various types of volcanic activity appear to be, they are all connected together in one continuous series. In all cases, steam of very high temperature and under enormous pressures is an important agent, while the differing results are due to varying degrees of pressure, quantity of imprisoned steam, amount of resistance to be overcome, the character of the lava, and similar factors. The intermediate, or Vesuvian, type of eruption is the most frequent.

Submarine Volcanoes. — Several instances of submarine eruptions have been actually observed, and there is much reason to believe that the number of vents on the ocean-floor is very large. Volcanic islands are merely submarine volcanoes which have built their cones above sea-level and these represent a great proportion of the vents now active. The durability of volcanic islands depends upon the materials of which they are constructed. Cones built of loose masses, or of ash and tuff, are speedily destroyed by the sea when the activity ceases, and cut down into reefs and shoals, while masses of solidified lava resist destruction for very long periods.

F

New Volcanoes. — During historic times a considerable number of new volcanoes have been formed, both on land and in the bed of the sea, the latter resulting in the birth of new islands. Aside from certain newly formed volcanoes, the origin of which has been recorded by ancient writers of Greece, Rome, and Japan, a few more modern instances may be cited.

Near Pozzuoli is a hill called Monte Nuovo, 440 feet high, which is hardly distinguishable from the other low volcanic cones

FIG. 17. — Monte Nuovo, near Pozzuoli, formed in 1538

among which it stands, and which are mentioned by several classical writers. Monte Nuovo was formed by an eruption which broke out September 29, 1538, the ground swelling up and bursting, leaving a fissure, which disclosed glowing lava, and which ejected great masses of blocks, sand, and ash. The activity lasted for a week and has not since been repeated.

Jorullo, in Mexico, was formed in 1759, the eruptions continuing for several years and then dying out. Immense quantities of lava flowed forth and several " cinder cones " were built up, one of which is 1300 feet high.   Isalco, a volcano on the west coast of Central America, north of the city of San Salvador, was first formed in 1770, and has been in almost uninterrupted activity ever since, sometimes with great violence.   A cone of more than 2000 feet in height has been built up.   In 1831 a new island appeared off the southwest coast of Sicily, where previous soundings had

FIG. 18. — Another view of the crater-floor and walls of Kilauea.   (U, S. G. S.)

shown a depth of 100 fathoms, and in the course of a few weeks grew to a height of 200 feet above the sea and a diameter of a mile at sea-level.   The activity soon ceased and the island, composed of loose materials, was swept away by the sea.

In the Greek archipelago the group of islands known as Santorin has been the scene of repeated operations for more than 2000 years. The outer islands are evidently fragments of an old crater ring, within which are several small islands, which were formed in 186 B.C., 1573, 1707, and 1866, respectively.

Finally may be mentioned an especially interesting group of three new islands, the Bogoslofs, north of the line of the Aleutian islands, Alaska. The first of the islands, Old Bogoslof, was formed by a submarine eruption in 1796, which was observed by a Russian trader. New Bogoslof was formed in 1883 and was first seen in September of that year, but the exact date of origin is not known. A third and very large island, composed of jagged lava, was seen between the older ones on May 28, 1906, " giving off clouds of steam and smoke from any number of little craters scattered all over it." (C. H. Gilbert.)

CHAPTER III

VOLCANOES (cont.) — INTERNAL CONSTITUTION OF THE EARTH

Volcanic Products. — These form the most important part of the subject from the geological point of view, because they contribute

FIG. 19. — Edge of Hale-mau-mau, showing the ropy forms of the highly fluid lava, when cooling. (Photograph by Libbey)

largely to the permanent materials of the earth's crust. We meet with such materials of all geological ages, sometimes developed on a vast scale. The study of volcanic products is the key which enables us to comprehend the great group of rocks which are called igneous, though, as we shall see later, by no means all of these were poured out on the surface of the ground.

Volcanic products are of three kinds: (1) *lava*, or molten rock; (2) *fragmental material*, including blocks, lapilli, bombs,

FIG. 20. — Ropy lava, Vesuvius

the so-called volcanic ashes, cinders, and the like; (3) *gases and vapours*.

(1) *Lava*. — A lava is a more or less completely melted rock; the degree of fluidity varies greatly in different lavas, but is rarely, if ever, perfect. Instead of being a true liquid, a lava ordinarily consists of larger and smaller crystals, embedded in a pasty mass, which is saturated with steam and gases. The degree of fluidity

depends upon several factors, the most obvious of which is temperature; the more highly heated the mass is, the more perfectly will it be melted. The quantity of imprisoned gases and vapours present has also an important effect, and some lavas appear to owe nearly all their mobility to these vapours. A third and most significant factor is the chemical composition. Those lavas which contain high percentages of silica (SiO_2), the *acid lavas*, are much

FIG. 21. — Sunset Butte, Arizona. An extinct volcano, with scoriaceous block-lava in foreground. (Photograph by A. E. Hackett, Flagstaff, Ariz.)

less readily fusible than the *basic lavas*, in which the percentage of silica is lower. The difference in the proportion of silica present is associated with other chemical differences which have a similar effect upon fusibility, the basic kinds having much more lime, magnesia and iron in them, which act as fluxes.

The experiments of Barus on the fusibility of lavas, which he divides into three groups, resulted as follows: (1) Certain lavas fuse readily (2250° F.); these are of basic composition and are made up of lime-soda felspars, the augitic and allied ferro-magnesian minerals, and iron oxide, but rarely have quartz. (2) A second group is of medium fusibility (2520° F.), and is made up of lime-soda felspars, augitic or hornblende minerals, and frequently quartz. (3) The third series melts with difficulty (2700° F.), and remains pasty at even 3100° F. These are acid lavas, and are composed of potash felspars, with quartz, hornblende, or mica. Lavas

FIG. 22. — Lava-tunnel, and "Spatter-cone" formed by escaping steam, Kilauea.
(Photograph by Libbey)

which, like those of the Sandwich Islands, are notably fluid, are always of basic composition.

When a lava stream reaches the surface of the ground, the imprisoned vapours immediately begin to escape and the surface of the molten mass to cool and harden. The surface layers are blown by the steam bubbles into a light, frothy or slaggy consistency, forming "scoriæ" or cindery masses. The motion of the lava breaks up this thin crust into loose slabs and blocks, and on the advancing front of the stream these loose masses rattle down over one another in the wildest confusion. The less perfectly fused lavas are soon covered with heaped-up cindery blocks, while the more completely fluid lavas are characterized by curiously twisted, ropy surfaces, such as may be observed in the slag from an iron furnace.

The front of a lava stream advances, not by gliding over the

FIG. 23. — Lava stalactites and stalagmites in lava-tunnel, Kilauea. (Photograph by Libbey)

ground, but by *rolling*, the bottom being retarded by the friction of the ground and the top moving faster, so that it is continually rolling down at the curved front end and forming the bottom. Thus, the scoriæ, though formed mostly on the top of the stream, are rolled beneath it, and the whole is enclosed in a cindery envelope. Or the flow may be checked by the mass of cinders, until the fluid lava bursts through them in a fresh stream. The

scoriaceous mass is a non-conductor of heat, and greatly retards
the cooling of the interior mass, which may remain hot for many
years. The arched surface of cindery blocks may become self-
supporting, and then the still fluid mass will flow away from
beneath it, leaving long tunnels or caverns. These tunnels are
especially well shown in Iceland and the Sandwich Islands.

The distance to which lava streams extend and the rapidity
with which they move are determined by the abundance and
fluidity of the lava and the slope over which it flows. Some lavas
are so liquid that they flow for many miles, even down moderate
slopes, while others are so pasty that they stiffen and set within a
short distance of the vent, even on steep grades. Ordinarily the
motion soon becomes very slow, though thoroughly melted masses
pouring down steep slopes may, for a short time, move very swiftly.
One of the lava floods from Mauna Loa moved fifteen miles in
two hours, and for shorter distances much higher rates of speed
have been observed; but this is very exceptional.

The cooling of the surfaces of the lava stream takes place rapidly, while the interior cools but slowly, and great thicknesses
require very long periods of time to become entirely cold. The
differences in the rate of cooling produce very strongly marked
varieties in the appearance and texture of the resulting rock. The
portions which have chilled and solidified very quickly are glassy
and form the volcanic glass, *obsidian*. If the swiftly cooling portions have been much disturbed by the bubbles of steam and
vapours, they are made light and frothy; in some cases, as in
pumice, they will float upon water. Otherwise, the glass is solid
and is usually very dark in colour, resembling an inferior bottle
glass in appearance. Microscopic examination shows minute,
hair-like bodies in the glass, which are called *crystallites*, and
represent the incipient stages of crystallization.

Passing inward from the surface of the lava stream, we find
the steam bubbles becoming rarer, until they cease altogether, the
vapours having escaped while the lava was still so soft that the
bubble holes soon collapsed. At the same time the glassy texture

of the rock is replaced by a stony character, which the microscope shows to be due to the formation of crystals too minute to be recognized by the unaided eye. Still deeper in the rock the stony texture passes gradually into an obviously crystalline one; and the slower the cooling, the larger will these crystals be, though in lava streams which have cooled on the surface of the ground, the whole mass, even of the deeper parts, is never coarsely crystalline.

FIG. 24. — A hand-specimen of obsidian, showing the glassy lustre and fracture

Large crystals are, it is true, very often found in lavas, but these were formed before the ejection of the mass from the volcano. Such crystals frequently contain enclosures of glass, which indicate that the crystallization went on while the surrounding mass was still fluid. The edges and angles of these crystals are often corroded by the action of the melted portion of the lava, and the motion of the stream often cracks them. These facts go to prove that the large crystals were complete when the lava, as a whole, was still fluid and in motion. Stromboli ejects great numbers of perfect crystals of augite, which must have existed in the molten lava of the vent. The lavas which contain large crystals embedded in a fine stony or glassy base are said to be of a *porphyritic texture.*

It is important to remember that all these various textures may be found in one continuous rock mass, and bear witness to the circumstances under which each part cooled and solidified. These textures also recur again and again in ancient rocks and enable us to determine their volcanic origin. The processes of rock destruction and removal have in many cases laid bare deep-seated masses which were plainly once melted like true lavas, but which have cooled very slowly and under great pressures. In such rocks the texture is usually coarsely crystalline and shows no traces of glass or scoriæ. Between the surface lava flows and such deep-seated reservoirs every form of transition may be traced, often in continuous rock masses.

Where several successive lava flows issue from one vent, at intervals which allow one stream to be consolidated before the next is poured out over it, a rough bedding or stratification results, each flow being perfectly distinguishable when seen in section. Deceptive resemblances to the true stratification of sedimentary rocks may thus arise, especially when the exposed section is short. But the wedge-like form of the sheets, the absence of bedding within the limits of each flow, and the nature of the rock itself, always enable us to distinguish these masses from the sediments which have been stratified by the sorting power of water.

A mass of lava, when it cools and solidifies, necessarily contracts, and since the cohesion of the mass is insufficient to allow it to contract as a whole, it must crack into blocks, separated by fine crevices, which are called *joints.* The mutual relations of the jointing planes, and the consequent shape of the blocks, are determined largely by the grain of the lava and its degree of homogeneity. In fine-grained (and some coarse-grained) homogeneous lavas the jointing is apt to be very regular, and to give rise to prismatic or columnar blocks, which are usually hexagonal. This shape is due to the fact that the formation of hexagons requires less expenditure of work than other figures, and is produced by the intersection of systems of three cracks, radiating from equidistant points at angles of 120°. The long axes of the prisms are at right angles to the cooling surface. Starch and fire-clay, which shrink on drying, joint in the same way. The coarser and more heterogeneous lavas usually break up into blocks of irregular size and shape.

FIG. 25. — Stream gorge, island of Hawaii; displaying modern columnar lava. (Photograph by Libbey)

It must not be inferred that the joints of all rocks are due to shrinkage on cooling. It will be shown in a subsequent chapter that such is very far from being the case.

Not all the lava produced in and around a volcanic vent can reach the surface. Some of it may be forced horizontally between the beds of the surrounding rocks, thus forming *intrusive sheets*, which, when exposed in section, may be readily distin-

guished from surface flows by the fact that they have consolidated
under pressure, and hence have no slag or scoriæ associated with

FIG. 26. — Obsidian Cliff, Yellowstone Park. Hexagonal jointing. (U. S. G. S.)

them. Other portions of the lava will fill up vertical fissures in the volcanic cone or in the underlying rocks, and, solidifying in these fissures, form *dykes*. Such a fissure, twelve miles in length and filled with molten lava, was observed by Sir Charles Lyell in the neighbourhood of Ætna. In the great eruption of Skaptar Jokul (Iceland) in 1783 lava was poured out at several points along a line two hundred miles long, and doubtless this was a great lava-filled fissure which consolidated into a dyke.

We thus see that the molten masses may not all well up through the crater of a volcano, but will seek egress along the line of least resistance, wherever that happens to be, breaching the walls of the volcanic cone, rising up through vertical fissures, or foreing their way as intrusive sheets between the beds of preëxisting rocks. In these various situations the different rates of cooling produce many varieties of rocks, though the original molten mass may have been nearly or quite identical in all of them.

Lavas which flow into the sea from a terrestrial vent, or are poured out from a submarine one, show, as a rule, but little difference from those which solidified on land, because the rapid formation of a cindery crust will protect the hot lava from contact with the water. Sometimes, however, the sudden chill will cause the lava to disintegrate into a mass like black sand.

The lavas which flow from a given vent do not always remain constant in character and composition, but change at successive periods of activity. It has frequently been observed, for example, that a series of lavas, at first intermediate in chemical composition, then acid and finally basic, have been successively ejected from the same volcano. It does not appear, however, that there is any definite law of succession in the kinds of lava emitted.

It should also be noted that neighbouring vents may simultaneously produce lavas of different composition. Thus, in the Lipari Islands, the lava of Stromboli is basic, while that of Vulcano is highly acid.

(2) *Fragmental Products.* — This division includes all the materials which are ejected from the volcano in a solid state. These

G

are of all sizes and shapes, from huge blocks weighing many tons, down to the most impalpable dust, which the wind will carry for thousands of miles. The very large blocks are commonly fragments of the older rocks through which the volcanic vent has burst its way, tearing a great hole and scattering the fragments widely. For fifteen miles around the lofty volcano of Cotopaxi in Ecuador lie great blocks of this nature, some of them measuring nine feet in diameter.

FIG. 27. — Volcanic bomb, showing scoriaceous texture; about ⅔ natural size

More important and much more extensively formed and widely spread are those fragmental products which are derived from the lava itself and are due to the sudden and explosive expansion of the vapours and gases with which the molten mass is intimately commingled and saturated. The more violently explosive the eruption, the greater the proportion of the lava that will be blown into fragments. In such eruptions as that of Krakatoa, all of it is thus dispersed and none remains to form lava flows. Cindery fragments thrown out of the vent are called *scoriæ*, while portions of still liquid lava thus ejected will, on account of their rapid rotation, take on a spheroidal form and are called *volcanic bombs*. *Lapilli* are smaller, rounded fragments, and *volcanic ash* and *dust* are very fine particles, though with a wide range of variation in size. The term *ash* is so far unfortunate that it implies combustion, but nevertheless it accurately describes the appearance of these masses.

In the immediate neighbourhood of the vent fragments of all sizes accumulate, but the farther we get from the volcano, the smaller do the fragments become. The coarser masses around the vent form a *volcanic agglomerate*, in which the fragments are of all shapes and sizes, heaped together without any arrangement. More regular sheets of large angular fragments form *volcanic breccia*, and these may be seen on a grand scale in the Yellowstone National Park, and in many other parts of the Rocky Mountain region. The finer accumulations of ash, formed at a greater distance from the vent, are roughly sorted by the air and often quite distinctly divided into layers, while, as already explained, the muds on drying set into quite a firm rock, called *tuff* or *tufa*.

As volcanoes so generally stand in or near the sea, and as the lighter fragments, such as pumice, often drift for months upon the water before they sink, while the finer dust is carried vast distances by the wind, it would naturally be expected that volcanic materials should have a very wide distribution upon the sea-bottom. Such, indeed, proves to be the case, and this kind of material, laid down in the sea, has formed important rock masses in nearly all the recorded ages of the earth's history. The exact character of the rock formed in this fashion will be governed by various circumstances, such as the fineness and abundance of the material, whether it is showered into quiet waters or along a wave-beaten coast, whether and in what proportion it is mingled with sand or mud. When the volcanic ash preponderates, a tuff is formed, very much like those which accumulate on land, but more regularly stratified.

G

The fragmental volcanic products, whether coarse or fine, retain their characteristic texture and appearance, so as to be readily recognizable, though perhaps only with the microscope. The great bulk of these materials consists of lava shattered by the steam explosions and quickly chilled. The coarser fragments display the frothy and vesicular nature of scoriæ, while the finer particles are glassy or crystalline. Mere comminution of the mass does not change its essential texture.

It will be readily imagined that lavas very rarely contain fossils. Though the flows often overwhelm living beings, the intense heat at once destroys them, seldom leaving a trace behind, though charred tree-trunks are sometimes recognizably preserved. In tuffs, on the other hand, fossils, especially those of plants, are frequently well preserved, and tuffs formed under water have fossils as abundantly as any other aqueous rocks.

(3) *The Gaseous Products* are important as agents of the eruptions, in promoting the crystallizing of the lavas, and in altering the rocks with which they come in contact. · The most abundant is steam. Carbon dioxide is common, especially when the action is failing, and often continues after all other signs of activity have died out. Sulphur dioxide (SO_2) is very characteristic and is the source of many other compounds. Sulphuretted hydrogen (H_2S) is a common volcanic gas, as is also hydrochloric acid (HCl). Several solids are vapourized, such as the chlorides of ammonium, iron, calcium, etc., but these are of little significance.

It is important to emphasize the vast quantity of material which, in many volcanic eruptions, is brought from the interior of the earth and deposited on the surface. Thus, the eruption of Skaptar Jokul in Iceland, in 1783, produced an amount of lava which is calculated as exceeding six cubic miles in volume. The fragmental materials derived from the great explosion of Krakatoa, in 1883, are estimated at 4.3 cubic miles, while for that of Temboro, in 1815, Verbeek gives the almost incredible figures of 28.6 cubic miles.

Volcanic Cones are built up by the material which the volcanoes eject, and vary in shape according to the character of those materials and to the violence of the eruptions. Those vents which yield only lavas build up cones of solid rock, the steepness of which corresponds to the degree of fluidity of the flows. The remarkably liquid lavas of the Sandwich Islands have formed cones of exceedingly gentle slope, 3° to 10° (see Fig. 28, the cone of Mauna Loa). Very stiff lavas which consolidate rapidly form very steep-sided cones. The cones which are constructed principally out of fragmental materials are steep (30°); the more so, the coarser

FIG. 28. — Mauna Loa, seen from a distance of 40 miles. (Photograph by Libbey)

the fragments which compose them, and often beautifully symmetrical, as in the noble mountains of our Pacific States, such as Mt. Shasta, Mt. Hood, and Mt. Rainier. Most cones are built up of scoriæ, ashes, and lava flows, while the fissures that radiate from the crater are filled by dykes, greatly strengthening the mountain, as in the case of Vesuvius. The latter is noted for its double head, Monte Somma being part of an ancient crater ring, one side of which was destroyed by an explosion before the cone of Vesuvius was built up. It is usually stated that the explosion which destroyed part of Monte Somma was that of 79 A.D., but it is not improbable that the destruction took place at a much earlier date.

Violent explosions occurring within a volcano blow off more or less of the top, thus producing the truncated cones and crater rings so often seen among volcanic mountains.

Volcanoes, like other mountains, are subject to the destructive activity of the atmosphere, of rivers and of the sea, and, when eruptions have ceased, this destruction may go on with great rapidity, especially in the case of cones made up of loose materials.

FIG. 29. — Mt. Shasta, California.   (U. S. G. S.)

Very ancient cones can seldom be found, for this reason, and often the lava-filled pipe is the only record left of an ancient volcano.

Fissure Eruptions. — There is much reason to believe that the mode of volcanic eruption from a single vent, described in the foregoing pages, is not the only method by which molten lava may reach the surface.   It would seem that in past times lava has welled up through great fissures and overflowed immense areas in successive floods.   As an example of this may be mentioned the vast fields of lava which occur in Idaho, Oregon, and Washington, covering more than 100,000 square miles to the depth of several hundred feet. On an even larger scale is the lava plateau of the Deccan in India, while similar but smaller lava fields occur in Patagonia, Iceland, Scotland, and other regions.

Eruptions of this type are rare in modern times and are best displayed in Iceland, where lava wells out through great fissures, some of which are 20 miles in length, and, in some cases, repeatedly

FIG. 30. — Vesuvius and Monte Somma

through the same fissure. Small craters which eject scoriæ are ranged along the fissures. At Schemakha, near the west coast of the Caspian Sea, the earthquake of 1902 was accompanied by the formation of a fissure, through which lava was extruded, very unexpectedly because igneous rock had been previously unknown in that area.

The Causes of Volcanic Activity. —Many theories have been advanced to explain the causes of vulcanism, but, it must be can-

didly admitted, none of them is satisfactory. In an elementary
book, like this, no adequate discussion of this most difficult problem can be given, but merely a brief sketch of some of the ways
in which its solution has been attempted. This problem is intimately connected with those concerning the origin of the Solar
System and the planetary evolution of the earth, which are astronomical rather than geological in their nature.

FIG. 31. — Mt. Wrangel, Alaska. (U. S. G. S.)

The principal questions for which an answer must be found in
any complete and adequate theory of vulcanism are as follows:
(1) What is the depth of the reservoir whence the volcanic materials are derived? and, consequently, what are the relations of
the different vents, near and remote, to one another? (2) What
causes the high temperature of volcanic materials? (3) What is
the origin of the steam and other vapours and gases? (4) What
produces the ascensive force of the lava? (5) Why should volcanic action be so generally intermittent? (6) The past and .

present distribution of volcanoes should be explained, as also the shifting of activity, which dies out in one region and appears in another.

(1) No certain answer can yet be given to any of these questions, chiefly because we can observe only what goes on at the surface of the earth and still remain ignorant concerning the physical condition of the interior. Hence, nothing is known as to the

FIG. 32. — Truncated tuff cone, island of Oahu. (Photograph by Libbey)

depths from which the volcanic materials rise. According to one view which is quite widely held, the reservoirs of lava are local and comparatively superficial, which would explain the fact that vents which are quite near together may be entirely independent of each other and eject very different materials. This view is further confirmed by the speedy exhaustion of many volcanic vents, a large number of which have had but a single eruption. According to this hypothesis, the length of time during which a volcanic region remains active is determined by the size of the reservoir which supplies it.

On the other hand, it is maintained by many students of vulcanism that the material is derived from a deep-seated layer of actually or potentially fused material, which everywhere underlies the surface of the earth. In support of this opinion, it is pointed out that often widely separated volcanoes are evidently connected in some manner, and that the volcanic products of all regions are closely similar. These two hypotheses are not altogether contradictory, for it is quite possible that some volcanoes may be supplied from shallow reservoirs which are soon exhausted, and others from a deeper and general source of supply.

(2) The high temperature has been explained in two principal ways. If we accept the nebular hypothesis of the origin of the Solar System, we must grant that the earth was once a globe of glowing gas, which subsequently condensed, in part at least, to a molten globe, and then solidified on the surface and to unknown depths. One explanation of volcanic heat is that it is due to the originally high temperature of the earth, not yet lost by radiation, whether in local reservoirs or in a universal, deep-seated layer. By those who accept the latter view, it is generally assumed that the interior of the earth is exceedingly hot, but solidified by pressure, and that when, by folding or fracturing of the overlying rocks, this pressure is partially relieved, the highly heated masses become liquefied along the line of diminished pressure.

In the second class of hypotheses on this subject of temperature, it is assumed that the earth never was in a molten condition, or that it has already so far cooled that its proper heat is no longer sufficient to produce fusion of rock. From this point of view, the great heat is believed to be generated mechanically, by the friction of internal masses under compression and contraction, or, with much less probability, to be due to chemical processes, or even to radio-activity.

Similar divergences of opinion obtain with regard to the nature and origin of the lavas ejected by volcanoes. The view most commonly held is that they are, for the most part, the original, unaltered material of the globe, whether this has always remained fluid, or has been remelted by release of pressure, or otherwise. According to another opinion, volcanic products are formed from the fusion of sedimentary material which was laid down under water, but has been deeply buried within the crust of the earth by subsidence. A third view recognizes both sources of supply.

(3) The problem as to the origin of the steam which plays so important a part in volcanic eruptions is likewise very differently solved by different investigators. One opinion is that the steam, like the lava itself, is primordial and was absorbed from the atmosphere (which then contained all the waters of the sea) when the surface of the globe was still molten. Melted substances will, it is known, absorb many times their own volume of steam and gases, when in contact with them under pressure. From this it is inferred that the lava has contained the steam ever since the first cooling of the surface crust. A second opinion derives the water from the surface of the earth, supposing that it descends partly through fissures and partly through the pores of the overlying rocks by capillarity. The nearness of most volcanoes to the sea is looked upon as favouring this view. Others, again, employ both methods of explanation, regarding the ordinary steam which impregnates all lavas as primordial, but believing that the violently explosive eruptions are caused by the sudden access of large bodies of water to the lava masses. The evidence of known facts is at present distinctly in favour of the view that the steam is essentially primordial.

(4) The causes of the ascensive force of the lava column are sought by various writers in several different agencies. Some find an all-sufficient cause in the steam pressure, while others maintain that some other force must be at work and find this partly in the weight of overlying masses, especially in the case of sinking blocks, and partly in the unequal contraction of the earth, and consequent pressure upon the molten or plastic layer beneath. It has been calculated that a radial contraction of one millimetre " would

suffice to supply matter for five hundred of the greatest known volcanic eruptions." (Prestwich.) That steam is an essential factor in this, as in other volcanic phenomena, appears to be well established.   Steam and other gases and vapours, under great pressures and at high temperatures, have a remarkable penetrating power and, when suddenly released, will perforate metal or rock like a projectile.   Even under a pressure of only a few hundred atmospheres, superheated steam corrodes and abrades like the sand-blast, as a substitute for which it is now frequently employed.

(5) The intermittency of volcanoes and their mode of distribution add to the difficulty of the whole subject, but any complete theory must explain them.   The views which bring volcanic action into relation with the mechanical changes in the crust are those which seem most consonant with the known facts of the past and present distribution of the vents.

Here, for lack of space, we must leave the subject.   Enough has been said to show how far we still are from understanding the mystery of volcanoes.

THE INTERNAL CONSTITUTION OF THE EARTH

The interior of the earth is completely beyond the reach of direct observation and what is known, or may be reasonably inferred, as to its physical constitution, is derived from various lines of indirect evidence.   The deepest boring ever made is less than $\frac{1}{3000}$ part of the earth's radius, and we have no experience with such enormous pressures as obtain within the mass of the globe, and can therefore form but imperfect conceptions of their effects.

From observations with the pendulum and plumb-line it is calculated that the specific gravity of the earth as a whole is 5.6, while the average specific gravity of the rocks which form the accessible parts of the crust is only 2.6.   It follows that the interior of the globe is composed of much denser materials than the superficial portion, and this fact, together with the phenomena of terrestrial magnetism, has led many to the belief that the earth is substantially a globe of iron.   This inference is also supported by the occurrence of native, or unoxidized, iron in certain igneous rocks.

Temperature of the Earth's Interior. — Volcanoes, which eject molten and white-hot lavas, and thermal springs, which pour out floods of hot and even boiling water, plainly indicate that the interior of the earth is highly heated, at least along certain lines. Direct observations tend to prove that this high temperature is universally diffused through the earth's mass.   For a short distance below the surface of the ground the temperature varies, like that of the air, though not so greatly, between day and night. Farther down, the daily variation ceases, but there is a seasonal variation, also with a less extreme range than the seasonal differences of the air-temperatures.   The almost constant temperature of deep cellars and ice-houses is a familiar fact.   At a still greater depth is reached a level where the temperature remains the same throughout the year, and is but slightly in excess of the *average annual* temperature of the air above ground at a given locality. Evidently the temperature at the level of no variation is determined by the solar heat and other climatic factors, and its depth depends upon the range of temperature changes in the air.   In the tropics, with their uniform degree of heat, the level of no variation is only three or four feet below the surface, and much the same is true of the polar regions, where the ground is permanently frozen to a depth of several hundred feet, but in the temperate zones this level is much deeper; generally speaking, the depth increases with the latitude and at New York is about fifty feet, but the level again rises toward the surface as the polar regions are approached.

Even at the level of no variation the inherent heat of the earth makes itself apparent, and below this level the temperature increases with the depth, though at very different rates in different places.   Thus, in Great Britain the rate of increment varies between 1° F. for every 30 feet of descent and 1° for every 90 feet. Increasing heat with increasing depth is observed in all deep borings, tunnels and mines, and often has completely checked any further penetration. The levels at which the great tunnels under the Alps were placed were determined chiefly by considerations of temperature, as it was necessary to avoid a degree of heat in which men could not work.

The deepest borings in the world are those in Prussia, some of them considerably exceeding a mile in depth. Observations made in these borings give an average increment of about 1° F. for every 60 feet. Should this rate be continued regularly, it would reach, at a depth of 35 miles, a heat sufficient to melt almost any known rock, *at atmospheric pressure*. However, the available observations are much too superficial to permit the formulation of any general law, farther than the establishment of the fact of the universal increment of temperature with descent into the earth.

Physical State of the Earth's Interior. — Opinions concerning the internal constitution of the earth differ very radically and only within the last few years has evidence begun to accumulate which permits the drawing of certain inferences with a considerable degree of probability.

Many hypotheses as to the condition of the earth's interior have been proposed, of which the following are the most important: (1) That the earth is a molten globe, covered only by a relatively thin crust. (2) That it is substantially a solid body. (3) That the interior passes gradually from a solid crust to a gaseous core, heated beyond the critical temperature and yet under such enormous pressure that the core is as rigid as a solid body, but still a gas in molecular condition. According to this theory, the temperature of the earth at the centre is about 180,000° F. and the pressure 3,000,000 atmospheres. (4) That it has a very large solid nucleus surrounded by a layer of fused material, upon which the crust floats in equilibrium.

In the present imperfect state of knowledge, it is not possible to decide definitely between these conflicting hypotheses, but, as mentioned above, evidence has been obtained which seems to point clearly to certain conclusions.

(1) The first, or " thin crust " hypothesis, is now almost entirely abandoned, for there is really no evidence in its favour and very much against it. The velocity and character of the earthquake waves which traverse the mass of the globe and the astronomical relations of the earth as a planet, especially the tidal phenomena, are strongly opposed to this view.

(2) That the earth is substantially a solid body, is the opinion held at present by many geologists and astronomers. In support of it may be cited the astronomical evidence just mentioned, and the earthquake waves, the speed of which requires a medium more rigid than steel, while the very transmission of the transverse or distortional waves would seem to require a solid medium.

(3) Between the second and third hypotheses the distinction is one not easy to explain in an elementary manner, and there are many modifications of the latter. According to Arrhenius, " the rigidity of the earth is greater rather than less than that of steel, but the interior forms an extremely viscous mass, with qualities somewhat like those of asphalt at a low temperature, of pitch, sealing-wax and glass." These bodies behave under forces of deformation, which act quickly or with constantly changing direction, like solids; but under slow, long-continued pressures, acting in a constant direction, they behave like fluids. Observations and records of very distant earthquakes show that when the path of the mass-waves penetrates to a depth of more than three-fifths of the earth's radius, the transverse waves of distortion are either extinguished or greatly retarded. This points to a change in the character of the medium and decidedly supports the notion of a gaseous core postulated by this hypothesis.

(4) The fourth hypothesis, which assumes the presence of a fused layer between the crust and the solid nucleus, with gradual transitions from one to the other, is believed to avoid the astronomical objections to a molten globe, as well as certain geological difficulties in accepting the hypothesis of an entirely solid earth. The earthquake observations, so frequently cited, are decidedly op-

posed to the belief that a layer of actually fused matter can exist
at a moderate depth below the surface.

It is thus probable that below the superficial crust, only a few
miles in thickness, the great mass of the earth is composed of very
dense material, which transmits elastic waves like a very perfectly
elastic solid, and yet is so highly heated and under such enormous
pressure that it is potentially fused and liquefies upon sufficient
release of pressure, and yields plastically to slow, long-continued
stresses which act in a constant direction. Furthermore, there is
evidence that a core, two-fifths of the earth's diameter and composed of matter in a different state of aggregation, which may be
gaseous, occupies the centre.

In this connection something should be said concerning the
important theory of *isostasy*, which may be thus defined: "The
earth is composed of heterogeneous material which varies considerably in density. If this heterogeneous material were so arranged that its density at any point depended simply upon the
depth of that point below the surface, . . . a state of equilibrium
would exist, and there would be no tendency toward a rearrangement of masses.

"If the heterogeneous material composing the earth were not
arranged in this manner at the outset, the stresses produced by gravity would tend to bring about such an arrangement; but as the
material is not a perfect fluid, . . . the rearrangement will be
imperfect. . . . The excess of material represented by that portion of the continent which is above sea-level will be compensated
for by a defect of density in the underlying material. The continents will be floated, so to speak, because they are composed of
relatively light material; and, similarly, the floor of the ocean will,
on this supposed earth, be depressed, because it is composed of
unusually dense material. This particular condition of approximate equilibrium has been given the name isostasy." (Tittmann
and Hayford.)

The recent very extensive and exact operations of the United
States Coast Survey have brought strong confirmation of the

theory of isostasy. "The United States is not maintained in its position above sea-level by the rigidity of the earth, but is, in the main, buoyed up, floated, by material of deficient density." (Tittmann and Hayford.)

It should be noted that isostasy, being a condition of approximate equilibrium, is conservative in tendency and does not explain the active movements of elevation and depression of the crust.

The whole subject of the temperatures and physical state of the earth's interior has been complicated and obscured by the discovery of radio-activity, and already some very far-reaching inferences have been drawn from the distribution of radio-active substances in the rocks. At present, however, it would be premature to give any extended discussion of this problem.

Summary. — The study of the subterranean, or igneous, agencies has proved to be very unsatisfactory in the way of explaining the phenomena and referring them to the operation of understood physical agents, because so little is really known and so much remains to be discovered. Nevertheless, we have learned much that is of great importance in geological reasoning. We have seen that the earth contains within itself a great store of energy, and that its interior, in whatever physical state that may be, is highly heated, and possesses great quantities of material which is either actually or potentially molten, and is permeated with superheated steam and other gases. This molten material is often forced upward, and is either poured out at the surface, or fills up fissures and cavities in the rocks, or pushes its way between them. Cooling under various circumstances, the molten masses consolidate into a great variety of characteristic rocks, frothy, glassy, or crystalline. Explosive discharges of steam blow the melted rock into fragments of all grades of fineness, and these fragments likewise accumulate either on the land or under water, and form rocks, the nature and origin of which may be readily recognized.

We have further seen that the operation of these subterranean forces produces shocks and jars in the interior, which are propagated to the surface as earthquakes, and there bring about permanent changes, associated with the fissuring and dislocation of the rocks, landslips, alteration in the course of rivers, formation of lakes, and the like. The frequency of earthquakes, their wide geographical range, and the constant tremor of the ground detected by delicate instruments, led us to infer that the crust of the earth is decidedly unstable.

This conclusion we found strengthened by the oscillations of level between land and sea, which, though extremely slow, are seen to be still in progress. Historical geology will show us that these changes of level have, in the course of ages, been effected on the grandest scale. Almost all the great continents are composed of rocks which, for the most part, were laid down in the sea and still contain the fossils of marine animals, and this shows that these continents have been under the sea. Not that all parts of any continent were submerged at the same time, but now one part and now another was overflowed and again emerged, until nearly all have been covered in their turn.

In brief, the principal geological functions of the subterranean agencies are two: (1) they bring up from below and form at the surface, and at all depths beneath it, certain characteristic kinds of rocks; and (2) they tend to increase the inequalities of the earth's surface, and thus to counteract the agencies which are cutting down the land and steadily tending to reduce it to the level of the sea.

SECTION II

SURFACE AGENCIES

THE superficial, or surface, agents, all of which are manifestations of solar energy, are those which act upon or near the surface of the ground; only one, circulating water, is able to penetrate to considerable depths within the earth. The work of the surface agents may all be summed up in two categories, the *destruction* and *reconstruction of rock*. These two processes are complementary; for, since matter is indestructible, and can have only its position and physical and chemical relations changed, it is obvious that what is removed in one place must be laid down in another. Neither process, therefore, can go on without the other, and reconstruction necessarily implies antecedent destruction to furnish the materials. Ceaseless cycles of change are everywhere in progress, new combinations continually formed, and older rocks worked over into newer. It is this circulation of matter upon and within the crust of the earth, which we have already compared to the physiological changes in the body of a living organism.

The work of destruction and reconstruction is in a continuous series of changes, beginning with the mechanical *disintegration* or chemical *decomposition* of an older rock, followed by the *transportation* of the material thus supplied, for longer or shorter distances, its *deposition* in a new place, and finally, if the series is complete, the *consolidation* of the loose débris into rock.

The processes of rock destruction and removal, which are grouped together under the general name of *denudation*, or *erosion*, are chiefly confined to the land surfaces, while those of reconstruction take place principally under bodies of water and, most

H   97

of all, in the sea. An important work of reconstruction is also performed on the land, but on a less extensive scale than in the sea.

The surface agents all act both destructively and reconstructively according to circumstances, but with very different degrees of efficiency. Certain agents are preëminently destructive, others as preëminently reconstructive, while others again operate most efficiently as agents of transportation. Again, as was pointed out in the Introduction, the depth below the surface at which operations take place has a very important bearing upon the effect produced. From this point of view, we may regard the earth's crust as being composed of a number of concentric shells, of somewhat irregular thickness and rather indefinite, or even fluctuating, boundaries. The superficial shell, which extends from the surface down to the level of the ground water (see p. 124), is called the *shell of weathering* (Van Hise's " belt of weathering "), and is characterized by the oxidation, carbonation, and hydration of minerals, and great quantities of material are removed in solution. As a result of these operations the rocks are decomposed, becoming soft and friable; the minerals produced are few in number, of simple composition, and are usually imperfectly crystallized. The second shell, that of *cementation*, which extends downward from the ground-water level to a varying depth with undetermined lower limit, is largely saturated with water, and hence has but a limited supply of oxygen and carbon dioxide. Therefore, while oxidation and carbonation occur, they are less important than hydration, and the resultant minerals are more crystalline than in the shell of weathering. Solution goes on, but deposition becomes very important and fills the openings in the rocks with mineral matter, and the general effect of the various processes is to harden the rocks.

The work of the surface agents, in its threefold aspect of erosion, transportation, and deposition, is profoundly affected by the diastrophic movements of the earth's crust. In a given case the effects produced will vary greatly in accordance with the elevation, subsidence, or stationary character of the region. In general, elevation favours denudation, and subsidence favours deposition of transported material.

In studying the work of the surface agents, the logical order of treatment requires that the destructive operations be considered first. The agencies to be examined are: (1) the atmosphere, (2) running water, (3) ice, (4) lakes, (5) the sea, (6) animals and plants. Of these various agents the work is principally mechanical, but water, in its various forms, is a slow but extremely efficient agent of chemical changes.

CHAPTER IV

DESTRUCTIVE PROCESSES. — THE ATMOSPHERE

THE atmospheric agencies are by far the most important of the destructive or denuding agents, because no part of the land surface is altogether exempt from their activity. Their work is described by the general term *weathering*, and is shown at once by the different appearance of freshly quarried stone from that which has been long exposed in the face of a cliff, or even in ancient buildings. While such agents as rivers and the sea do work that is much more apparent and striking than that of the atmosphere, yet they are more locally confined, and even in their operations the atmosphere renders important aid. Though no part of the land surface is entirely free from the destructive activity of the atmosphere, the rapidity and intensity of this activity vary much in different places. There are, in the first place, the great differences of climate to be considered, differences in the amount and distribution of the rainfall, of temperature, and of the winds. In the second place, the resistance offered by the various kinds of rocks to the disintegrating processes differs very greatly, in accordance with the differences of hardness and chemical composition. Again, the presence or absence of a covering of protective vegetation has an important influence upon the amount and character of the destruction effected.

The outcome of all these varying factors is to produce very irregular land surfaces. While the tendency of the atmospheric agencies is gradually to wear down the land to the level of the sea, yet in that process some parts are cut away much more rapidly than others; and hence the *first effect* of denudation is an increasingly irregular surface. The overlying screen of soil conceals many of these irregularities, especially the minor ones; and were that screen removed, the rock surface would be seen to be much more irregular and rugged than the actual surface of the ground.

So long as the land surface is varied by hill and valley, it is said to possess *relief;* and when it has all been planed down to a flat or gently sloping surface, raised but slightly above the level of the sea, it is said to have reached the *base-level of erosion*, or to be *base-levelled*.

The atmospheric agents may be conveniently divided into (1) rain, (2) frost, (3) changes of temperature, (4) wind.

1. RAIN

The work of the rain, which is both chemical and mechanical, varies greatly in accordance with climatic factors. The annual precipitation of two regions may be the same in amount, but in one the rainfall may be in very frequent, gentle showers, and in the other in less frequent, but far heavier and more violent downpours; under such different conditions the destructive work of the rain will be very different. Effects of still another kind are produced in those regions which have regularly alternating rainy and dry seasons. Temperature also modifies the work of the rain to an important degree, so that results are brought about in warm countries quite different from those observed in temperate and cold regions. Thus, each climatic zone exhibits the work of the rain with characteristic differences.

Perfectly pure water would act upon rocks with extreme slowness, but such water is not known to occur in nature. The raindrops, formed by the condensation of the watery vapour of the atmosphere, absorb certain gases which very materially increase the solvent power of the water. Of these gases the most important are oxygen (O) and carbon dioxide (CO_2), and all rain-water contains them.

It was formerly supposed that rain-water in percolating through the soil acquired additional destructive efficiency by absorbing certain products of vegetable decomposition called humous acids. Recent exact investigations, however, have thrown grave doubts upon the existence of these acids in the soil, and the effects which had been ascribed to them are now referred to carbon dioxide acting out of immediate contact with the atmosphere.

One of the first and simplest effects of atmospheric moisture consists in the *hydration* of the minerals exposed to it. Hydration, the taking up of water into chemical union, is an important agency of decay; it causes an increase of volume and thus greatly increases the pressure in the lower parts of rock masses which contain hydrating minerals. In the District of Columbia " granite rocks have been shown to have become disintegrated for a depth of many feet, with loss of but some 13.46 per cent of their chemical constituents. . . . Natural joint blocks brought up from shafts were, on casual inspection, sound and fresh. It was noted, however, that on exposure to the atmosphere, such not infrequently fell away to the condition of sand." (Merrill.)

Oxidation affects especially the iron minerals and thus brings about conspicuous colour changes, for iron compounds form the principal colouring matter of the rocks and soils. Ferrous compounds give little colour, but the rocks in which they occur are apt to have a blue or grey tint, due to other substances, both organic and inorganic. When such rocks are exposed to the action of air and water, the ferrous compounds are oxidized, producing ferric oxide and ferric hydrates, the former giving a red colour and the latter various shades of yellow and brown.

When fired in a kiln, a blue clay will yield red bricks, by the conversion of $FeCO_3$ into Fe_2O_3. In nature, rain-water effects a similar change, and the contrast between the superficial and deep-seated parts of the same rock is often as great as between blue clay and red brick. Weathered blocks stained red on the outside are often blue, grey, or nearly black on the inside, because the change has not affected the whole mass.

An especially important and wide-spread change is *carbonation*, due to the carbon dioxide which all natural waters contain in greater or less quantity. The silicates are attacked and decomposed in a manner that will be explained below, and, under certain circumstances, the insoluble ferric hydrates are converted into soluble ferrous carbonate.

Finally, *solution* plays a highly important rôle in the destructive work of the rain. All rocks contain some soluble material, and when this soluble material is removed, the rock crumbles into a friable mass, which, on complete disintegration, forms soil. This may be illustrated by a block of frozen earth, which is as hard as many rocks, being cemented by the ice crystals, which bind the particles of soil together. When the ice is melted, the mass immediately becomes incoherent. So, in the rocks, the removal of even a small quantity of soluble material often causes the whole to crumble. Except vegetable moulds, all soils are derived from the decay and disintegration of rocks.

The chemical composition of the rock-forming minerals varies so much, that the processes which destroy them must vary correspondingly, differing in the case of the igneous rocks, on the one hand, from that of the sedimentary rocks, on the other.

Most igneous rocks are made up of crystals of some kind of felspar (see p. 13), associated with such minerals as augite (p. 16), hornblende (p. 16), and quartz (p. 12). In granite, for example, which is composed of orthoclase felspar, quartz, and mica or hornblende, rain-water slowly attacks the orthoclase by dissolving out the potash, probably as a carbonate, and also a considerable proportion of the combined silica. The aluminous silicate, which forms the residue, is hydrated as kaolinite (Al_2O_3, $2 SiO_2$, $2 H_2O$), forming clay, while the quartz is little or not at all affected. The mica, if present, is very slowly attacked, the hornblende more readily.

The decomposition of granite, then, results in the formation of clay, through which are scattered flakes of mica (if mica were originally present) and the unaltered grains of quartz. In the other igneous rocks the manner of decomposition is essentially similar; the complex silicates are broken up into simpler compounds, clay being derived from the aluminous silicates, especially the felspars, while the quartz, if present, is broken up into fragments and forms sand. The bases, potash, soda, lime, magnesia, iron, etc., are removed in solution, chiefly as carbonates, and more or less of the silica is also dissolved and carried away. Even when an igneous rock is yet firm and hard and, to the naked eye, appears to be quite unchanged, the microscope often reveals the first stages of decay.

FIG. 33. — Boulders of weathering, Eldon Mt., Arizona. (Photograph by A. E. Hackett, Flagstaff, Arizona)

In most tropical regions, where there is a long dry season, followed by a wet season of violent rainfall, the manner of decay is characteristically different. The felspars and other aluminous silicates lose nearly all of their silica, so that the residue is the hydrated oxide of aluminium, not the silicate, and the iron is oxidized, forming nodules and masses and staining the whole a deep red. This characteristic warm-country soil is called *laterite*.

Very many igneous rocks weather into heaps and masses of rounded boulders, which are often mistaken for the deposits made by glaciers. The spheroidal shape is due to the more rapid decay of the edges of the original joint blocks, which are attacked on both sides at once. As the edges and angles are removed more quickly than the broad faces of the blocks, a rounded form results. Once acquired, the round shape is long retained, because then decay penetrates at a nearly equal rate from all parts of the surface.

Rocks which are themselves composed of substances derived from the decay of older rocks are attacked in their turn and yield material for new formations. These derivative rocks, such as sandstones, slates, and limestones, are affected in characteristic ways by the rain.

Sandstones are composed of grains of sand (quartz, SiO_2) cemented together; the cementing substance may be silica itself, some compound of iron, such as Fe_2O_3, or carbonate of lime ($CaCO_3$), and the dissolving away of the cement causes the rock to crumble into sand. In a sandstone with siliceous cement the action is excessively slow, atmospheric waters having very little effect upon silica, but underground it is slowly attacked. Ferric oxide (Fe_2O_3) is likewise unchanged by rain-water, but beneath the soil it is converted into the soluble carbonate and removed. The uppermost layers of red sandstone are often thus completely disintegrated into loose sand, bleached by the removal of the iron which gave it its colour. Carbonate of lime is very soluble in water containing carbon dioxide, as all rain-water does, and in sandstones with calcareous cement, disintegration is rapid. In sandstones and slates it is the cementing substance which is removed, leaving the grains of sand or particles of clay unchanged. This is because the materials of these rocks were, for the most part, originally derived from the decomposition of the igneous rocks, and the minerals which compose them are already of a very simple and stable character.

The sandstones are largely employed for building materials, and their value and permanence for such purposes depend principally

upon the character of the cementing substances in them. For
this reason, the siliceous and ferruginous sandstones are the most

FIG. 34. — Soil originating in place by the decomposition of sandstone

durable, those with calcareous cements usually yielding with comparative rapidity to the attacks of the weather.

Slates and shales, by removal of their soluble constituents, crumble down into clay.

Limestones are among the few rocks which are chiefly or entirely made up of soluble material, the carbonate of lime ($CaCO_3$). This is attacked by the rain-water, dissolved and carried away in solution, while the insoluble impurities contained in the rock remain to form soil. The proportion of such impurities varies greatly in different limestones, and hence the residual soil will vary, but it is generally a clay, since that is much the commonest of the impurities in limestone. Sand also occurs in limestones, either with or without clay. When the sand forms a coherent mass, out of which the calcareous material has been dissolved, it is called *rotten stone*.

The gradual formation of soil by the disintegration of rock may. be easily observed in excavations, even shallow ones, such as cellars, wells, railroad cuttings, and the like. At the surface is the true soil, which is usually dark-coloured, due partly to the admixture of vegetable mould, partly to the complete oxidation and hydration of its minerals. Next follows the subsoil, which, owing to the absence of vegetable matter and the less complete oxidation and hydration, is of a lighter colour. The subsoil is frequently divided into distinct layers, and often contains unaltered masses of the parent rock, which have resisted decomposition, while the surrounding parts have become entirely disintegrated. By imperceptible gradations the subsoil shades into what looks like unaltered rock, but is friable and crumbles in the fingers; this is *rotten rock*. From this to the firm, unchanged rock the passage is equally gradual.

In the northern portions of the United States the soil is, in most localities, of only moderate depths, because at a late period (geologically speaking) this region was covered with a great ice-sheet, which swept away much of the accumulations of ancient rock decay. In the parts of the country where the ice-sheet did not come, the soil is much deeper, and in tropical lands it attains remarkable depths. In our Southern States the felspathic rocks are often found thoroughly disintegrated to depths of 50 or 100 feet, while in Brazil the soil is often 200 to 300 feet deep.

The mechanical effect of rain is less extensive, perhaps, than its chemical work of disintegration, but is very important, nevertheless. Under ordinary conditions, this mechanical work consists in the washing of soil from higher to lower levels. How considerable is the movement of soil that has thus been brought about, may be imagined when one sees, after a heavy rain, the rills which run over the slopes, muddy and charged with sediments, and how turbid the streams become with the soil which the rain washes into them. Bare soil is rapidly torn up and washed away by the action of rain, but a covering of vegetation, and especially of the elastic and matted stems and roots of grasses, much retards the action.

Other things being equal, the rapidity with which the rain sweeps away the soil depends upon the steepness of the slope upon which the soil is formed; for gravity largely determines these movements. On cliffs and steep hillsides the soil is quickly removed, and in such places it is thin or quite lacking, while in the valleys it often accumulates to great depths. Even on gentle slopes and almost level stretches the rains slowly wash it downward, and eventually into the streams which carry it to the sea. The soil is thus not stationary, but under the influence of the rains and streams is slowly but steadily travelling seaward. Disregarding the alluvial deposits made by rivers, and soils accumulated by the action of ice or wind, the soil of any district is thus a residual product, and its quantity represents the surplus of chemical disintegration over mechanical removal.

The mechanical action of rain is greatly increased by extreme violence and volume of precipitation; a single "cloud-burst" will do far more damage than the same quantity of rain falling in gentle showers. Those who know only the temperate regions can form but imperfect conceptions of the violence of tropical rains. On the southern foot-hills of the Himalayas, for example, the rainfall is exceedingly great (in some localities as much as 500 inches per annum), and almost all of it is precipitated in six months of the year; especially remarkable is the quantity which often falls in a single day. "The channel of every torrent and stream is swollen at this season, and much sandstone and other rocks are reduced to sand and gravel by the flooded streams. So great is the superficial waste, that what would otherwise be a rich and luxuriantly wooded region is converted into a wild and barren moorland." (Lyell.)

FIG. 35. — Bad lands of South Dakota.   (U. S. G. S.)

The action of rain is thus by no means uniform, the results depending upon so many and such varying factors, that we may find marked differences in closely adjoining regions, and even in one and the same mass of rock. One of the most remarkable monuments of rain erosion is exhibited by the curious districts in the far Western States known as the *bad lands*, which cover many thousands of square miles in the Dakotas, Nebraska, Wyoming, Utah, etc. The bad-land rocks are mostly rather soft sandstones and clays, with prevailingly calcareous cements, and formed in nearly horizontal beds or layers. The rainfall is light, though torrential showers sometimes occur; but the absence of vegetation is favourable to its efficiency, and the present aridity of the climate is not of very long standing, from a geological point of view. The chemical action of the rain has disintegrated the rocks by dissolving out the calcareous cement, and then the débris so formed has been mechanically washed away.

At the present time the action of the rain is very slow, because the débris which covers the sides of the cliffs and slopes is almost

FIG. 36. — Bad lands near Adelia, Nebraska. The rock in the middle distance is sandstone formed in a stream-channel and the bluffs are flood-plain deposits. (U. S. G. S.)

impervious to water, and holes left after the excavation of fossil skeletons often remain visible for many years; but where the bare rock is exposed, the disintegration often proceeds with extraordinary rapidity, and a single shower will produce notable effects. The different layers of rock resist decay differently, and even in the same bed some parts are much more durable than

others.   This differential weathering has resulted in that remarkable variety and grotesqueness of form, resembling the ruins of
gigantic towers and castles, for which the bad-land scenery is
famous.   The sculpture of the rain produces this variety in accordance with the arrangement of the more and less durable
layers.   The varying arrangement of these layers produces a
fantastic topography.   A variant of bad-land topography is given

FIG. 37. — Bad lands in Wyoming, with talus slopes.   (U. S. G. S.)

by the pillars of Monument Park, Colorado, which are due to
weathering, the capping of hard rock protecting the soft sandstone
below..   (See Fig. 38.)
    The mechanical wash of rain is greatly retarded when the ground
is covered with dense vegetation, which protects the soil against
the impact of raindrops and the wear of rain rills.   The removal
of the vegetation is often speedily followed by disastrous results,

and especially the reckless and wanton destruction of the forests, which has gone on in this country ever since its settlement by Europeans, has been followed by the loss of valuable soil on a vast scale. Speaking of the soil destruction in the old fields of

FIG. 38. — Monument Park, Colorado.   (U. S. G. S.)

southern Mississippi, McGee says that they are washed away, " leaving mazes of pinnacles divided by a complex network of runnels glaring red toward the sun and sky in strong contrast to the rich verdure of the hillsides never deforested. . . . Whole villages, once the home of wealth and luxury, are being swept away at the rate of acres for each year."

It is to be hoped that the work of the United States Bureau of Forestry, in endeavouring to check this terrible destruction, may receive the support it so well deserves from every intelligent lover of his country.

2. FROST

The term *frost*, in this connection, is restricted to the freezing of water. Water is one of the comparatively few substances which expand considerably on solidifying. This expansion amounts to about one-eleventh of the original bulk of the water, and, exerting a pressure of somewhat more than 2000 pounds per square inch, takes place with irresistible power, bursting thick iron vessels like egg-shells.

Excepting loose, incoherent masses, like sand and gravel, no rocks are formed of continuous sheets of material, but are rather to be considered as masses of blocks, divided by *joints* (see pp. 5 and 136). In addition to these visible clefts, the blocks are traversed by minute crevices, rifts, and pores, all of which openings take up and retain quantities of water, as may readily be seen by examining freshly quarried stone. When exposed to a low temperature, the water freezes and forces out the large blocks and shatters them into pieces of smaller and smaller size. The fragments thus formed are called *talus*, and great accumulations of such blocks are found at the foot of cliffs in all regions where the winters are at all severe. Talus accumulations are also formed by other agencies, as will be seen in the sequel. Alternate freezings and thawings not only break up rocks, but cause the broken fragments and soil to work their way down slopes. Each freezing causes the fragments to rise slightly at right angles to the inclined surface, and each thawing produces a reverse movement; hence the slow *creep* down the slope.

The action of frost is, of course, practically absent in the lowlands of the tropics, but in high mountains and in all countries which have cold winters, frost is an agent of great importance in the mechanical shattering of rocks and slow destruction of cliffs. The hardest rocks are shivered into fragments and, dislodged from their places, the fragments roll down the mountain side till they

FIG. 39. — Shales " creeping " under the action of frost. (U. S. G. S.)

come to rest, perhaps thousands of feet below. Immense accumulations of frost-made talus are to be found in such places as the foot of the Palisades of the Hudson, the abrupt southern slope of the Delaware Water Gap, and wherever cliffs or peaks of naked rock are exposed to severe cold. Many mountain passes are so bombarded by falling stones as to be extremely dangerous; in the

Sierra Nevada of California, talus slopes as much as 4000 feet high are reported, all the work of frost.   At Sherman, where the Union Pacific Railroad crosses the "continental divide," the ground is covered for miles with small, angular fragments of granite broken up by the frost.

In the polar regions frost is probably the most important of the disintegrating agents.   In Spitzbergen Beechy found that in summer the mountain slopes absorb quantities of water, which freezes

FIG. 40. — Cliff and talus slope, Delaware Water Gap, N.J.

in winter with very destructive effect.   "Masses of rock were, in consequence, repeatedly detached from the hills, accompanied by a loud report, and falling from a great height, were shattered to fragments at the base of the mountain, there to undergo more rapid disintegration."   Similar phenomena are reported from the Aleutian Islands of Alaska.

The action of frost is, in itself, purely mechanical; no chemical change is occasioned by it, and the smallest fragments into which a block may be riven are sharp and angular, and the minerals have

unaltered and shining faces. But, on the other hand, frost pre-
pares the way for the more rapid action of rain and percolating
waters. The effects of these agents are produced upon the surface
of the rocks and the walls of the crevices which run through them.
By breaking up the blocks, the frost greatly increases the surface
and thus facilitates the work of the rain. A breaking up of one
cubic foot into cubic inches multiplies the exposed surface by 12.

Frost is an extremely superficial agent and acts effectively only
a few feet below the surface. In the polar regions the ground
is permanently frozen to a depth of several hundred feet, but the
shattering of rocks requires alternate freezings and thawings.

Rain and frost are agents whose effects are most important in
regions of moist climate and abundant rainfall, for both are forms
of the activity of water. Few regions of the earth's surface are
altogether rainless, but nearly all of the continents have great
desert areas in which atmospheric precipitation is very light. It
might seem that in such deserts the work of rock disintegration
must be practically at a standstill, and that the circulation of material must be so slow as to be hardly distinguishable from complete stagnation. Even in these regions, however, the rain
accomplishes something, and it is aided by other agencies which in
moist climates play a much more modest rôle; these are the changes
of temperature and the wind.

3. Changes of Temperature

In moist and equable climates these temperature changes are of
very subordinate importance as a destructive agent and act chiefly
in giving an easier passage to percolating waters. In arid regions,
on the other hand, especially on high mountains and plateaus,
where there are great differences of temperature between day and
night, this agency becomes much more important. During the
day the naked rocks are heated very hot by the full blaze of the
sun, while at night the rapid radiation which occurs in dry and

thin air chills the outer layers of rock very quickly, and they attempt to contract upon the still heated and expanded interior. Thus, stresses are set up which the rock cannot resist, and pieces, great and small, are split off from the surface. In this manner great talus slopes, like those due to frost action, accumulate at the foot of cliffs and on mountain slopes in all dry regions which have hot days and cool nights. Even when the rocks are not shattered to pieces, their crevices and fissures are slowly widened.

FIG. 41. — Smooth exfoliated surface of granite, Matopos Hills, Rhodesia, South Africa

Certain rocks, notably granites, *exfoliate* under extreme temperature changes, that is, the surface portions split off in large sheets, which may be of almost any thickness, and are either flat or, more commonly, are curved. In this way are produced the granite domes which are found in so many parts of the world, such as those of the Yosemite, Stone Mountain in Georgia, the Matopos Hills in South Africa. The smooth slopes, due to exfoliation, are often deceptively like those worn and smoothed by glaciers, a resemblance which is heightened by the large boulders, remnants of exfoliated masses, which often occur upon these slopes.

The effect of temperature changes is frequently the disintegration of rocks into minute fragments. This extreme effect is especially noteworthy in those igneous rocks which are coarsely crystalline. A rock of this kind is made up of several different minerals, each

FIG. 42. — Slope of exfoliating granite, Matopos Hills

of which has its own particular rate of expansion and contraction, and thus the particles are subjected to stresses which gradually separate them and cause the rock to crumble. In Egypt one may pick up granite fragments from the ancient monuments which will break into small pieces upon very slight pressure. The Egyptian obelisk in New York seemed, when first brought to this country, to be perfectly sound and fresh, but the severe winters of our Atlantic seaboard speedily showed how the granite had been rifted and weakened by the centuries of exposure to temperature changes in the dry climate of Egypt.

Changes of temperature do work of a purely mechanical kind, as does frost, and are even more entirely superficial than the latter, for a thin covering of débris suffices to put an end to their efficiency.

FIG. 43. — Exfoliation of glaciated granite, Sierra Nevada. (U. S. G. S.)

4. WIND

The wind, of itself and unassisted by hard particles, can accomplish but little in disintegrating firm rocks, but on high mountain crests and " knife-edges," where the wind blows with great velocity, it may accomplish considerable destruction. When, however, the wind is able to drift along quantities of sand and fine gravel, it becomes a disintegrating agent of importance. Except on sandy coasts, this agency is of small efficiency in regions of ordinary rainfall, because in these the soil is protected and held together by its covering of vegetation. On sandy coasts we may often observe the abrading effects of wind-driven sand. In a Cape Cod light-house a single heavy gale so ground a plate-glass window as to render it opaque and useless, and on that same coast window-panes are sometimes drilled through by the sand flying before a storm. Fragments of glass lying on the sand dunes are soon worn as thin as a sheet of paper.

FIG. 44. — Exfoliating granite dome, Yosemite Valley, California. (U. S. G. S.)

In arid regions, and more especially in sandy deserts, high winds sweep along much sand and fine gravel, which are hurled against any obstacle and gradually cut it away.

Very hard rocks yield but slowly to the cutting action of wind-driven sand, and in them the chief effect to be observed is a scratching and polishing of the surface. The same principle is employed in the sand-blast, which is a jet of sand, driven at a high velocity and used to engrave glass, polish granite, and do other work of the kind. Soft rocks are quite rapidly abraded and

cut down by the drifting sand, and go to increase the mass of
cutting material. The softer parts are cut away first, leaving the

FIG. 45. — Wind-sculptured Sandstone, Black Hills, South Dakota. (U. S. G. S.)

harder layers, streaks, or patches standing in relief. In this way very fantastic forms of rocks are frequently shaped out; pot-holes and caverns are excavated by the eddying drift, and archways cut through projecting masses. (See Frontispiece and Figs. 45–46.)

As the wind does not lift the harder and heavier particles to any great height, the principal effect is produced near the level of the ground, and thus masses of rock are gradually undermined

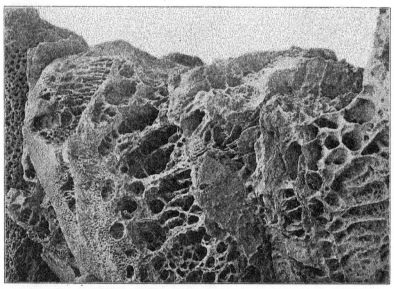

FIG. 46. — Honey-combed rock, due partly to wind erosion and partly to the Solvent action of rain.   (U. S. G. S.)

and fall in ruins, which in their turn are slowly abraded. Isolated blocks are sometimes so symmetrically cut away on the under side, that they come to rest upon a very small area and form rocking stones, which, in spite of their size and weight, may be swung by the hand.

The fine particles abraded from the rocks by drifting sand have undergone no chemical change, the process being entirely mechanical.

The abrading effects of wind-driven sand may be observed in any desert region where naked rocks are exposed, as, for example, in the arid parts of Utah and Arizona. One very characteristic effect of this natural sand-blast is found in the appearance of the pebbles shaped by it. Pebbles of very hard and homogeneous materials, such as quartz or chalcedony, are highly polished. Those made from igneous rocks have the softer minerals worn away, leaving the harder to stand in relief in curious patterns, while limestone is carved into beautiful arabesques.

The wind-driven sand, which does the work of abrading, is itself abraded and grows finer, the longer the distance which it traverses.

We have seen that the rain is slowly shifting the soil seaward, and in dry countries the wind acts in similar fashion. Strong winds, blowing steadily in one direction, carry great quantities of dust and fine sand with them, sometimes directly into the sea or other bodies of water, sometimes into rivers, or again to moister regions, where it comes under the influence of the rain.

Slowly as they work, the wind and temperature changes prevent any complete stagnation in the circulation of material, and thanks to them, the processes of disintegration of rock and transportation of soil are kept up even in the dryest deserts.

CHAPTER V

DESTRUCTIVE PROCESSES. — RUNNING WATER

THE source of all running water, whether surface or underground, is atmospheric precipitation. All springs and streams are merely rain (or snow) water collected and fed from reservoirs. The rain-water which falls upon the land is disposed of in three ways: one part is returned to the atmosphere by evaporation; another part flows over the surface to the nearest watercourse. The remainder sinks into the soil to a greater or less depth, and though some of it is returned to the surface in springs, yet a great part must reach the sea by subterranean channels. The surface flow, together with the supply from springs, constitutes the " *run-off*."

The relative proportions of these three parts of the total precipitation vary much in accordance with the climate and with the topography of the land surface. In a moist climate with heavy rainfall the run-off may amount to one-half of the precipitation, and the loss by evaporation is at a minimum. In arid regions, where evaporation is very great, the run-off is from one-fifth to zero. Climatic factors being equal, run-off increases with the steepness of the slopes and is thus relatively less in large drainage basins than in small ones.

1. THE GROUND WATER

Within the soil the movement of water is in different directions according to circumstances. In climates of considerable rainfall which have no long dry season, the movement is chiefly downward, due to gravity, but if there are long periods of drought, evaporation from the surface causes an upward movement of the water by capillarity; and in the tropics this upward movement produces certain important and characteristic effects.

At a depth below the surface, which varies greatly at different times and places, the soil and rocks are saturated with water, which is called the *ground water*. Near the sea, or other bodies of surface water, the ground water may be very little below the surface of the ground, while in arid regions, with irregular topography, it may sink to great depths. In the eastern United States the ground water is encountered at depths of 1-100 feet, as is shown by the countless wells which are supplied by it. In the limestone plateau of eastern Kentucky and Tennessee the ground water is from 200-300 feet below the surface and is determined by the level at which the surface streams flow, that is, the *drainage level* of the region. In the plateau of the Colorado River, which is dissected by profoundly deep cañons, the ground water is, in places, nearly 3500 feet from the surface.

The level of the ground water is thus highly irregular and depends upon the amount of precipitation and upon the topographical features. As a general rule, the level of ground water is at that of the streams and rises toward the divides, but less steeply than the surface of the ground. Similarly, the ground water level fluctuates with the rainfall, rising in wet seasons and sinking in dry, as is shown by the failure of wells after a long drought.

It is usual to regard the ground water as everywhere penetrating to great depths, and, from this point of view, it is frequently called the " sea of ground water," but there is reason for much hesitation in accepting this belief. In a large number of very deep mining shafts in various parts of the world, and in both humid and arid regions, water is found only in the upper levels, within 2500 feet or less of the surface, while below the mines are dry, even dusty. Such shafts frequently encounter water in the lower levels, when they intersect large fissures, and this indicates that water descends to great depths principally through such fissures. The character of the rocks themselves has a great effect upon the depths to which water can penetrate, — some rocks being porous and with such open joints as to permit a free passage of water, while others are almost impervious. "It is probable that the universal presence of ground water is characteristic of a comparatively shallow surface belt, below which the water which has not been again drawn off at the surface, at a lower level, or has not been used up in hydration processes, is concentrated into the larger fissures." (Spurr.)

Aside from the extremely slow movement of water through the mass of porous rock, underground waters follow the larger openings, such as joint-cracks, bedding planes, etc. The inclination of the stratified rocks, the alternation of porous and impervious beds, and the character of the joints and fissures are thus the factors which determine the direction of flow, especially in the shell of weathering, where the rocks are not saturated with water. In soluble rocks, such as limestones, the water may dissolve out its own channels. Surface topography has but a subordinate effect upon the course of underground waters, and it often happens that, for considerable distances, the surface and subterranean movements of water are in exactly opposite directions.

The factors which determine the movement of underground waters are of great practical importance in all problems of drainage and water supply. Serious evils have followed from carelessly taking for granted that the underground flow would be in the same direction as that on the surface.

As the movement of underground waters is almost always excessively slow, their mechanical work is trifling, but chemically they bring about important changes. The water, making its way downward through the joints and bedding planes of the rocks, exerts its solvent and decomposing action upon the walls of these crevices, in the manner already described in connection with the work of rain. Down to the level of the ground water, or in the shell of weathering, percolating waters are the great agent of decomposition and therefore always contain more or less mineral matter in solution, the nature and quantity of which depend upon the character of the rocks traversed. Below the ground water level in the shell of cementation, the effects are more reconstructive than destructive, though solution and alteration of minerals continue at these lower levels.

In passing through limestones in the shell of weathering, percolating waters dissolve channels, great and small, through the rock. Pipes and sink-holes are dissolved downward from the surface, and in the mass of the rock great caverns are formed by the solvent power of the carbonated waters. Such caverns, as the

FIG. 47.—Sink-hole in limestone, near Cambria, Wyoming. (U. S. G. S.)

Mammoth Cave of Kentucky, for example, are often many miles in extent and have considerable rivers flowing in them. Indeed, in limestone regions the smaller streams generally have a longer or shorter underground course. The lower level of the caverns is determined by the general drainage level at which the surface streams flow. In the shell of cementation the movement of water is very much slower and its solvent effects are much lessened. The beds of rock-salt, which would long ago have been dissolved away by moving waters, are found at depths which may be reached by mining or boring.

When underground waters become highly heated through con-

FIG. 48. — Cañon and lower falls of the Yellowstone River.   (U. S. G. S.)

tact with hot volcanic masses, or by descending to great depths along channels which permit a return to higher levels, their solvent efficiency is greatly increased.   Rocks penetrated by such *thermal*

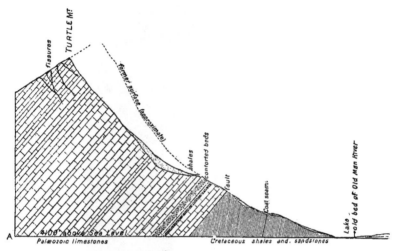

FIG. 49. — Profile of Turtle Mt., showing the amount of material removed in the Frank rock-slide. (Brock)

waters are profoundly altered in character and composition. The complex minerals of the igneous rocks are decomposed; the felspars become opaque from the formation of kaolin, or are altered to hydrated micas; minerals containing magnesia and iron give rise to talc, chlorite, serpentine, and the like, while the lime compounds are converted into the bicarbonate and carried away in solution. Some of the minerals are altered in place, and others are deposited in the crevices of the rocks. Thermal waters also alter minerals by bringing in new material in solution. In the Yellowstone Park the lavas of the great volcanic plateau, which has been deeply trenched by the Yellowstone River, are profoundly decomposed and altered by the hot waters which traverse it.

Except in caverns, underground waters flow too slowly to accomplish direct mechanical erosion, but indirectly they may bring about important mechanical changes. Masses of soil or talus, lying on steep slopes, saturated by long-continued, heavy rains, may have their weight so increased and their friction so reduced,

K

FIG. 50. — Rock-slide of 1903 at Frank, Alberta.   The lake in the foreground due to the damming of a stream by the mass of débris.   (Brock)

as to glide downward in land-slips, which are sometimes disastrous. Of this kind was the great land-slip of 1826 in the White Mountains of New Hampshire.

Rock-slides occur when the rocks forming a slope become saturated with water, until they can no longer support themselves. The movement is much facilitated by underlying beds of clay, or clay rocks, which become very slippery when lubricated with water. Mountain valleys in all parts of the world show plain evidence of such rock-slides, and often a vast quantity of rock is thus displaced. At Elm, Switzerland, in 1881, more than 12,000,000 cubic yards of rock were carried down for a distance of 2000 feet. In 1903 a great rock-slide occurred at Frank in the Canadian province of Alberta, when the entire face of Turtle Mountain fell and rushed across the valley in a huge avalanche of rock fragments, estimated at 40,000,000 cubic yards. The causes of this great rock-slide were several, but an unusual amount of ground water and a severe frost following warm weather were the chief agents.

2. SPRINGS

Springs are the openings of the ground water upon the surface, and could not be formed were the land perfectly free from irregularities, for gravity controls the movement of underground waters, and the source of a spring must be higher than its mouth. It must be remembered, however, that a subterranean stream is often confined as in a pipe, and that the pressure to which it is subjected may seem to make it flow upward, as when a spring rises from a deep fissure, or bursts out upon the top of a hill. But these are not real exceptions, and here also the source, which may be many miles distant, is above the spring, and it is this which produces the necessary pressure.

The commonest type of spring is formed when a relatively impervious bed of rock (usually clay in some form) overlaid by porous rocks, crops out on a hillside. The ground water saturates the lower layers of the porous beds, until its descent is arrested by the impervious bed, and then the water follows the upper surface of the latter. When, by some irregularity of the ground, the impervious bed comes to the surface, the water will issue as a

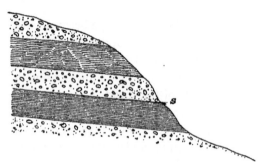

spring, or a line of springs (see Fig. 51).

A second class of springs are those which rise through a crack or fissure in the rocks. Inclined porous beds, enclosed between more impervious ones, allow the

FIG. 51. — Arrangement of strata which causes hillside springs. The lower close-lined bed impervious

water to follow them downward, until in its lower course such water is under great pressure, or " head " (Fig. 52). On reaching a fissure opening upward, the water will rise through it and, if under sufficient pressure, will come to the surface.

An artesian well is an "artificial fissure-spring." It is a boring which taps a sheet or stream of the ground water, when the water is under sufficient pressure to rise to the surface or even spout high above it.

In limestone districts depressions of the surface may intersect the course of considerable underground streams, which thus reach the surface in springs of unusual volume. Very striking and beautiful examples are the Giant Spring in the cañon of the upper Missouri, near Great Falls, Montana, and Silver Spring, Florida, which is navigable for steamboats.

Springs, as such, do little in the way of rock disintegration, but they accomplish something by undermining the rocks at the point where they issue, and thus working their way backward. This process is known as the *recession of spring-heads.* The underground streams, of which springs are the outlets, have often effected much in the way of dissolving rock material, and hence spring-water always contains dissolved minerals, principally the carbonates and sulphates of lime and magnesia, and the chlorides of magnesium and sodium. In mineral springs the quantity of dissolved materials is larger and perceptible to the taste.

Thermal Springs are those whose temperature is notably higher than that of ordinary springs in the same region, and they range from a lukewarm to a boiling state. This increase of temperature may be caused in either of two ways: (1) In volcanic regions, water coming into contact with uncooled masses of lava is highly heated and reaches the surface as a hot spring. Of this class are the in-

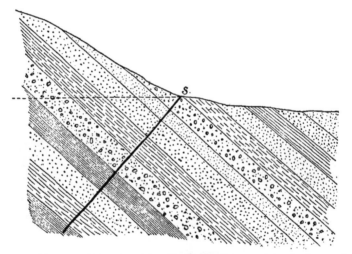

FIG. 52. — Diagram of fissure-spring. The heavy line represents the fissure along which the water rises

numerable thermal springs of the Yellowstone Park. (2) Wherever the disposition of the rocks is such that water may descend to great depths within the earth and yet return to the surface by hydrostatic pressure, thermal springs appear. These conditions are found only in regions where the rocks have been much folded and fractured. In this case the temperature of the water is raised by the interior heat of the earth, which, as we have seen, increases with the depth. Springs of this class occur numerously along the Appalachian Mountains, and in larger numbers and of higher temperatures they accompany the various ranges of the Rocky Mountains and Sierra Nevada.

Geysers are thermal springs which periodically erupt, throwing up hot water in beautiful fountains, accompanied by clouds of steam. Though of great scientific interest, geysers are not important geological agents, because of their rarity, since they occur only in Iceland, the Yellowstone Park, and New Zealand.

FIG. 53. — An artesian well.   (U. S. G. S.)

The destructive effects of thermal springs are principally accomplished below the surface, and have already been considered under the head of underground waters. The high percentages of dissolved materials which such springs usually contain are evidence of the important work of rock disintegration which they perform.

3. RIVERS

Erosion by Rivers. —The destructive work of rivers, including in that term all surface streams, is far less extensive, in the aggregate, than that of the atmospheric agencies, but because the work of a stream is concentrated along its narrow course, it appears much more striking and impressive.

FIG. 54. — The "Bottomless Pit," Arizona.. The stream disappears in a limestone cavern and is not known to reappear. (Photograph by A. E. Hackett, Flagstaff, Ariz.)

A certain amount of solution and decomposition is performed by rivers upon the rocks of the bed, and in limestones this may be considerable, especially if the water be charged with organic acids from a swamp or peat-bog. Limestone regions are characterized by a paucity of surface streams, most of which pass into caverns and underground channels which they have made by dissolving the limestone. Such subterranean streams may or may not reappear on the surface, according to circumstances.

The mechanical work of a river is much greater than the chemical, and is dependent upon the velocity of the current, varying directly as the square of that velocity. The velocity of a stream is the rather complex resultant of several factors, the chief of which is gravity; the steeper the slope of the bed, the swifter the flow of the water. A second factor is the volume of water, the velocity varying as the cube root of the volume. That is to say, if one of two streams which flow down the same slope has eight times as much water as the other, it will flow twice as fast. Other factors enter into the result, but slope of bed and volume of water are much the most important.

Pure water can do little to abrade hard rocks, though it can wash away sand, gravel, and other loose materials. When the Colorado River broke into the Salton Sink in southeastern California in 1905, it cut a deep trench with incredible rapidity through the soft alluvial soils. Streams also take advantage of the joint-blocks, into which all rocks are divided, and often loosen and carry down such blocks. This process is called *plucking* and is important in the destructive work of glaciers and the sea. As in the case of the wind, the stream merely supplies the power; the implement with which the cutting is performed is the sand, pebbles, and other hard particles which the water sets in motion. These abrade the rocks against which they are cast, just as the wind-driven sand does, but more effectively, because of the ceaseless activity of the stream, and because many rocks are rendered softer and more yielding by being wet. The cutting materials are themselves abraded and worn finer and finer by continued friction against the rocks and against one another. In the case of complex minerals this abrasion is accompanied by more or less chemical decomposition, as has been shown experimentally by rotating crystals of felspar in a drum half filled with water. When the felspar was ground down to mud, the water showed the presence of potash and soda in solution. Angular blocks are speedily worn into cobblestones and these into pebbles of spheroidal or flat, discoidal form. A process of selection goes on, by which the softer materials are ground into mud, the harder remaining as pebbles and sand.

An example of exceedingly rapid wear of hard rock by running water, under favourable conditions, is given by the Sill tunnel in Austria, which is provided with a pavement of granite slabs more than a yard thick. Great quantities of débris are swept over this pavement at a high velocity and so rapid is the abrasion, that it was found necessary to renew the granite slabs after a single year.

A river which is subject to sudden fluctuations of volume, being now a rushing torrent and again almost dry, is a much more efficient agent, both of erosion and of transportation, than is one which carries nearly the same quantity of water at all times, or which fluctuates only slowly.

The velocity of a stream differs much in its various parts, diminishing, as a rule, from the head waters to the mouth. In very many cases there are also local variations of speed, falls, rapids, and eddies alternating with quiet reaches. In eddies and at the foot of cascades the water acquires a rotary motion, which is transmitted to stones lying on the bottom. In a rocky bed these revolving stones excavate cylindrical holes, often of remarkable regularity, called *pot-holes*, or giant kettles. The diameter and depth of pot-holes are determined by the volume and velocity of the water and by the length of time during which the eddy or fall remains at the same point.

Since the velocity of a stream is so largely dependent upon gravity, it is obvious that the deeper a stream cuts its channel, the less steep does its slope become, and that so long as the region is neither upheaved nor depressed, the river performs its vertical erosion at a constantly decreasing rate. Unless, therefore, the work is done under very exceptional conditions, as in the case of the Niagara, we cannot reason from the present rate of excavation to the length of time involved in cutting out a given gorge.

Unless the region through which a river flows is upheaved, and

FIG. 55. — Undermined pot-hole, Little Falls, N.Y.   The arrow points to the upper opening and the partly concealed figure stands in the lower part.   (Photograph by van Ingen)

thus, by increasing the fall, renewed power is given to the stream, a stage must sooner or later be reached when the vertical cutting of the stream must cease. This stage is called the *base-level of erosion*, or *regimen* of the river, and it approximates a parabolic curve, rising toward the head of the stream. Elevation of the country will start the work afresh, until a new base-level is reached,

FIG. 56. — Pot-hole in stream, Mill Creek, Oklahoma. (U. S. G. S.)

while depression will have a contrary effect and may put a stop to vertical erosion where it was in active progress before. When the base-level is reached, the river cuts laterally, undermining its banks (see Fig. 55) and working like a horizontal tool upon the country-side.

As valleys are also excavated by other agents, it is important to note the characteristic features of river-formed valleys. Unassisted by other agencies, a river cuts a narrow, steep-sided trench or gorge, the possible depth of which depends upon the height above base-level at which the river began its work, disregarding any subsequent elevation of the land. As soon as the gorge begins to form, its sides are attacked by the atmospheric destructive forces and a process of widening is begun; but this is very slow and to widen out the gorge or canon into a broad valley, with gentle slopes, requires a very long period of time, determined by the activity of the climate and the resistant power of the rocks. Even in the gorge stage, a river valley tends to have a V-shaped cross-section, because the upper part of the gorge, having been longest exposed to weathering, has suffered the greatest loss.

A river valley is rarely straight for any considerable distance, but takes a sinuous course, with rocky spurs projecting alternately from opposite sides of the stream, and these spurs have a continuous (or a terraced) slope from top to bottom. This is true even of swift streams flowing through hard rocks, and the tendency is much exaggerated when the velocity of the current is diminished and the river-bed is in soft materials, as in the lower Mississippi. Under such conditions the stream meanders to an extraordinary degree and often, by cutting through the narrow neck of a meander, abandons part of its channel and leaves an " ox-bow lake " to mark its former course.

Still another feature of river valleys is the accordant relation between the main valley and its tributaries. Normally, the tributaries lower their beds at the same rate as the main stream and enter the latter on the same level. Exceptions to this rule do occur, as in the case of the tributaries below a great cataract in the trunk river, which enter high up on the gorge walls, but such exceptions admit of a ready explanation.

When a river enters the sea or a lake, its velocity is checked and it is no longer able to excavate a channel, so that the continuation of the river channel across the sea-floor, like that of the Hudson (see p. 34), is a proof that the lower course of the stream has been submerged under the sea. There are, however, two known instances of the excavation of a stream channel in the bed of a lake, the upper Rhone in Lake Geneva and the upper Rhine in Lake Constance. These streams are both glacial streams of very cold and dense water and charged with great loads of coarse sediments. Discharging into bodies of warmer and lighter water, the river currents are able to maintain themselves for some distance from shore and thus to cut trenches. Other examples will, no doubt, be found under similar conditions, but must be very uncommon.

FIG. 57. — A meandering stream; ox-bow lakes at the right: Alashuk River, Alaska. (U. S. G. S.)

Having learned the general character of river erosion, we may illustrate it with a few concrete examples.

1. A particularly interesting case is that of the little river *Simeto* in Sicily, since the history of its gorge is so well known. In 1603 a great lava flood from Ætna was poured out across the course of the stream, and, when cold, solidified into a barrier of

the hardest rock. When Sir Charles Lyell visited the spot in
1828, he found that in a little more than two centuries the stream

FIG. 58. — The Au Sable Chasm, N.Y. (Copyright by S. R. Stoddard, Glens Falls, N.Y.)

had cut a gorge through this barrier of 40 to 50 feet deep, and varying in width from 50 to several hundred feet. The lava which had thus been trenched is not porous or slaggy, but homogeneous, dense, and very hard.

2. In the northern parts of the United States the great ice-sheet, which in late geological times covered the country, brought down with it vast quantities of drift, that filled up the channels of many streams and quite revolutionized the drainage of certain dis-

FIG. 59. — Old, high-level channel of the Niagara River, below the present falls. (U. S. G. S.)

tricts. Since that time the displaced streams have cut out new channels for themselves, often through hard rocks, and many now flow in quite deep gorges, with nearly vertical walls. Au Sable Chasm, New York, is an example of these geologically modern river gorges, the atmosphere not having had time to widen it.

3. The Niagara is an exceptional case, the gorge being cut, not only by the direct abrasion of the running water, but also by the action of the spray and frost at the falls. In the ravine the

upper rock is a hard, massive limestone, which is underlaid by a soft clay shale. The latter is continually disintegrated by the spray of the cataract and by the severe winter frosts undermining the limestone, which, when no longer able to bear its own weight, breaks off in tabular masses. Thus the falls are steadily receding, leaving behind them a gorge, which is deepened by the river and especially by the plunging masses of water at the foot of the cataract.

4. One of the most remarkable known examples of river erosion is seen in the cañons of the Colorado. The Grand Cañon is over 200 miles long and from 4000 to 6500 feet deep, with precipitous walls. It is extremely probable that the river has been rendered able to cut to such profound depths by the gradual uplifting of the whole region, which is now a lofty plateau, in places more than 8000 feet above the sea. The erosive power of the river has thus been continually renewed and a more or less uniform rate of excavation secured.

5. Finally, an extremely curious example of river erosion is the gorge of the Zambesi River in South Africa, at and below the Victoria Falls. Above the falls the river is more than a mile wide, and at the cataract it plunges into a narrow chasm which is *transverse* to the course of the river. From the chasm the only outlet is by a narrow gorge of only 50–60 yards in width, and below this gateway the gorge continues for many miles in a series of sharp zigzags, which are highly exceptional in such a hard rock as the basaltic lava which the river has trenched to a depth of 400 feet. This remarkable result is due to the fact that the excavating work of the river has been controlled by the lines of joints in the rock.

Transportation by Rivers. — The main importance of rivers as geological agents is not so much their work of erosion, but lies rather in what they accomplish as carriers of the results of their own destructive activity and that of the atmosphere, comprising both the materials which are mechanically swept along in suspension and those which are carried in solution.

Materials mechanically Carried. —The transporting power of running water is dependent upon the velocity of the current, and both mathematical and experimental treatment of the problem brings out the surprising result that the transporting power varies directly as the sixth power of the velocity. If the rapidity of a stream be doubled, it can carry 64 times as much as before. The destructiveness of sudden and violent floods is thus explained. In the terrible flood which overwhelmed Johnstown, Pennsylvania, in 1889, great locomotives and massive iron bridges were swept off, it is hardly an exaggeration to say, like straws, and huge boulders carried along like pebbles. The formula as to the relation of velocity to transporting power refers more particularly to the coarser materials which are pushed along the bottom of the stream. No relation has yet been determined for very fine particles of silt and clay, some of which remain suspended indefinitely even in still water. Transporting power also increases as the temperature of the water decreases.

It obviously follows from the relation obtaining between velocity and transporting power, that a slight increase in the rapidity of a stream will largely augment the load which it carries, provided the stream obtains as much material as it can transport, while a slight reduction of velocity will cause the deposition of a large part of that load. The buoyancy of water adds, in an important degree, to its ability to sweep along sediment, because when any substance is immersed in water, it loses weight to an amount equal to the weight of an equal bulk of water. The specific gravity of most rocks is from two and one-half to three, so that when immersed they lose from one-third to two-fifths of their weight in air. The shape of the fragments is likewise a factor in determining the velocity requisite to move them; the larger the surface of the fragment in proportion to its weight, the more easily it is carried in suspension. Thus flat grains or scales are carried farther than round ones; while, on the other hand, rounded fragments are more easily rolled along the bottom, when too heavy for the current to lift.

The greater part of the débris or sediment which a stream carries is furnished to it by the destructive activity of the atmosphere; the rains wash in the finer materials, while frost and land-slips bring in the larger masses which are carried down by mountain torrents. To this material the river adds that which is derived from its own work in the cutting away of its banks and bed.

Materials in Solution. — In addition to what the river carries down mechanically in suspension or sweeps along the bottom, there is a third class of material; namely, that which is dissolved in the waters of the stream. Dissolved matters are always present in greater or less quantity, and are the same in kind as those which we have already found to occur in spring-waters, whence they are, for the most part, derived by the rivers. River-water is, however, usually more dilute than that of springs, because of the rain which falls into it, or pours in from the banks. In very dry regions, where this additional rain supply is at a minimum, and where the streams are concentrated by continual evaporation, they are frequently undrinkable, on account of the quantity of matter in solution which they contain. Examples of this are the salt and so-called " alkali " (a very comprehensive term) streams of the arid West, which contain a great variety of dissolved minerals.

The quantity of material which rivers are continually sweeping into the sea is enormously great. Every year the Mississippi carries into the Gulf of Mexico nearly 7,500,000,000 cubic feet of solid sediment, either in suspension or pushed along the bottom, an amount sufficient to cover one square mile to a depth of 268 feet. In addition to this is the quantity brought down in solution, which is estimated at 2,850,000,000 cubic feet annually.

Different rivers vary much in the proportion of suspended and dissolved materials which they carry and discharge into the sea; a roughly approximate average makes the amount of material removed equal to about 11,400 cubic feet (600 tons) of annual waste for every square mile of the land surface of the globe, that is, *under existing conditions* of slope, temperature, rainfall, etc. How great a difference in the result a change in these factors may produce, will be seen from a comparison of the Mississippi and the Ganges. The amount of suspended matter discharged by the former represents a lowering of the surface of the entire drainage area at the rate of one foot in 4920 years, while in the case of the Ganges it is one foot in 1880 years, or more than twice as fast. The amount of material carried by the Amazon has not been determined, but there can be little doubt that it is far greater than that discharged by the Mississippi. The area drained by the Amazon is less than twice as large as the drainage basin of the Mississippi, and yet it brings to the sea five times as much water as does the great river of North America.

CHAPTER VI

DESTRUCTIVE PROCESSES. — SNOW AND ICE, THE SEA, LAKES, ANIMALS AND PLANTS

Avalanches are great masses of snow which descend from the mountain tops at a' very high velocity, and are frequent in all high mountains with heavy snowfall, and occur, though less commonly,

FIG. 60.—Summit of Mt. Blanc, Switzerland, showing the great accumulations of snow

on mountains of medium height. Winter avalanches of dry and powdery snow do comparatively little destructive work, but in thawing weather, when the snow is heavily charged with water,

148

great masses of earth and rock are brought down in the avalanche, which sweeps everything before it. Though acting only occasionally, avalanches are efficient agents in the removal of material from higher to lower levels.

On a small scale, snow-slides remove unprotected soil from slopes. In the bad lands (see p. 109), where the rain wash produces comparatively little effect upon the débris-covered buttes, sliding masses of snow strip off the covering of soil and expose fresh surfaces of rock to the destructive action of the water.

Glaciers are much the most important form of ice as a geological agent. A glacier is a stream of ice which flows *as if* it were a very tough and viscous fluid, and does not merely glide down a slope, as snow slides from the roof of a house. Glaciers play a very important part in keeping up the circulation of the atmospheric waters, and produce geological results of an extremely characteristic kind. Their contribution to the sum total of rock destruction and reconstruction is, it is true, relatively small, but it often becomes important to trace the former extension of glaciers, which, in its turn, has a wide bearing upon some of the most far-reaching of cosmical problems.

As we ascend into the atmosphere from any point on the earth's surface, we find that it becomes continually colder with increasing height. In this ascent a level is eventually reached where the temperature of the air never rises for any length of time above the freezing-point, and above this level no rain, but only snow, falls. This level is called the limit of perpetual snow, or simply the snow-line. While the height of the snow-line above the sea-level is, like climate in general, much affected by local factors, yet, speaking broadly, its elevation is determined by latitude. In the tropics the snow-line is 15,000 or 16,000 feet above the sea, — descending more and more, as we go toward the poles, and coming down nearly to sea-level within the polar circles, but does not actually reach that level at any known point in the northern hemisphere.

Were there no means of bringing the snow which accumulates above the snow-line to some place where it may melt, it would evidently gather indefinitely, and at last nearly all the moisture of the earth would be thus locked up.　As a matter of fact, there is no such indefinite accumulation.　In very dry regions the excess of snow is disposed of by direct evaporation, and on high mountains avalanches carry the snow down to lower levels, where it melts.

FIG. 61.—Two valley glaciers descending Mt. Blanc, showing the terminal moraine at the foot of each.　On account of the foreshortening the glaciers appear to be unduly steep

In places where the excess of snow cannot be disposed of in either of these ways, glaciers are formed and thus keep up the circulation of the waters, by carrying the surplus snow down to lower levels at which it can melt, or by entering the sea and in the shape of icebergs (which are fragments of glaciers) being floated to warmer latitudes.

Though even at the present time there are in various parts of the world great tracts of glacier ice, they cannot be called common and are found only where certain conditions concur. The nature of these conditions will be best understood by examining the process of glacier formation.

Snow is made up of minute, hexagonal crystals of ice, which are intimately mixed with air and thus separated from one another. Though the individual crystals are transparent, snow is white and opaque, as always results when a transparent body is intimately mixed with a gas, as in the foam on water, or in powdered glass. Ice is composed of the same kind of crystals as is snow, but they are in contact with one another, not separated by air. To convert snow into ice, therefore, it is only necessary to expel the air and bring the crystals into contact, for which pressure alone is not ordinarily sufficient.

The first step in the transformation is the partial melting of the upper layers of snow, for which a change of temperature is necessary, though the change need not warm the air, but may be due to the direct rays of the sun. Glaciers are rare in the tropics because of the constancy of the temperature, and the small area which extends above the snow-line, which seldom permits the formation of extensive snow-fields. Sometimes, however, the conditions of glacier formation are fulfilled even in the equatorial zone; for example, there is a glacier on one of the peaks of Ecuador.

When the surface layers of snow have been partially melted, the water thus formed trickles down into the snow beneath, expelling much of the air. This underlying snow has still a temperature much below the freezing-point, and the percolating water is soon refrozen into little spherules of ice. This substance, midway between snow and ice, is called *névé*, and may be seen every winter wherever the snow lies for any length of time. The hardened " crust " which forms by the refreezing of partly melted snow is névé. The air, which is now in the form of discrete bubbles, is largely expelled by the increasing pressure of the overlying snow masses, which are continually added to by renewed falls, and the névé is thus converted into ice.

The structure of glacial ice is characteristically different from that produced by the freezing of a body of water. The latter is made up of parallel crystals with optical axes perpendicular to the surface of the water. Glacial ice, on the other hand, consists of crystalline grains, which increase in size toward the lower end of the glacier, with optical axes disposed irregularly. The banded structure of the glacier, often so conspicuous, is a kind of stratification and is derived from the successive snow layers of the névé.

The temperature of the interior of a glacier corresponds at every depth to the melting-point of the ice for the pressure at that depth. The melting-point of ice is lowered by pressure, and therefore pressure changes within and at the bottom of the glacier cause melting and refreezing without corresponding temperature changes. This fact that glacial ice is so nearly at the melting-point indicates that the maximum thickness of the glacier cannot exceed 1600 feet, which in truth appears to be the thickness of the Antarctic ice-cap. A greater thickness would cause melting by pressure of the bottom parts.

It follows from their mode of formation that glaciers can be formed only where the snow accumulates to great thicknesses, and cannot be disposed of by either melting or evaporation. Hence, glaciers are rare or absent in dry regions, as in most of the Rocky Mountains within the limits of the United States. It also follows that the ground upon which the snow lies must be so shaped as to allow great masses of it to gather.

A glacier moves in much the same way as a river, but at a very much slower rate. The middle portion moves faster than the sides, because the latter are retarded by the friction of the banks, and, for the same reason, the top moves faster than the bottom. While behaving like a plastic substance under pressure, ice yields readily to tension, and even a slight change in the slope of the bed will cause a great transverse crack, or *crevasse*, to form, which, like an eddy in a stream, seems to be stationary, because always formed again at the same spot. Other systems of cracks, the marginal

FIG. 62. — A hanging glacier, Cascade Pass, Wash. Note the terminal moraine and the crevasses. (U. S. G. S.).

crevasses, are formed along the sides of the glacier, and are due
to the more swiftly moving middle pulling away from the retarded
·sides.

The rate of glacier movement depends upon the snow supply,
upon the slope of the ground, and the temperature of the season.
The comparatively small glaciers of the Alps move at rates varying
from two to fifty inches per day in summer and at about half that·
rate in winter, while the vastly larger glaciers of the polar lands

FIG. 63.— Moraine-covered surface of the Malaspina Glacier, Alaska. (U.S.G.S.)

have a correspondingly swifter flow. The great stream of ice
which enters Glacier Bay in Alaska has a summer velocity of
seventy feet per day in the middle. ,

Southeastern Alaska is a region where glaciers are developed on
a very extensive scale. The Malaspina is an immense ice-sheet,
having an area of 1500 square miles, which is formed at the foot of
the St. Elias Alps by the confluence of several great glaciers from
the neighbouring mountains. Parts of this vast accumulation of

ice are stagnant and deeply covered with rock débris, upon which there is a luxuriant growth of vegetation, with not less than 1000 feet of ice beneath it.

In Greenland and the Antarctic continent the accumulations of ice are on a scale not elsewhere found, and these regions present conditions of great geological interest. Greenland, except for a narrow strip along the coasts, is buried beneath a vast ice-sheet, from which great glaciers descend to the sea. In the interior only

FIG. 64.—Nunatak rising through the ice-cap, Greenland. (Photograph by Libbey)

a few isolated mountain peaks, or *nunataks*, rise through the ice mantle; except for these, nothing is visible but illimitable fields of snow. The snowfall is not very great; but so little of it is disposed of by evaporation or melting, that there is a large excess which goes to the growth of the ice-sheet, and keeps up the supply for the innumerable glaciers which flow to the sea.

The Antarctic ice-cap is estimated to be nearly seven times as large as that of Greenland.

The source of a glacier is always above the snow-line, but the

ice-stream itself may descend far below that line, slowly melting
and diminishing in thickness as it flows.   The lower end is at the
point where the rate of melting and the rate of flow balance, so
that changes in the temperature of the seasons or in the amount
of the snow supply will cause the glacier to advance or retreat, as
one or other of these factors prevails.   Thus the Alaskan glaciers
have retreated notably within the last century, while some of the
Norwegian ones are advancing.   From the lower end of a glacier

FIG. 65. — Edge of the Greenland ice-sheet, with a glacier descending from it.
The dark line is a medial moraine.   (Photograph by Libbey)

there always issues a stream of water, which flows under the ice,
often in great volume, and even in winter, for the thick ice is a
non-conductor and protects the stream from the intense cold of the
air.

There are various forms of moving bodies of land ice corresponding to bodies of water.   We have (1) *Alpine glaciers*, of which
those in the Alps are types, and are relatively small streams occu-

pying narrow mountain valleys, each connected with a particular basin or gathering ground of snow. (2) *Hanging glaciers*, which descend but little below the snow-line and are small glaciers occupying steep clefts near the mountain tops. In some cases they are not connected with snow-fields, but are fed by avalanches. The glaciers of the northern Rocky Mountains and the Sierra Nevada are mostly of this· class. (3) *Ice fields*, exemplified in Scandinavia; these are extensive and continuous areas of thick ice,

FIG. 66. — The Columbia Glacier, Alaska. (U. S. G. S.)

with gently curved surface, from the margins of which numerous, but mostly small, glaciers descend through rocky gorges. (4) *Piedmont glaciers*, like the Malaspina of Alaska. These are great accumulations or lakes of ice which form at the foot of mountains, by the coalescence of numerous glaciers of the Alpine, or valley, type. (5) *Continental glaciers* are those which cover enormous areas of land, such as the ice-sheet under which nearly all of Greenland is buried and that which covers the Antarctic land. This is a type of especial interest and significance to the geologist, because of the light which it throws upon the often mysterious operations of the ice-sheets which once covered large portions of North America and Europe.

Glacier Erosion is highly characteristic, and enables us to detect the former extension of ice-streams which have greatly shrunken and their former presence in regions whence they have long van-

FIG. 67.— Glaciated surface, Sierra Nevada, Cal.   The angular recesses indicate places where blocks have been removed by plucking.   (U. S. G. S.)

ished. The erosive capabilities of moving ice have been and still are the subject of much dispute, but the researches of the last ten or fifteen years in many parts of the world have brought together a great body of evidence which strongly supports the view that glaciers, under favouring conditions, are extremely efficient and powerful agents of erosion.   Just as in the case of water, the destructive power of ice depends upon the velocity with which it moves and the pressure exerted on its bed, so that a glacier may erode actively at one part of its course, and little or not at all at another. It is not surprising, therefore, that advancing glaciers have been observed to override loose masses of gravel without moving them.

When acting effectively, newly formed glaciers remove the soil, talus, and other loose materials from the surface, which thus, in the first instance, is rendered more irregular than before, because

FIG. 68. — Steeply inclined strata, with edges roughened by glacial plucking, overlaid by glacial drift, Iron Mt., Mich. (U. S. G. S.)

of the varying depths to which the effects of weathering penetrate (see p. 101). Bare rocks are eroded by a double process: (1) The joint-blocks are torn away (*plucking*) by the advancing ice, an operation which is much facilitated by the continual liquefaction and regelation of the ice at the bottom of the glacier, owing to changes of pressure. As has been already remarked, the temperature of the ice within and at the bottom of the glacier is near the melting-point for the pressure. Thus, slight changes of pressure, due to the motion of the glacier and inequalities of the bed, cause the ice now to melt and again to solidify, and in this manner the joint-blocks of the bed-rock are loosened. (2) The bottom of the glacier is a mass of ice mingled with rocks, pebbles, boulders, sand, and débris of all sizes, and by their means the bed-rock is worn down, smoothed, polished, and scored with parallel marks,

FIG. 69. — Glacial striæ on limestone, overlaid by drift; Pillar Point, Lake Ontario. (U. S. G. S.)

in a fashion which forms the unmistakable autograph of the glacier. The rock fragments, firmly held by the immense weight of the ice, are slowly pushed over the rocky bed and cut grooves corresponding to the size of the fragments, from hair-like scratches to deep troughs. The scorings are, of course, in the direction of the movement and keep parallel often for considerable distances.

The smaller particles act as a polishing powder and smooth the bed, and, if the bed-rock is sufficiently hard, it receives a high polish. Hummocks of rock, over which the ice has flowed, are worn and rounded into the form called "*roches moutonnées*," with the upstream side gently sloping and polished, but with the downstream side abrupt and often rough.

FIG. 70. — Ancient glacial striæ (Permian); Riverton, on the Vaal River, South Africa. (R. B. Young)

As in the case of the river, the abrading material is itself abraded in its journey, and much of it is ground to the rock powder which heavily loads the stream flowing from the foot of the glacier. The pebbles and boulders are scratched with parallel or intersecting lines, smoothed and polished, but, as they are not rolled over and over like river pebbles, they are not spheroidal in shape, but are more or less angular and sometimes facetted, with smooth faces which meet at a distinct angle. This peculiar shape is due to the

M

shift̤g or turning of the pebble in the ice, so that, after one side
is wẏn flat, another is similarly worn, and this may be several
timẹrepeated.

Th amount of débris produced and carried by a glacier is often
veryʓreat. In summer the glaciers of the Justedal in Norway
togeter bring down nearly 2,400,000 cubic yards of material daily.

If laciers are powerful eroding agents and not merely a means
of ronding and polishing the rocks, we should expect to find that

_____ Glacial grooves on sandstone cliff; Delaware Water Gap, Pa.

_____ display topographical features not to be found
_____ is actually the case. Not only are rounded and
_____ forms produced, in marked contrast to the
_____ shapes due to ordinary weathering, but the
_____ excavated valleys is extremely significant,
_____ with that of river valleys (see p. 140).
_____ a glacial valley is U-shaped, with broad
_____ vertical sides. If the upper _____ of the

valley was not occupied by the ice, the slopes may be gradual down to the former level of the ice, where they become abrupt. (2) Glacial valleys are often straight and open for long distances; they may have spurs alternating from the opposite sides, but these spurs are truncated by the ice, and, if the action persisted sufficiently long, the spurs will have been entirely removed. (3) The tributary valleys do not enter the main valley at grade, as

FIG. 72. — U-shaped glacial valley; Kern Cañon, Cal. (U. S. G. S.)

they normally do in the case of rivers, but enter at the sides of the main valley much above its bottom and hence are called *hanging valleys*. The explanation of this peculiar arrangement is that the smaller, tributary glaciers could excavate their beds much less rapidly than the trunk ice-stream. " Although there is no uniform height at which these side valleys enter the main trough, in general it is true that, the smaller the tributary valley, the shifting or turning of the pebble in the ice, so that, after one side is worn flat, another is similarly worn, and this may be several times repeated.

The amount of débris produced and carried by a glacier is often very great. In summer the glaciers of the Justedal in Norway together bring down nearly 2,400,000 cubic yards of material daily.

If glaciers are powerful eroding agents and not merely a means of rounding and polishing the rocks, we should expect to find that

FIG. 71. — Glacial grooves on Sandstone cliff; Delaware Water Gap, Pa.

glaciated regions display topographical features not to be found elsewhere, and this is actually the case. Not only are rounded and flowing outlines and forms produced, in marked contrast to the craggy and angular shapes due to ordinary weathering, but the character of glacially excavated valleys is extremely significant, when compared with that of river valleys (see p. 140).

(1) In cross-section a glacial valley is U-shaped, with broad bottom and steep or vertical sides. If the upper portion of the

valley was not occupied by the ice, the slopes may be gradual down to the former level of the ice, where they become abrupt. (2) Glacial valleys are often straight and open for long distances; they may have spurs alternating from the opposite sides, but these spurs are truncated by the ice, and, if the action persisted sufficiently long, the spurs will have been entirely removed. (3) The tributary valleys do not enter the main valley at grade, as

FIG. 72. — U-shaped glacial valley; Kern Cañon, Cal. (U. S. G. S.)

they normally do in the case of rivers, but enter at the sides of the main valley much above its bottom and hence are called *hanging valleys*. The explanation of this peculiar arrangement is that the smaller, tributary glaciers could excavate their beds much less rapidly than the trunk ice-stream. " Although there is no uniform height at which these side valleys enter the main trough, in general it is true that, the smaller the tributary valley, the

higher its mouth lies above the main valley bottom." (Tarr.)
Hanging valleys may be formed in other ways than by glaciers,
but while they are common in glaciated regions, elsewhere
they are only occasional. (4) Glaciers may erode their valleys
below sea-level, because they exclude the sea water, as rivers
cannot do.

Glaciers flowing from high mountains. head usually in large
amphitheatres, or " cirques," caused by the lateral plucking of the
rocky walls by the fields of ice and snow. Each cirque slowly retreats upward, thus often reducing the divides between them to
extremely sharp " knife-edges," as is conspicuously displayed in
the Sierra Nevada of California.

The part of a glacier which descends below the snow-line is in
summer exposed to continual melting and may be more or less
completely covered with running water. Surface streams flow
in channels, which they melt for themselves, until they meet with
a crevasse, down which they pour in cataracts. As crevasses are
continually formed at the same spot (see p. 153), such a cataract
may remain stationary for a long period and wear out a cylindrical pot-hole, exactly as is done by a cascade in a stream.

Glacier Transportation. — The transporting power of a glacier
is not determined by its velocity, at least so far as the material
carried on its surface is concerned. This is because the rocks
may be regarded as floating bodies with reference to the ice, and
thus a rock weighing many tons is carried with as much ease as a
grain of sand. The masses of material transported by a glacier
are known as *moraines*. The moraines which are carried on the
top of the glacier are derived from the cliffs and peaks which
overhang the ice, and the action of frost and land-slips is continually showering down earth, sand, and rocks of all sizes, from
small blocks up to masses the size of houses. This material is
heaped up along the sides of the glacier in disorderly array, and
here forms the *lateral moraines*. When a. glacier is composed of
branch streams, it will have a corresponding number of *medial
moraines* (see Fig. 65), in the middle of the glacier. When two

branches unite, their coalesced lateral moraines form a single medial moraine.

The quantity of material thus carried on the top of the glacier depends upon the amount of rock surface which extends above the level of the ice and is subject to the action of the ice and the atmosphere. In the Alps, where the glaciers flow in deep ravines, the moraines are large, and some of the great Alaskan

FIG. 73. — Front of Bowdoin Glacier, Greenland. The dark bands are made by englacial drift. (Photograph by Libbey)

glaciers have their lower reaches so covered with rubbish, that the ice is visible only in the crevasses. In Greenland, on the contrary, the inland ice-cap has very little material on its surface, because only scattered nunataks rise above it.

The bottom part of the glacier is a confused mass of ice, stones, etc., and this débris is the *ground moraine*, which is to be regarded, not as so much material pushed along between the ice and the rocky bed, but as an integral part of the glacier.　At the foot or end of the glacier is the *terminal moraine* (see Fig. 62), where all the materials carried are dumped in a promiscuous heap, except so much as is swept away by the stream of water.　Besides the moraines proper, there is a certain amount of *englacial drift*, carried in the body of the ice.　This is derived from débris that comes from the surface, but does not work its way entirely to the bottom, as well as from that which gathers upon the surface of the snow or névé and is covered up by subsequent snowfalls.　The materials carried by a glacier are as characteristic as the marks left upon the rocks over which the ice has flowed.　Aside from the substances swept along by the subglacial stream, the various fragments are not rounded and water-worn, as is the sediment of rivers.　The moraines on the top of the ice (lateral and medial) are little or not at all abraded, but are deposited as angular blocks and fragments.　The ground moraine, on the other hand, is abraded in the peculiar way already described.　In all this work of glacial denudation the process is entirely mechanical, — chemical decomposition plays no part in it.

Certain other forms of transportation by ice may be conveniently mentioned here.

Ground Ice forms in rivers and ponds on the bottom, freezing around stones and boulders, and when broken up by thaws, this ice may float for long distances, carrying with it burdens far greater than the stream which transports the ice could carry unassisted.　The shores of the St. Lawrence River are fringed with lines of large boulders which have thus been brought down.

Lake Ice produces some curious effects in northern regions. When the lake is covered with cakes of ice as the result of an early thaw, refreezing, by expanding the water between the floating blocks, causes the ice to press strongly upon the shore.　In case the lake beach is covered with boulders, the push of the ice heaps up the boulders into a ring wall.

Coast Ice. — In Arctic regions the shallow water along the coast is frozen in winter into a broad shelf of ice called the *ice-foot*.　In the spring land-slips cover the ice with débris, while the bottom is studded with stones and pebbles. When the ice-foot is broken up in summer, part of it is drifted away and transports its load of rock for long distances. Other parts are worked backward and forward by the waves and tides, scoring the rocks of the coast and grinding and polishing the fragments of rock frozen in the ice, in much the same fashion as glacial pebbles are scored and ground. Over comparatively limited areas the marks of coast ice often have a deceptive resemblance to those left by glaciers.

Icebergs. — When a glacier enters the sea, it ploughs along the bottom until the buoyant power of the water breaks off great fragments of it, which float away as icebergs. These are often of gigantic size, veritable islands of ice, and huge as they appear, only about one-ninth of their bulk is above water. As icebergs are derived from glaciers, they carry away whatever débris the parent glacier had upon or within it.

2. THE SEA

The destructive work of the sea is accomplished mainly by means of the waves which the wind raises upon its surface and by wind and tidal currents. The great ocean currents are, as a rule, so far from shore, and flow in such deep water, that their erosive power is comparatively small. The Gulf Stream is said to scour the bottom in the Florida Straits and off the Carolina coast, but this is exceptional.

Waves act continually upon all coasts, but with very different force at different times and places. According to observations made for the Scotch Lighthouse Board, the average wave pressure on the coast of Scotland is for the five summer months 611 pounds per square foot, and for six winter months 2086 pounds. These are average figures and are greatly exceeded in storms, when the force of the breakers often rises to many tons per square foot.

The effect produced by this great force depends upon the character of the rocks of the coast, its height, and the angle at which it rises out of the water; also, in the case of stratified rocks, upon the attitude of the beds, whether they are horizontal, or inclined toward or away from the sea. When the coast is high, steep, and rocky, the waves continually wear away its base, partly by dislodging the blocks into which all consolidated rocks are divided, and partly by using as projectiles the blocks which it has dis-

FIG. 74. — Wave erosion; Étretat, France

lodged, or which have been loosened by the frost. In heavy gales great masses, weighing tons, it may be, are hurled with tremendous violence against the base of the cliffs, cutting them into caverns, which are further excavated by the ordinary surf. Eventually, the cliff is undermined, and the unsupported masses above fall in ruins.

Waves are not so entirely dependent for their effectiveness as rivers are upon the hard materials which they dash upon the coast for their efficiency as destructive agents. The force of the mere blow given by a storm breaker is very great, and the hydrostatic pressure which first forces the water into every fine crevice of the rock, and then withdraws it, together with the sudden compression and reëxpansion of the air contained in these fissures, assists materially in the loosening of the blocks.

FIG. 75. — Wave-cut arch, coast of California. (U. S. G. S.)

Along coasts which are composed of hard rocks the work of cutting back the land by the sea is comparatively slow, but when the rocks are soft and yielding, and yet rise abruptly from the ocean, the waste is so rapid as to attract every one's attention. The coast of Yorkshire in England is washed away at an average rate of nearly seven feet per annum. The island of Heligoland, near the German coast, has suffered great loss from the attacks of the sea within historic times; the small eastern island was cut off from the larger island, Heligoland proper, by a great storm in 1720.

At Long Branch, New Jersey, the sandy bluffs must be artificially protected against the attacks of the sea; yet in spite of such protection, almost every severe gale does considerable damage.

Sandy coasts which are low-lying and flat often suffer less from the inroads of the sea than rocky and precipitous ones, especially as they are apt to be lines along which material is accumulating. Even such coasts may, however, be rapidly cut back, as is shown

FIG. 76. — Wave erosion, strata dipping seaward; Orkney Islands, Scotland

in the familiar example of Coney Island, where great damage has been done of late years. When the sea is eating away a sandy shore, the homogeneous material prevents the occurrence of such irregularities of the coast-line as occur in rocky districts. So long as the coast is neither elevated nor depressed, the surf cuts it back at a continually decreasing rate, because the retreat of the coast-line leaves a shelf covered with shallow water, in passing over which the waves are retarded by friction and strike the shore with greatly diminished force. Just how far such a coast may be cut back is not definitely known, but it probably does not exceed a few miles, at most. On the other hand, if the land in question be slowly sinking, the sea gains a great advantage and may continue its destructive work indefinitely. Indeed, several high authorities are of the opinion that this is the only method by

FIG. 77. — Wave erosion, strata dipping landward; Duncansby Head, Orkney Islands

which great areas can be planed down to an approximately level surface. Again, it should be noted that when the sea is advancing over an ancient land surface, it finds ready to hand an immense body of soft materials which are speedily removed. The deep decay of the rocks into soil and the deposits made by the wind, rivers, lakes, etc., have all prepared the way for the erosive action of the

sea. When the coast is elevated, the sea cuts a succession of
terraces.

Besides cutting back its shores, the sea continually grinds up the
material which is brought into it by the rivers, and that which it
obtains by its own wear of the coast. The great blocks on the
shore are rolled about in storms, and worn into rounded boulders,
which are gradually reduced to smaller and smaller size. All the

FIG. 78. — Joint-block partly dislodged by the surf on wave-cut terrace; Wick,
Orkney Islands

minerals softer than quartz are rapidly ground into fine particles
and swept away by the undertow into deeper and quieter waters,
leaving the larger quartz fragments to form the pebbles and sand
of the beach.

The action of the waves is limited vertically, ceasing to be effective in quite shallow water, not far below the low-tide mark. In
violent storms the waves often accomplish much destruction far

above high tide, but the principal work of the waves is confined to a belt extending from a little above high tide to somewhat below low tide. However, Graham Island, near Sicily (see p. 67), which was thrown up in 1831, has been so completely removed by the waves that not even a shoal remains. Below the low-water mark the wave work is often efficiently supplemented by tidal currents, which under favourable circumstances acquire great velocity and depth, scouring away loose materials and even cutting into solid rock. When an island of considerable

FIG. 79. — Igneous rock, corroded by sea-water, about ½ natural size. (Photograph by van Ingen)

extent is exposed to the incoming tide, the latter travels around the island in both directions, and if the shape of the mainland is favourable, one of these currents will be much higher than the other, which will produce a " race " between the island and the mainland. Hell Gate, New York, is an example of this; the tide advances through New York Bay and Long Island Sound, being higher at flood, lower at ebb, in the sound than in the bay. The consequence is a swift current into the bay at flood-tide and into the sound at ebb. The floor of the British Channel, over which the tidal currents run very rapidly, is swept bare of sand, which is carried into the North Sea. By such means as this, the sea cuts away the land to depths much greater than unassisted waves can effectively reach.

Rocks are also attacked chemically by the solvent and decomposing action of sea-water. The silicates, such as the felspars, augite, hornblende, etc., are attacked much more rapidly than in fresh water. In shoal water and on the shore this action is obvious only in spots sheltered from the direct assault of the waves, because the products of decomposition are immediately removed by the surf and the mechanical work is so much more striking. Limestone coasts suffer from solution by sea-water, and are characterized by long caverns and tunnels, though sea caves are worn by the surf in all classes of rocks In the profound depths of the oceanic basins, where the water is never disturbed and where its motion is extremely slow, chemical activity becomes relatively very important. Calcareous shells are completely dissolved, and the volcanic débris, which covers the sea-bottom over vast areas, is disintegrated into a characteristic red clay.

3. LAKES

In comparison with the long life of the earth, lakes must be regarded as merely temporary bodies of water, which will sooner or later disappear, either by being drained of their waters or by being filled up with the sediments which are washed into them. The general term *lake* is employed for any inland body of water, which does not form part of the sea, but lakes are formed in very different ways and have correspondingly different histories. Most lakes occupy depressions below the general drainage level of the country, whether these depressions be due to movements of the earth's crust, to glacial excavations, to unequal decomposition by the atmospheric agencies, or to some other factor. Others, again, are held back by dams, such as lava streams, glacial moraines, or the glaciers themselves, by the débris of land-slips, or by the deltas of tributary streams which bring in more material than the main river can dispose of. Others still are enlarged basins cut out by rivers. Great lakes that persist for long periods of time are contained in basins, often of great depth, which were formed by movements of the earth's crust; the other kinds are more evanescent and usually of rather small size.

FIG. 80. — Wave-cut bluff on Lake Ontario. (U. S. G. S.)

Small lakes accomplish very little in the way of rock destruction, but are rather places of accumulation. The waves, even in storms, are not heavy enough effectively to cut back the shores, while the current of water through the lake is too slow and the sediment transported too small and light to erode the bottom as a river does. In great lakes, such as those which drain into the St.

Lawrence, storms develop a very heavy surf, and such lakes eat into their shores as the ocean does, but the very small tide confines the work of the waves within narrower limits, and the lighter breakers are less effective. Lakes are subject to various accidents which cause great fluctuations of the water-level. Deserted shore-lines are marked by beaches and terraces. The method of denudation by lakes is the same as that of the sea, but the modes of accumulation of material are characteristically different.

4. Organic Agencies

The organic agencies are animals and plants, both living and after death. In some respects these agencies tend to counteract the destructiveness of others, and the protective effects may be taken up first.

(1) *Protective Effects.* — These have already been considered in part, in connection with the processes of weathering (see p. 100). The protective effects of organisms are almost entirely those of plants, since animals, on land at least, are not sufficiently abundant to be of any importance in this connection. A thick covering of vegetation, especially the elastic, matted roots of grassy turf, protects the soil against the mechanical wash of rain. How complete this protection often is, may be seen in the different effects produced by a heavy fall of rain upon a grass field and on the adjoining ploughed lands, or even on the roads. The roads may be so washed out as to be impassable, while the grass fields have not suffered at all. In certain of the western bad lands, the efficient protection given by grass is very well shown; where the grass has established itself thickly, the country is gently rolling, but where it is absent, the wild and broken bad lands are developed.

Forests also are very important conservers of the soil, especially on mountain sides and other steep slopes. The removal of forests only too often is followed by calamitous results.

Vegetation, especially grass, protects loose, light soils from the wind, and often this is the only means by which sand dunes can be held in place and prevented from overwhelming valuable lands. Even the banks of rivers and the seacoast may be efficiently protected by plants. Dense masses of seaweed growing on the rocks form an elastic buffer against the surf, and along low-lying tropical

FIG. 81. — Erosion following removal of forest; Great Smoky Mts., Tenn.
(U.S. Bureau of Forestry)

coasts the mangrove trees, with their interlacing aërial roots, so break the force of the waves that they cannot wash away even fine mud.

The only protection afforded by animals that requires mention

N

is in the case of coral reefs, which, thrown up along or parallel to the coast, shield it from the heaviest surf.

(2) *The Destructive Effects* of the organic agencies are decidedly subordinate to those of the other classes which have so far been considered, but they are not without importance. The products of vegetable decomposition in bogs and in beds of clay, muds on the sea-bottom, etc., are efficient means of chemical change, and

FIG. 82. — Soil destruction due to removal of forest; Mitchell Co., N.C.
(U. S. Bureau of Forestry)

observations show that the decay of animals in the deep sea is an agent of no mean importance in promoting the chemical changes which there take place. But even living animals and plants do an important work in disintegrating rocks, that should not be overlooked. Bacteria play a considerable, but not yet fully known, part in the surface decomposition of rocks and soils. Certain of these microscopic plants have the power of fixing the atmospheric nitrogen and converting it into nitric acid, while others are the indispensable agents of organic decomposition. Seeds germinating in the crevices of rocks, or the roots of trees which invade such crevices from above, wedge the rocks apart with the same irresistible power as is displayed by frost, and often large areas of rock are thus most effectively broken up. The roots of living plants also secrete an acid, which dissolves out some of the soluble constituents of rock, thus adding a chemical activity to the wedge-like mechanical effects of growth.

Many marine animals bore into rocks, even the hardest, and cause them to crumble, and on the land great numbers of animals continually bore and tunnel through the soil, allowing a freer access of air and water. In the tropics the soil is fairly alive with the multitude of burrowers. Earthworms are among the most important agents in work of this kind, and the last of Mr. Darwin's books was a most interesting one upon the geological work of worms. The worms swallow quantities of earth, for the sake of the organic matter which it contains, and grind it exceedingly fine in their muscular gizzards. This ground-up soil is always deposited on the surface, in the form of the coiled " worm-castings," so abundant in grassy places. Worms are thus continually undermining the soil, bringing up material from below and depositing it on the surface, while, by the collapse of the old burrows, the first surface gradually sinks. In England the material thus yearly brought to the surface varies from seven to eighteen tons per acre, which means an average annual addition of one-tenth to one-sixth of an inch. By this means the surface of the ground is constantly changed, and substances spread over the ground in the course of years make their way down into it, forming well-defined layers beneath the surface. In the tropics ants and termites (so-called white ants) are even more active than worms in tunnelling the soil, and in many semi-arid plains burrowing mammals in incredible multitudes are continually working over the soil to great depths, as in the prairie-dog villages of our western plains. The occasional heavy rains thus penetrate to depths which could not otherwise be reached.

Summary of Destructive Action. — The surface of the land is everywhere attacked by the universally present atmosphere at a rate which differs much in different regions, depending upon climate, elevation above sea-level, and the resistant power of the rocks. The rain chemically decomposes the rocks, converting them into soil, and mechanically washing this soil to lower levels and into the streams. Frost shatters the rocks into smaller and smaller fragments. In arid regions the extreme changes of temperature break up the rocks much as does the expansive force of freezing water, while the wind transports immense volumes of sand and dust, which cut and carve and wear away the exposed rocks. Underground waters, especially when heated, do an important work of solution and decomposition, and, under favourable circumstances, cause the dislodgment of great masses of earth and rock in land-slips and rock-slides. Rivers excavate valleys and serve as the great agents of transportation, bearing the waste of the land to the sea, and glaciers do similar work in a highly characteristic manner. The sea cuts into its coasts by the action of waves, deepening its bed in shallow places by tidal currents, and in the case of a slowly sinking land may plane down great areas to a flat, gently sloping surface. Animals and plants add an important quota to the general work of destruction.

The annual waste of the land at the present time is estimated at 20 cubic kilometers (Penck), and, in past times, an incalculably great amount of material has been removed from the land. The Appalachian Mountain system has thus lost thicknesses of rock which vary in different regions from 8000 to 20,000 feet, and it is altogether probable that the average waste of all the continents amounts to several thousands of feet. The figures given for the basins of the Mississippi and Ganges show that such waste implies enormously long periods of time.

CHAPTER VII

RECONSTRUCTIVE PROCESSES. — CONTINENTAL DE-POSITS, LAND, SWAMP, AND RIVER

WE have now to inquire what becomes of the material which is derived from the decomposition and disintegration of the rocks. At the present time, it is estimated, about one-half of the waste of the land is carried directly into the sea, while the remainder is arrested in its journey and deposited upon the land. It must be remembered, however, that when the sea advances over the land, these deposits are, to a large extent, rapidly worked over by the waves and converted into marine deposits. The accessible rocks of the earth's crust are more largely composed of marine deposits than of those laid down in other ways, yet the non-marine sedimentary rocks are also extensively represented. It is only quite lately that the importance of this latter class of rocks has been appreciated. Furthermore, their importance is not merely quantitative, but lies also in the help which they give in the determination of ancient land surfaces, lake beds, river channels, ice-fields, and the like. It is therefore necessary to study all the methods by which rock reconstruction is effected, on however small a scale.

The most natural primary division of the sedimentary accumulations is into the marine and the continental, including in the latter the deposits which are made upon the land, or in such bodies of water as are not parts of the sea. Between these two principal classes there is a transitional series, consisting of deposits laid down in bodies of salt water which are in tidal connection with the sea, such as estuaries, almost closed bays and sounds, or seas, like the Baltic, which are partly brackish, as well as the littoral, or

seashore, which by the movement of the tides is alternately a
land surface and a sea-bottom. These distinctions are sufficiently
obvious, yet they are not always easy to apply, especially in the
absence of fossils; hence great differences of opinion continually
arise concerning the interpretation of certain rock masses.

Stratification. — It is an almost universal characteristic of sedimentary accumulations, whether modern deposits or ancient rocks,
that they are *stratified*, that is, divided into more or less parallel
layers or beds.   Indeed, the terms *secondary, derivative, sedimentary,* and *stratified* rocks are but different names for the
same thing.   Stratification is due to the *sorting power* of water, or
of the wind, by which, so long as conditions remain the same, particles or fragments of similar size and weight are thrown down at
the same spot.   If sand, gravel, mud, and clay are shaken together
in a jar of water and then allowed to stand, the various materials
will settle to the bottom in the order of their coarseness, the finest
coming down last.   Yet the change from one kind of material
to another will be so gradual that no well-defined layers will
appear, and thus no true stratification results.   Layers clearly demarcated from one another may be produced in either one of two
ways: (1) by such a change of conditions that the material deposited changes abruptly, though perhaps only as a mere film of a
different substance, or (2) by a pause, however brief, in the process
of deposition.   In the latter case, each layer represents a time
of deposition broken by an interval which allows the surface particles to arrange themselves somewhat differently from the position
they would take were the deposition continuous.   The planes of
contact between the successive layers, which may be indistinct or
very sharply defined, are called the *bedding* or *stratification planes,*
and each one of these formed the surface of the lithosphere, either
as a land surface or the bottom of some body of water, for a short
time.   The thickness of each layer indicates the length of time
during which the deposition of similar material went on without
interruption, and varies from hundreds of feet to a small fraction of an inch.

The power of ordinary winds to transport material is much less than that of water, and wind-borne débris is, on the average, much finer than water-borne sediment, and furthermore the winds are less constant in direction and subject to greater and more sudden changes of velocity. Consequently, stratification by the wind is, as a rule, less even and regular than that which is due to water; but still wind-made deposits are stratified, and it is not always practicable to distinguish with certainty between the two classes. Fine volcanic ash and dust may be spread by the winds over immense areas and in very regular beds or strata.

The sorting power of water or wind results in the concentration of similar material, so that, as a rule, each bed is made up of some predominant substance in a state of greater or less purity, such as gravel, sand, clay, etc., and thus heterogeneous material is separated into its constituent parts, though the separation is rarely quite complete, and sometimes there is hardly any separation at all. On examining a thick series of deposits, we find that the materials are apt to change both vertically and horizontally. Changes in the vertical direction imply changes of conditions, in accordance with which different kinds of material are successively laid down over the same area, so that gravel is deposited on sand, sand on mud, or vice versa. Such changes are usually abrupt, so that each stratum is sharply demarcated from the one above and the one below it. On the other hand, changes in material in the horizontal direction are usually gradual, and a bed of sand may pass by imperceptible transitions into one of gravel or of mud. This is because of the gradual change in the velocity of the transporting agent and therefore of its carrying power. In the sea or a large lake the material on the bottom grows finer outward from the shore, while a river, whose velocity diminishes from head waters to mouth, lays down material of decreasing coarseness, from the boulders and cobbles of the head waters to the fine silt of the lower course.

Each agent of reconstruction, or deposition, has its own characteristic manner of accumulating material, and, in typical instances, it is easy to distinguish between them, but there are also many similarities and, as we have already learned, it is sometimes exceedingly difficult to determine which of several possible agents was the actual means of forming a given series of deposits. If no fossils (*i.e.* recognizable traces of animals or plants) are present, it is not always easy to determine, for example, whether a given sandstone was laid down in the sea, or in a lake, or heaped up by the winds in a desert. This uncertainty is, however, largely due to our ignorance concerning all the minute details of structure which characterize the work of each agent, and may be expected to disappear as knowledge of these details advances. Of late years great progress has been made in these matters, and systematic study, it may reasonably be hoped, will remove an ignorance which is owing chiefly to a neglect of the subject and to certain preconceptions inherited from the early days of geology.

A. CONTINENTAL DEPOSITS

The continental deposits may be classified in several different ways, each one of which has its advantages according to the object aimed at. Our present purpose will best be served by arranging these accumulations, in general, in accordance with the agency by which they are made. However, it is not feasible, nor even desirable, to carry out this scheme with rigid consistency, for so many deposits are formed by two or more agents acting together, wind and rain, ice and water, rivers and the sea, etc. Then, too, the various agents so often have shifting boundaries: on the seashore the tides, especially the spring and neap tides, make the limits of land and sea somewhat indefinite, while rivers, now confined to their channels, again are flooded so as to form great temporary lakes; the rare but violent rains of the desert may cover with a sheet of shallow water great areas which are ordinarily baked and cracked by the blazing sun. Owing to this shifting of limits and the alternation of agencies, continental deposits seldom display such uniformity over wide areas as obtains on the sea-bottom, where the conditions are so much more constant.

The land is the scene both of denudation and deposition, and which of these two processes shall prevail in any area depends upon the topography and the climate of that area. As is shown in the diagram, Fig. 83, only about one-fifth of the land surface is raised more than 1200 meters (about 4000 feet) above the level of the sea, and this fifth includes the regions of most active denudation; three-fifths, at successively lower levels, are areas of progres-

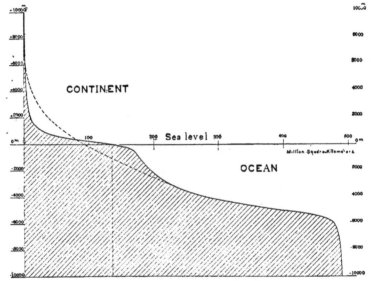

FIG. 83. — Diagram showing the relation between height and area of land above sea-level and of water in ocean basins. Vertical columns of figures indicate heights and depths in meters; on horizontal line, millions of square kilometers. (Penck)

sively less effective erosion, as we descend from higher to lower ground, while the remaining fifth receives deposits upon it. Nearly a fifth of the land of the globe is comprised in interior continental basins which have an arid climate and are without an outlet to the sea; some, indeed, like the lower Jordan valley and the Dead Sea, are far below the ocean level. Probably one-half of this desert and semi-desert area is the seat of extensive deposition. The areas of

denudation and those of deposition are thus determined by climate
and topography, and shift as those factors change or are modified
by diastrophic movements of the earth's crust.

I. TERRESTRIAL DEPOSITS

Under this head are included those accumulations of the mechanical and chemical waste of preëxistent rocks which are formed
on land surfaces and not in permanent bodies of water. Deposits
made by ice are considered in a separate section. The principal
agencies which form deposits of this class are rain and wind and
springs, and the great variety of them is due to climatic factors, the
velocity and constancy of the winds, the quantity and seasonal
distribution of the rainfall, the amount and rapidity of temperature
changes. Hence we find different kinds of deposits in deserts and
humid regions, in cold, temperate, and tropical climates, near the
seashore, and in the interior of the continents. The necessity of
considering and emphasizing these differences lies in their value
for historical studies. Every rock bears within it a record of its
history, could we only decipher it.

Residual Accumulations; Soil. — As we have already seen, the
disintegration and decay of rock results in the formation of soil,
which is the residuum after the removal of more or less of the parent rock. In humid climates there is usually a distinct subsoil
which is less thoroughly oxidized and hydrated and is lighter in
colour and much less fertile than the top soil, which is largely due
to the washing downward of the fine clay particles by percolating
rainwater. In arid climates there is less kaolinization of the
aluminous silicates, a much deeper top soil, and little or no distinct
subsoil. Under the influence of wind, rain, and other agencies,
the soil travels down the slopes and accumulates, often to great
depths, in valleys and depressions, and is carried in enormous
volume by the rivers. Very little soil, as such, is built into the
rocks of the earth's crust, but sometimes it is buried under a lava
stream or depressed beneath the sea or a lake in such a manner

as not to be washed away, but immediately covered by other deposits. Such an old soil, or " dirt bed," may be recognized by its texture and appearance and by the fossil roots and stems of land plants with which it is apt to be filled.

Laterite is a peculiar soil very widely spread in the tropics and of a deep red colour, caused by the presence of Fe_2O_3. It differs from ordinary soils in the fact that much of the alumina is present as an oxide, instead of the silicate, and is frequently characterized by lumps and nodules of Fe_2O_3, which are produced by a chemical process of concentration.

Chemical Deposits. — In the tropics, which so largely have a regular alternation of rainy and dry seasons, and in arid regions, where the rain often falls in torrential showers, followed by long periods of drought, the movement of water through the soil is frequently reversed in direction. During the rains the movement is downward; in the dry period evaporation from the surface and capillarity cause a slow ascent of the water through the soil. Often this ascending water is charged with material in solution and this material is deposited on or near the surface as the water evaporates. In deserts and semi-deserts the surface is often white with salt, the sulphate or carbonate of soda, borax, and other soluble compounds. The iron nodules of laterite are produced in this manner, and sometimes these nodules are cemented into continuous sheets of crude hæmatite. Where the soil and underlying rocks contain the carbonate of lime abundantly, the water concentrates them at the surface, it may be, as in South Africa, in very extensive sheets of hard limestone. These terrestrial chemical deposits may cover very wide areas, but never in any great thickness.

Mechanical Deposits are made on land surfaces by various agencies and form quantitatively much the most important series of the class.

Talus and Breccia. — As has been pointed out (pp. 114 and 118), great masses of angular blocks of all sizes accumulate at the foot of cliffs and on mountain slopes as talus, which shows an imperfect division into layers and is slowly but continually creeping

downward. By the deposition of some cementing material (usually CaCO₃) in the interstices of the talus the blocks may be bound

FIG. 84. — Loess deposits; North China. (Photograph by Willis)

into a solid mass, called *breccia*, of which the peculiarity is that the
fragments composing it are angular, not rounded.

Loess. — In arid regions the wind often carries the finer parts of the soil to immense distances and deposits them where they are less exposed to the wind, and where there is vegetation enough to hold them. In Central Asia the sun is often darkened for days by these dust-storms, and after they are past, a fine deposit of yellow dust is found over everything. Loess is a deposit formed in this way, and it is found in many lands. One of the largest known accumulations of it is in northern China, where it covers an immense area,

FIG. 85. — Sand dune with wind-ripples, River Terraces in distance; Biggs, Oregon. (U. S. G. S.)

to depths of 1000 to 1500 feet. It is not stratified, but cleaves vertically, and thus the ravines and valleys excavated in it have very abrupt sides. Loess also occurs in Europe, and the Pampas of the Argentine Republic are covered with a great thickness of it. The loess of the Mississippi valley, though of rather exceptional character, is yet probably of aeolian origin.

Blown Sand. — Wherever a sandy soil occurs unprotected by vegetation, as in deserts or along the seacoast, the wind drifts the sand and piles it up into hills or sand dunes. The dunes are roughly divided into layers, the thickness and inclination of which depend upon the force and direction of the wind, and often imitate the confused arrangement of sands piled up by waves and currents under water. The sand-grains of the dunes are, however, more rounded by the abrasion which they have undergone and, especially in deserts, they are apt to be smaller. When the sands

FIG. 86. — Sand dune; Beaufort Harbor, N.C. (U. S. G. S.)

are mixed with pieces of shells and other calcareous material, per-colating waters, by dissolving and redepositing the $CaCO_3$, may cement the sands into firm rock. This is the more conspicuous when the whole material is calcareous, as in the shell sands of Bermuda. This substance, ground up by the surf, is transported inland by the wind and piled up into dunes. Rain-water cements the loose grains together, and by the alternate accumulation by wind and cementing by rain is formed the stratified æolian or drift-sand rock.

Spring Deposits. — As our knowledge of microscopic plants increases, many processes which were believed to be purely chemical, are found to be dependent upon the activity of minute plants. At present, it is not possible to distinguish accurately, in all cases, between the two kinds of processes.

Many springs precipitate carbonate of lime, on coming to the surface. The quantity of $CaCO_3$ which a given volume of water will dissolve depends upon the amount of CO_2 contained in that water, and the quantity of dissolved gas, again, is determined by the pressure to which it is subjected. When the spring-waters

FIG. 87. — Ideal section through Mammoth Hot Springs, showing the water rising through limestone. (Hayden)

reach the surface, the pressure is relieved, much of the CO_2 immediately escapes, and more or less of the $CaCO_3$ is deposited as *travertine* in the neighbourhood of the spring, often in masses of considerable extent and thickness. The process is not always entirely chemical. The beautiful calcareous terrace formed by the Mammoth Hot Springs, in the Yellowstone Park, is, in part at least, due to the separation of the lime salt from the water by a jelly-like plant, which grows in the hot water and is spread in bright coloured layers over the surface of the terrace. The parts of the terrace where deposition is no longer in progress can be at once distinguished by their white colour.

Siliceous deposits are much less common than the calcareous, because of the rare conditions under which silica is dissolved

in any considerable quantity, hot solutions of alkaline carbonates being necessary for this purpose. In the Yellowstone Park,
especially on the Firehole River, are great terraces and flats of
hard white siliceous sinter, or geyserite, which have been formed
and are still being added to by the innumerable hot springs and
geysers. The silica is deposited partly by the evaporation of the
water and partly by the action of *Algæ* (minute plants) which

FIG. 88. — Travertine terrace of the Mammoth Hot Springs, Yellowstone Park

flourish in hot-water pools. Similar deposits are found in the
geyser regions of Iceland and New Zealand.

Iron deposits are formed by the springs known as *chalybeate*,
which contain the carbonate of iron ($FeCO_3$) in solution. Contact with the air speedily converts the soluble carbonate into the
insoluble Fe_2O_3, which forms brown stains and patches on the
channels leading from such springs, and considerable quantities of
it collect in pools. Here again, organic agency may supplement

FIG. 89. — Crater of Castle Geyser, Yellowstone Park. The crater is composed of silica in the form of geyserite

the chemical work, for certain atoms extract iron from the water, as other Algæ extract lime and silica.

Certain mineral springs are of importance, as indicating a way in which mineral veins my have been formed (see p. 430). The Sulphur Bank Springs in the Coast Range of California are an especially instructive example i this activity. Below the depths to which the atmospheric influences penetrate, the fissures in the rocks are filled with hydrated sica, which is as soft as cheese and contains more or less cinnabar(sulphide of mercury). In other places the silica is hardened to chalcedony, and deposits of cinnabar mixed with iron pyrites fi up the crevices. The hot waters which build up these deposits re alkaline, charged with certain acids and alkaline sulphides. Near Virginia City, Nevada, hot alkaline springs rise through a sries of fissures, in which they have deposited linings of silica, amorphous and chalcedonic, with some quartz, containing minute crysals of iron pyrites and traces of copper and gold. On the surfac the springs have formed a thick layer of geyserite.

Phosphate Deposits are the olv strictly terrestrial organic formations which require notice. These are principally derived from guano, which is the accumulate excrement of birds (in caves, of bats), and contains phosphate in large quantity. In rainless regions, such as the Peruvian csts and islands, the guano may accumulate to great thickness without loss of its soluble matters, but in rainy districts these are largely carried away by percolating waters. Should the underlying rock be a limestone, it will be gradually converted from a carbnate into a phosphate of lime. Such is believed to be the mode i origin of the phosphatic rock of Florida and the West Indies. On the other hand, the phosphatic nodules of South Carolina are regarded as due to the action of swamp water upon underlying sell rocks, though the source of phosphoric acid is not well understood.

Cave Deposits. — The chemically formed cave deposits are due to the solution and redeposition f carbonate of lime. Caves are very generally found in limestoes, and the percolating waters

which make their way through the roof of a limestone cavern always have more or less $CaCO_3$ in solution. A drop of such water, hanging from the cavern roof, will lose some of its CO_2, upon the presence of which the solubility of the $CaCO_3$ depends, and deposit a little ring of the lime salt. Successive depositions will lengthen the ring to a tube, and then the tube will be built up by layers on the inner side, until it becomes a cone. At first, the deposit is white, opaque, and very friable, crumbling at a touch, but repeated depositions fill up the interstices of the porous mass and convert it into a hard, translucent stone, which assumes a crystalline structure through the development of calcite or aragonite crystals. The masses, thus formed, that depend from the roof of the cavern, are called *stalactites*. After hanging for a time from the roof, the drop of water falls to the floor of the cave, and there, in similar fashion, deposits a little layer of $CaCO_3$, which gradually grows upward into a cone. This is a *stalagmite*, and differs from the stalactite only in the fact that it grows upward from the floor, instead of downward from the roof. The stalagmite is, of course, exactly beneath the stalactite, and as long as the water continues to follow the same path, the two cones are steadily, though very slowly, increased both in height and thickness, until they meet, unite, and form a pillar extending from floor to roof of the cavern.

These deposits form the most curious and beautiful features of limestone caverns. The stalactites assume all manner of shapes, determined by the way in which the water trickles over them, and the abundance or scantiness of the water supply. Fantastic and beautiful shapes of every description, fringes of crystal spar, and curtain-like draperies hang from the roof and cover the walls of the chambers, while grotesque shapes rise from the floor, which is itself often a solid mass of the same deposit, and the pillars, once formed, are ornamented with every variety of fringe and sculpture. The constancy of the paths by which the water descends through the roof of the cavern, insures that the process shall continue uninterruptedly for very long periods of time. The Luray Caverns of Virginia are famous for the bizarre beauty of their formations, but limestone caves everywhere have more or less of the same deposit to show.

This process may be readily observed in any masonry arch, through which rain-water percolates, as a bridge, for example. The lime of the mortar is converted, in course of time, by contact with moist air, into $CaCO_3$, and this again is partially dissolved by the rain. When the rain-water trickles through the arch, it leaves icicle-like deposits, or thin sheets of calcareous matter, fringing the under side.

In a cave, it frequently happens that angular fragments fall from the roof and are cemented into a breccia by deposits of stalagmite. In caves connected with the surface by openings, sand and gravel, or fine soil and loam, are washed in by streams, or by the rain, and form the characteristic deposit known as *cave earth*. In ancient caverns, no longer subject to this wash, the whole deposit of earth may be sealed in by a covering of stalagmite. Cave earth has, in many instances, yielded great quantities of bones, which were washed in with the earth, or dragged in by the carnivorous animals which inhabited the cavern. The Port Kennedy cave in Pennsylvania is almost filled up by the bones of extinct animals which were washed into it, and many such cases are known, especially in Europe.

II. PALUSTRINE OR SWAMP DEPOSITS

The most important of the swamp and bog deposits are the vegetable accumulations, for the preservation of which a certain amount of water is necessary. The vast quantities of coal which occur in so many parts of the world, testify to the significance of the part which bog and swamp accumulations of vegetable matter have played in the earth's history. The nearest approach to coal that we have in process of formation at the present day, we find in the peat-bogs, which are especially abundant and extensive in cool, damp climates, as in Ireland, Scandinavia, and the northern parts of North America. In northern regions the peat is formed principally by mosses, and especially by the bog moss, *Sphagnum*; elsewhere, as in the Great Dismal Swamp of Virginia, the leaves of trees and various aquatic plants are the sources of supply.

The processes of organic decomposition depend upon the activities of bacteria, but, for the sake of simplicity, we may treat the subject *as if* the processes were chemical only. Vegetable matter consists of carbon, hydrogen, oxygen, and nitrogen, with a certain proportion of mineral matter, or ash. When decaying on the ground, exposed to the air, the plant tissues are completely oxidized, and form such simple and stable compounds as CO_2, H_2O, NH_3, and the more complex humous acids, and thus hardly any solid residue is left. In forests the accumulation of leaves for many centuries results only in a shallow layer of vegetable mould. Under water, where the supply of oxygen is very limited, vegetable decomposition is much less complete. Some CO_2, H_2O, and CH_3 (marsh gas) are formed, but much of the hydrogen and nearly all of the carbon remain; the farther decomposition proceeds, the higher does the percentage of carbon rise, and the darker does the colour of the mass become. Peat frequently forms in small lakes and ponds, aquatic plants growing out from the edges and on the surface, until they gradually fill up the basin and convert the pond into a bog.

The Great Dismal Swamp of Virginia and North Carolina probably more nearly reproduces than do most existing peat-bogs the conditions of the ancient coal swamps. The swamp, which measures thirty miles by ten, is a dense growth of vegetation upon a water-covered soil of pure peat about fifteen feet deep and with no admixture of sediment. The swamp cypress grows abundantly in the bog, and prevents, by its dense shade, the evaporation which would take place in summer, could the sun's rays penetrate to the wet soil. The shallow layer of water which covers the ground receives the fallen leaves, twigs, and branches, and sometimes even the trunks of fallen trees, preventing their complete decomposition, while the dense covering of mosses, reeds, and ferns which carpet the ground, add their quota to the mass of decaying vegetable matter.   At the bottom of the bog, it is of interest to observe, is a layer of fire-clay, which, by its imperviousness, tends to hold the water and prevent its draining away.   Peat swamps, formed in a similar manner, also occur at the mouths of great rivers, such as the Mississippi.

FIG. 90. — Great Dismal Swamp, Virginia.   (U. S. G. S.)

The bogs of northern latitudes are due principally to the bog moss *Sphagnum*, which forms dense and tangled masses of vegetation, dead and decaying below, green and flourishing above. As these mosses hold water like a sponge, they will develop bogs in any shallow depression, or even on a flat surface, where they may get a foothold.   The depth of peat is sometimes as much as fifty feet, and its density and fineness of grain increase with the depth and the length of time it has been macerating in water.

Fire-clay is frequently found at the bottom of peat-bogs, and is directly connected with the processes of vegetable decomposition, though not itself of organic origin. Fire-clay contains a large admixture of siliceous sand, but is free from lime, magnesia, the alkalies, and any high percentage of iron; it is thus a mixture of nearly pure clay and sand, which may be heated very highly without melting or crumbling. The iron, alkalies, and alkaline earths are gradually leached out of the clay by the action of the peaty water, which is charged with organic acids, and thus an ordinary clay is converted into a fire-clay. Fire-clay occurs frequently beneath coal seams; as the percentage of silica becomes very high, fire-clay passes over into *gannister*, which is largely used for the lining of iron furnaces.

Bog Iron-ore is another substance which is indirectly due to the decay of plants; it is found at the bottom of bogs, or lakes, in deposits which are sometimes many feet thick. Iron is a very widely disseminated substance, occurring in almost all rocks and soils, though usually in very small quantities; by the action of the bog water the oxide is converted into the soluble carbonate ($FeCO_3$). Solutions of $FeCO_3$ accumulate under peat-bogs and deposit their mineral by concentration; but when the iron-bearing waters evaporate in contact with the air, the carbonate is reconverted into the red oxide, by the loss of CO_2 and absorption of O.

III. FLUVIATILE OR RIVER DEPOSITS

In a preceding chapter we learned that the power of a stream of water to transport sediment depends upon its velocity, which, in its turn, is determined by the slope of the ground and the volume of water. Further, we discovered the very surprising fact that, for the coarser material which is pushed along the bottom, the transporting power increases as the sixth power of the velocity ($T = V^6$). It follows from this that a slight decrease in the swiftness of a stream will cause it to throw down the greater part of its load of sediment, while a slight increase will cause it to carry off what it had before deposited. Thus, great rivers, like the Mississippi, which flow in soft, easily moved deposits, are preëminently whimsical and treacherous. As the volume and velocity of the stream are much subject to change, there will obviously be corresponding changes in the scour and deposition at any given point, but there are certain places where deposition is so constant that extensive accumulations

FIG. 91. — Manti Creek, Utah; flood of August, 1901

may be formed there. As we trace a river downward from its source in a mountain region, we find that in the upper stream, which is a torrent in swiftness, only large stones remain at rest, everything else being swept along. Farther down stream, as the slope of the bed diminishes, the coarse gravel is thrown down, next the coarse sand is deposited, and in the lower reaches of a river, which, like the Mississippi, flows over land that has a very gentle

slope, and is raised but little above the sea-level, only the finest silt
gathers on the bottom.   The exact limits of the different kinds of
deposit will vary with the stage of water.

At points where the velocity of the stream meets a constant check,
there will be constant deposition, and thus bars and islands are
built up in the channel, which will be permanent unless some change
of conditions is brought about.   In the sand-bars and gravel-spits

FIG. 92. — Effects of flood; Black Hills, S.D.   (U. S. G. S.)

the up-stream side is a gentle slope, ending abruptly on the downstream side, the bar or spit advancing by having sand or gravel
pushed up the gentle slope by the current and dropped over the
steep face, where it forms inclined layers.· Flattened and elongated pebbles arrange themselves so as to offer the least resistance
to the current, in a slanting position, with their tops down stream.
When the stream is subsiding, the material tends to assume a

more horizontal direction, giving an irregular and confused
stratification to these deposits.

Alluvial Cones or Fans. — Where a swift torrent, descending a
steep slope, debouches on a plain or wide valley, its velocity is
greatly diminished, and a large part of the material which it carries
is thrown down and spread in a fan shape from the opening of the
ravine in which the torrent flows. The thickness of the cone is

FIG. 93. — Sand deposits, North Platte River, Nebraska. (U. S. G. S.)

greatest at the mouth of the ravine, while its breadth increases outward from that point. Where several such torrents open on the
plain near together, their fans may coalesce and form a continuous
fringe along the base of the mountain. The slope of the cone's
surface diminishes with the size of the stream; in small streams it
may be as steep as 10°. These cones are formed on much the
same principle as deltas, and might, with propriety, be called terrestrial deltas. Very large alluvial cones are found in the Rocky

Mountain and Great Basin regions, generally in the forelands which front high mountains. In the western part of the Argentine Republic, along the front of the Andes, the temporary rivers formed by the melting snow bring down enormous quantities of mud and fine sand and spread it out over the plain. Where the rivers discharge into the sea, this process of upbuilding is limited, except where there is slow downwarping of the foreland, but in interior

FIG. 94. — Alluvial cone, trenched by stream, with secondary cone below.
(U. S. G. S.)

arid basins, without outlet, it may accumulate very great thicknesses of river-made sediments.

Flood Plains. — Rivers, as is well known, are subject to floods when the volume of water is enormously increased and can no longer be contained in the ordinary channel, but spreads out over the level ground on each side. By this spreading, which may be for many miles in both directions, the velocity of the water is much

FIG. 95. — Flood plain, Genesee River, N.Y. (Photograph by van Ingen)

diminished, and over the flooded area (flood plain) large quantities of material are thrown down, while the unchecked velocity in the channel may cause a scouring and deepening there or, under other conditions, the channel and flood plain may both be built up, especially if the river flow through a slowly subsiding region. The nature of the material deposited over the flood plain will depend on the character and swiftness of the flooded stream, and varies from the coarsest gravel to the finest silt. The latter is more usual, for the flood plain is widest along the lower course of the river. Flood-plain deposits attain great importance in interior basins which have no drainage outlet and consequently retain all the material which is washed into them from the surrounding mountains by the rain or rivers. In such basins the rivers end in salt lakes or die out in the sands, but at intervals, it may be only rarely or during an annual wet season, the rivers are flooded and immense areas of the desert are converted into shallow seas. Near the mountains is formed a fringe of alluvial cones, which may coalesce into a continuous belt, and over the central parts of the basin is spread the finer material in even and regular stratification. The wind may carry away all this material and remove it beyond the limits of the basin, leaving only stony wastes, or the deposits may accumulate to a great thickness, according to circumstances.

Even in climates of heavy rainfall great interior basins, due to downwarping, are found, and though they drain to the sea, they may become filled to great depths with river deposits. The interior of South America, drained by the Orinoco, Amazon, Paraguay, etc., is an example of such basins where river deposition is actively progressing on a very large scale.

In climates with abundant rainfall (*pluvial* climates) the flood plains of rivers are covered with vegetation which protects the flood deposits, but in arid climates the flood plains are bare of vegetation for the whole or most of the year, and the river deposits are exposed to the sun. Great areas of mud and silt, thus exposed, shrink in drying and crack in deep fissures, which enclose polygonal areas, as may be seen in any mud puddle which is drying in the sun. The cracks thus formed are called *sun* or *mud cracks* and may be preserved indefinitely in rocks which are formed from flood-plain accumulations. In such deposits there is apt to be a difference in the material thrown down in the earlier and later stages of the flood, because of the difference in the velocity with which the waters move. After the river ceases to rise, the water over the flood plain becomes almost stagnant and lays down very fine

FIG. 96. — Sun cracks in Newark shale, about ¼ natural size

material, which thus forms the cracked surface. When the next flood arrives, it carries coarser material, frequently sand, which fills up the cracks and thus preserves them. Mud cracks are formed under other conditions, as will be seen in the following pages, but probably nowhere on such an extensive scale as on the flood plains of rivers which flow through arid regions. Footprints of land animals may be preserved in the same manner by being baked hard in the hot sun and then buried under the deposits of the next flood. Such cracks and footprints, and even the impression of rain-drops, are frequently found in the rocks and give valuable assistance in determining the conditions under which those rocks were formed.

In ancient flood-plain deposits, which the rivers that made them have long since deserted, the old channels are indicated by coarser material, sands and gravel cemented into sandstone and conglom-erate. When these old channels are plotted on a map, their sinuous course, great length in proportion to width, and their ramifying tributaries clearly mark them out as the records of an ancient system of drainage. Such channels and the accompanying broad and regularly stratified flood-plain deposits cover very extensive areas in South Dakota, Nebraska, and others of the Western States.

River Terraces and Old Gravels. — The lower courses of many rivers, including most of those in the northern United States, and some in the southern, are bordered by a succession of terraces that rise symmetrically on the two sides of the stream. Sometimes, as in many English rivers, the terraces are at different levels on oppo-site sides. The formation of these terraces is due to a twofold activity of the river; the combined deepening of the channel and building up of the flood plain at length make the trough of the river so deep that floods no longer fill it, especially if the velocity of the current be maintained or increased by an elevation of the region drained by the river. Then the energy of the current is partly employed in widening the channel and forming a new flood plain, cutting back the edges of the old flood plain, which it can no longer overflow, thus converting it into a terrace, which is the remnant of an old flood plain. The process may be repeated many times, and thus successive terraces rise, one above another, as we recede from the river.

It necessarily follows from this account that the highest terrace is the oldest, and the lowest is the last formed. This seems to be a violation of the rule that, in any series of sedimentary deposits, the oldest must be at the bottom and the newest at the top; but the violation is only apparent, not real. Were the river to flow at a constant level, no terraces could be formed, and the deposits would follow the rule, just as they do now in each successive flood plain and terrace. Because, however, the stream flows at successively lower levels, the lower flood plain is made up of the newer deposits. It should further be observed that the older gravels do not actually overlie the newer ones, but are merely at higher levels.

FIG. 97. — River terraces; Chelan River, Wash. (U. S. G. S.)

Unsymmetrical terraces, which are either confined to one side of the river, or if present on both sides, are on different levels, are formed when a stream is widening its valley by steadily cutting away the bank on one side, shifting the channel toward that side, and at the same time deepening it. This will result in the formation of terraces representing the former positions of the stream. If the lateral movement be all in one direction, the terraces will all be on the side away from which the channel is shifting; if it be alternately in opposite directions, terraces will be formed on both sides, but at different levels.

Another method of terrace formation should be mentioned. If a river which has excavated a deep valley, have its velocity checked by a slow subsidence of the country, it will commence to fill up its valley with gravel or other sediment, and may thus

FIG. 98. — Terrace on deserted channel; central New York.   (U. S. G. S.)

accumulate material of great thickness and extent.   Should a re-elevation of the country now occur, the river will acquire new destructive power and cut a terraced channel down through its own deposits.   In such a case the material is a continuous mass, and the gravels of the higher terraces are newer (not older) than those of the lower.   The rivers Mersey and Irwell in England are believed to be examples of this mode of terrace formation.

P

Rock terraces in river valleys are the result of erosion, not of deposition, and are due to harder ledges of rock which are exposed by the cutting of the river. Rock terraces may likewise be occasioned by diastrophic movements and by long-period fluctuations in the volume of the stream. In all cases terraces indicate the successive levels at which the river has flowed, but they do not imply, as would seem at first sight to be the case, that the river was once of sufficient volume to fill up the space between and below the terraces.

Deltas are accumulations of river deposits at the mouths of streams, land areas which the rivers have recovered from the body of water into which they flow. The factors which determine the formation of a delta are not altogether clear. The presence or absence of a strong tide is evidently one of these factors, for in lakes and in seas with little or no tide, almost all streams form deltas, while those rivers which empty into the open ocean almost invariably do so by means of estuaries, in which the sea encroaches on the land. In North America the rivers which discharge into the Gulf of Mexico form deltas, while the Atlantic streams nearly all have estuaries. In Europe the delta-forming rivers empty into the Mediterranean, the Baltic, and the Black and Caspian seas. Nevertheless, the tide is evidently not the sole factor in determining the presence of a delta. The Ganges and Brahmapootra have formed a vast delta in spite of the powerful tide of the Bay of Bengal; the Thames and the Rhine discharge into opposite sides of the North Sea, yet, while the latter has built up a delta, the former opens into a wide estuary. If the sea-bottom is subsiding faster than the river deposit is built up, no delta will be formed, but an estuary; while, on the other hand, slow and moderate subsidence is favorable to delta formation.

When a stream loaded with sediment flows into the relatively stationary waters of a lake or sea, its velocity is checked and the greater part of its load very rapidly thrown down. Deposition takes place much more rapidly in salt water than in fresh, because the dissolved salts reduce the cohesion of the water, and hence diminish the friction which retards the settling of silt. The excessively fine particles of clay, which in fresh water remain suspended for weeks, are thrown down in salt water in a few hours; hence the great mass of the sediment falls to the bottom in the vicinity of the stream's mouth. Such rapid accumulation obstructs the flow of the river and causes it to divide and seek new channels, especially in time of flood, and form a network of sluggish streams

FIG. 99. — Delta of Rondout Creek in the Hudson River; Rondout, N.Y.
(Photograph by van Ingen)

meandering across the low flats. The height of the delta is increased by the spreading waters of the river, when in flood, and the growth of vegetation assists in raising the land. Though the Mississippi delta is an area of subsidence, two-thirds of its surface is above water, when the river is in its ordinary stages. But for the levees, however, most of it would be inundated in times of flood, when the unconfined waters of the river would form a lake

600 miles long, 60 miles wide, and with an average depth of 12½ feet.

FIG. 100. — Settling of clay in salt and fresh water, after 24 hours; the jar on the left contains salt water and is clear, while the fresh water is still turbid.

The nature of the materials of which deltas consist varies according to circumstances. When mountains are near the coast, the streams flowing from them may descend into the sea with sufficient velocity to build a delta of cobblestones and coarse gravel. Usually, however, deltas formed in seas are composed of very fine materials, because the lower course of most rivers is through flat plains, and the stream can carry only very fine silt. Even in such cases, there will be differences in the coarseness and fineness of the material, corresponding to the seasons of high and low water in the river.

The material composing a delta is stratified in characteristic ways, which vary according to circumstances. Three different kinds of beds may be distinguished: (1) the *bottom-set* beds, which consist of fine material, spread out in regular, nearly horizontal layers over the sea-bottom; (2) the *foreset* beds, which are made up of slightly coarser sediment in layers which have a decided seaward inclination, the steepness of which depends upon the depths of water in which the débris is thrown down; (3) the *topset* beds, horizontal layers, which are deposited by the river upon the advancing foreset strata, as the latter shoal the water, and are usually, for the most part, of subaërial origin. As a rule, the topset and bottom-set beds cover the wider area, but the foreset make up the greater volume of the delta.

When a rapid subsidence is going on, the whole delta may be submarine, and the same result may be effected by powerful wave and tidal action, as in the case of the Amazon, which has a submerged delta extending about 125 miles to sea and covered with less than 10 fathoms of water. When the rate of sinking is less than that of accumulation and the power of waves and tides is relatively small, most of the area of the delta, neglecting the bottom-set beds, belongs to the region of continental sedimentation.

The rate of delta growth depends upon the quantity of sediment supplied by the river, the depth of the sea or lake into which it discharges, the power of the waves, tides, and currents which distribute the sediment, and the stationary or subsiding character of the sea-bottom. The Mississippi delta is advancing into the Gulf at the rate of a mile in 16 years, and that of the Rhone has been

built out more than 14 miles into the Mediterranean since the beginning of our era.   The coast of the upper Adriatic is fringed with
delta deposits which have widened from 2 to 20 miles since Roman
times.   The combined delta of the Ganges and Brahmapootra
measures about 50,000 square miles and is still gaining, despite the
powerful tides and waves of the Bay of Bengal.   On the other
hand, there seems to be a limit to delta growth in the sea; the
Nile delta has advanced very little in the last 2000 years, for a strong
current sweeps along the sea-front and carries away the sediment.
Débris from the Indus extends out 800 miles from the mouth of the
river, covering an area of more than 700,000 square miles.

CHAPTER VIII

RECONSTRUCTIVE PROCESSES. — CONTINENTAL DEPOSITS, LAKE AND ICE

IV. Lacustrine or Lake Deposits

THE term *lake* is a comprehensive one and includes all continental bodies of water, not in tidal communication with the sea, in which the water is relatively stationary and not actively running like a stream. In lakes which have an outlet there is a movement of the water toward the outlet, but this movement is extremely slow, and in the Lake of Geneva a given particle of water requires more than 11 years to pass through the length of the basin. (Forel.)

Lakes are formed in a great variety of ways, of which it is necessary to mention here only a few of the more important ones. We have: (1) *Tectonic lakes*, due to movements of the earth's crust, whether warping or faulting, by which basins, subsequently filled with water, are formed. In this class might be included volcanic lakes, which occupy ancient craters. (2) *Erosion lakes*, in basins which have been excavated by one or other of the erosive processes. In both of these classes the lakes occupy basins which are below the general drainage level of the country. (3) *Barrier lakes*, in which the water is retained by a built-up dam or barrier, such as a landslip, a lava stream, glacial moraines, a delta formed by a tributary in the main stream, etc. Such lakes are frequently above the general drainage level.

As compared with the sea, lake basins are but small and shallow, and, from the geological point of view, they are short-lived and ephemeral, because in course of time they are either drained by the outlet's cutting through the retaining barrier, or by filling up with the sediment brought in by tributary streams. The material transported by the Mississippi, which would require 11,000,000 years to fill the Gulf of Mexico, would fill the basin of Lake Superior in 66,000 years. (Barrell.)

From another point of view we may speak of temporary and permanent lakes, the former usually in arid climates. Such temporary lakes are called *playas* and are formed occasionally or periodically, according to the amount and distribution of the rainfall, but the distinction between playas and river floods is rather arbitrary, and hence playa deposits have already been described in connection with flood plains.

Lakes are important places of sedimentary accumulation, for they act as settling basins and retain all the sediment brought in by streams. However turbid the inflowing streams may be, the outlet is beautifully clear, as is exemplified very strikingly by the Rhone, which enters the Lake of Geneva a muddy stream and leaves it in a state of exquisite clearness and brilliancy. The Yellowstone and Niagara rivers are other examples of the same kind. Occasional exceptions to this rule may occur when a shore current washes some sediment into the outlet, as happens in Lake Huron and the St. Clair River.

1. **Fresh-water Lakes.** *a. Mechanical Deposits.* The mechanical sediment which accumulates in a lake basin is of two kinds, (1) that which is brought in by tributary streams, and (2) that which the lake itself acquires by cutting back its shores; of these the former is much the greater in volume. Almost without exception, rivers entering lakes form deltas, which spread out fan-like from the stream mouths, and, if sufficiently numerous, may fringe the entire lake shore with delta deposits. Part of the materials will be distributed by waves and currents, but the coarser material remains to form the foreset beds, the inclination of which depends upon the depth of the lake at the stream mouth and upon the coarseness of the débris. If deposited in deep water, the beds may be inclined at a considerable angle; if in shallow water,

they form a very gradual slope. In small lakes the coalescence of deltas, or the advance of a single one, will eventually fill up the basin, forming first swamps and then smooth, grassy meadows, through which flow the streams, keeping their own channels clear. Such filled-up lakes are common in many mountain valleys. In large lakes the process is, of course, much slower.

FIG. 101. — Gravel beach, Lake Ontario. (U. S. G. S.)

Away from the deltas the combined action of the waves and currents fringes the lake with coarse deposits of boulders, gravel, and sand, which form the beach, the sand extending some distance out into shallow water. In large lakes the heavy surf cuts a terrace on the shore and the débris thus obtained builds out the terrace, which is therefore said to be " cut and built." A succession of terraces indicates the various levels of the water, for lakes are often subject to great fluctuations of level. The finer materials are carried out into deeper water and deposited in successive layers over the whole lake bottom, the finest materials in the

centre. The coarse and fine sediments grade into each other, .
dovetail and overlap, because in heavy storms, or when the streams
are in flood, the coarser sediments are carried farther out and deposited on the fine, and .these changes of material in any given
vertical section, not too far from shore, may be often repeated.
Special lines of accumulation for the coarse substances also occur
in the form of shoals, spits, embankments, and the like. If the
lake is subject to fluctuations of its level, with the water much

FIG. 102. — Outlet of Lake Bonneville, Utah.   (U. S. G. S.)

higher at one time than another, even more wide-spread changes
in the character of the deposits will occur. The deposits now forming in the great Laurentian lakes, which occupy a relatively small
drainage basin and receive no large tributaries, are principally
blue muds and clays, partly made up of kaolinite and partly of the
débris of other minerals in an extremely fine state of subdivision,
but not decomposed chemically. In Lake Superior the clay has
generally a pinkish tinge.

Owing to the way in which the materials are arranged, lake

deposits betray the form of the basin in which they were laid
down.   Around the old shore line are masses of coarse materials,
with deltas interspersed, to mark the mouths of streams, while
towards the middle of the basin, quantities of fine mud and clay
have accumulated.   An excellent example of such a deserted lake
basin is that known as Lake Bonneville in Utah, of which Salt
Lake is the shrunken remnant.   The drying up of this lake, which
was once fresh and had an outlet northward to the Snake River,

FIG. 103. — Terraces of Lake Bonneville, Utah.   (U. S. G. S.)

is an event geologically so recent, that its form and size, its shores
and islands, its high and low stages, in short, its history, can be
made out with great clearness, as has been admirably done by
Mr. Gilbert of the United States Geological Survey.   At its time
of greatest extension, Lake Bonneville had an area of 19,750
square miles and a maximum depth of 1050 feet, while Salt Lake
(which is variable) had in 1869 an area of 2170 miles and an
extreme depth of 46 feet.   Around the ancient shores are beauti-

fully preserved the terraces, embankments, and deltas of the various stages of water, with the gravels and sands appropriate to the shallow water. The principal part of the basin is a level plain, filled to a great but unknown depth with beds of clay and marl (Fig. 104).

In still more ancient lakes the terraces, embankments, and other shore features have been swept away by the processes of denudation, but the outline of the lake may frequently be reconstructed from the character of the deposits.

b. *Chemical Deposits* are not common, nor of much importance in fresh lakes. In a few, chemically precipitated carbonate of lime is found, and more abundant is limonite ($2 Fe_2O_3, 3 H_2O$). This is carried into the lake by streams that contain dissolved ferrous carbonate ($FeCO_3$), which, becoming oxidized and hydrated, is no longer soluble, and accumulates on the bottom. In Sweden ores of this kind are dredged out of the lakes and employed as a source of iron.

c. *Organic Deposits* are seldom important in large lakes, but often decidedly so in small ones. As we have already seen, peat often forms to such an extent as to choke up the lake and convert it into a bog. Siliceous accumulations are made on an extensive scale by the minute plants, *diatoms*, which though of microscopic size, yet multiply with extraordinary rapidity; their tests of transparent flint gather on many lake bottoms in a fine deposit, as white as flour, and variously called Tripoli or polishing powder, or infusorial earth. Calcareous accumulations are formed by the shells of fresh-water molluscs, often in masses of considerable thickness. The lower layers of this *shell marl* have generally been so much disintegrated by the water as to be without any obvious organic structure. Such marls are frequently found under peat bogs and indicate that the latter were originally lakes, and in the marl often occur the bones of extinct animals.

2. **Salt Lakes** are especially characteristic of arid climates, in which the rainfall is light and evaporation great. They may be formed in either of two ways: (1) through the separation of bodies of water from the sea. This is exemplified by the Salton Sink in

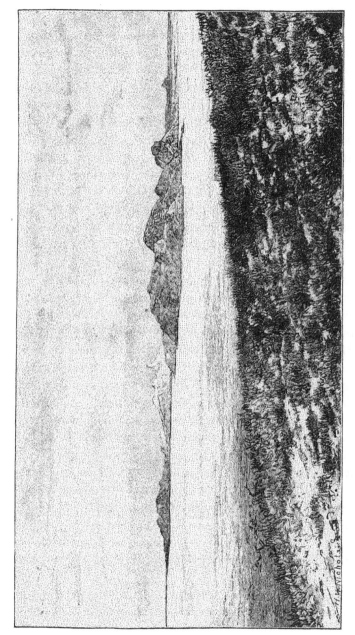

FIG. 104. — Mountains nearly buried under old lake deposits; plain of Salt Lake, Utah. (Gilbert)

the Colorado desert of southeastern California; the bottom of the Sink is considerably below sea level and has recently been converted into a lake by the influx of the Colorado River. Originally the Salton Sink was the head of the Gulf of California and old beaches, shell banks, etc., still remain to indicate this. The Colorado River, which enters the Gulf from the east some distance below the head, built its delta across the Gulf and thus cut off the upper portion into a salt lake, which subsequently disappeared by evaporation. Beds of salt, which demonstrate the lacustrine stage, occur in the Sink. (2) By the long-continued concentration of river water in basins that have no outlet, where the influx of water is disposed of by evaporation from the surface of the lake. In either case an arid climate is requisite to maintain the salinity; in a moist region the large rainfall and slower evaporation would cause the lake to rise until it found an outlet, and then the water, if originally salt, would become fresh. The history of Lake Bonneville exemplifies the change from fresh to saline conditions. As long as the water level was maintained above the outlet, the lake was fresh, but when the advancing aridity of the climate diminished the rainfall and increased the rate of evaporation, the water level sank until it fell below the outlet. Then the lake became saline, reaching its maximum salinity in the intensely bitter waters of Salt Lake, which is the remnant of the large lake.

All river water contains greater or less quantities of dissolved substances, and of these one of the commonest is ordinary salt (NaCl). When such waters are evaporated, the solids remain behind, and thus the water becomes more and more saline till it reaches saturation. The salt is, however, not entirely derived from the rivers, for salt is a very wide-spread constituent of desert soils and surfaces, and in the Pampas of Argentina, salt crusts form repeatedly. Wind and rain thus bring into the lake quantities of salt in addition to that carried by the tributary streams. Other substances occur also, as will be seen below.

The mechanical deposits formed in salt lakes do not differ in any very important manner from those of fresh lakes. The finer clays settle more rapidly in brine than in fresh water, which makes strongly saline lakes extraordinarily clear and limpid. The organie deposits of salt lakes are practically nothing, for brackish water is not favourable to many organisms and in dense brines very few animals or plants can exist, and those that can are not the kinds which give rise to peat, or to siliceous or calcareous deposits. For the same reason, the deposits of whatever kind, laid down in salt lakes, are almost barren of fossils, except of land animals and plants, such as are washed into the lake by flooded streams.

The *chemical deposits* are much the most interesting and characteristic of the accumulations gathered in salt lakes. These chemical precipitates differ much in the various lakes, according to the nature of the rocks which form the drainage basins, but while some of the substances are rare and restricted in extent, others are extremely common and wide-spread. Several changing factors combine to vary the order of precipitation of the salts, which is a highly complex process, but, in general, it follows the inverse order of solubility, the least soluble material being deposited first and the most soluble last. Comparatively little chemical reaction appears to take place in these lakes; the substances are, for the most part, thrown down merely by the evaporation of saturated solutions and are the same as those carried in very dilute solutions by the tributary springs and streams. If the precipitation of salts is slow and occasional, the chemically and mechanically formed deposits are mingled together; but if such precipitation be rapid, then thick and nearly pure masses of the salts are thrown down in their proper order, as the concentration by evaporation proceeds.

The first substances to be deposited from solution are the carbonate of lime and oxide of iron ($CaCO_3$ and Fe_2O_3), and in moderately saline lakes this is about the limit of precipitation. These same materials are thrown down in fresh lakes also, and their deposition is principally due to the loss of the solvent CO_2. The ancient Lake Lahontan, which formerly occupied part of northwestern Nevada, was the seat of calcareous deposition on a magnificent scale, and every crag and island which its waters touched is sheathed in thick masses of calcareous sinter. Pyramid Lake, a remnant of Lahontan, has a remarkable island of calcareous tufa; and Mono Lake, California, is famous for similar deposits, which have assumed curious and whimsical shapes.

As the concentration of the lake waters proceeds, the next substance to be precipitated is gypsum ($CaSO_4$, $2 H_2O$), which,

FIG. 105. — Island of calcareous tufa, Pyramid Lake, Nevada. (U. S. G. S.)

though much more soluble than the carbonate of lime, is yet only sparingly so. After all the gypsum in solution has been thrown down, there follows a pause in the deposition, until a further stage of concentration has been reached, and then common salt is precipitated, which deposition continues steadily as concentration proceeds, but at an advanced stage the salt is mingled with the sulphate of magnesia ($MgSO_4$), should that be present. The highly soluble salts, such as the chlorides of magnesium and calcium ($MgCl_2$, $CaCl_2$), remain in solution until the water is completely evaporated to dryness, hence they are rarely found in beds of rock salt; yet when they do occur, as at Stassfurt in Prussia, they are mingled with salt, which is thus precipitated till the very end of the process.

FIG. 106. — Salt deposit, El Paso, Texas. (U. S. G. S.)

Various circumstances may change the order of precipitation just given. In seasons of high water the flooded rivers dilute the waters of the lake, checking the chemical precipitation and, at the same time, increasing the mechanical deposition; thus beds of sand and mud are thrown down upon the beds of gypsum and salt, alternating with them, as the influx of fresh water or evaporation predominates. Changes of temperature also have an effect upon the order of precipitation. Thus, in cold weather, Salt Lake

Q

washes up on its shores quantities of sulphate of soda (Na_2SO_4), which is formed at low temperatures by the double decomposition of NaCl and $MgSO_4$.

Besides the chemical deposits already mentioned, others occur on a smaller scale. On the western side of the Great Basin, in Nevada, California, and Oregon, are several lakes which contain large proportions of carbonate of soda, and in some of them the

FIG. 107. — Deposits in an alkali lake.   (U. S. G. S.)

concentration is sufficiently advanced to cause precipitation, while others form deposits of borax.

Much the most abundant of the chemical deposits made in salt lakes are gypsum and rock salt, and the enormous scale on which the latter was formed in past ages of the world's history is demonstrated by the vast bodies of rock salt which are found embedded in the rocks in so many parts of the world.   Near Berlin, at Sperenberg, an artesian well was sunk through such a deposit for nearly 4000 feet, without reaching the bottom. In various regions of the United States, notably in New York and Kansas, large bodies of salt are found, but not on such a scale as in Europe.

Salt bodies of such immense extent and thickness, with little or no interstratified, mechanically deposited material, are not explained by the usual operations of a salt lake as above described. The key to the understanding of these enormous deposits of salt appears to be given by a gulf from the eastern part of the Caspian Sea, known as the Karibogas. This gulf, which is connected with the Caspian only by a very narrow channel, is situated in an extremely hot and almost rainless region, and evaporation is very rapid. On the bottom masses of salt have been deposited uninterruptedly for a long time past, and in a very deep or a slowly subsiding basin the process might continue almost without a limit.

It should be noted that the chemical deposits made in salt lakes are crystalline and at the same time stratified. This association is not the usual one, as stratified rocks are ordinarily not crystalline, and crystalline rocks are mostly unstratified.

V. Ice Deposits

Deposits made, directly or indirectly, by the agency of ice are very characteristic, and though some are formed on land and some under water, it is desirable to consider them together in a single section. The peculiar features of ice formations may be much obscured by the action of water, either at the time of their deposition or at some subsequent period. Ice deposits play but a small part quantitatively in the construction of the earth's crust, but the light which they throw upon changes of climate and similar questions, lends them an unusual degree of interest. Only very recently has the great importance of the part played by glaciers in former ages of the earth's history been appreciated.

Glacial Deposits. — We have already learned that glaciers carry with them great masses of débris, either in the form of lateral

and medial moraines upon their upper surfaces, or frozen in the
interior of the ice, or pushed along beneath it. When the ice
reaches the end of the glacier, where the rates of motion and
melting balance each other, all the burden which it is transporting
is deposited in a great mound or ridge, the *terminal moraine.* The
terminal moraine is composed of material which was carried upon
and within the ice and that which was pushed along beneath the

FIG. 108. — Fluted ground moraine; Columbia Glacier, Alaska.   (U. S. G. S.)

glacier.   It is gradually built up around the end of the glacier and
extends up along its sides so far as the conformation of the ground
will permit the material to gather, and is thus more or less crescentic,
with concave side directed upward.   Moving ice does not sort the
material which it carries, as flowing water does, because in a glacier
there is no such definite relation between velocity and transporting
power.   Hence, the terminal moraine is *unstratified* and is composed of materials of all sizes, from dust and sand up to great

boulders weighing hundreds of tons, all mingled together in confusion. In the case of a glacier which carries the principal part of its burden upon its upper surface, the terminal moraine is chiefly made up of angular blocks that have undergone little or no abrasion, together with earth, sand, gravel, and whatever kind of material the overhanging cliffs may have delivered to the moving ice. Mingled with these materials, however, will be found more or fewer

FIG. 109. — Glacial moraine, Montauk Point, L.I. (Photograph by B. N. Mitchill)

of the characteristically worn and grooved glacial pebbles and boulders, which have been dragged along under the ice, and scored and polished by the rocky bed. There will also be found some, at least, of the sand and fine rock flour which the glacier's own movement produces and which have escaped the washing of the subglacial stream.

When a glacier is retreating, it may build up a new terminal

moraine at each point of arrested withdrawal, or if the retreat is
gradual and steady, the ground in front of the ice will be covered
with moraine material, spread out in a sheet, not heaped up in a
moraine or mound. The retreat of the glacier may leave behind
it isolated masses of ice deeply buried in the débris of the terminal
moraine; when such masses melt, they form depressions in the
mound and give rise to the " kettle moraines." A shrinking glacier

FIG. 110. — Glacial pebble.   (U. S. G. S.)

will contract laterally and in depth, as well as longitudinally, and
in this way the blocks of the lateral moraine will be left stranded at
intervals over the former glacial bed. Such blocks and boulders
are known as *erratics*, or perched blocks, and when their parent
ledge can be discovered, it is easy to determine the distance to
which they have been carried. Sometimes a great boulder is
lowered so gradually and gently by the retreating ice, that it is

FIG. 111. — Striated glacial boulders from Permian of South Africa

FIG. 112. — Kettle moraine, Alaska. (U. S. G. S.)

exactly balanced, and may be moved backward and forward by the hand. This is a " rocking-stone," though it must not be supposed that all rocking-stones are glacial (see p. 122).

The *ground moraine* consists of the débris which is deposited underneath moving ice and that which is left from the bottom of a retreating glacier. Deposition beneath moving ice is much less in amount than erosion, but may occur in areas sheltered from ice

FIG. 113. — Perched block near the Yellowstone Cañon, National Park. (U. S. G. S.)

pressure, or where the ice thins rapidly from melting. The ground moraine is not stratified, or but imperfectly so when deposited partly from water, and is of very different thickness in different places; it consists of fine material containing boulders and pebbles, many of which show the characteristic facetting and striæ, and is called *till*, or *boulder clay*. The general term for a sheet of glacial deposits is *drift*.

FIG. 114. — Glacial drift, Bangor, Pa. (U. S. G. S.)

FIG. 115. — Glacial drift in Permian of South Africa. (Photograph by Rogers)

The water deposits which are made in the neighbourhood of and in association with a glacier, are also characteristic and should be noticed in this connection. The streams which flow beneath and from the foot of the glacier are loaded to their utmost capacity with débris, and usually build up their beds by rapid deposition of the coarser sediment. From an alpine glacier this deposit, which is stratified, forms a *valley train*, but when the glacier ends, as do some of those in Iceland, upon a more or less flat surface, a broad *overwash plain* is the

FIG. 116. — Glacio-fluvial deposits, Yahtse River, Alaska.   (U. S. G. S.)

result. These accumulations may usually be distinguished from ordinary river and flood plain deposits by their upward extension into moraines and by the evidently glacial origin of their materials. *Eskers* (or *Åsar*) are long, winding ridges of gravel, which often ramify like the branches of a stream and were formed by streams which flowed in channels upon or in tunnels beneath the ice. *Kames* are hillocks, or short ridges of stratified gravel and sand heaped up by subglacial streams, as they escape from the margin of the ice, and *drumlins* are elliptical mounds of ground moraine, sometimes with some stratification, formed near the margin of the ice.

Very instructive examples of this combined action may be observed about the great Malaspina Glacier in Alaska. This is an immense ice-sheet, with an area of 1500 square miles, which is formed at the foot of the St. Elias Alps by the confluence of several

FIG. 117. — Gravel flood plain of glacial stream, Alaska. (U. S. G. S.)

great glaciers. All the outer borders of the glacier are covered with sheets of moraine matter, and upon the stagnant portion of this is a luxuriant growth of bushes, beneath which is a thickness of not less than 1000 feet of ice. About the margin of the ice-sheet, small lakes are formed, the water being held in place by the ice barrier, but these lakes are subject to great fluctuations, and often their

FIG. 118. — Esker, central New York.   (Photograph by Fairchild)

FIG. 119. — Kame moraine, central New York.   (U. S. G. S.)

waters escape through tunnels in the ice. In some of these lakes stratified deposits are made by the inflowing streams. Innumerable streams, some of them quite large, rise from under the glacier, and many others flowing from the north pass under the free margin of the ice by means of long tunnels. All of these streams are loaded to their utmost capacity with sediment, gravel, and boulders; by

FIG. 120. — Drumlin near Newark, N.Y. (U. S. G. S.)

blocking up their own openings from the ice, they likewise cause the deposition of sand, gravel, and boulders within their tunnels, which, when the glacier retreats, will be left standing as eskers, while conical mounds, or kames, are built up where the streams burst from under the ice, and sometimes, owing to the great press-ure, rise like fountains. This kind of deposition is characteristic of retreating ice-sheets, such as the Malaspina; in advancing glaciers denudation will prevail over deposition, or, if the advance be not so great as to sweep away all the previous accumulations, purely glacial deposits may be laid down upon the stratified fluvio-glacial material.

Iceberg Deposits. — When a glacier flows into the sea great masses are broken off from the foot and float away as icebergs. Icebergs are thus seen to be, as indeed they always are, derived from land ice and not from the freezing of sea-water. The ice-

FIG. 121. — Drift-covered surface of the Malaspina Glacier, Alaska. (U. S. G. S.)

berg will, of course, carry with it whatever parts of the glacial débris are contained within or upon that particular fragment of the glacier, and drops this load over the sea-bottom, as the berg gradually melts. As the Greenland icebergs sometimes drift as far south as the Azores, glacial boulders are scattered all over the bed of the North Atlantic, and thus we see how large blocks may be embedded in stratified deposits very far from the place where they were torn from their parent ledges.

Coast Ice Deposits. — In high latitudes with intensely cold winters, great fields of ice (the ice-foot) are formed by the freezing of sea-water along the shore. The ice-foot becomes loaded with great masses of rock, part of which is thrown down from over-hanging cliffs by the action of frost, part picked up from the shore-line by the ice forming around it. In summer the coast ice breaks up and floats away with its load of blocks and boulders, distributing

FIG. 122. — River issuing from the Malaspina Glacier, Alaska.
(U. S. G. S.)

them over the sea-bottom just as icebergs do. In storms great masses of coast ice are often driven on the shore, where they may pile up to heights of fifty feet or more, carrying some of the boulders above the levels at which they were picked up. The coast of Labrador is covered for long distances with boulders thus trans-ported, as are many other Arctic shores. Great masses of rock are thus transported in the Baltic, and the divers report that in the Copenhagen Sound the sunken wrecks of vessels are covered with ice-borne blocks.

CLIMATIC RELATIONS OF CONTINENTAL DEPOSITS

This subject, though of great importance in deciphering the earth's history, is extremely complex and can be but briefly outlined here.

FIG. 123. — Deposit partly made by stranded ice, west coast of Greenland. (Photograph by Libbey)

1. Polar Regions. — The activity of frost destruction in polar lands results in the accumulation of great talus masses of sharply angular blocks and fragments, while the low temperature, even of summer, is unfavourable to chemical decomposition of the rocks and the consequent formation of soil. River deposits are not extensively formed, for rivers in the true polar lands are comparatively few and small, while glacial accumulations, on the other hand, assume great importance. Small lakes are frequently found among the moraines, but their deposits are not made on a large scale. On the other hand, peat is very largely accumulated in the illimitable "tundras," or swamps, of the lower polar latitudes, the low temperatures retarding the work of decomposition.

2. **Temperate Regions.** — The temperate regions are distinguished from the tropics on the one hand, and the polar lands on the other, by their frequent and wide changes of temperature and by the very great variety of conditions which obtain. In temperate climates with normal rainfall, river deposits, and sometimes lake deposits, assume the most important place, while glacial drift and moraines are local only and confined to mountain areas sufficiently high to extend above the snow-line. Such mountains are, in fact, extensions of polar lands into the temperate regions, and the deposits characteristic of the former accompany them. The temperate regions of pluvial climate are densely covered with vegetation, so that, except on sandy coasts, wind deposits are of small importance and the material laid down in the flood plains of rivers is not extensively sun-cracked, while vegetable accumulations are very extensively formed in swamps and bogs, especially in the cooler and moister parts. Pluvial climates, again, are highly favourable to the decomposition of the rocks and the formation of deep soils which are prevailingly brown in colour from the presence of hydrated iron oxide, or limonite.

In the semi-arid parts of the temperate regions wind-made accumulations of loess are developed very extensively, and in the more arid parts, river and playa and salt lake deposits become conspicuous. Sun cracks are highly characteristic of playa muds and around the margins of fluctuating salt lakes.

3. **The Desert Zones.** — In both the northern and southern hemispheres, between the temperate and tropical regions, is an irregular belt of desert, encircling the earth. The deserts are not entirely rainless, but so nearly so that they have no drainage outlet. The Nile and the Colorado flow across deserts, but their sources lie outside of the desert zone, and water plays but a subor-

R

dinate part in desert accumulations. Talus masses, due to the rapid changes of temperature, are very common in true deserts, where " all the mountains rise like islands out of an almost flat sea of talus " (Walther). Wind-blown sand is, of course, very characteristic of deserts, which lies in chains of dunes, instead of with a level surface; the sand grains are small, rounded and polished by attrition, often fractured by the sun's heat, and prevailingly of a more or less reddish colour. Salt lakes can exist only in arid climates, so that salt and gypsum are among the most characteristic of desert deposits.

4. **The Tropics** have a constantly high temperature with but small daily or seasonal range, and a very heavy rainfall, which, however, is usually confined to a part of the year. Chemical decomposition is thus very active and complete, and the soil accumulates to great depths. Especially characteristic of the tropics is laterite (see p. 104), the red colour of which tinges the river silts. Where the topography is favourable to such accumulations, enormous masses of river deposits are formed, as in the interior basins of South America, while the alternating wet and dry seasons, by reversing the movement of water in the soil, occasion the concentration and deposition at the surface of iron and limestone. Despite the great luxuriance of tropical vegetation, the climate is much less favourable to the formation of peat than is that of cooler latitudes, because of the rapidity and completeness of decomposition. It is almost needless to say that the climatic zones pass into one another by gradual transitions.

CHAPTER IX

RECONSTRUCTIVE PROCESSES. — MARINE AND ESTUARINE DEPOSITS

B. MARINE DEPOSITS

THE sea is the great theatre of sedimentary accumulation, and rocks of marine origin form the larger part of the present land surfaces. Important as other classes of deposits may be, they are less so than those laid down in the ocean and the waters immediately connected with it. There is great variety in the sedimentary deposits made in the sea, which change in accordance with climate, the depth of water, the nature of the coast rocks, the force of winds and tides, and the nearness or remoteness of the mouths of rivers and, in a very important degree, with the elevation and relief of the adjoining land masses; very different materials are supplied by bold and rocky shores from those derived from flat, sandy coasts. Large land-locked seas, like the Gulf of Mexico and the Mediterranean, again, have deposits more or less different from those of the open ocean, a difference which is largely due to the absence or insignificance of the tide, and the reduced force of the waves.

It is important to remember that the actual line of meeting of sea and land is not the structural margin of the continent, for the water may cover a broad submerged shelf of the latter. For 100 miles east from the coast of New Jersey the water deepens very gradually to the 100-fathom line, whence it shelves very steeply to the profound oceanic abyss. Shoal water, less than 100 fathoms deep, surrounds all coasts, sometimes as a narrow belt, again as a very broad zone. The transition from shoal water to the deep sea is by steep slopes, with well-defined upper margin or edge; as a rule, these steep slopes of the continental mass begin at the 100-fathom line, but there are frequent departures from this rule. For example, in the Gulf of Guinea the steep descent begins at the 40-fathom line, while off the west coast of Ireland the descent is gradual almost to a depth of 200 fathoms and then becomes steep. The bed of the Atlantic off the Carolinas displays no well-marked

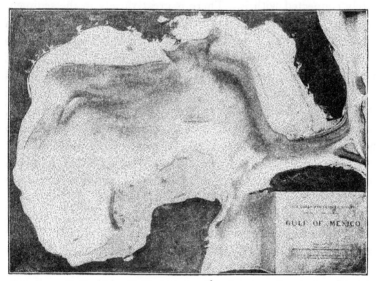

FIG. 124.—Basin of the Gulf of Mexico, showing the submerged margin of the continental platform and the steep descent of the bottom at the 100-fathom line. Vertical scale much exaggerated. (From a model by the U. S. Coast Survey)

edge of the continental shelf. The ocean thus fills its own basin and overflows the margins of the continental platforms to a greater or less extent; this submerged shelf constitutes the shallow sea (Fig. 124).

Marine deposits may be classified primarily in accordance with the depth of water in which they were laid down, one of the most valuable guides to the history of ancient rocks, and secondarily in accordance with the nature of the material of which they are composed, and the processes by which they were accumulated. The following table gives, in a somewhat modified form, the classification of Murray and Renard, founded upon the great collections of modern marine deposits made by the " Challenger " expedition.

MARINE DEPOSITS

1. Littoral Deposits, between high and low water marks.	Sands, Gravels, Muds, etc.	I. Terrigenous Deposits, material derived from the land in suspension (except the calcareous masses).
2. Shoal-water Deposits, between low-water mark and 100 fathoms.	Sands, Gravels, Muds, Calcareous Accumulations, etc.	
3. Aktian Deposits, laid down on the continental slope.	Coral Mud. Volcanic Mud. Green Mud and Sand. Blue Mud. Red Mud. Calcareous Deposits.	
4. Abysmal Deposits, laid down on the ocean floor.	Foraminiferal Ooze. Pteropod Ooze. Diatom Ooze. Radiolarian Ooze. Oceanic Red Clay.	II. Pelagic Deposits, formed in deep water far removed from land.

The material brought into the sea by rivers, or washed from the shore by waves, is partly mechanically suspended and partly in a state of solution; the former is deposited when the water is no longer able to transport it, while some of the latter is extracted by animals and plants, and some remains permanently dissolved. The sorting power of the water arranges the mechanically borne sediments according to the coarseness and fineness of their constituent particles, at the same time separating them according to their mineralogical composition, a separation which is usually imperfect, but sometimes very complete. Marine deposits are thus typically stratified, though when deposition continues long

and uninterruptedly, thick masses, not obviously divided into layers,
may be accumulated, but this is exceptional in those parts of the
sea where deposition is most rapid.

1. **Littoral Deposits** are laid down between low and high-water
marks, and by heavy storms and exceptional tides, somewhat
above the former. · Thus, the accumulations of this class grade into
continental deposits on one side and into those of the shallow sea,

FIG. 125. — Gravel beach and wall, Conception Bay, Newfoundland. (U. S. G. S.)

on the other, and are themselves alternately covered with water
and exposed to the sun and wind. The material of the littoral
beds varies much on different coasts; on rocky shores boulders,
coarse shingle and gravel form the beach, but gravel and more particularly sand are the most widely distributed. Boulders and shingle
may be composed of any kind of hard rock, but as the process of
attrition continues, the greater hardness of quartz has its effect,
so that gravel and sand generally consist of that mineral, the softer

minerals being ground into fine particles and swept out into deeper water. Locally, sand of other composition occurs, as, for example, around the Bay of Naples sands of olivine, felspar and other volcanic minerals, are found.

The waves cast material upon the beach, throwing the coarsest parts up in storms as a beach wall or embankment, above their ordinary reach, while the undertow carries back the finer particles,

FIG. 126. — Gravel beach, Long Island, N.Y. (U. S. G. S.)

thus washing the sands and gravels clean of other minerals. Even in the littoral belt, fine sand and mud may gather in sheltered spots, but the material is preponderatingly coarse. At any given time the littoral is a narrow belt, measuring at present about 62,000 square miles, but its breadth at a particular place depends upon the amount of tidal rise and fall, and the slope of the bottom. On a stationary coast, or one where accumulation is more rapid than sinking, littoral deposits may form a broad area by building out the land at the expense of the sea. When sinking and building are about equal, great thicknesses of littoral beds may be formed.

Aside from the coarseness of the material, littoral deposits are apt to retain certain characteristic marks of their exceptional mode of formation. *Ripple Marks* are formed by the wind, or by the rippling movement of water, and may be seen on any sandy beach;

FIG. 127. — Ripple-marked Sands, low tide; Mont St. Michel, France

they occur especially in sands, in shoal-water deposits, as well as in those made on flood plains and in lakes and on sand dunes. They are found in rocks of all geological periods, and though most frequent in sandstones, occur in other kinds of rocks. *Wave Marks* are formed by waves washing up on the beach after they have broken, and are preserved by the deposition of thin layers of sand on the edges of the waves; they are confined to the littoral zone. *Rill Marks* also are peculiar to littoral deposits and are made by the excavating action of rills of water trickling over the sand or mud, as the tide ebbs. *Sun Cracks* (also called mud cracks and shrinkage cracks) are formed where flats of fine mud or silt exposed to the drying action of the sun, harden and crack in

FIG. 128. — Ripple-marked Sandstone. (U. S. G. S.)

more or less regular patterns. As we have already learned, such cracks form over vast areas of flood-plain and playa deposits, and the littoral are the only truly marine deposits which display them, though on a limited scale. They do not form in the clean sands and gravels which make up the greater part of littoral sediment, but only in fine silt, and in pluvial climates probably only in

such areas as are not reached by the ordinary tides.   In very hot
and dry climates, as on parts of the Red Sea coast, cracks may
develop in the course of a few hours.

FIG. 129. — Steeply inclined beds of ripple-marked shale; near Altoona, Pa.
(U. S. G. S.)

Rain Prints are little pit-like marks made by light showers; the prints are circular where the raindrops fall vertically, or oval and with edge raised on one side where the rain falls obliquely before the wind. *Tracks of Land Animals* are made by the animals walking upon the soft sediment, which is yet fine enough to retain the footprints and is afterwards hardened by exposure to the sun and air. Rain prints and footprints are not so common in

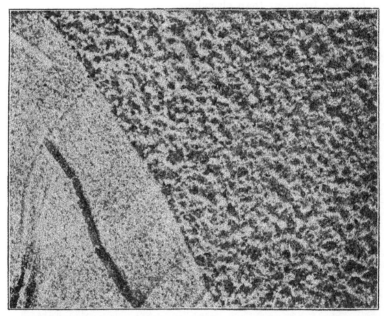

FIG. 130. — Wave mark and rain prints, modern sandy beach. (U. S. G. S.)

littoral deposits as in those of flood plains and playas, but where they do occur in rocks which contain marine fossils, they prove the littoral origin of those rocks, for only in the littoral zone can such marks be made.

It might seem incredible that such slight marks could be preserved for ages in the solid rocks, were it not for the fact that we actually find them so often. The explanation is that surfaces which are capable of preserving these marks are those of accumulation and that each layer with its marks is hardened by the sun and wind before the next layer is deposited upon it.

Climatic differences are not well marked in the littoral zone, the character of which is chiefly determined by the elevation and topography of the adjoining land. ' Only in the polar regions is a special character given to the littoral by the activity of frost and coast ice, so that block and boulder beaches are more common, and sandy beaches less frequent than elsewhere.

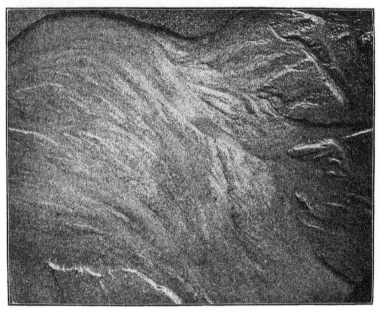

FIG. 131. — Rill marks on modern sandy beach.   (U. S. G. S.)

2. **Shoal-water Deposits.** — The material of the littoral zone is continued out beyond low-water mark to distances which vary according to several circumstances. Where, for long distances, no large rivers enter the sea and the material is all derived from the wear of the coast, the arrangement of coarse and fine deposits is quite regular, and gravel beds may extend as far as ten miles from land. Waves and currents sweep sediment not only toward the

shore, but parallel with it, and tend to simplify the coast-line by
building barriers and spits across the mouths of bays, which the

FIG. 132.—Sun cracks in limestone, Rondout, N.Y. (Photograph by van Ingen)

waves mȧy pile up above high tide, as is seen all along the eastern
coast of the United States. Behind these barriers streams bring

in sediments, filling up the bays and converting them into salt
marshes and eventually into land.

While gravel and, in sheltered or deeper spots, mud are found
in the shallow sea, the most abundant and characteristic material
of this zone is quartz sand. If the bottom shelves very gradually
and the continental margin is far from land, the sand will extend
100 to 150 miles out, growing finer and finer with the increasing

FIG. 133. — Cross-bedded Sands, Bennett, Nebraska.   (U. S. G. S.)

depth of water.   Further, the sand travels along the shore for long
distances from its place of origin, as on the Atlantic coast of Florida,
where there is a belt of siliceous sand that cannot have been derived from the peninsula.

Throughout the whole of the shoal-water zone wave action is
exerted upon the bottom, though to a very insignificant extent in
the deeper parts.   Very near shore currents produce irregularities
of stratification, especially the structure known as *cross bedding*

(also called current or false bedding) in which the separate layers are inclined at a considerable angle to the horizontal. (See Figs. 133-4 and Frontispiece.) This structure is due to the heaping up of bars and ridges on the sheltered side of which sand or gravel is dropped in inclined layers. Frequently we find horizontal strata built up of inclined layers, the latter all truncated by a horizontal bedding plane. Cross bedding occurs in shoal water of all kinds, the sea, lakes, and rivers, wherever the bottom is frequently stirred

FIG. 134. — Cross-bedded Sandstones, Arizona.   (U. S. G. S.)

up by currents, and the foreset beds of deltas are a typical example of it. In the rocks it is most frequently observed in consolidated sands and gravels, which are called respectively sandstones and conglomerates. Ripple marks are also extremely common in deposits of the shoal-water zone and tracks of marine animals; tracks of land animals and sun cracks are of course wanting.

Much less widespread than sand or gravel on the bottom of the shallow sea is mud or clay. When these occur, their presence may be due either to holes and depressions, where the bottom water

is less disturbed and therefore deposits finer material, or to a large
supply from the neighbouring land.   For example, a large triangular patch of clay invades the sand area south of Block Island,
and mud-holes are found along the New Jersey coast near the
entrance to New York Bay.

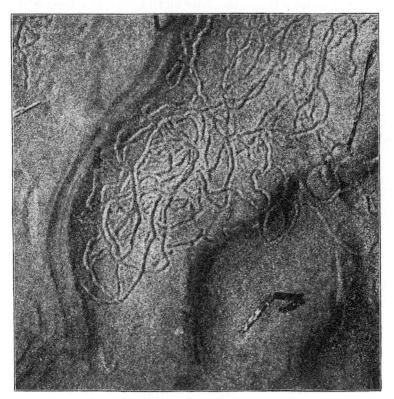

FIG. 135. — Markings by marine worms, modern

It is manifest that a great thickness of shoal-water deposits
can be formed only upon a sinking sea-bottom, for otherwise the
water would be filled up and the coast-line pushed out to sea.
If the subsidence be very slow, deposition may shoal the water
and thus extend the coarse materials seaward ; if it be rapid, deepen-

ing the water, fine sediment will be thrown down upon coarse, while, if the rate of deposition and subsidence be nearly equal, the coarser material will form long, narrow bands, running parallel with the coast. Thus, in the same vertical line may be accumulated many different kinds of sediment, corresponding to the different depths of water at the same spot. When traced laterally, beds of any given kind of material will eventually give way to those of another kind, either by gradual transition, or by thinning to an edge and dovetailing with the thin edges of the other beds. The dovetailed structure is caused by shifting conditions, a succession of heavy storms sweeping coarser material out to unusual depths, and long periods of calm occasioning the deposition of fine sediment unusually near shore.

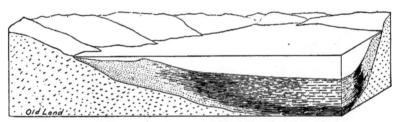

FIG. 136. — Diagram showing dove-tailed deposition on the sea-floor

Organic deposits are much less common in shallow water than are the terrigenous, and yet under favourable conditions they are developed on a very extensive scale. The most important of such conditions is an abundant supply of food. Even in the far North limestone accumulations are formed, but this work is most extensively done in the warm waters of tropical and subtropical seas. The sea is constantly receiving from the land materials in solution, of which the most important are the carbonate and sulphate of lime. Many classes of marine animals extract the $CaCO_3$ from the sea-water and form it into hard parts, either as external shells and tests, or as internal skeletons. There is also good reason to believe that some, at least, of these organisms are able to convert the sulphate into the carbonate.

s

The classes of marine organisms which at present or in times
past have played the most important part in the accumulation of
calcareous material are: the Foraminifera, Corals, Echinoderms,
and Molluscs; but other groups, such as Bryozoa, worms and
calcareous seaweeds, contribute extensively to the same result.
The Foraminifera do not accumulate with sufficient rapidity to add
largely to the calcareous deposits of shallow water, and will therefore be considered in connection with the deep-sea formations.

Mollusca. — The ordinary shell-fish (Mollusca) supply a very
large amount of calcareous material for the formation of shallow-

FIG. 137. — Modern shell limestone (coquina), Florida

water limestones, especially in the neighbourhood of the coasts,
and are found in warm, temperate, and even in Arctic seas. The
shells accumulate in great banks, frequently, though not always,
mingled with more or less sand and mud, and when gathered below
the limit of violent wave action, they are entire, embedded in finer
material, which may be calcareous or not. More commonly the
shells are ground by the waves into fragments, making shell sand
and mud, which is then cemented into a more or less compact mass.
The coquina rock of Florida is an example of a recently made
shell limestone (though it is forming no longer), and among the

rocks of the earth's crust are many immense limestones which were accumulated in this way. In the formation of shell-banks carnivorous Crustacea and fishes play an important part, for they grind up even quite thick shells and produce an angular calcareous sand, which may be deposited by itself, or constitutes the finer material in which the entire shells are embedded. The shell-banks thus form lens-shaped limestone masses of greater or less extent and thickness, which are intercalated in areas of terrigenous sediment, more particularly of sand.

Echinodermata. — This group of marine animals, which includes the starfishes, sea-urchins, crinoids or sea-lilies, etc., is made up of forms which all secrete skeletons of calcareous plates, and which contribute largely to the formation of marine limestones. At the present day, however, they seldom build up any extensive masses unassisted, but in former ages of the world's history they did so on a great scale. This is particularly true of the crinoids (sea-lilies or feather-stars), which have now become comparatively rare, but many ancient limestones are composed almost entirely of their remains, and especially of their hard and heavy stems.

Limestone Banks. — In favourable situations immense submarine plateaus or banks are built up in shallow waters by the accumulated remains of all sorts of lime-secreting animals, corals, echinoderms, molluscs, worms, and Foraminifera. These are well exemplified in the Gulf of Mexico and the Caribbean Sea by the great banks along the west coast of Florida, the Yucatan Bank, and the plateau which extends from the coast of Nicaragua almost to Jamaica. On these banks the luxuriance and fulness of life are astonishing, myriads of animals flourishing in the warm waters, and abundantly supplied with food by the great ocean currents which sweep over the banks. Innumerable molluscs, echinoderms, and calcareous worms are continually dying and adding their hard parts to the sea-floor; the waves and tides sweep calcareous sand and mud from the coral reefs over the flats, and all of these masses are rapidly consolidated into rock.

An example of a limestone bank in moderately deep water is

the Pourtalès Plateau, which extends southward from the Florida Keys, and is covered by 90 to 300 fathoms of water. "The bottom is rocky, rather rough, and consists of a recent limestone, continually, though slowly increasing from the accumulation of the calcareous débris of the numerous small corals, echinoderms, and molluscs, living on its surface. These débris are consolidated by tubes of serpulæ; the interstices are filled up by Foraminifera and further smoothed over by nullipores. — The region of this recent limestone ceases at a depth varying from 250 to 350 fathoms, and beyond it comes the trough of the straits." (A. Agassiz.)

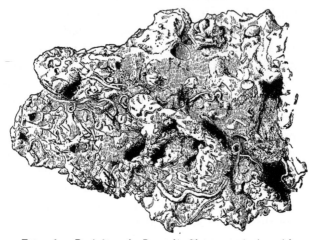

FIG. 138. — Rock from the Pourtalès Plateau. (A. Agassiz)

It is not known how thick these modern limestone banks are, but some indication of their thickness is given by the raised terrace of modern limestone in northern Yucatan, composed of the same species of animals as still abound in the adjoining seas. In this rock are caverns more than 400 feet deep, which do not reach the bottom of the mass.

Corals. — The animals of this group show great variety of form, size, and habit of growth, and by no means all of them are important as rock-makers. The solitary corals, which are widely distributed, even in the deep sea, are never sufficiently abundant to form deposits by themselves. The corals which do accumulate in great masses and are called " reef-builders," form compound colonies or stocks, consisting of hundreds and thousands of individuals. The adult corals are sedentary, but the newly hatched young are worm-like, free-swimming larvæ. When the young animal has established itself in a suitable place, preferably upon a rock or other fixed foundation, it develops into a *polyp*, or fleshy sack, with rows of tentacles around the mouth, and then by budding or partial division (*fission*) gives rise to great numbers of other polyps, which are connected by a tissue common to them all. In this compound mass is secreted a skeleton of carbonate of lime, which reproduces the form of the colony and, in most cases, displays cells for the individual polyps. The great variety of form shown by these compound colonies is determined by the mode of budding or fission and the relative position of the newer to the older polyps. Thus, some are like trees, others like bushes; some form flat, irregular plates, while others grow into great dome-like masses.

The reef corals have, at present, a restricted distribution, and can flourish only where several favourable conditions are found united. They are preëminently shallow-water animals and can live only in depths of less than twenty fathoms. They also require a high temperature, and they cease wherever the average temperature of the water for the coldest month is below 68° F.; this is the minimum, and for full luxuriance a higher temperature is necessary. Another requisite is sea-water of full salinity and uncontaminated with mud; hence, few corals can live at the mouth of a river, which, even if it brings down no sediment, freshens the water and is thus fatal to the polyps. Another condition favourable to the growth of corals is the presence of ocean currents, not too rapid, which bring abundant supplies of food, and they flourish best in the broken waters of heavy surf, which gives the necessary oxygen and prevents the smothering of the polyps in the calcareous silt and débris of the reef. In short, the reef corals are tropical, marine, shallow-water animals, and their reefs are widely spread throughout the warmer seas of the globe, but they do not always occur where we should naturally expect to find them.

A coral reef is not built, as many people imagine, by the industry of the polyps — these furnish the material by extracting lime salts from the water and forming solid skeletons; the actual construction is largely the work of the waves and other agencies.

FIG. 139. — Corals on the Great Barrier Reef of Australia.   (Savile Kent)

The coral colonies are scattered over the sea-bottom, much like vegetation on the land, scantily in some places, thickly in others, and in still others they are absent. The waves, especially in storms, break up the masses of coral, which are much weakened by the borings of many kinds of marine animals, and the surf grinds them down to fragments of all sizes, from large blocks to the finest and most impalpable mud. The process is the same as with the ordinary rocks of the coast, only the material differs, and thus are formed boulders, pebbles, sand and mud, all of coral fragments. The many animals which feed upon coral greatly facilitate this work, partly by boring into the masses, partly by grinding the

FIG. 140. — Various forms of modern coral limestone. (Savile Kent)

smaller fragments into fine powder. Considerable masses of calcareous débris are added by the shells and tests of the various animals which live on and about the reef, and the coral-like seaweeds, called *Nullipores*, contribute an important quota, while shell sand often makes up as much as one-half of the volume of a reef. All of this material is ceaselessly ground up by the waves, distributed by tides and currents, and brought to rest in quiet waters. A single deposit of two or three inches in thickness has been observed to form between tides after a gale along the Florida reefs, and in storms the water is often discoloured and turbid for miles around the reef. The sea-water dissolves and redeposits $CaCO_3$, cementing the fragments into a firm rock, which, especially after exposure to the air, may become very hard.

By these processes several varieties of rock are formed, corresponding, in all but the material, to the ordinary marine deposits. In one form the standing and unbroken colonies are filled up with calcareous débris and enclosed in solid masses. This is perhaps the most important kind of rock, at all events, in many reefs, for the branching corals retain the shell sand and other calcareous débris and prevent the waves from washing it away. Reefs of this kind have many and deep holes penetrating them, where the colonies are not in contact and the sand has not filled up the interspaces. Coral conglomerate or breccia is a cemented mass of coral pebbles or angular pieces, or is made up of fragments of an older coral rock. Reef rock is the dense and solid mass formed by the cementing of the finer débris which accumulates in quiet water. It is important to notice that even under the microscope reef rock frequently shows no trace of organic structure, and is a definite proof that the absence of such structure is not a sufficient reason for denying the organic origin of a rock. The interior of growing masses which are still alive on the outside, and have never been broken up, may be so crystallized by the action of the sea-water that the organic structure is obscured or destroyed. On the beach is formed a curious rock called *oölite*, which is made up of minute spherules of $CaCO_3$ cemented into a mass not unlike fish-roe in appearance. This is due to the deposition of $CaCO_3$ from solution around tiny grains of calcareous sand, until the spherules are built up and cemented together.

The growth of coral ceases when the reef extends up a little above low-water mark, but the waves continue their work and throw up débris and build up a platform, upon which they establish a beach of calcareous sand. The latter may be further piled up by the winds into dunes and solidified by the cementing action of percolating rain-water. According to circumstances, the new platform may be an extension of the shore or an island like the Florida Keys.

Coral reefs are classed according to their relation to the shore, and are of three kinds. (1) Fringing reefs are those attached directly to the land, though the exposed part may be at some distance out from the shore and separated from it by a shallow channel with coral bottom. The width of a fringing reef is determined by the slope of the sea-bottom, being narrower on a steep grade, broader on a gentle one. (2) Barrier reefs are farther out from shore, to which the reef is parallel in a general way, and separated by a broad and often quite deep channel. The distinction between the two kinds of reefs is not very sharply drawn, for the same reef may be fringing in parts of its course and a barrier in others. Even at the present time barrier reefs are sometimes constructed on an enormous scale. A great barrier reef runs parallel to nearly the whole north shore of Cuba, while the barrier reef of Australia, the largest known, extends, with some breaks, for over 1200 miles along the northeast coast of Australia, from which it is distant 20 to 80 miles; its breadth varies from 10 to 90 miles, though but little of this width is exposed above water; its sea-face is in some places more than 1800 feet high (*i.e.* above the sea-bottom, not the surface). (3) Atolls are coral islands of irregularly circular shape, which usually enclose a central lagoon and frequently, as in the Pacific, rise from the profoundest depths. The way in which such islands have been built up is still a subject of much controversy. No doubt, atolls have been formed in various ways, especially those which arise from small depths, but probably the most important method is by a slow subsidence of the sea-bottom, with which the growth of the reef can keep pace. Such subsidence is the only explanation of great thicknesses of coral rock, as of any other kind of shoal-water deposits. Borings in the Hawaiian Islands, and especially in Funafuti, an island of the South Pacific, have demonstrated the existence of immensely thick coral limestones formed in the modern period.

Coral reefs in shoal water frequently have gentle slopes, but those which rise from the deep sea have very steep faces, sometimes as much as 65.°, and thus reefs may occur as lens-shaped areas, or steep-sided masses of limestone, in which stratification is very obscure, or absent, in the midst of well-stratified fragmental sediments.

Dolomitization. — A process has been observed in the closed lagoons of certain atolls which is significant as throwing light upon a very difficult problem, that of the formation of dolomite or magnesian limestone. In the closed lagoon, shut off entirely from the sea, the isolated body of sea-water becomes considerably concentrated by evaporation. All sea-water contains chloride of magnesium ($MgCl_2$), and this percolating into the coral rock, by double decomposition with $CaCO_3$, forms $MgCO_3$. The change occurs more readily when the $CaCO_3$ is in the form of aragonite, as is the case in many shells and corals.

Chemical Deposits. — It is not known just how important a part is played by chemical precipitation in the formation of marine deposits, but probably a greater one than has been generally supposed. Rivers which bring in quantities of $CaCO_3$ in solution may so overload the sea with this substance (for sea-water will dissolve little of it) that more or less is precipitated in the neighbourhood of the land. On the coast of Asia Minor, for example, are large areas of sandstone and conglomerate, formed within recent times by the precipitation of $CaCO_3$ in masses of sand and gravel, binding them into hard rock. Similar examples are known elsewhere. There is also some reason to believe that the decay of marine animals evolves sufficient carbonate of ammonia to convert the sulphate of lime into the carbonate by double decomposition, and to precipitate the latter in some quantity.

3. Aktian Deposits. — The 100-fathom line is by Murray and

Renard regarded as the boundary between shallow and deep water, for it generally marks the edge of the continental shelf, from which the bottom rises very gently to the land, but slopes abruptly down to the oceanic depression. The great bulk of the material derived from the waste of the land is thrown down upon the continental shelf, within the 100-fathom line, but the finer particles are carried farther out and subside in deeper and quieter water. A considerable quantity of the finest sedimentary particles remains long suspended in sea-water, especially in the cold water of the polar seas. On the continental slopes, extending from the 100-fathom line to the bottom of the great oceanic abysses, are laid down most of the very fine materials derived from the land, which are grouped together under the somewhat indefinite term, *mud.* Mud is a mixture of minerals in a state of extremely fine mechanical subdivision, but not chemically decomposed, thus differing from clay.

(1) *Blue Mud.* — The materials of this deposit, which are principally, though not altogether, derived from the land, are very heterogeneous. Quartz grains in an excessively fine state of subdivision are very abundant; clay is often a considerable ingredient, and then the mud is plastic when wet, but it is usually more earthy than clay-like. Minute particles of other terrigenous minerals, like felspar, hornblende, augite, etc., are common. $CaCO_3$ is almost always present, averaging 7%, and in some instances rising to 25%; this is due chiefly to the foraminiferal shells, both of those species which live at the surface and those which live on the bottom. Siliceous organisms are also present to the average amount of 3%, and are principally diatoms, radiolarians, and spicules of sponges. Glauconite is found in nearly all the samples. The blue colour of this mud is due to the sulphide of iron and the organic matter which prevents the oxidation of the sulphide. Of the terrigenous deep-sea deposits blue mud is the most extensively developed; it is estimated as covering 14,500,000 square miles of the sea-bottom, and surrounds almost all coasts, and fills enclosed basins like the Mediterranean and even the Arctic Ocean. The depths at which blue mud is found range from 125 to 2800 fathoms.

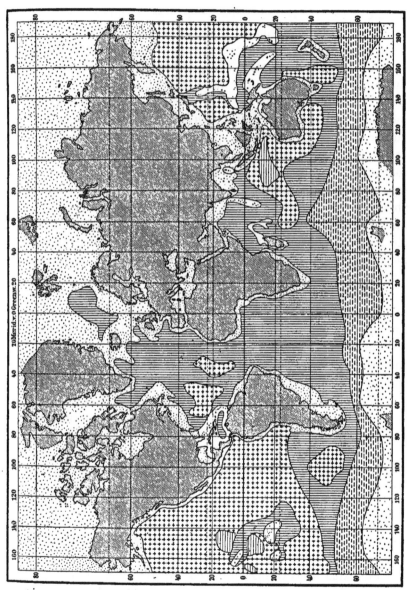

FIG. 141.— Map of marine deposits. (Kayser after Murray and Renard) Dotted area
= terrigenous; vertical lined = foraminiferal ooze; horizontal broken-lined = dia_
tom ooze; crosses = oceanic red clay; white = radiolarian ooze

(2) *Red Mud* is a local development, which occurs principally upon the Atlantic coast of Brazil, and in the Yellow Sea of China. Silt of this character, the red colour of which is due to Fe_2O_3, contained in laterite, is brought down in large quantities by the Amazon and the Orinoco. Foraminiferal shells are abundant; radiolarians very rare. Probably a more minute examination of the continental slopes will show that red mud has a wider distribution in tropical seas than is here indicated.

(3) *Green Mud* is much the same in character as the blue mud, but owes its green colour to the higher percentage of glauconite which it contains.

(4) *Green Sand* is granular in appearance, and is made up largely of grains of glauconite and casts in that material of the interior of foraminiferal shells, together with nearly 50 % of $CaCO_3$. The green sands occur in shallower water than the muds, and often within the 100-fathom line, as in the case of a deposit of this kind which is now forming off the coast of Georgia and the Carolinas. The estimated area of the green muds and sands is 1,000,000 square miles.

(5) *Volcanic Muds.* — In the deeper water surrounding volcanic islands are deposits of fine mud made from the disintegration of volcanic rocks, mixed with considerable clay, and also calcareous matter derived from organisms.

4. The **Abysmal Deposits** are those the materials of which are not directly derived from the land, but consist of matters carried to the sea in solution and extracted from the sea-water by the agency of organisms, together with volcanic substances in a more or less advanced stage of decomposition. Only rarely are terrigenous materials found in these deposits, as, for example, off the west coast of Africa, where fine sand, carried by the wind from the Sahara, is found in deep water, and ice-borne fragments are common in high latitudes. The pelagic deposits are found far from land, mostly in the deepest oceanic abysses, where the rate of accumulation is almost inconceivably slow, and the remains of extinct animals still lie exposed upon the ocean floor.

(1) *Foraminiferal Ooze.* — The Foraminifera are minute animals, each one a tiny speck of jelly, most of which, in spite of their
extreme simplicity of structure, have the power of secreting very
beautiful and complex shells of $CaCO_3$. The species which are
of importance in this connection are ·those which live in infinite
multitudes at the surface of the ocean, and the most abundant at
the present time are those which belong to the genus *Globigerina*,
whence this deposit is frequently called Globigerina ooze. These
surface Foraminifera flourish best in warm water and follow the
warm currents, often into quite high latitudes. Their shells, which

FIG. 142. — Foraminiferal ooze, × 20.   (Agassiz after Murray and Renard)

drop to the bottom as the occupants die, are present in almost all
marine deposits, but near land the terrigenous materials prepon-
derate to such a degree that the Foraminifera make up but a slight
proportion of the deposit. In deeper water, where the wash from
the land does not come, the foraminiferal shells become relatively
much more abundant, and when 30 % or more of a given sample
of the bottom consists of them, it is classed as a foraminiferal ooze.
Other organisms which secrete calcareous shells or tests always
contribute more or less to these oozes (coral mud, echinoderms,
molluscs, nullipores, etc.). The deposit is purest and most

typical in the medium depths of the ocean, far from any land; in such places the ooze may contain as much as 90 % $CaCO_3$ and is white, while nearer land the slight admixture of terrigenous minerals gives a pink, gray, brown, or other colour to the mass. Below the depth of 2500 fathoms the proportion of $CaCO_3$ becomes much diminished, owing to the increasing percentage of CO_2 in the seawater, which attacks and dissolves these delicate shells.

The foraminiferal oozes have a vast geographical extent, estimated at 49,520,000 square miles, and are especially developed in the Atlantic, though they are largely present in all except the polar seas, and range in depth from 400 to 2900 fathoms.

FIG. 143. — Pteropod ooze, × 4. (Agassiz after Murray and Renard)

(2) *Pteropod Ooze.* — The thin and delicate shells of the molluscan groups known as the pteropods and heteropods abound at the surface of the warmer parts of the ocean, but their dead shells are found only in depths of less than 2000 fathoms. In shallow water (and even in greater depths near land) the shells are concealed by other kinds of material, but in moderate depths, far from any land, these shells sometimes become so frequent in the foraminiferal ooze as to give it a special character. In its typical development this pteropod ooze has been found only in the Atlan-

tic, where it covers some relatively small areas, in depths of 400
to 1500 fathoms.

(3) *Radiolarian Ooze.* — The organisms which we have so far
considered secrete only shells or tests of $CaCO_3$, but this is not the
only substance which is very extensively extracted from sea-water
by living beings.   Silica is also dissolved in sea-water, and various
organisms construct their tests of that substance.   The *Radiolaria*
are, like the Foraminifera, a group of microscopic, unicellular
animals, which secrete siliceous tests of the most exquisite delicacy
and beauty; they live both at the surface and at the bottom of the
sea.   Radiolarian tests may be detected in all sorts of marine
deposits of both deep and shallow water, but it is only in very profound depths that they occur in quantity sufficient to give character
to the deposit.   When 20 % or more of a bottom deposit consists
of radiolarian tests, it is called a radiolarian ooze, but clay and
volcanic minerals make up most of the materials.   This ooze has
been found only in the Pacific and Indian oceans, where, it is estimated, it covers 2,290,000 square miles of the bottom, at depths
of 2350 to 4475 fathoms.

(4) *Diatom Ooze.* — In our study of fresh-water deposits we
learned that the siliceous cases of the microscopic plants known as
diatoms form considerable accumulations in lakes and ponds, and
they also flourish abundantly in brackish water and in the sea.
Diatoms are found in many marine deposits, but in relatively small
quantities.   In the Antarctic Ocean, however, is an immense
belt of ooze, believed to cover 10,880,000 square miles and extending around the globe, which is largely made up of their frustules.
Besides the great Antarctic zone, an area of some 40,000 square
miles is known in the North Pacific.   The diatom ooze entirely
resembles the fresh-water deposit, but may be distinguished by the
presence of foraminiferal and radiolarian shells and tests.   The
depths at which this ooze is found are from 600 to 2000 fathoms.

(5) *Red Clay.* — The profoundest abysses of the ocean, far
from any land, are covered with a deposit of red clay, which,
though varying much in composition and colour, is yet of a quite

uniform character.   In these vast depths the foraminiferal shells are almost all dissolved by the carbonated sea-water, but some $CaCO_3$ is very generally present, averaging about 6%, and diminishing in quantity as the depth increases.   In the less profound abysses the red clay passes gradually into the foraminiferal oozes, the number of shells increasing until the ooze-like character is attained.   The clay is derived from the disintegration and decay of volcanic substances, especially pumice, which floats upon water, often for months, and drifts long distances in the ocean currents. The greater part of these volcanic materials is believed to be derived from terrestrial volcanoes, but the submarine vents doubtless contribute largely; particles of undecomposed volcanic minerals and glasses are also common.   In some regions the clay is coloured chocolate-brown by the oxide of manganese, and many separate nodules of this substance are found.   The excessive slowness with which this abysmal deposit is formed, is shown by the occurrence, in recognizable quantities, of meteoric iron, which reaches the earth in the form of meteorites, or " shooting stars," and by the presence of the remains of animals which have long been extinct.

Of all the oceanic deposits the red clay is the most widely extended, covering 51,500,000 square miles of the bottom.   Almost four-fifths of this vast area are in the great depths of the Pacific; the shallower Atlantic has much more of the foraminiferal ooze than of the red clay.   The observed range in depth is from 2225 to 3950 fathoms.

Comparing the marine deposits now accumulating in the sea with the rocks of evidently marine origin which form most of the land, we find that the great bulk of these rocks, the sandstones, slates, and limestones, are such as are formed in water of shallow and moderate depths, while only rarely do we discover a rock that implies really deep water.

Climatic Relations of Marine Deposits. — The shoal-water deposits of cold and temperate seas are much alike, both having a preponderance of sand and gravel, with occasional limestones which are more frequent in lower latitudes, while in the deeper waters of the continental slopes, blue mud is laid down. On the floor of the oceanic abysses, however, is a difference in that the terrigenous material is more widely distributed, while foraminiferal ooze and oceanic red clay are absent from the polar seas, and in the Antarctic Ocean is a complete zone of diatom ooze, encircling the earth. The tropical seas are characterized by red muds, and by great development of limestones, especially of the coral reefs, which are almost entirely confined to the tropical waters. The deposits of the truly deep sea are essentially the same in the tropical and temperate zones, and are determined almost entirely by the depth of water; foraminiferal and radiolarian oozes and red clay cover almost the whole floor of the oceanic basins in these regions. As conditions are more uniform in the sea than on the land, climatic differences are less clearly marked in marine than in continental deposits.

Estuarine Deposits

An estuary is a wide opening at the mouth of a river into which the sea has penetrated by the depression of the land. In such bodies of water the tide often scours with much force. Estuaries abound along our Atlantic coast, Delaware and Chesapeake Bays and the mouth of the Hudson being excellent examples of such. The water in them is brackish, and unfavourable to abundant aquatic life, for only a limited number of marine animals, and fewer fresh-water ones, flourish in brackish water.

Estuarine deposits are, in general, much like those of the sea, except that they are apt to be of a finer grain for a given depth of water; muds are abundantly laid down, especially in the more sheltered nooks and bays, with fine and coarse sands and gravels in the more exposed situations. The sands are apt to show a confused stratification from the conflicting currents and eddies in which they are deposited, but with horizontal layers formed at slack water. Extensive mud-flats often surround an estuary, especially if the rise and fall of the tide be great. On these flats, exposed during low tide to the sun and air, in dry, hot climates, sun cracks are formed on the drying surface, and these, together with the prints of raindrops and the tracks of land animals, will be preserved when the incoming tide, advancing too gently to scour the slightly hardened surface of the flat, deposits a fresh layer of sediment upon it. In pluvial climates, the mud-flats of estuaries do not dry with sufficient rapidity to permit the formation of shrinkage cracks, except when the flats are exposed to the air for a longer time than the ordinary interval between tides. This longer exposure occurs when part of the flats is covered only by spring tides, or the general level of the water is raised by a storm. If the estuary be the opening of a large river, considerable deposits of river sediment will, in times of flood, be laid down upon the other beds, producing an alternation of fresh and brackish water beds. On the coast of North Carolina the low sand-spits thrown up by the waves enclose extensive shallow sounds, into which the tide enters by only narrow openings, but which have numerous streams flowing into them. At high water the incoming tide acts as a barrier, damming back the river waters, checking their velocity, and causing them to deposit their burdens of sediment. In course of time, the sounds must be silted up by the rivers, first converted into salt marshes and then into land. The great areas of salt marsh along our Atlantic coast have, for the most part, been formed in this way.

For reasons that we have already discussed, estuaries are not favourable to either fresh-water or marine organisms, and hence estuarine deposits do not contain any great variety of remains of either group. These remains may, however, represent numerous individuals, sufficient sometimes to form limestone layers, as is especially true of oyster banks. Diatoms may also accumulate in great quantities, as in one of the Baltic harbours, where they form 18,000 cubic feet of deposit annually, which necessitates continual dredging to keep the harbour open. On the other hand, estuaries are often favourably situated for the reception and preservation of the remains of land animals and plants which are swept into them by streams and buried in the soft silt of the mud-flats.

The Consolidation of Sediments (Diagenesis)

The processes of deposition upon the land and beneath the water, which we have so far been studying, result, for the most part, only in the bringing together of great masses of loose and incoherent material, which in the case of marine deposits are saturated with salt water. If such masses are properly to be compared with the hard rocks of the earth's crust, it will be necessary to show that loose sediments may be consolidated and rendered hard and firm, like the latter. This is not difficult, for we have abundant evidence to prove that such consolidation actually does take place, and in a variety of ways.

(1) *Consolidation by Weight of Sediment.* — When deposited on a sinking sea-bottom, sediments often accumulate in masses of great thickness, and in such cases the lower portions must tend to consolidate from the weight of the overlying masses. Of course, such a process cannot be directly observed in modern accumulations, because only the surface of them is accessible, but from the analogy of observed facts we may safely infer that this weight is not without effect.

(2) *Consolidation by Cement.* — Sediment is often penetrated by percolating waters, which carry in solution various cementing substances, such as SiO_2, $CaCO_3$, $FeCO_3$, etc., and the deposition of these materials in the interstices of the loose sediment will bind the particles into a firm rock. This process we have already had occasion to observe in several instances, as in the coral reefs, the drift-sand rock of Bermuda, the modern sandstones on the coast of Asia Minor and Brazil, and many others. In all of these cases the cementing substance is $CaCO_3$, but other modern rocks are known in which Fe_2O_3, formed by the oxidation of $FeCO_3$, plays the same rôle, as in Florida where the waters from ferruginous springs bind the grains of calcareous sand into a hard rock, the modern date of which is proved by the presence in such rock of the bones of Indians. Both of these substances are very common as cements

among the ancient rocks. The deposition of silica in the interstices of sand has also been observed, where the original sand grains can with difficulty be detected with the microscope, the rock appearing to be a mass of crystalline quartz grains. A cementing effect may also be produced by reactions within the mass of the sediment itself, as is seen in the solidification of volcanic ash mingled with water to form tuffs.

(3) *Consolidation through Heat.* — This may be local, as in the neighbourhood of volcanoes, or general and due to the internal heat of the earth. For sediment to reach great thickness it must subside, and this subsidence brings the lower parts of the mass deep down into the crust, where they are invaded by the earth's interior heat, and baked as bricks are burnt in a kiln. This process is likewise one which cannot be directly observed, but the effects of molten lava upon loose sediments may be watched, and the consolidating power of heat has been tested experimentally.

(4) *Consolidation by Lateral Pressure.* — This is probably the most widely acting and important agency of consolidation. Though it acts so gradually and at such depths that we cannot see it in operation, yet the inference is, none the less, a safe one. We shall see later that very many of the stratified rocks are no longer in the nearly horizontal position in which they were first laid down, but have been folded and fractured through the operation of great lateral pressures. The more intensely folded and compressed any rock has been, the harder has it become, not only through the mechanical pressure, but by the heat and the chemical changes which such compression generates. In addition to this, we know from experiment that loose materials may be consolidated by powerful compression. Certain exceptional rocks of very ancient date are known, which are almost as incoherent as when first accumulated, but these all retain their original horizontal position and have not been compressed. It must not be supposed, however, that only compressed sediments have become hard, for great areas of scarcely disturbed rocks are found, which are perfectly solid and firm; here some other solidifying agent has been at work.

There are certain other features in which the loose modern sediments differ from the older and harder rocks, such as joints, and cleavage which divides many rocks into thin plates, independently of the planes of stratification. These may be shown, however, to be structures which the rocks have acquired after their formation, and therefore need not be discussed here.

The parallel is now complete between the sediments which we may observe to-day in the process of accumulation, and the hard stratified rocks which make up by far the largest part of the dry land. For all these ancient rocks we may find a counterpart in sediments now forming, and we may conclude with perfect confidence that the ancient rocks were formed by the same agencies as the modern accumulations. Every rock contains a more or less legible record of its own history.

Summary of the Reconstructive Processes. — The destructive agencies supply a great mass of material, of which, *under existing conditions* of climate, topography, etc., about one-half is arrested in its journey to the sea and the remaining half completes that journey; the former moiety constitutes the continental deposits, and the latter moiety the terrigenous marine deposits.

Continental deposits are of great variety, and their nature is determined chiefly by the factors of climate and topography. In the arid and desert regions we have great accumulations of drift-sands, of angular talus, of flood-plain and playa sands and muds, which are characteristically sun-cracked and more or less impregnated with various salts. Deposits from salt lakes, such as salt, gypsum, soda, borax, etc., are confined to arid climates and are not formed in humid climates. In pluvial climates of the temperate zones, rain-wash, deep soils, lacustrine deposits from fresh-water lakes, and river deposits on flood plains and in channels are characteristic. In such climates sun cracks do not form over great areas, as they are largely prevented by the dense covering of vegetation. Peat bogs are the seats of great vegetable accumulations, especially in the cooler and moister regions. In the polar regions, glacial deposits and frost talus are the principal modes of accumulation, and in high mountains these also penetrate deep into the temperate and even the tropical zones. In the tropics we find extremely deep soils, which contain or are made up of the red laterite, and surface deposits of iron oxide and chemically formed limestone are extensively made. Immense masses of river alluvium gather in interior basins, but vegetable accumulations are less abundant than in temperate lands, and lakes are not common in the tropics. The absence of lakes, however, is not determined by temperature, but by the *antiquity* of land surfaces. It cannot be inferred from the fact that only half of the annual land-waste finds its way to the sea, that such should be the proportion between continental and marine deposits among ancient rocks, for a transgression of the sea over an ancient land surface, deeply buried under continental deposits, would rapidly rework the latter into marine deposits. At present, we observe that material derived from the land and in mechanical suspension laid down in the sea is distributed by the waves and currents, sorted into layers according to the fineness of the material and, more or less incompletely, according to its mineralogical composition. The most important factors which determine the character of the deposit at any given point on the sea-floor are the depth of water and the topography and elevation of the adjoining land. The coarser material, gravel and sand, are laid down upon the beach and in shoal water, the sand generally extending to the 100-fathom line, while on the continental slope are deposited the various muds, and on the floor of the ocean basins the organic oozes and the oceanic red clay, derived chiefly from the decay of volcanic minerals. Limestone banks are formed by the extraction of the dissolved lime-salts through organic agencies, a process which goes on most extensively in warm seas of shallow and moderate depth. Climatic differences also have their effect upon marine deposits, but less markedly than in the case of the continental accumulations. The loose sediments accumulated on land or under water are, under favouring conditions, consolidated into hard rocks, thus making the parallel with the ancient sedimentary rocks complete, and finishing the cycle of destruction and reconstruction from rock back to rock.   All these various kinds of deposits, continental and marine, are forming simultaneously, but one kind of deposit does not gather indefinitely at one point, except perhaps on the floor of the deep ocean-basins.   Conditions shift and change, so that one kind of material is laid down upon another, and in the same vertical section we may discover many different beds, each one recording the conditions at that point, for the time during which the bed formed the surface of the lithosphere.   All these changes we have studied in order to obtain a key to the record contained in the rocks, and we have found that the processes now at work do furnish a partial key.   However, before a systematic history of the earth can be attempted, we have first to study the ways in which the rocks are arranged and the disturbances which they have undergone; this constitutes *structural* or *tectonic geology.*

PART II

STRUCTURAL OR TECTONIC GEOLOGY

CHAPTER X

THE ROCKS OF THE EARTH'S CRUST — IGNEOUS ROCKS

In the first section of this book we made a study of the processes and agencies which are still at work upon and within the earth, tending to modify it in one or other particular. We there found that slow but ceaseless cycles of change take place and that a continual circulation of material is going on.

We have now to take up the second branch of our subject, that of *structural geology*, which deals with the materials of the earth's crust, their mode of occurrence, and their arrangement into great masses. Structural geology is, however, not merely a descriptive study; hand in hand with the examination of the rock-masses must go the attempt to explain their structure, and to show how they have come to be as we find them. Dynamical principles must be continually called in to interpret the facts of structure, and many of the observations concerning the construction, destruction, and reconstruction of rocks find their application in the study of structure.

This application cannot, in all cases, be made with confidence, both because a given structure may often be referred, with equal probability, to different processes, and because certain of the great dynamical agencies are so slow and gradual in their mode of operation, that no one has ever been able to observe them at work. In

this latter class of cases the agency must be inferred, not from anything which we have actually seen accomplished, but from the
traces which it has left in the structure.   Under such circumstances,
it need not surprise us to find that the explanation is not always
obvious, but may be very problematical, and that great differences
of opinion may arise concerning the rightful interpretation of a
complex region.

Here, as in all other provinces of geology, the historical standpoint is the dominant one.   Our object is to learn, not only the
agencies which have produced the structures and the way in which
they operated, but also the successive steps by which the structures
originated, the order in which they occurred, and their geological
date.   Thus they may be coördinated into the great history of the
earth, which it is the main problem of geology to construct.

ROCKS

The distinction between a rock and a mineral is not always an
easy one for the beginner to grasp, yet it is essential that he should
do so.   A *Rock* is any extensive constituent of the earth's crust,
which may consist, though rarely, of a single mineral, but in the
great majority of cases is a mechanical mixture of two or more
minerals.   A rock thus has seldom a definite chemical composition, or homogeneous internal structure.   An examination
with the microscope almost always shows that a rock is an aggregate of distinct mineral particles, which may be all of one kind,
or of many different kinds, in varying proportions.   Rocks, then,
are mechanical mixtures, and their properties vary in proportion
to their various ingredients, while minerals are chemical compounds (see p. 6).

In ordinary speech the term *rock* is held to imply a certain
degree of solidity and hardness, but in geological usage the word
is not so restricted.   Incoherent masses of sand and clay are regarded as being rocks, quite as much as the hardest granites.

The classification of rocks is a very difficult and obscure problem, and would be so, even were our knowledge much more complete and exhaustive than it is. There are, therefore, great diversities in the various schemes of classification which have been proposed and which are still in use, and all such schemes require modification to meet continually advancing knowledge.

Bearing in mind the principle, already emphasized so often, that geology is primarily a historical study, the most logical scheme of classification is obviously one that, so far as possible, is *genetic*, that is to say, one which expresses in brief the history and mode of formation of the rocks. Other criteria, such as texture and chemical and mineralogical composition, must be employed for the minor subdivisions. On this genetic principle we may divide all rocks into three primary classes or groups.

A. *Igneous Rocks*, those which were melted and have solidified by cooling. Texture glassy or crystalline.

B. *Sedimentary Rocks*, those which have been laid down (most commonly) under water, by mechanical, chemical, and organic processes. Rocks composed of more or less rounded and worn fragments, seldom crystalline.

C. *Metamorphic Rocks*, those which have been profoundly changed from their original sedimentary or igneous character, often with the formation of new mineral compounds in them. Texture fragmental or crystalline.

IGNEOUS ROCKS

The igneous rocks have a deep-seated origin and have either forced their way to the surface, or have cooled and solidified at varying depths beneath it. Though rocks of this class, there is every reason to believe, were the first to be formed, they have been made all through the recorded history of the earth, and, as volcanoes show, are forming now. They are thus the primary rocks and all the others have been derived, either directly or indirectly, from them. The products of the chemical decomposition or mechanical abrasion of the igneous rocks have furnished the materials out of which the sedimentary rocks were formed, at least in the first instance.

The igneous rocks are massive, as distinguished from stratified, and though sometimes presenting a deceptive appearance of stratification, may always, with a little care, be readily distinguished from the truly stratified rocks. The term *massive* is, indeed, frequently used for these rocks in the same sense as igneous, and *eruptive rocks* is another term meaning the same thing, though eruptive is also employed in a more restricted sense. Still another term which should be defined is *magma*, by which is meant a continuous molten mass before solidification.

Characteristic differences appear between those igneous masses which have solidified deep within the earth and have been brought to light only by the denudation and removal of the overlying rock-masses, and those which have cooled at or near the surface of the ground. The former are called *plutonic* (abyssal, or intrusive) and the latter *volcanic* (or extrusive). Between the two may be found every gradation, and the term *hypabyssal* is sometimes employed for rocks which are transitional between the typical plutonic and the typical volcanic kinds. These terms, plutonic, hypabyssal, and volcanic, are used to describe the character of *rock-masses*, not as terms of classification.

Texture. — The *texture* of an igneous rock means the size, shape, and mode of aggregation of its constituent mineral particles. Texture is a very important means of determining the circumstances under which the rock was formed, and hence great attention is paid to it. Since texture responds so accurately to the circumstances of solidification, rate of cooling, pressure, etc., all the varieties shade into one another by imperceptible gradations and form a continuous series. Nevertheless, it is necessary to distinguish and name the more important kinds.

Among the igneous rocks are found four principal types of texture, with several minor varieties : —

1. *Glassy*. — Here the rock is a glass or slag, without distinct minerals in it, though the incipient stages of crystallization, in the

form of globules and hair-like rods, are often observable with the microscope. (See Fig. 24, p. 75.) When the glass or slag is made frothy by the bubbles of escaping steam and gas, the texture is said to be *vesicular, scoriaceous,* or *pumiceous* (see Figs. 27, 210), according to the abundance of the bubbles. These are varieties of the glassy texture, though other kinds may also be vesicular. A vesicular rock in which the steam-holes have been filled up by the subsequent deposition of some mineral is called *amygdaloidal,* a term derived from the Greek word for almond.

2. The *Compact* (or Felsitic) texture is characterized by the formation of exceedingly minute crystals, too small to be seen by the unassisted eye, giving the rock a homogeneous but stony and not glassy appearance. If the crystals are too minute to be identified even by the aid of the microscope, the rock is said to be *cryptocrystalline,* and when such identification can be made, it is called *microcrystalline.*

3. *Porphyritic.* — In rocks of this texture are large, isolated crystals, called *phenocrysts,* embedded in a ground mass, which may be glassy or made up of small crystals. The phenocrysts may have sharp edges and well-formed faces, or they may have irregular and corroded surfaces. The porphyritic texture indicates two distinct phases of crystallization. The first is the formation of the phenocrysts, which remain suspended in the molten mass, or *magma,* and are often corroded and partially redissolved (resorbed) by it. These crystals are said to be of *intratelluric* origin, because formed before the eruption of the lava, and such crystals are showered out of certain active volcanoes at the present time. Stromboli (see p. 75), for example, ejects quantities of large and perfect augite crystals. There is reason to believe, however, that not all phenocrysts are thus intratelluric, but that the first phase of crystallization sometimes takes place after the ejection of the molten mass. The second phase consists in the formation of the ground mass, which may be glassy, finely crystalline, or both. Mineral particles having distinct crystalline form are called *idiomorphic.*

4. *Granitoid.* — In this texture the rock is wholly crystalline,

without ground mass or interstitial paste. The component grains, which may be fine or very coarse, are of quite uniform size, and as the crystals have interfered with one another in the process of formation, they have rarely acquired their proper crystalline shape. Such grains are said to be *allotriomorphic*.

An additional texture which should be mentioned is the *fragmental*. This is represented by the accumulations of the fragmental products ejected by volcanoes (see p. 79), agglomerates,

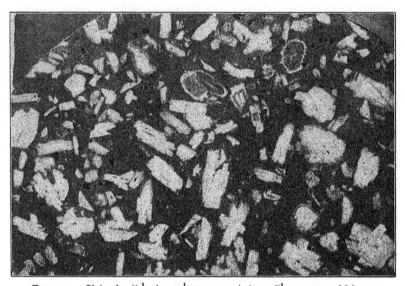

FIG. 144.—Slab of polished porphyry, natural size. Phenocrysts of felspar

bombs, lapilli, ashes, etc. Many such materials accumulate in bodies of water and are there sorted and stratified and, it may be, mingled with more or less sand and mud and other sedimentary material. Rocks formed in this manner partake of the nature of both the igneous and sedimentary classes, and may be regarded as a series intermediate between the other two and in a measure connecting them. These rocks will here be treated as a special subdivision, under the name of pyroclastic rocks.

In our studies of the products of modern volcanoes, we saw that the same molten mass will give rise to rocks of very different appearance in its different parts, according to the circumstances of rapidity of cooling, pressure, etc. We may now express this in somewhat more general form and say that the texture of an Igneous rock is determined by the several factors which affect the molten mass during consolidation. Of such factors may be mentioned the chemical composition, temperature, rate of cooling,

FIG. 145. — Hand Specimen of granite, natural Size

degree of pressure, and the quantity present of dissolved vapours and gases, which are called *mineralizers*.

Solidification. — Chemical composition determines the fusibility of a rock at a given temperature. The least fusible rocks are, on the one hand, those which contain large quantities of silica, 60 to 75 %, and, on the other, those which contain less than 40 % of silica. The most fusible rocks are those with an intermediate percentage of silica, and among these the fusibility increases, as the percentage of silica diminishes, until the lower limit is reached. The effect of chemical composition upon texture is seen in the rapidity with which the less fusible rocks chill and stiffen, and therefore the greater frequency with which they form glasses.

Chemical composition is, however, important in this connection chiefly through its effect upon the rate of solidification. We have already learned (p. 9), that solidification very generally takes place by a process of crystallization, and this requires time. Hence, very rapid cooling results in a glass, but the microscope reveals the incipient stages of crystallization in many of even the glassy rocks. A somewhat slower rate of solidification produces a cryptocrystalline rock, and successively slower rates bring about the porphyritic, microcrystalline, and granitoid textures. Large crystals form slowly, and other things being equal, the larger the component crystals of a rock, the more slowly has it consolidated.

Pressure is of importance in preventing the rapid escape of the vapours and gases contained in the molten mass, and hence frothy, scoriaceous, and vesicular textures cannot be produced under high pressures. Pressure is also believed to be necessary for the formation of many phenocrysts in porphyritic rocks.

The mineralizers, such as steam, hydrochloric acid, and other vapours, determine the crystallization of many minerals, which refuse to crystallize in the absence of such vapours. Variations in the quantity of mineralizers present in different parts of the same mass occasion corresponding differences in the local textures. The well-known Obsidian Cliff, in the Yellowstone National Park, is formed by a great lava-sheet, made up of alternating layers of glassy and microcrystalline rock, a difference which is referred to varying proportions of mineralizers present in different parts of the molten mass.

It must not be supposed that a molten magma consists merely of a number of fused minerals, mechanically mixed together and having no effect upon one another. If such were the case, the

minerals in cooling should all crystallize in the order of their fusibility, the least fusible forming first, and the most fusible last. This
is not what we find, and many facts which cannot be discussed
here have led petrographers to the belief that a molten magma
is a *solution* of certain compounds in others, and that crystallization occurs in the order of solubility, as the point of saturation for
particular compounds is successively reached by the cooling mass.

Similar phenomena may be observed among the metals. If
strips of copper be thrown into a vessel of melted tin, the latter
will dissolve the copper at a temperature far below that at which
the copper would melt alone.

In a rock magma the crystallization of the more and more
soluble minerals will proceed regularly, provided the pressure and
rate of cooling continue constant. As these conditions are, however, subject to variation, it frequently happens that the more
soluble minerals begin to crystallize before the less soluble have
all been formed, and thus the periods of formation of two or more
kinds of minerals partly overlap.

Usually, the order of formation of the different kinds of minerals in a solidifying magma is as follows. First to form are apatite,
the metallic oxides (magnetite, ilmenite), and sulphides (pyrite),
zircon, and titanite. " Next come the ferro-magnesian silicates,
olivine, biotite, the pyroxenes, and hornblende. Next follow the
felspars and felspathoids, nepheline and leucite, but their period
often laps well back into that of the ferro-magnesian group. Last
of all, if excess of silica remains, it yields quartz. In the variations
of pressure and temperature, it may and often does happen that
crystals are again redissolved, or resorbed, as it is called, and it
may also happen that after one series of minerals, usually of
large size and intratelluric origin, have formed, the series is again
repeated on a small scale, as far back as the ferro-magnesian
silicates. Minerals of a so-called second generation thus result,
but they are always much smaller than the phenocrysts and are
characteristic of the ground mass.

" It results from what has been said that the residual magma is

U

increasingly siliceous up to the final consolidation, for the earliest crystallizations are largely pure oxides. It is also a striking fact that the least fusible minerals, the felspars and quartz, are the last to crystallize." (Kemp.)

A very considerable number of minerals are found in the igneous rocks, but comparatively few in any large quantity. It thus becomes necessary to distinguish between the *essential* minerals of a rock and the *accessory* ones. The essential minerals are those which characterize a given kind of rock, while the accessory minerals are those which occur in small quantities and which may be present or absent, without materially affecting the nature of the rock. The distinction is necessary and useful, but is rather arbitrary.

Another necessary distinction is that between *original* and *secondary* minerals. Original minerals were formed with or before the rock of which they are constituents, and secondary minerals are produced by the alteration or reconstruction of the original ones.

With comparatively few exceptions, the igneous rocks are made up of some felspar or felspathoid, together with one or more of the pyroxenes, amphiboles, micas, olivine, or quartz. Magnetite is also very common.

Differentiation. — Different parts of the same continuous rock-mass frequently display chemical and mineralogical variations, resulting from a process of differentiation, or segregation, of the magma. How this is brought about, is far from certain, but there can be little or no doubt as to the fact. "When large areas of eruptive rocks are carefully investigated, it is found that there is a perfect and gradual transition of one kind into another — all intermediate varieties existing — and that quantitatively no special part of the series is universally predominant, although there are often immense masses of nearly uniform character, and there may be smaller bodies of quite variable composition." (Iddings.) It is further found that in a given volcanic district, or *petrographical province*, the rocks erupted at a particular geological period have certain peculiarities which distinguish them from those of other provinces. It thus appears probable that the igneous rocks of such a province were derived from the same magma, and the relationship between the various kinds of rocks of the province is called *consanguinity*.

The existence of these petrographical provinces does not imply that the rocks of each one differ from those of all the others. In fact, similar or identical groups of rocks are found in many parts of the world, but each province differs more or less from the surrounding ones. Thus, rocks which from the genetic point of view are closely related, are, by any scheme of chemical or mineralogical classification, often placed in widely separated groups.

Assimilation. — When we examine in the field the igneous rocks in their relations to other rock-masses, we frequently find cases where it is exceedingly difficult to account for the presence of the igneous mass, except upon the assumption that the magma made way for itself by fusing and incorporating the rocks which must formerly have occupied its present position, the surrounding rocks showing no evidence of being merely pushed aside by the ascending magma. In certain instances, such a melting and incorporating of opposing rocks would seem to be clear, as when a sheet of magma has made its way into a series of strata, parallel with the bedding planes, without increasing the thickness of the series. This incorporating of freshly fused material with the intruding magma is called *assimilation*, but, save on a very small scale, its reality is a subject of much dispute, and some of the highest authorities altogether reject it. Nevertheless, many observed facts strongly favour this assimilation theory, which has a most important bearing upon some of the fundamental problems of geology. According to this view, the ascending magma is at an extremely high temperature and very fluid, and it forces its way upward partly through crevices and fissures, partly by detaching joint-blocks, which sink into the molten mass and are dissolved by it, thus greatly modifying the chemical constitution of the magma. Subsequently, the magma becomes differentiated, so that the different varieties of rock separate from one another in the manner already described. Perhaps, as Daly has suggested, there is a universal subcrustal magma, of basic composition, which, owing to the pressure of the overlying crust, is only potentially fluid, liquefying along lines of a partial release of pressure. The overlying solid rocks are, on the average, more acid than the subcrustal magma and thus, by assimilation followed by differentiation, the many varieties of igneous rock are formed.

Classification. — What was said above with regard to the difficulty of classifying rocks, applies more especially to the igneous group, because of the way in which the various kinds shade into one another, since even the same molten mass may differentiate into several species, showing not only differences of texture, but marked changes of chemical and mineralogical composition. In an elementary work, like the present, only a meagre outline of the subject can be attempted, for the microscopic study of rocks, or petrography, has now become an independent science of great scope and interest and cannot be compressed into a few pages.

The classification of the igneous rocks now most generally adopted is made upon a threefold method, according to texture, and chemical and mineralogical composition. In the following table (modified from Kemp's) the textures are given in vertical order, while transversely the arrangement is mineralogical, chiefly in accordance with the principal felspar. In this manner the acidic rocks come at the left side of the table and the basic at the right side. The percentages of silica are given on a lower line of the table.

The acid rocks are so called because they are rich in silica, but they have only small quantities of lime, magnesia, and iron; hence they are very infusible, of low specific gravity, and generally of light colours. The basic rocks, thus named because of the predominance of the bases, have much smaller percentages of silica and higher ones of lime, magnesia, and iron; the latter substances act as fluxes, making the basic rocks much more fusible, as well as giving them a higher specific gravity and darker colour. The

CLASSIFICATION.

ACIDIC →   → BASIC

Texture / Occurrence	ACID GLASSES: Obsidian, Perlite, Pumice, Pitchstone	CHIEF FELSPAR ORTHOCLASE (Biotite and Hornblende or Augite)			ANDESITE OBSIDIAN — CHIEF FELSPAR PLAGIOCLASE (Biotite and Hornblende)		(Pyroxenes)		Nepheline Leucite	BASIC GLASSES, SCORIÆ, TACHYLITE, BASALT OBSIDIAN — NO FELSPAR (Augite and Hornblende and Biotite or)	
		+Quartz	−Quartz	Leucite Nepheline	+Quartz	−Quartz	−Olivine	+Olivine	−Olivine / +Olivine	−Olivine	+Olivine
Glassy (Surface Flows)	Obsidian, Perlite, Pumice, Pitchstone				Andesite Obsidian					Basic Glasses, Scoriæ, Tachylite, Basalt Obsidian	
Porphyritic with few Phenocrysts — Surface Flows, Thin Sills, Dykes, Laccoliths		Rhyolite (Felsite, Quartz-Porphyry)	Trachyte (Felsite)	Phonolite (rare) Leucite Rocks (Very rare)	Dacite (Felsite)	Andesite (Felsite)	Augite-Andesite, Basalt (Diabase)	Olivine-Basalt, Olivine-(Diabase)	A series of Nepheline and Leucite Basaltic Rocks, very rare in ɑa	Augitite	Limburgite
Porphyritic, Compact, or Glassy with abundant Phenocrysts — Dykes, Sills, Laccoliths		Rhyolite-Porphyry	Trachyte-Porphyry	Phonolite-Porphyry	Dacite-Porphyry	Andesite-Porphyry	Augite-Andesite-Porphyry	Olivine-Basalt-Porphyry		Augitite-Porphyry	Limburgite-Porphyry
(Dykes, Sills, Laccoliths)		Granite-Porphyry	Syenite-Porphyry	Nepheline-Syenite-Porphyry	Quartz-Diorite-Porphyry	Diorite-Porphyry	Gabbro-Porphyry	Olivine-Gabbro-Porphyry		Pyroxenite-Porphyry	Peridotite-Porphyry
Granitoid — Batholiths, Sills, Laccoliths / Batholiths		Granite	Syenite	Nepheline-Syenite, Leucite-Syenite (Very rare)	Quartz-Diorite •	Diorite •	(Diabase) Gabbro •	(Olivine-Diabase) Olivine-Gabbro •	Theralite (exceedingly rare)	Pyroxenite	Peridotite •
SiO₂		80—65%	65—55%	60—50%	70—60%	65—50%	50—45%		50—40%	55—30%	
		Granite fam.	Syenite family		Diorite family		Gabbro family			Peridotite family	

distinction between acid and basic rocks cannot be very sharply drawn, because the two kinds are connected by every variety of intermediate gradation. The same is true, however, of all the divisions given in the table, which is apt to produce a false impression of sharply distinguished groups of rocks, such as do not occur in nature.

As a general rule, the glassy and porphyritic textures characterize those rocks which have solidified at the surface of the ground, or not very far below it, while the granitoid types have cooled slowly and at great depths; but there are exceptions to both statements. Between the glassy and porphyritic textures at one end of the series and the granitoid at the other comes the felsitic which represents an intermediate rate of cooling and intermediate depths within the earth as the place of solidification (hypabyssal rocks).

The division of the igneous rocks into families is made primarily in accordance with the mineralogical composition, with subdivisions according to texture. This method gives us five principal groups.

I. The Granite Family

The molten magma, which on solidification gives rise to the rocks of this group, is very rich in silica (65 to 80%) and has from 10 to 15% of alumina; the quantity of alkalies (Na and K) is relatively large (6 to 8%), and there are small amounts of iron oxides (2 to 4%), magnesia (1 to 2%), and lime (1 to 4%). In the process of consolidation the principal minerals formed are orthoclase and quartz, with smaller amounts of oligoclase, iron oxide, and of the ferro-magnesian minerals, biotite or hornblende. Differences of texture, produced in the manner already described, give rise to rocks of totally different appearance, which it is difficult to imagine are of similar or identical composition.

Obsidian is a volcanic glass, which is usually black or dark brown or green (but sometimes blue, red, or yellow). It breaks with a shell-like fracture, and in very thin pieces is translucent.

The microscope shows " crystallites," the incipient stages of crystals, which are present in great numbers. The name *obsidian* is used for the various kinds of volcanic glass in which the percentage of water is small, and so for exact description a prefix is necessary, such as *rhyolite obsidian, andesite obsidian*. Though the glasses are of varying composition, by far the greater number of them belong to the granite family. When the glass is divided by concentric cracks, due to shrinkage on cooling, so as to form onion-like spherules, it is called *Perlite*.

Pitchstone has much the same appearance as obsidian, but contains from 5 to 10 % of water.

Pumice is a glass blown up by the bubbles of escaping steam and other vapours into a rock froth, so light that it will float upon water. A very similar substance is produced when a jet of steam is blown through the melted slag from an iron furnace.

It not infrequently happens that, in course of time, the volcanic rocks become *devitrified*, losing their glassy texture and assuming a stony one. The homogeneous rock becomes converted into a mass of extremely minute crystals of quartz and felspar, and the original glassy texture is then shown only by the lines of flow, or by the perlitic character, which are not affected by the change. Devitrification has also been observed in artificial glasses, especially when the glass, owing to insufficient annealing, has been subject to internal stress.

Rhyolite ordinarily occurs as the lava outflow of a granitic magma, cooled rapidly, but yet more slowly than obsidian. The texture is porphyritic, the phenocrysts being chiefly quartz and the glassy form of orthoclase known as sanidine, while the ferro-magnesian minerals are present in very much smaller quantities, and of these the commonest is biotite. The phenocrysts are embedded in a ground mass of minute felspar crystals and a varying proportion of glass. Other names used for rhyolite are *liparite* and *quartz trachyte*. The rhyolites are exceedingly common in the western part of the United States. The *Felsites* are very dense, fine-grained, and light-coloured rocks, in which phenocrysts are

absent or scanty; they are rocks which have been formed in different ways, by the devitrification of obsidians and rhyolites, by the
recrystallization of tuffs, and by original cooling from fusion.

Quartz Porphyry shades imperceptibly into rhyolite or felsite
on the one hand, and into granite on the other; it is made up
of phenocrysts of quartz, or of quartz and orthoclase, in a crystalline ground mass of the same minerals. If the phenocrysts
are very abundant and the ground mass rather coarse grained, the
rock is called *granite porphyry*. Syenite porphyry, Diorite porphyry, etc., bear similar relations to the other members of their
respective families and need no further description.

Granite. — The granites are thoroughly crystalline rocks, of
typically granitoid texture, to which they have given the name,
and without any ground mass. The grains have not their proper
crystalline shape, the separate minerals interfering with each other
in the process of crystallization. The characteristic minerals are
quartz, orthoclase, some acid plagioclase, muscovite, biotite, and
hornblende; magnetite and apatite are always present, though in
small quantities. The variations in granite are principally in the
ferro-magnesian minerals. Thus we have *muscovite granite*, with
white mica only; *granitite*, with biotite only; *hornblende granite*,
the hornblende replacing the mica, or in addition to biotite; *augite
granite*, with augite and biotite. Those in which the percentage
of soda is high are called *soda-granites*. When the ferro-magnesian
silicates are all absent, the rock is called a *binary granite*, or *Aplite*.

The colour of granite is dark or light in accordance with the
proportion of dark silicates present, while the shades of the felspar
determine whether the rock shall be red, pink, or white. The
texture of granite varies from fine to very coarse, and in some
cases becomes nearly porphyritic. A very coarse-grained granite
is called *Pegmatite*, or *giant granite*.

II. THE SYENITE FAMILY

In this family the magma much resembles that of the granite
group, except that the quantity of silica is less (50 to 65 %); hence

it is nearly or quite taken up in the formation of silicates, leaving little or none to crystallize out separately as quartz, and orthoclase is thus the chief mineral. The two families are connected by many transitional rocks.

Syenite Obsidian is indistinguishable, except by chemical analysis, from the glasses of the preceding family, but it is much less common.

Trachyte is a volcanic rock, consisting of phenocrysts of sanidine in a ground mass of minute felspar crystals, but having little or no glass, together with more or less biotite, amphibole, or pyroxene, according to which we get the varieties *mica, amphibole*, or *pyroxene trachyte*. In America the trachytes are very much less abundant than the rhyolites.

Phonolite differs from trachyte in the higher percentage of soda which it contains, and in the presence of the felspathoid nepheline or leucite, or both. The name is derived from the ringing sound which thin plates of the rock give out when struck with a hammer. Phonolites are quite rare rocks, and in this country the best-known locality for them is the Black Hills region of South Dakota.

Syenite is a thoroughly crystalline rock, without ground mass, and much resembling granite in appearance, but having no quartz. It is composed typically of orthoclase and hornblende, with plagioclase, apatite, and magnetite as accessories.   When the hornblende is replaced by biotite, the rock is called *mica syenite,* and when by augite, *augite syenite.* The name *syenite* is sometimes given to the rock we have called " hornblende granite."

Nepheline Syenite is marked by the presence of nepheline, and bears the same relation to phonolite as syenite does to trachyte, being the granitoid crystallization of the same magma.

The syenites occur just as do the granites, but are not nearly so frequent.

III. THE DIORITE FAMILY

The magma of these rocks has about the same silica percentages (50 to 65 %) as have the syenites, but the quantity of alkalies is less, while that of the lime and magnesia is greater. Hence orthoclase is absent or much less important, and the principal mineral is a soda-lime felspar. The textures display the usual variety from glassy to granitoid.

The glasses of this family (andesite obsidian) can be distinguished from those of the preceding groups only by chemical analysis, but they are rare.

Andesites are dark-coloured lavas of porphyritic or compact texture, composed of a glassy plagioclase felspar and some ferromagnesian mineral, embedded in a ground mass of felspar needles and glass. In accordance with the nature of the predominant ferro-magnesian mineral, we have *hornblende andesite, biotite andesite*, and several varieties of *pyroxene andesite*. These rocks are very common in the western United States and along the Pacific coast of both North and South America; they are named from the Andes.

The **Dacites** differ from the andesites in having quartz, and therefore a higher percentage of silica.

The **Diorites** are the plutonic equivalents of the andesites and dacites, having granitoid texture. The ferro-magnesian mineral is usually green hornblende, but augite and other pyroxenes and biotite occur in the different varieties. Most diorites have a little quartz; but when this mineral becomes abundant, it gives a *quartz diorite*, which is related to the dacites as the typical diorite is to the andesites.

IV. The Gabbro Family

In the magmas of this series the percentage of silica is much less than in the preceding groups (40 to 55%), and the quantity of alkalies is small, while that of iron, magnesia and lime is much greater. They are heavy, dark-coloured rocks. The principal minerals are a plagioclase felspar, rich in lime (labradorite or anorthite), some kind of pyroxene, magnetite, and frequently olivine. There is a wide range of mineralogical composition and many varieties of rock occur in this family.

Tachylyte is a basaltic glass, which is not at all common.

Basalt is a name of wide application covering many varieties. The basalts are very common volcanic rocks, and most of the active volcanoes of the present day extrude basaltic lavas. In texture the basalts are ordinarily porphyritic, but they may be without phenocrysts, and consist of a finely crystalline mass. The ground mass is made up of tiny crystals, mingled with a dark glass.

The basalts are closely related to the andesites and connected with them by a number of transitional forms, but in the andesites the phenocrysts are principally felspars, which is not the case in the basalts. Those basalts which contain olivine in notable quantities are called *olivine basalt;* while those in which the felspar is replaced by leucite or nepheline are called *leucite* and *nepheline basalt*, respectively.

Trap is a useful field name for various sorts of dark, granular rocks, which cannot readily be distinguished by inspection. The term is often applied to diorite and especially to diabase.

Dolerite is a coarsely crystalline basaltic rock, which is either porphyritic or granitoid in texture.

Diabase is a rock of peculiar texture; the felspar crystals are long, narrow, and lath-shaped, and contain the dark minerals in their interstices. The trap rocks of the Palisades of the Hudson, and many localities in the Connecticut valley, New Jersey, Maryland, Virginia, and North Carolina, are diabase.

Gabbro is a term which is now used comprehensively to include the coarse-grained, plutonic phases of the various basaltic rocks, which are typically composed of plagioclase and pyroxene. *Olivine gabbro* and *hornblende gabbro* are names that explain themselves. *Norite*, or *hypersthene gabbro*, contains orthorhombic pyroxene. *Anorthosite* is nearly pure labradorite in large crystals, with little or no pyroxene; great masses of it occur in Canada and the Adirondack Mountains of New York. Gabbros are present on a great scale in the Adirondacks, and occur in the White Mountains, on the Hudson, near Baltimore, around Lake Superior, in California and various parts of the West.

V. Peridotite Family

These rocks have no felspars, and in most of them the quantity of silica is below 45 %, while that of magnesia is from 35 to 48 %; they are composed almost entirely of ferro-magnesian minerals.

Limburgite is made up of crystals of augite, olivine, and magnetite, embedded in a glassy ground mass.

Augitite is a similar rock, but without olivine.

Pyroxenite is a holocrystalline, plutonic rock composed of one or more varieties of pyroxene.

Hornblendite is a similar rock made up of hornblende.

The **Peridotites** are likewise plutonic rocks which are principally composed of olivine, with iron ore and some of the pyroxenes or hornblende.

The **Serpentines** are products of alteration, and many of them have been formed from peridotites, though some have been derived from augitic rocks, such as gabbro, and others from hornblendic rocks. In rarer instances they have arisen from the alteration of acid rocks.

APPENDIX

THE PYROCLASTIC ROCKS

These rocks are formed out of the fragmental materials ejected from volcanoes. The materials are of course igneous, but the rocks themselves differ from the typical igneous rocks in several important respects. They have not been formed in their present state of aggregation by cooling from a molten mass, and in many cases they are more or less distinctly stratified. It seems best, therefore, to group them separately, under the name *pyroclastic*.

Volcanic Agglomerate, or **Breccia**, is a mass of angular blocks of lava, with which may be mingled fragments of sedimentary rocks, which the volcano has torn off from the sides of its chimney. The blocks may be loose or cemented together into hard rock by a filling of finer materials. Ordinarily the breccia is formed only near the vent, but sometimes it is developed on a great scale, as in the eastern part of the Yellowstone Park.

Tuffs are masses of volcanic ashes and dust, which accumulate in beds, either on the land or in bodies of water. Even in falling through the air, the particles are sorted, in some degree, in accordance with their size, and the tuffs are thus usually stratified, and frequently have fossils in them. When accumulated under water, the ashes are, of course, stratified and may be mingled with more or less sedimentary débris. Such subaqueous tuffs pass into the ordinary sedimentary rocks, by the gradual diminution of the volcanic material. When examined under the microscope, even the finest tuffs are found to consist of crystals and particles of glass.

The volcanic breccias and tuffs may best be classified in accordance with the nature of the component fragments. Thus, we find rhyolite tuffs and breccias, andesite tuffs and breccias, basaltic tuffs and breccias, and the like. ·

CHAPTER XI

THE SEDIMENTARY ROCKS

THE materials of which the sedimentary rocks are composed were, in the first instance at least, derived from the chemical decay or mechanical abrasion of the igneous rocks, and hence they are often called derivative or secondary. They have been laid down under water, or on land, and are therefore almost always stratified and, for the most part, are composed of rounded fragments, seldom crystalline.

Almost all the minerals which we have found in the igneous rocks also occur, in a more or less worn and comminuted condition, in the sedimentary class. However, with the exception of quartz, the great bulk of the sedimentary materials consists of simpler and more stable compounds than the igneous minerals, from the decomposition of which they have been derived. The principal minerals which compose the sedimentary rocks are quartz (SiO_2), kaolinite (Al_2O_3, $2\ SiO_2$, $2\ H_2O$), and calcite ($CaCO_3$).

Quartz is a very simple and stable chemical compound, and hence, in the ordinary process of rock decay, it remains unchanged further than being broken up into smaller pieces and rounded by the action of wind or running water. Kaolinite is derived principally from the decay of the felspars, and the lime of calcite from the complex silicates containing lime, which are so frequent in the igneous rocks. These rocks also yield the iron oxides which are so widely diffused in the sedimentary class, though comparatively seldom in any very great quantity. Very many varieties of rocks are produced by the mixture of the siliceous (quartz), argillaceous (clay), and calcareous (lime) materials in varying proportions. The sorting out of material by water, according to its chemical nature, is usually imperfect (although siliceous and calcareous concentrations are often remarkably pure), and changes from point to point, so that the sedimentary rocks have an even less definite chemical composition than have the igneous.

It is, unfortunately, not yet practicable to apply to the sedimentary rocks the arrangement employed for modern continental and marine deposits, and the most useful classification at present of the sedimentary rocks is, primarily, according to the mode of their formation, and secondarily, according to their composition. This gives two principal divisions: I, *the Aqueous Rocks*, or those laid down under water ; II, *the Æolian Rocks*, those which were accumulated on land, which are of more limited extent and importance.

The aqueous rocks may be further divided into three classes: 1, Mechanical Deposits; 2, Chemical Precipitates; 3, Organic Accumulations.

I. Aqueous Rocks

The rocks laid down under water form the larger and more important part of the sedimentary series.

I. MECHANICAL DEPOSITS

These have resulted from the accumulation of débris derived from the destruction of preëxisting rocks, carried in mechanical suspension by moving water, whether waves, currents, or streams, and dropped when the velocity of the moving water was no longer sufficient to carry them. The study of the dynamical processes has already taught us that such accumulations are forming to-day in all kinds of bodies of water, and an examination of the rocks will show that similar accumulations have been made since the beginning of recorded geological time. Mineralogically, the mechanical deposits are of two principal kinds, the *siliceous* and the *argillaceous*. The sorting power of water has been sufficient to separate them more or less completely; though we find mixtures of the two in all proportions.

a. Siliceous Rocks

In these rocks the principal component is quartz in fragments of greater or less size, either angular, or more or less rounded by wear. Of the common rock-forming minerals quartz is the hardest and the one which best resists chemical change. Small quantities of other minerals, such as magnetite, mica, felspar, garnet, etc., are generally present.

Sand is made up of fine grains of quartz, not compacted together, but forming a loose, incoherent mass. River sands and those formed by the atmospheric disintegration of rocks commonly have angular grains, due to the splitting up of the quartz fragments along preëxisting flaws, though desert and wind-blown sands are apt to be fine-grained, rounded and pitted by abrasion. Beach sand is somewhat rounded, due to the constant wash of the surf.

Sandstone is a rock of varying degrees of hardness, the grains of sand being held together by a cement. The most important cementing substances are carbonate of lime, the oxides of iron and silica. The sandstones with calcareous cement usually yield quickly to the action of the weather, because of the solubility of the cement. Those with ferruginous cement are much more durable and more highly coloured, being of various shades of red, yellow, and brown. Most durable of all are the siliceous cements.

Varieties of sandstone are produced by the conspicuous admixture of other minerals; thus, *micaceous sandstone* has abundant flakes and spangles of mica deposited along the planes of stratification. *Argillaceous sandstone* is composed of a more finely grained sand than the more typical sandstones, contains considerable quantities of clay, and is, in general, more thinly bedded. The flagstones, so largely used for pavements, are examples of such a rock, and split readily into slabs of almost any desired size.

Arkose is a sandstone containing considerable quantities of felspar in a mechanically subdivided but undecomposed state.

Gravel is composed of rounded, water-worn pebbles, varying in size from a pin-head up to cobblestones and boulders. The coarser kinds are often called *shingle*. Gravel may be composed of almost any kind of rock material, but the commonest pebbles are of quartz, because of its greater resistance to wear. Masses of quartz will be only rounded into pebbles, when other substances are ground into fine silt, or chemically disintegrated, and so washed into deeper water.

Conglomerate is a cemented gravel. Different names are given to conglomerate, according to the character of the pebbles, as *quartz conglomerate, flint conglomerate, limestone conglomerate, granite conglomerate*, etc.

FIG. 146. — Hand specimen of conglomerate, natural size

b. *Argillaceous Rocks*

Clay — Mud. — Clay consists of kaolinite nearly always with large admixtures of other substances, such as exceedingly fine grains of

x

quartz, felspathic mud, and the like. When moist, clay is plastic, differing in this respect from mud. The particles of clay and mud are extremely fine and are carried for long distances before settling to the bottom. Hence the muds and clays are distributed over wider areas than the gravels and sands, and deposits of them indicate quieter and, usually, but not always, deeper waters than the conglomerates and sandstones.

Clay is found in very different conditions of purity. *Kaolin*, or porcelain clay, is nearly pure, while *Potter's* and *Brick Clay* contain finely divided quartz, and the latter in addition, lime, magnesia, iron, and alkalies. Clays with considerable percentage of iron burn red on firing.

Fire-clay is a nearly pure mixture of sand and clay, with only traces of iron, magnesia, or lime, and therefore burns to white or buff-coloured bricks, which will resist very high temperatures. Fire-clays occur frequently beneath coal seams, representing the ancient soil in which the coal plants grew. Such ancient fire-clays are often hard rocks, and must be ground up before using.

Mudstone is a rock which is composed of solidified clay or felspathic mud, or a mixture of the two, and which crumbles rapidly into mud when exposed to the action of the weather.

Shale is a finely stratified or laminated clay rock, formed from the solidification of mud and silt. In some of the paper shales there are as many as thirty or forty laminæ to the inch, each representing a separate stage of deposition. Shales ordinarily contain more or less sand, and as this increases in quantity, they shade gradually into arenaceous shales and argillaceous sandstones, or by the increase of calcareous matter into limestones. *Bituminous shale* is coloured very dark or black by the carbonaceous matter with which it is saturated. When distilled, the bituminous shales yield hydrocarbons, and are of considerable economic importance; the carbonaceous matter may be of either animal or vegetable origin. Shales of this class grade into coals.

Marl is clay containing carbonate of lime, which rapidly crumbles on exposure to the weather.

2. CHEMICAL PRECIPITATES

Rocks which have been principally or entirely formed by chemical processes are, for the most part, of locally restricted extent, and are not at all comparable to the great masses of mechanical and organic sediments. This arises from the fact that the chemical processes occur in a conspicuous way only around the mouths of certain classes of springs (p. 191), and in closed bodies of water without outlet and subject to evaporation.

The chemical precipitates may be classed under the following heads: *a*, Precipitates of the alkalies and alkaline earths; *b*, siliceous precipitates; *c*, ferruginous precipitates.

a. Precipitates of the Alkalies and Alkaline Earths

Calcareous Tufa or **Sinter, Travertine, Stalactite, Onyx Marbles,** are all forms of carbonate of lime deposited from solution, either around the vents of springs, or by percolating waters in limestone caverns, or in lakes and streams. These deposits are made of calcite (or aragonite), are often very pure, and usually white, and more or less translucent, though they may be stained by other substances dissolved with the lime. In structure they are banded and show rings of growth, which distinguishes them from the organic limestones. The so-called "Mexican onyx" or "onyx marble" is a beautifuly banded travertine derived from ancient spring deposits.

Oölite is a limestone composed of minute spherules of carbonate of lime, cemented into a more or less compact mass, somewhat resembling fish-roe, whence is derived the name, meaning "egg rock." The spherules are made up of concentric layers of carbonate of lime, deposited from solution around some nucleus, it may be a particle of sand or dust, or a calcareous fragment. The beach rock of a coral reef (p. 264) is made in this fashion, and calcareous sinter often has a similar structure. When the spheres are larger, resembling peas in size and shape, the rock is called *pisolite*.

Gypsum ($CaSO_4.2H_2O$) is deposited from solution in salt lakes and lagoons, in which evaporation balances the influx of water (p. 224). When pure, gypsum is white, but it is often coloured gray, brown, or red, by iron stains, and it may even be black. It forms compact, crystalline, or fibrous beds, looking like limestone, but much softer and not effervescing with acid; portions of the beds may consist of transparent selenite crystals. The mineral sometimes occurs in the form of anhydrite ($CaSO_4$), but it is not known under what conditions the anhydrous sulphate has been deposited from solution.

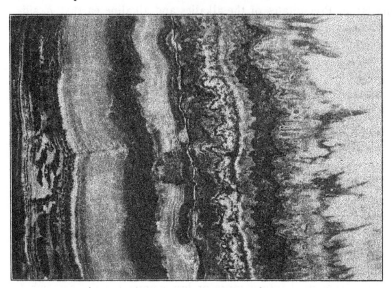

FIG. 147. — Piece of banded travertine polished, natural size

Rock Salt (NaCl) is precipitated by evaporation from the dense brine of salt lakes and lagoons, following the deposition of gypsum, which explains the very common association of the two rocks in successive beds. The salt may be present only as an ingredient of shale (saline shale), or may form thin layers, indicating brief periods of deposition, followed by freshening of the water. Again, it may occur in enormously thick masses, the result of long-continued precipitation. One such mass, near Berlin, exceeds 4000 feet in thickness. Rock salt is often very pure, and then it is transparent and colourless; but it is frequently stained by iron, or mingled with dust blown into the lake or lagoon which deposited the salt, or mixed with clay and other mechanical sediments.

b. Siliceous Precipitates

These are much less common and extensive than the calcareous, and are formed under exceptional conditions.

Geyserite, or **Siliceous Sinter**, is deposited in dense and hard masses around the mouths of geysers, partly by the evaporation of the water which holds the silica in solution, and partly by the action of Algæ (see p. 192). Large terraces of this rock have been built up by the geysers of the Yellowstone Park. Geyserite also occurs as an uncompacted white powder.

Chert (**Flint** or **Hornstone**) forms exceedingly dense and fine-grained masses, which the microscope shows to be made up of very minute grains of chalcedony mixed with more or less amorphous silica and crystals of quartz. The mode of origin of these masses is not at all well understood, but is believed to be by precipitation from sea-water.

c. Ferruginous Precipitates

Bog and **Lake Iron Ore** results from the oxidation and consequent precipitation of iron circulating in solution in the soluble ferrous condition. The deposits often have a concretionary structure, not uncommonly becoming oölitic, and consist of impure limonite, sometimes mingled with siderite.

3. ORGANIC ACCUMULATIONS

The organically formed rocks are those whose materials were accumulated by living beings, on the death of which more or less of their substance was preserved, added to by successive genera-

tions, and finally compacted into rock. In preceding chapters
we have read of these processes as going on at the present time,
in peat bogs, in the coral reefs, shell-banks, limestone plateaus,
and organic oozes of the ocean. Similar processes have been
at work in all the recorded ages of the earth's history since the
first appearance of living things, and very extensive rocks have
thus been built into the solid crust of the globe. An exact
classification would require us to place certain of these rocks
among the mechanical sediments, because the actual work of
accumulation was performed by mechanical agencies, such as
waves and currents. But it will be more convenient to examine
together all those rocks which are principally made up of organic
materials, especially as it is not always easy to distinguish the
results of one mode of formation from those of the other.

a. Calcareous Accumulations

Limestone is a very abundant, important, and widely distributed
rock, the commonest of the organic accumulations. It is composed of carbonate of lime in varying degrees of purity, hardness,
fineness of grain, and crystalline texture. . Sand or clay is frequently present as an impurity, and by an increase in these materials, the limestones pass gradually into sandstones and shales.
In some varieties of limestone the organic nature of the rock is
most obvious, shells, corals, crinoid stems, and the like being conspicuously shown, especially on weathered surfaces. In other
kinds the microscope is required to make this organic nature
clear; while in others, again, the calcareous materials have been so
ground up by the action of the waves, or so completely modified
by crystallization, that all traces of organic structure have disappeared. The example of the reef rock now forming in many coral
reefs (p. 264) is a warning that the absence of even microscopic
structure in a limestone cannot be relied upon as a proof that the
rock is not of organic origin.

The great limestones are almost entirely of marine origin, though quite extensive fresh-water limestones are known.   The chemically formed ones are never very widely extended, though they may form quite thick masses.   As a rule, the limestones are deposited in deeper water than the sandstones and shales, but not necessarily so, freedom from large amounts of terrigenous sediments being more important than depth of water.   This is shown by the great calcareous banks of the Gulf of Mexico and the Caribbean Sea (p. 259), and coral reefs are always formed in water of less than twenty fathoms in depth.

The classification of the limestones is very difficult, and cannot be readily made on any single principle;   mode of formation, purity, texture, and nature of organic material, all being employed for the purpose.

Shell Marl is an incoherent and crumbling rock, formed principally at the bottom of fresh-water lakes and ponds, by the accumulation of shells;   it frequently occurs beneath peat bogs, and is an indication that the bog arose from the choking up of a lake by vegetable growth.   When the shells are cemented into a hard rock they form a *fresh-water limestone.*

Chalk is a soft limestone of friable, earthy texture, and frequently very pure;   in colour it may be snowy white, pale gray, or buff.   The microscope reveals the fact that chalk is principally composed of the shells of Foraminifera, and closely resembles the foraminiferal oozes forming to-day at the bottom of the sea (p. 270). A chalky deposit may, however, be formed from the débris of corals ground up by the waves.

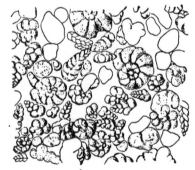

FIG. 148.—Chalk from Kansas × 45. (Drawn from a photograph by the Geological Survey of Iowa)

The ordinary massive marine limestones are named from the character of the organic material which predominates in them.   Thus, we have *coral lime-*

stone, *foraminiferal limestone,* made up of the shells of very large
extinct forms of the Foraminifera (Fusulina, Nummulites, Orbitolites, etc.), *crinoidal limestone, shell limestone,* and the like.

Though much the larger part of the limestones is of animal origin,
yet certain seaweeds contribute extensively to the formation of these
rocks, and there is much reason to believe that chemical precipitation is of greater or less importance in nearly all varieties of the
rock. Many of the massive limestones, which show little or no
sign of disturbance, are quite completely crystalline, due to the
action of water upon them. Calcite recrystallizes with the greatest
ease, and the interior of coral-masses, which are still alive on the
outside, may be so crystallized as to obliterate all traces of their
original structure.

Dolomite, or **Magnesian Limestone,** is a compact, granular rock
of white, gray, or yellow colour, composed of the carbonates of
lime and magnesia. Nearly all limestones contain some carbonate
of magnesia, but the name *dolomite* is given only to those with a
considerable percentage of that substance (5 to 20 %). How far
this rock is made up of the mineral dolomite, and how far it is
merely a mixture of the two carbonates, is uncertain, as is also the
way in which the rock was formed. Dolomite contains a much
larger proportion of magnesia than the shells or tests of any known
animals, and this ingredient must therefore have been added after
the accumulation of the calcareous organisms. Opinions differ as
to just how this has been accomplished, but probably the magnesia
has been derived from the strong brine of lagoons and salt lakes.
The frequent association of dolomite with gypsum gives additional
probability to this view. A similar process has been observed in
the lagoons of coral reefs at the present time (p. 266), and it has
been shown that dolomitization takes place much more readily
when the $CaCO_3$ is in the form of aragonite, as is the case in the
shells and tests of many marine animals.

Green Sand is not strictly a calcareous deposit, but has a natural
connection with that series of rocks. . Green sand is seen by the
microscope to be largely composed of internal casts of foraminiferal

shells in the mineral glauconite (p. 19). The dead foraminiferal shells which lie upon certain areas of the ocean floor are gradually filled up with glauconite, and then the shells are dissolved, leaving the grains of the mineral, which retain the form into which they were moulded. This process is still going on, and has been observed at several points (p. 269). Glauconite also forms on the sea-floor in nodules, quite independently of foraminiferal shells.

b. Siliceous Accumulations

The siliceous deposits of organic origin are very much less common and less extensively developed than the calcareous, because of the relatively small amount of silica which is in solution in ordinary waters, and of the comparatively few organisms which secrete shells or tests of it. Nevertheless, these beds are of sufficient importance to require mention.

Infusorial Earth is a fine white power composed of the microscopic tests, or frustules of the minute plants called diatoms. The fineness and excessive hardness of the particles make this an excellent polishing powder. Beds of this earth occur in both marine and fresh-water deposits. At Richmond, Virginia, is a celebrated deposit of this kind.

Siliceous Oozes are exceedingly rare as rocks of the land; they consist of the tests of Radiolaria, such as are now accumulating in the deeper parts of the ocean (p. 172). The only land areas in which such deposits have been found occur in certain of the West Indian Islands (Barbadoes, Cuba, and others).

Flint or **Chert** occurs in nodules or beds, especially in marine limestones, though it is also found among the sands and clays of certain fresh-water formations, as in Wyoming. Microscopic examination sometimes reveals the presence of sponge spicules and other siliceous organisms, but this is by no means always the case. As we have seen, the structureless cherts are believed to have been formed by chemical precipitation.

c. Ferruginous Accumulations

The iron deposits which can be referred to the activity of *living* creatures are of small extent and importance, but certain of the bog-iron ores are believed to be due to the agency of diatoms, Bacteria and Algæ, which extract the iron from its dissolved state.

d. Carbonaceous Accumulations

The rocks of this group are formed, almost entirely, by the accumulation of vegetable matter and its progressive, though incomplete, decay under water. This decay is cf such a nature that the gaseous constituents diminish, while the carbon is removed much less rapidly, consequently the *proportion* of the latter substance steadily rises. All the varieties of carbonaceous rocks pass into one another so gradually, that the distinction between them seems somewhat arbitrary. From fresh and unchanged vegetable matter to the hardest anthracite there is an unbroken series of transitions.

Peat is a partially carbonized mass of vegetable matter, brown or black in colour and showing its vegetable nature on the most superficial examination, though the parts which have been longest macerated are often as homogeneous and as fine grained as clay, and reveal their true nature only under the microscope.

Lignite or **Brown Coal** is a brown or black mass of mineralized and compressed peat, and though still plainly showing its vegetable nature, it does so less obviously than peat, being more carbonized. It is an inferior fuel, though often very valuable in regions where other fuel is scarce or entirely wanting.

Coal is a compact, dark brown or black rock, in which vegetable structure cannot be detected by the unassisted eye, though microscopic inspection seldom fails to reveal it. Coal is found in beds or strata, interstratified with shales, sandstones, and, less commonly, limestones. The different kinds of coal vary much in hardness and chemical composition, but they are all connected by intermediate gradations. *Bituminous Coal* has (neglecting the ash) 70 to 75 % of carbon and 25 to 30 % of volatile matters, chiefly hydrocarbons, which are driven off on destructive distillation. Under the term *bituminous* are included many varieties of coal, which differ much in their behaviour and in their value for different purposes. *Anthracite* is a hard, lustrous coal, that is nearly pure carbon (aside from the ash) and has little or no volatile matter; it burns without smoke or flame and gives an intense heat. *Semibituminous* or *Steam Coal* is intermediate in character and composition between the bituminous and anthracite varieties.

Cannel Coal does not belong in the series of coals above enumerated, but forms a very distinct variety. It occurs in lenticular patches, not in beds, and is very compact, though not very hard or heavy. This coal has from 70 to 85 % of carbon and the high proportion of 6 to 7 % of hydrogen, giving off large quantities of gas when heated, and burning with a white, candle-like flame. Even with the microscope, it is difficult to detect the vegetable structure of cannel, so thoroughly has the material been macerated. Evidently, cannel is an exceptional coal and has been formed in a somewhat peculiar way. While the ordinary coals evidently represent ancient peat bogs, which by subsidence allowed the sea, or other body of water, to overflow them and were thus sealed up and buried under sedimentary deposits, cannel was formed in pools of clear water, in which vegetable matter was accumulated and very completely disintegrated. This is shown not only by the shape of the coal patches, but also by the fossil fish not infrequently found in cannel.

The following table (from Kemp) displays the composition of the typical varieties of coal, not including the ash: —

	C.	H.	O.	N.
Wood .	50	6	43	1
Peat .	59	6	33	2
Lignite	69	5.5	25	0.8
Bituminous Coal .	82	5	13	0.8
Anthracite	95	2.5	2.5	trace

The Hydrocarbons. — The great economic importance of these substances requires that brief mention of them be made, though they can hardly be considered abundant as rocks. The natural hydrocarbons of the earth's crust belong principally to the methane series, with the general formula C_nH_{2n+2}. The most abundant are *marsh gas* (CH_4), *petroleum*, a mixture of several hydrocarbons, which are liquid at ordinary temperatures, and *asphalt*, which is solid or extremely viscous, and results from the oxidation of hydrocarbons. The hydrocarbons impregnate porous or shattered rocks, which they have invaded from below, and are frequently retained under great pressure by overlying impervious beds. Natural gas and petroleum tend to collect in the upward arches (anticlines) of folded beds, and when these reservoirs are tapped by the drill, the oil and gas rise in spouting wells which may continue to flow for many years.

While certain eminent chemists have maintained the inorganic origin of the hydrocarbons, there is no evidence that they actually were formed in this way, and nearly all geologists are agreed that they have been derived from the fatty and oily parts of organic accumulations, both animal and vegetable, at high temperatures and pressures. That such a mode of generation is at least possible has been demonstrated experimentally, and the geological mode of occurrence of these hydrocarbons renders the hypothesis of their derivation from organic substances extremely probable. Petroleum is found in rocks of a very wide range in geological time, and the various oil-fields of the United States are of very different geological dates.

Asphalt is found in beds interstratified with ordinary sediments or in cavities and fissures of the rocks, or impregnating porous limestones and sandstones.

II. ÆOLIAN ROCKS

The rocks formed on dry land form less of the earth's crust than do the aqueous rocks, but they have a special importance because of the hints which they often give as to the physical geography of the place and time of their formation.

Blown Sand is heaped up by the wind into dunes, and displays an irregular kind of stratification. The sand-grains, abraded by their contact with hard substances, are smaller, more rounded, and less angular than the grains of river or even beach sands.

Drift-sand Rock (also called æolian rock) is the consolidated sand of dunes. If the sand contains any considerable quantity of calcareous matter, the solution and redeposition of this by percolating waters binds the sand into quite a firm rock. The calcareous sands of Bermuda give an often quoted example of this.

Talus gathers at the foot of cliffs in large masses, and in many deserts it forms great sheets.

Breccia is a rock composed of angular fragments cemented by deposition of material, commonly $CaCO_3$, in the interstices; the fragments may be any kind of rock. Breccia is also found in zones of fracturing and shattering of the rocks along fault-planes, and is then called a *fault-breccia*.

Soil. — In Chapter IV it was shown that soil is mainly the residual product left by the atmospheric decay of rocks, and that its surface layers contain more or less organic matter and are filled with the roots of plants. Soils may be buried under aqueous deposits by floods, or after subsidence marine deposits may be built up upon the soils, which are then interstratified with marine rocks. Ancient soils have been frequently preserved in this manner, filled with fossil roots, and sometimes with the stumps of trees still standing upon them.

Loess. — A very fine grained terrestrial deposit, usually unstratified and with a vertical cleavage. It is quite firm and may even become hard and stony (see p. 190).

In logical order, the Metamorphic Rocks would next come up for consideration; but since we have, as yet, learned nothing of the processes by which these rocks are formed, it will be best to defer the study of this class to a future chapter, when the rocks and their mode of formation will be examined together.

CHAPTER XII

THE STRUCTURE OF ROCK MASSES—STRATIFIED ROCKS

In the preceding chapter we have studied the rocks which make up the crust of the earth, so far as that is accessible to observation. It remains for us to inquire how these rocks are arranged on a large scale, and to what displacements and dislocations they have been subjected since the time of their formation. Examined with reference to the simplest and broadest facts of structure, we find that rock masses fall into two categories: (1) *Stratified Rocks*, and (2) *Unstratified* or *Massive Rocks*. A very brief examination will show us that these two categories correspond respectively to the sedimentary and igneous divisions of the classification according to mode of origin, neglecting, for the present, the metamorphic class.

We shall begin our study of rock masses with the stratified series, because their structure and mode of occurrence are, on the whole, the simplest and most intelligible, and tell their own story. The unstratified series, on the other hand, can be understood only by determining their relation to the former.

The stratified rocks form more than nine-tenths of the earth's surface, and if the entire series of them were present at any one place, they would have a maximum thickness of about thirty miles, but no such place is known. The regions of greatest sedimentary accumulation are the shallower parts of the oceans, while those regions which have remained as dry land, through long ages, may not only have had no important additions to their surfaces, but have lost immense thicknesses of rock through denudation. The great oceanic abysses are also areas of excessively slow sedimentation, and thus the thickness of the stratified rocks varies much from point to point, a variation which has been increased by the irregularities of upheaval and depression and of different rates of denudation. Even with this irregularity in the formation and removal of the stratified rocks, it would be exceedingly difficult, if not impossible, to investigate the entire series of them, if they had all retained the original horizontal positions in which they were first laid down. In many places, however, the rocks have been steeply tilted and then truncated by erosion, so that their *edges* form the surface of the ground, and thus great thicknesses of them may be examined without descending below the surface.

Stratification, or division into layers, is the most persistent and conspicuous characteristic of the sedimentary rocks. In studying the sedimentary deposits of the present day (Chapter VII) we learned that by the sorting power of water and wind, heterogeneous material is arranged into more or less homogeneous beds, separated from one another by distinct planes of division, and the same thing is true of the sedimentary rocks of all ages. This division into more or less parallel layers is called *stratification*, and the extent to which the division is carried varies according to circumstances.

A single member, or bed, of a stratified rock, whether thick or thin, is called a *layer*, though for purposes of distinction, excessively thin layers are called *laminæ*. Each layer or lamina represents an uninterrupted deposition of material, while the divisions between them, or bedding planes, are due to longer or shorter pauses in the process, or to a change, if only in a film, of the material deposited. A *stratum* is the collection of layers of the same mineral substance, which occur together and may consist of one or many layers. However, the term is not always employed in just this sense and often means the same as layer. The passage from one stratum to another is generally abrupt and indicates a change in the circumstances of deposition, either in the depth of water, or in the character of the material brought to a given spot, or in both. So long as conditions remain the same, the same kind of material will accumulate over a given area, and thus immense thicknesses of similar material may be formed. To keep up such equality of conditions, the depth of water must remain constant, and hence the bottom must subside as rapidly as the sediment accumulates.

Usually, a section of thick rock masses shows continual change of material at different levels. Figure 149 is a section of the rocks in Beaver County, Pennsylvania, in which several different kinds of beds register the changes in the physical geography of that area. At the bottom of the section is a coal seam (No. 1), the con-

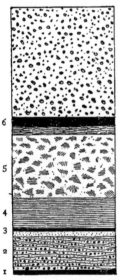

FIG. 149. — Section in coal measures of western Pennsylvania. (White)

solidated and carbonized vegetable matter which accumulated in an ancient fresh-water swamp. Next came a subsidence of the swamp, allowing water to flow in, in which were laid down mixed sands and gravels (No. 2). The accumulations eventually shoaled the water and enabled a second peat swamp to establish itself; this is registered in the second coal bed (No. 3), the thinness of which indicates that the second swamp did not last so long as the first. Renewed subsidence again flooded the bog, as is shown by the stratum of shale (No. 4) which overlies the second coal bed. Next, the water was shoaled by an upheaval, and argillaceous sands were laid down, which now form the flaggy sand-stones (No. 5) overlying the shale. The twenty-five feet of sandstone, aided by continued slow rise, silted up the water and allowed a third peat bog to grow, the result of which is the third coal seam (No. 6), while a repetition of the subsidence once more brought in the water, in which were laid down the seventy feet of gravel at the top of the section. In this fashion the succession of strata records the changes which were in progress while those strata were forming. Whether the beds, other than the coal seams, were laid down in fresh water, or in salt, by a lake, a flooded river, or the sea, may be determined from the fossils contained in those beds. In the absence of fossils it Is not always possible to make the distinction.

Somewhat similar changes in the strata may be occasioned by the steady lowering of a land surface through denudation. This diminishes the velocity of the streams, which, In its turn, changes the character of the materials which the rivers bring to the sea.

We have no trustworthy means of judging how long a time was required for the formation of any given stratum or series of strata, but it is clear that different kinds of beds accumulate at very different rates. The coarser materials, like conglomerates and sandstones, were piled up much more rapidly than the shales and limestones; so that equal thicknesses of different kinds of strata imply great differences in the time required to form them. Comparing like strata with like, we may say that the thickness of a group of rocks is a rough measure of the time involved in their formation, and that very thick masses imply a very long lapse of time, but we cannot infer the number of years or centuries or millennia required.

Geological chronology can be relative only. Such a relative chronology is given in the section that we have examined by the order of succession of the beds. Obviously the lowest stratum is the oldest and the one at the top the newest. This may be put as a general principle, that, *unless strata have lost their original position through disturbance or dislocation, their order of superposition is their order of relative age.* It is for this reason that in geological sections the strata are numbered and read from below upward.

Change in the character of the strata takes place not only vertically, but also horizontally, since no stratum is universal, even for a single continent. Our study of the processes of sedimentation which are now at work, showed us that the character of the bottom in the ocean or in lakes is subject to frequent changes, varying with the depth of water and other factors. The same is true of the ancient sea and lake bottoms, now represented by the stratified rocks of the land. Strata may persist with great evenness and uniform thickness over vast areas, and in such cases the bedding planes remain sensibly parallel. But sooner or later, the beds, whenever they can be traced far enough, are found to thin out to edges and to dovetail in with beds of a different character. When the strata are of constant thickness for considerable distances, and the bedding planes remain parallel, the stratification is said to be *regular.* In many cases these changes take place rapidly from point to point, and then the strata are plainly of *lenticular* shape, thickest in the middle, thinning quickly to the edges. Here the bedding planes are distinctly not parallel, and the stratification is *irregular.*

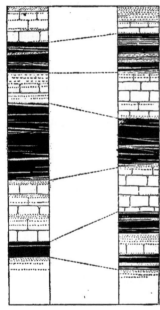

FIG. 150. — Parallel sections near Colorado Springs, Col. (Hayden)

An example of rapid horizontal changes is given in the two accompanying parallel sections (Fig. 150), taken through the same beds, only twenty feet apart. In these sections the differences of thickness of the coal seams and of the sands and clays which separate them are very striking.

The finer details of structure of the stratified rocks, such as crossbedding, ripple and rill-marks, rainprints, tracks of animals, and the like, likewise afford valuable testimony as to the circumstances under which the rocks were laid down.

Concretions, or *Nodules,* are developed after the formation of strata. They are balls or irregular lumps of a material differing from that of the stratum in which they occur. They are not pebbles, which are older than the stratum which contains them and which were embedded just as we find them, but are younger than the stratum and were formed subsequently. This is shown by the fact that the planes of stratification may often be traced through the concretion, and that fossils are sometimes found partly within and partly without the nodule. In shape the concretions vary greatly, from almost true spheres, to grotesque aggregations, but always with rounded form, and almost as great a variety of material is found among them. Very often a foreign body, like a fossil shell or leaf, forms the centre or nucleus of the nodule, which has

FIG. 151. — Concretions in Laramie sandstone, exposed by weathering. (U. S. G. S.)

been built up, often in concentric layers, around the nucleus. One form of concretion, known as a *septarium*, is divided internally by radial cracks, which were subsequently filled up with some mineral deposited from solution by percolating waters.

The agency which produces concretions cannot as yet be explained. The material of which they are made must have been scattered through the stratum and then gathered together at a later period.   Such nodules have been observed in the process of formation in modern sediments, and it has further been noticed that when finely powdered substances are mixed together, certain of them do segregate into lumps. These observations, however, merely confirm the conclusion that concretions are due to segregation of scattered material in the stratum; they give us no explanation of the fact.

FIG. 152. — Ironstone concretion, split open to show the fossil leaf which forms the nucleus; Mazon Creek, Illinois

The commonest concretions are those of clay in various kinds of rock, of flint and chert in limestone, and of ironstone in clay rocks.

DISPLACEMENTS OF STRATIFIED ROCKS

It is evident that the stratified rocks which form the land must have been changed, at least relatively, from the position which they originally occupied, since the great bulk of them were laid down under the sea. Originally they must have been nearly horizontal, for this is a necessary result of the operation of gravity. Just as a deep fall of snow, when not drifted by the wind, gradually covers up the minor inequalities of the ground and leaves a level surface, so on the sea-bottom the sediments are spread out in nearly level layers, disregarding ordinary inequalities. We must remember, however, that this original horizontality is not exact, and departures from it are not infrequent. On a large scale, these departures from the horizontal position are very slight, while those that are conspicuous are local.

Examples of such original deviations from horizontality are the following: (1) When a sediment-laden stream or current empties abruptly into a deep basin with steeply sloping sides, the sediment is rapidly deposited in oblique layers, which follow the slope of the sides (*i.e.* foreset beds). (2) Alluvial cones, or fans (p. 202), have steeply inclined layers, for a similar reason. Both of these cases resemble the artificial embankments which are built out by dumping earth or gravel over the end, until each successive section is raised to the necessary level. In such embankments the obliquity of the layers is often plainly visible. (3) Sand beaches often have a considerable inclination, as much as 8 %, and newly added layers follow this slope. (4) On a large scale, the great sheets of sediment that cover the sea-bottom generally have a slight inclination away from the land, with a somewhat increased slope along lines of depression. These slight original inclinations of sedimentary masses, either as a whole, or along certain lines, are called *initial dips*, and have an important bearing upon the results of subsequent movements of displacement.

The displacements to which strata have been subjected after their formation are of two principal kinds: (1) In the first kind, the strata have been lifted vertically upward, often to great heights, without losing their horizontality. Over great areas of our Western States and those of the Mississippi valley, the beds are almost as truly horizontal as when they were first laid down. In some of the lofty plateaus through which the Grand Cañon of the Colorado has been cut, almost horizontal strata are found 10,000 feet above the sea-level. (2) More frequent and typical are the displacements of the second class, by which the beds are tilted and inclined at

various angles, sometimes bringing the strata into a vertical position, and occasionally even overturning and inverting them.   In
the comparatively small exposures of strata which may be seen
in ordinary sections in cliffs and ravines, the rocks appear to be
simply inclined, and the strata themselves to be nearly straight.
But when the structure is determined on a large scale, it is often
found that this appearance is due to the limited area visible in one
view, and that the apparently straight beds are really portions of
great curves.   Such curves are called *folds*.

Dip. — The angle of inclination which a tilted stratum makes
with the plane of the horizon is called the *dip*, and is measured in
degrees.   The line or direction of the dip is the line of steepest inclination of the dipping bed, and is expressed in terms of compass
bearing.   For example, a stratum is said to have a dip of 15° to
the northwest. • The angle of dip is measured by means of an instrument called a clinometer, of which many kinds are in use.

Strike. — The line of intersection formed by the dipping bed
with the plane of the horizon is called the line of strike and is
necessarily at right angles to the line of dip.   (See Fig. 154.)   If a
piece of slate be held in an inclined position and lowered into a vessel of water, the wet line will represent the strike.   As long as the
direction of the dip remains constant, the line of strike is straight,
but as the direction of the dip changes, the strike changes also,
always keeping at right angles to the dip, and in such cases as the
Appalachian Mountains the lines of strike are sweeping curves.

Outcrop is the line along which a dipping bed cuts the surface of
the ground, and is, of course, due to erosion, which has truncated
the folds of strata.   Except in the case of fractured beds, which
will be considered in the following section, if there were no erosion,
there could be no outcrop.   When the surface of the ground is
level, outcrop and strike become coincident, because the surface
then is practically a horizontal plane.   With the dip remaining constant, the more rugged and broken the surface becomes, the more
widely do strike and outcrop diverge.   For a given form of surface,
outcrop and strike differ more when the beds dip at a low angle

than when the dip is steep, for when the strata are vertical, outcrop and strike again coincide, and the more nearly the strata approach verticality, the more closely do the two lines come together.

Having digressed to make these necessary definitions, we may now return to the subject of folds.

Folds present themselves to observation under many different aspects, all of which may be regarded as modifications of three principal types.

FIG. 153. — Symmetrical folds; anticline on left, and syncline on right. (U. S. G. S.)

(1) The *Anticline* is an upward fold or arch of strata, from the summit of which the beds dip downward on both sides. The curve of the arch may be broad and gentle, or sharp and angular, or anything between the two. The line along which the fold is prolonged is called the *anticlinal axis* and may be scores of miles in length, or only a few feet. This may be illustrated by an ordinary roof, which represents the two sides or limbs of the anticline, while the ridge-pole will represent the anticlinal axis.

Whether long or short, the fold eventually dies away, and thus the summit of the arch is not perfectly level, but more or less steeply inclined, and this inclination is called the *pitch* of the fold.   **In**

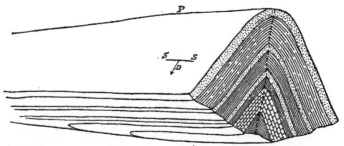

FIG. 154. — Model of anticline.   *P*, axis pitching to the left;   *S S*, line of strike; *D*, line of dip.   The dotted line is the plane of the axis.   (Willis)

accordance with the length of the axis and the steepness of the pitch the uneroded anticline is either short and dome-like, or elongate and cigar-shaped.

(2) The *Syncline* is the complement of the anticline, and in this the beds are bent into a downward fold or trough, dipping from both sides toward the bottom of the trough, which forms the

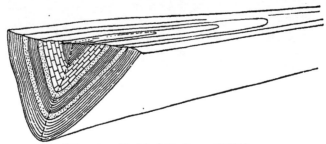

FIG. 155. — Model of syncline.   (Willis)

longitudinal synclinal axis.   As in the anticline, the axis may be long or short, with gentle or steep pitch, forming long, narrow, " canoe-shaped " valleys, or oval, even round, basins.   In section the syncline may be shallow and widely open, or with steep sides and angular bottom.

Domes and *Basins* are special cases of anticlines and synclines. The dome is an anticlinal fold in which the axis is reduced to zero, the dip of the beds being downward in all directions from the summit of the dome. As the dip changes, the strike changes, describing an oval or circle. Similarly, the basin is a syncline with axis reduced to zero, the beds dipping downward from all sides to the bottom of the basin, and the strike forming the edge

FIG. 156. — Anticline near Hancock, Md. (U. S. G. S.)

of the basin. ·The term *basin* is used in different senses, and it is necessary to distinguish carefully between a basin of folding and one which has been excavated by erosion.

It is rare to find a single anticline or syncline occurring by itself; very much more frequently they are found in more or less parallel series, each pair of anticlines connected by a syncline. At one end of the system we may find several axes converging and unit-

ing into a single fold, and they all die away sooner or later, the
pitch of the folds coinciding with the dip of the beds.

Anticlinorium and *Synclinorium.* — The system of roughly
parallel folds which are grouped together may be, when regarded as
a whole, either anticlinal, rising up into a great compound arch, or
synclinal, depressed into a great compound trough. The former
is called an anticlinorium, and the latter a synclinorium. The
secondary folds which compose one of these systems may themselves be compound and made up of many subordinate folds, the
smallest of which can be detected only with the microscope.

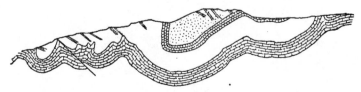

FIG. 157. -- Synclinorium, Mt. Greylock, Mass.   (Dale)

Geanticline and *Geosyncline.* — The folds and flexures which
we have so far examined are those which affect the strata at the
surface or at comparatively moderate depths. It is quite impossible that the whole crust can be involved in folds of such small
amplitude. The crust is, however, subject to flexures of its own,
which are characterized by their great width and gentle slope.
Such flexures have been named by Dana geanticlines and geosynclines, to express their importance for the earth as a whole.
The great thickness of sediments which form the Appalachian
Mountains (exceeding 25,000 feet) was laid down in an immense
geosynclinal trough, which through long ages slowly sank as the
sediments accumulated. The rate of subsidence so nearly equalled
the rate of deposition, that almost the entire thickness was accumulated in shallow water, as is indicated by the character of the rocks
themselves. Geanticlines are less easy to detect, but there is evidence to show that they do occur on an equally great scale.

Folds may be classified either in accordance with the relation

which their opposite limbs bear to each other, or with reference
to the degree of compression to which they have been subjected.
Using the first method, we may distinguish the following varieties.

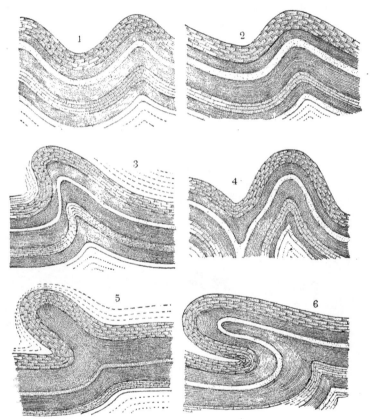

F... 158. — Diagrams of folds. (Willis.) 1. Upright or symmetrical open folds.
2. Asymmetrical fold, open. 3. Asymmetrical fold, closed and overturned.
4. Symmetrical fold, closed. 5. Closed anticline, overturned. 6. Closed anticline, recumbent

Upright or *Symmetrical*. — In this case the two limbs of the fold
dip at the same angle in opposite directions, the plane of the axis of
the flexure is vertical and bisects the fold into equal halves. In
asymmetrical, or *inclined,* folds the opposite limbs have different

angles of dip, the axial plane is oblique and divides the flexure into more or less dissimilar parts. When one limb has been pushed over past the perpendicular, the fold is said to be *overturned* or *inverted*, and when this has gone so far that one of the limbs becomes nearly or quite horizontal, the fold is *recumbent*.

FIG. 159. — Asymmetrical open fold, High Falls, N.Y. (Photograph by van Ingen)

According to the second mode of classification, we have a somewhat different series of terms; but both methods have their uses and must be employed. *Open folds* are those in which the limbs are widely separated; strata with open, gentle flexures are said to be *undulating*. *Closed folds* are those in which the limbs of the flexures are in contact and any further compression must be relieved by a thinning of the beds. *Contorted* strata are thrown into closed folds, which are connected by sharp, angular turns.

Plications are intense crumplings and corrugations of the strata. *Isoclinal folds* are those which have been so bent back on themselves that the limbs of the flexures are all parallel, or nearly so. When a series of isoclines has been planed down by erosion to a level, the strata show a continuous, uniform dip and present a deceptive appearance of being a simple succession of tilted beds. A still further compression of isoclinal folds produces *fan folds*.

FIG. 160. — Overturned sharp fold; Big Horn Mountains, Wyoming. The conspicuous white stratum is sharply bent on itself about the middle of the mountain face, in reversed Z-shape. (U. S. G. S.)

In this structure the anticline is broader at the summit than at the base and the syncline broader below than above, a reversal of the normal arrangement.

The isoclinal and fan folds may be upright, inclined, inverted, or recumbent. In the closed folds there has been such enormous compression that the same strata are of different thickness

FIG. 161. — Closed recumbent fold, East Tennessee.   (U. S. G. S.)

FIG. 162. — Plicated gneiss, Montgomery Co., Pa.   (U. S. G. S.)

in different parts of the flexure. This is especially marked in fan folding, in which the beds are much thinner on the limbs than at the summit, and sometimes the central beds in the folds have been actually forced to flow upward or downward, forming isolated masses cut off from their original connections.

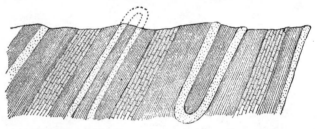

FIG. 163. — Inclined isoclinal folds, eroded. (Willis)

Besides the simple folds above described, there are frequently found complex systems of flexures, in which the compressing force has acted simultaneously or successively in different directions, producing highly complicated cross folds. These are, however, often extremely difficult to work out, and in an elementary book, intended for the beginner, it is not necessary to do more than mention them.

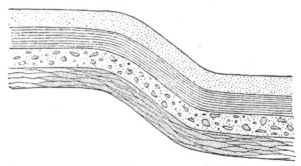

FIG. 164. — Diagram of monoclinal fold

(3) The *monoclinal fold* is a somewhat exceptional type, which can hardly be regarded as a modified form of the anticline. A mono-

FIG. 165. — Monoclinal fold, Selmas Valley.   (U. S. G. S.)

FIG. 166. — Monoclinal fold, Mead River, Alaska.   (U. S. G. S.)

clinal flexure is a single, sharp bend connecting strata which lie at different levels and are often horizontal except along the line of flexure. Folds of this character are very common in many parts of the West, especially in the high plateau region of Utah and Arizona.

CHAPTER XIII

FRACTURES AND DISLOCATIONS OF ROCKS

THE rocks are often unable to accommodate themselves by bending or plastic flow to the stresses to which they are subjected, and therefore break, usually with more or less dislocation. A simple fracture, not accompanied by dislocation, is called a *fissure*, and the strata on the two sides of the fracture are the same at corresponding levels, so that the crack was evidently made through continuous beds.

FAULTS

When the strata on one side of a fissure have been shifted in any direction relatively to the beds on the other side, so that the strata, which were once continuous across the fracture, are now separated by a vertical interval and lie at different levels, the structure is called a *fault*. We have learned that faulting is an accompaniment of many great earthquakes (see Chapter I), and these modern faults show that the movement may be in any direction, vertical, horizontal, oblique, or rotational. "Whenever the rocks of the earth's crust are subjected to strain, fractures take place in them as in any other body under similar conditions, and the different parts of the rock tend to move past one another along the fracture-planes, seeking to obtain relief from the strain and to accommodate themselves to new conditions. In this movement one part of the fractured rock-mass may move upon the other in any direction, up, down, sidewise or obliquely, according to the conditions, which are different in each instance." (Spurr.)

It is obvious that faulting displays highly complex phenomena, which cannot be adequately presented by diagrams, since three

338

dimensions are involved, and the simple cross-section may be altogether misleading. To add to the difficulty, movements frequently occur at intervals along the same fracture-planes, but, it may be, in entirely different directions, so a vertical movement may be succeeded by a horizontal one, or *vice versa*, and the final outcome may be the resultant of many different movements.

It is very unfortunate that several of the terms used in the de-

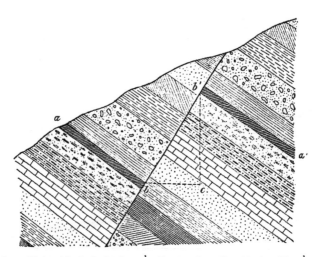

FIG. 167. — Normal fault, fault-plane hading against dip of beds; *b' c*, throw; *b c*, heave; *b b'*, stratigraphic throw, which in this case is measured along the fault-plane, because the latter happens to be at right angles to the bedding-planes. The angle *bb'c* is the angle of hade; *b'bc*, the angle of dip. Foot-wall to the right of the fault, and hanging wall to the left. — N.B. The line *b' c* should have been drawn from the *top* of the obliquely lined bed, slightly increasing both throw and heave

scription of faults, adopted from the miners by English geologists, should in American practice have acquired meanings quite different from those originally given to them, so that the student finds in different books the same terms employed in different senses. The following definitions are those commonly to be found in the text books: —

Faults are usually inclined, and the angle of inclination,

measured from a *vertical* plane, is called the *hade*, or *slope*, of the
fault, while the *dip* of the fault, like that of a stratum, is measured
from a horizontal plane, and is thus the complement of the hade.
For example, if the fault is vertical, the hade = 0, and the dip = 90°;
if the fault is horizontal, the hade = 90°, and the dip = 0, while a
hade of 45° gives a dip of the same amount. The side on which the
beds lie at a higher level than their continuations on the other side
of the fault-plane is called the *upthrow side* and the other is the
downthrow side, without reference to the actual direction of the
movement. Owing to the inclination of the fault, the rocks on one
side project over those on the other, and are hence called the *hang-*

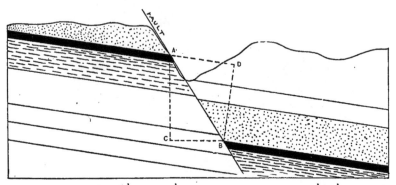

Fig. 168. — Normal fault hading with dip of beds. *DB*, stratigraphic throw; *AC*,
throw; *CB*, heave

ing wall, and the side which projects underneath the other is called
the *foot-wall*. Either the hanging or the foot-wall may be on the
upthrow or the downthrow side, according to the nature of the fault.
The vertical displacement between the fractured ends of a given
stratum is called the *throw* (*b'c*, Fig. 167) and the *heave*, or horizontal throw, is the horizontal distance through which one end of a
faulted bed has been carried past the corresponding end on the
other side of the fault-plane (*bc*, Fig. 167). When the movement
has been vertical, the heave is due to the obliquity of the fault and
therefore increases, in proportion to the throw, as the hade increases. A fault with plane perpendicular to the surface has no

heave, for it has no hade. *Offset* is the distance between the two corresponding ends of a faulted bed, measured on a horizontal plane and usually applied to the outcrop˙ (see Fig. 177, III). The *stratigraphic throw* is the thickness of beds which is included between the two fractured ends of a faulted stratum and is taken at right angles to the bedding-planes. (*DB*, Fig. 168.)

The throw of faults varies greatly in different cases, from a fraction of an inch up to thousands of feet. In those of small throw

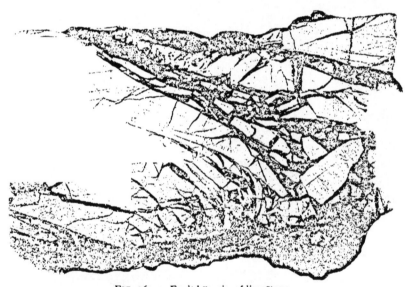

FIG. 169. — Fault-breccia of limestone

the plane of fracture is frequently a clean, sharp break; but in the greater faults the rocks in the neighbourhood of the fault are often bent, crushed, and broken, forming a confused mass of fragments, large and small, which may be cemented into a breccia, which is then called *fault-breccia* or *fault-rock*. In soft rocks the fault is always closed by the immense weights and pressures involved, but in rigid rocks it may remain partly open, especially if the break be not a plane, but of curved, warped, and irregular course, as is usually the case. The term *fault-plane* is thus rarely accurate, though it is constantly employed as a matter of convenience. **In** faults of considerable throw the ends of adjoining strata are apt to

*FIG. 170. — Vertical slickensides; Rondout, N.Y.   (Photograph by van Ingen)

be bent more or less sharply upward or downward, in accordance with the direction of movement.   This is *drag* (Fig. 178).

In the more rigid rocks the friction of the masses grinding against one another on the fault-plane grooves and polishes them, which produces the characteristic appearance known as *slickensides*. The grooves or striæ indicate the direction of the *last* movement along the fault-plane, for ordinarily this last movement obliterates the earlier striæ, but does not always do so, for we sometimes find cases in which two, or even three, sets of striæ are preserved, each demonstrating motion in a different direction.

FIG. 171. — Limestone faulted on bedding-planes, with vertical slickensides; Rondout, N.Y. (Photograph by van Ingen)

In stratified rocks faults usually break across the strata, separating each bed into two or more parts, according to the number of dislocations, yet sometimes the fault-planes coincide with the bedding planes, which are slickensided, pushing each stratum upon those above and below it, but without fracture.

The preceding discussion of faults deals only with those of stratified rocks, but this is merely because such displacements are the easiest to observe. As a matter of fact, dislocations may and do traverse rocks of all kinds, but it may be quite impossible to detect a fault, even one of great throw, in a thick, homogeneous, crystalline mass, for lack of any definite points of reference on the two sides of the fault-plane. On the other hand, in thinly laminated rocks

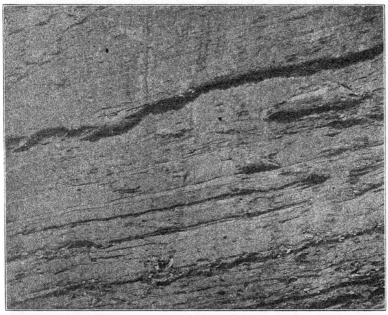

FIG. 172. — Minute vertical fault, of recent date, interrupting glacial striæ. (G. F. Matthew)

with well-defined colour lines the most minute displacements are strikingly apparent.

It is customary in geological literature to apply the term *fault* to any dislocation of the rocks, in which the broken ends of the beds are carried past one another, yet, used in this manner, it includes structures of very different significance, produced in dissimilar ways. It therefore seems advisable to distinguish between the two main classes of structure by removing *thrusts* altogether from the category of faults. Used in this restricted sense, faults are those dislocations which generally tend toward the vertical and occur in horizontal, inclined, or but slightly folded strata, which, to all appearance, have been subjected to tension rather than compression, though the latter frequently occurs locally, while thrusts tend to be horizontal and are found in regions of violent compression. The classification of faults is even yet the subject of vigorous discussion, and no general agreement has been reached, so that the following scheme is to be taken as merely tentative, though it departs but little from the customary plan, except in the complete separation between thrusts and reversed faults, which has been advocated by many writers. In the present state of knowledge, however, any scheme of classification has an undue appearance of exactness.

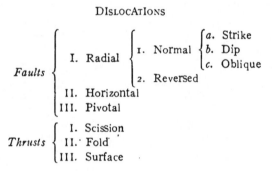

DISLOCATIONS

$$
\text{Faults} \begin{cases} \text{I. Radial} \begin{cases} \text{1. Normal} \begin{cases} a. \text{ Strike} \\ b. \text{ Dip} \\ c. \text{ Oblique} \end{cases} \\ \text{2. Reversed} \end{cases} \\ \text{II. Horizontal} \\ \text{III. Pivotal} \end{cases}
$$

$$
\text{Thrusts} \begin{cases} \text{I. Scission} \\ \text{II. Fold} \\ \text{III. Surface} \end{cases}
$$

I. **Radial Faults** are those in which the principal component of the movement has been upward, downward, or both, though subordinate movements of tilting and rotation frequently occur. However, it is not always possible to tell from the observed data, whether the chief movement was vertical or horizontal, for in certain circumstances the results are so deceptively similar. According to prevalent belief, the radial faults comprise the great majority of dislocations, and horizontal movements are comparatively rare, but recent exact studies indicate that this opinion is at least an exaggeration and that horizontal movements are far from uncommon.

1. NORMAL FAULTS (also called gravity faults) are those in which the fault-plane inclines or hades toward the downthrow side, which forms the hanging wall. " It seems best to use the term *normal* to cover those faults in which, using the horizontal plane as datum, the hanging wall has dropped relative to the foot." (J. A. Reid.) Locally, at least, a normal fault implies an extension of an arc of the earth's surface; for the beds occupy a greater space, measured across the fault, than they did before faulting occurred. The normal faults may be divided into three groups, as shown in the table.

FIG. 173. — Trough-fault of very small throw.   (U. S. G. S.)

a. Strike-faults are those which run parallel, or nearly so, to the strike of the beds. To this group belong the great normal faults, great both as to length and throw, though they may be extremely minute, or of any order of magnitude between these extremes. They may die out in a few yards, or run for hundreds of miles, and may be simple or compound, single or branching. A *compound fault* is made up of a number of parallel dislocations, placed close together, which may hade in the same, or in opposite directions, but in the latter case one hade prevails over the other. A series of parallel faults, wider apart than those of compound faults, and all hading in the same direction, are called *step-faults*. If two parallel dislocations are inclined toward each other, they form a *trough-fault* and include a wedge-shaped mass of rock, which is on the downthrow side of both displacements, while if they incline

Fig. 174. — Small faults in the roof of a mine, Idaho. Near the right end, a tiny Horst. (U. S. G. S.)

away from each other the inclined mass is on the upthrow side of both. For the latter structure, which is the converse of the *trough-fault*, there is no English term — many geologists have therefore adopted the German word *Horst*.

However long it may be, a simple fault sooner or later dies away by diminution of the throw, until it vanishes. This implies that the rocks are bent along the dislocation, upward on the upthrow side and downward on the downthrow side. It is comparatively sel-

dom that the upthrow side of a fault is left standing as a line of
cliff, or *fault-scarp*, which depends upon the length of time during
which the scarp has been exposed to denudation.   In the majority

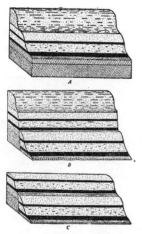

of instances the two sides of the fault
are worn to the same level or to one
continuous slope, so that there is no
feature of surface topography to indicate the existence of the fault, which
must be inferred from the effects of the
dislocation upon the outcrops of the
strata involved in it.   These effects
are determined by the direction and
throw of the fault, and by the attitude
and dip of the beds.   Strike-faults of
moderate throw which traverse horizontal strata, or strata inclined so that
the dip of the beds and the hade of the
fault are in opposite directions, repeat
the outcrop of the beds, bringing them
again to the surface, as shown in Fig.
175.   When dislocated by a series of
step-faults, a given stratum has a
number of outcrops greater by one

FIG. 175. — Effect of strike-fault
on outcrop.   *A*, before faulting; *B*, with fault scarp standing; *C*, with upthrow and
downthrow sides worn to a
continuous slope.   (Model by
Sopwith)

than the number of faults.   When the surface has been worn
down to one continuous slope,
such a repetition of the outcrops may be very deceptive.
In Fig. 176, for example, the
observer might easily be misled
into believing that seven seams
of coal were cropping out on the
hillside, whereas in reality there
are only two such seams with
outcrops repeated by faulting.

FIG. 176. — Effect of Step-faults in re-
peating outcrops.   (Model by Sopwith)

When the hade of a strike-fault is in the same direction as the dip of the beds, a certain number of the latter abut against the fault-plane and fail to reach the surface, their outcrops being cut out (Fig. 168). In great faults, with displacements of many thousands of feet, the beds cropping out on the two sides of the fault are entirely different. The deep-seated strata which are exposed by denudation on the upthrow side, are carried so far down on the downthrow side that they do not reach the surface at all, or, at least, do not crop out in the neighbourhood of the fault.

b. Dip-faults. These are, in general, parallel to the dip of the beds and therefore cross or branch out from the strike-faults of the same region, more or less at right angles; they are less important than strike-faults, having generally a smaller throw and less length. Dip-faults cut across the strike of the beds and interrupt the continuity by producing an offset in the outcrop. The outcrop of a given stratum ceases abruptly at the fault-line and when found on the other side is seen to be shifted for some distance along that line. How such a horizontal shifting is brought about by a vertical movement, is shown by the model (Fig. 177). In I is seen the model before faulting, the black band representing a dipping bed; in II the block has been faulted, the upthrow side remaining as a fault-scarp, while III shows the scarp removed by denudation. On the downthrow side the outcrop is shifted away from the dip of the beds and on the upthrow side toward the dip.

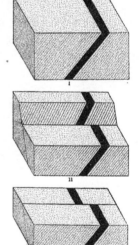

FIG. 177.— Model showing stratum offset by dip-fault, I, before faulting; II, with fault scarp standing; III, with scarp removed by erosion

When a dip-fault cuts across eroded folds, the distance between the outcrops of the same stratum in the two limbs of an anticline is increased on the upthrow side, diminishing on the downthrow; in the synclines this arrangement is reversed. This is due to the fact that, when both sides are planed down to the same level, the surface of the ground cuts the beds at a lower *stratigraphic* level on the upthrow than on the downthrow side, and as the limbs of an anticline diverge downward, the outcrops will be the more widely separated, the lower the level at which they reach the surface. , The limbs of a syncline, on the other hand, converge downward and the effect of the fault is therefore just the reverse of what occurs in the anticline.

FIG. 178. — Drag of strata near fault-plane.   (U. S. G. S.)· The hole is an artificial opening along the fault

   c. Oblique Faults. — Dip-faults do not always follow the dip, and strike-faults often deviate considerably from the strike of the beds, and sometimes the fault is neither one nor the other, but midway between the two, and then is called an oblique fault.   The outcrop of a given bed, obliquely faulted, has an offset, as in the case of a dip-fault, but if the fault inclines with the dip of the strata, there is a gap between the two adjacent ends of the outcrop, the gap widening as the line of fault approximates that of strike. If the fault hades in the opposite direction from the dip, the two ends of the outcrop overlap.

2. REVERSED FAULTS. This group, as usually defined, is made to include thrusts (*q.v.*), but the latter are here excluded and the term *reversed fault* comprises only those true, radial faults in which the hanging wall has been pushed up over the foot-wall and therefore forms the upthrow side. A reversed fault, which almost always coincides with the strike of the beds, implies a local compression, for the beds occupy less space than before dislocation. In a large faulted area, normal and reversed faults frequently occur together, compression in one place compensating for tension in another, and the two kinds of displacement appear to have been formed at the same time, or in close succession.

II. **Horizontal Faults** (or *Heave-faults*). — In displacements of this class the principal direction of movement is horizontal, and in horizontal strata may readily escape detection. When the strata are inclined, a horizontal displacement produces effects which in cross-section cannot be distinguished from those of ordinary normal and reversed faults, except when the striæ of slickensides remain to indicate the actual direction of movement. The deceptive appearance is exactly the counterpart of that which results from the vertical movement of a dip-fault, by which the offset of the outcrop is brought about. This is illustrated by the model, Figs. 179, 180, which shows that if the hanging wall is moved in a direction opposite to that of the dip of the beds, an apparently normal fault results, while if it is moved in the same direction as the dip, an apparently reversed fault is produced. Horizontal faults do not form scarps, for there is no vertical movement, but in certain cases, as shown by the striæ, the movement is obliquely upward. It is thus evident that what would ordinarily be regarded as normal and reversed faults of the typical kind may readily be formed by the same movement. For a long time these heave-faults were supposed to be very rare, but they are now known to be quite common, and doubtless very many faults,

which have been regarded as normal or reversed, will on further
study turn out to be heaves.

III. **Pivotal Faults.** — In faults of the preceding groups there

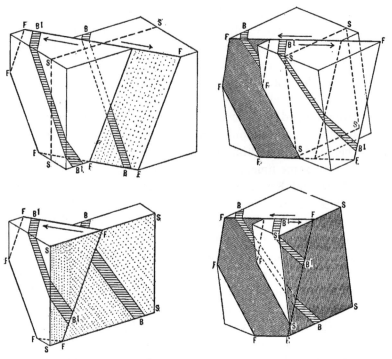

FIG. 179. — Model illustrating horizontal faulting, with hanging wall moved
against the dip and producing an
apparently normal fault. Upper figure (modified from Ransome) block
after dislocation. Lower figure, cross-
Section on plane *SSSS* : *BB*, *B'B'*,
Stratum of reference. *FFFF*, fault-
plane

FIG. 180. — Model illustrating horizontal faulting, with hanging wall moved
in direction of dip and producing an
apparently reversed fault. Upper figure (modified from Ransome) block
after dislocation. Lower figure, cross-
Section on plane *SSSS*. Lettering as
in Fig. 179.

is apt to be more or less rotation, because of unequal friction and
resistance of the walls, but in certain cases this movement of rotation is the principal one, exceeding any movement of translation,

and these are the pivotal faults. The result of the movement is that the hanging wall drops on one side of the axis of rotation, producing a fault of the normal type, and rises on the other side, forming a reversed fault. Thus, one and the same fault is " normal " in one part of its course and " reversed " in the other.

Systems of faults of different dates frequently traverse the same region, intersecting and crossing one another at all angles. An

FIG. 181. — Horizontal slickensides, Oklahoma.   (U. S. G. S.)

older fault crossed by a newer one is itself faulted and offset. The intersecting faults divide the rocks into large and small *fault-blocks*, which are generally tilted in different directions, but, as a rule, their beds are not strongly folded. As was pointed out in the discussion of earthquakes (Chapter I), this mosaic of fault-blocks is an important element in the production of seismic disturbances. Though faults often occur in regions of strata that are not folded,

2 A

there is, nevertheless, frequently a close connection between faults and folds, especially monoclinal flexures, which so often pass into faults, the strata bending along part of their course, fracturing and dislocating in another part.

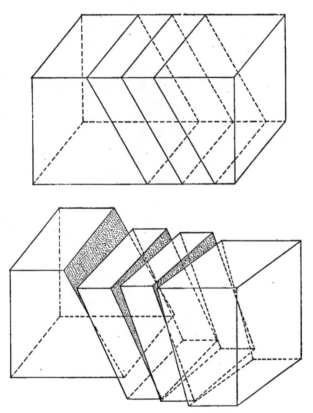

FIG. 182. — Model illustrating pivotal faulting. Upper figure, before dislocation. Lower figure, after dislocation. (J. A. Reid)

THRUSTS

A thrust is like a reversed fault in that it is the result of compression and that the inclination or hade of the fault is toward the upthrow side, which is the hanging wall, but differs in the tendency to a horizontal position of the plane of fracture and in the association with violent folding and plications. In his latest work on the subject Mr. Willis divides thrusts into the following three groups: —

I. **Scission-thrusts** are those in which the fault-plane is independent of any older structures, and occur chiefly in the crystalline

FIG. 183. — Fold thrust, near Highgate Springs, Vt. (U. S. G. S.)

schists (metamorphic rocks) and granite, and, as a rule, depart but little from horizontality. Thrusts of this kind are developed on a great scale in the southern Appalachians, especially in eastern Tennessee, where thrusts of 20 miles or more have been

observed.   On an even more gigantic scale they occur in the Highlands of Scotland and Norway, where the movement of translation
amounts to 75 miles.

II. **Fold-thrusts** are intimately connected with folds and occur
only among folded sedimentary rocks; they may arise by plication
and inversion, usually between an overturned anticline and the ad-

FIG. 184.—Steep fold-thrust. Big Horn mountains, Wyoming. (U. S. G. S.) Strata
of hanging wall, on left of thrust-plane, show drag

joining synclines. Thrusts of this character are very widespread and
common in regions of strongly folded and plicated strata and represent the breaking and dislocation of rocks in the process of folding.
The central and southern Appalachians, Arkansas, and Oklahoma,
and the northern Rocky Mountains are the regions of the United
States where great thrusts of this kind are most frequent.   In the
latter mountains, on both sides of the international boundary

line, great fold-thrusts have carried masses of strata at least eight miles to the eastward.   In fold-thrusts the plane of dislocation is somewhat steeper than in scission-thrusts and sometimes approximates the steepness of typical reversed faults. (See Fig. 49, p. 129.)

III. **Surface-thrusts**, as their name implies, are formed at the earth's surface, where a rigid, gently inclined stratum that crops

FIG. 185. — Surface-thrust of small displacement.   (U. S. G. S.)

out of the ground is subjected to lateral compression and thrust forward over the underlying beds.   Such a condition arises, for example, when an anticlinal fold has been planed down by erosion, so that some of the beds lying on the flanks of the fold are truncated and crop out freely; when renewed compression is applied to the fold, the more rigid bed will be pushed forward over the beds beneath, or it may be fractured and overthrust not far below the surface, as shown in the figure (Fig. 186).   Instances of surface-thrusts have not been identified in great numbers, though there is no reason to doubt that they are common, for they cannot always be distinguished from fold-thrusts without careful study.   They have been found in the southern Appalachians and, on a great scale, in Montana.

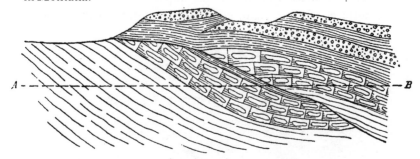

FIG. 186. — Surface-thrust, Holly Creek, Georgia.   (Hayes)

THE CAUSES OF FOLDING AND DISLOCATION

Like all processes which take place deep within the interior of the earth, the causes of crustal deformations are very obscure and there is much difference of opinion concerning them.   The view which is held as to the physical state of the earth's interior will necessarily condition the explanation of folding and faulting, which is but one special aspect of the general problem.   Any complete theory must of course contain a satisfactory solution of all the problems involved, but such a theory has still to be propounded and for the present we must be content with tentative hypotheses.

The first step in the inquiry is to determine the direction in which the folding force acted.   At first sight, it might seem natural to suppose that the direction of the force was vertically upward, acting with maximum intensity beneath the anticlines and with minimum intensity beneath the synclines.   But such an explanation could apply only to open, symmetrical, and simple folds, and even in these cases is not satisfactory.   Folded strata must either occupy less space transversely than they did before folding, or else they must have been stretched and made much thinner, but a comparison of continuous beds, in the flexed and horizontal parts of their course, shows no such thinning. Again, such an explanation is obviously insufficient to account for closed, inclined, and inverted folds, for contortions and plications, and for flexures of different orders, one within another.

If the folding force did not act vertically, it must have acted horizontally, and this is the explanation now almost universally accepted. A horizontally acting force would compress and crumple up the beds, producing different types of flexure in accordance with varying circumstances. Furthermore, the microscopic study of intensely folded rocks shows that they have actually been compressed and mashed, and the minutest plications are visible only under the microscope.

Assuming, then, that the folding force was one of compression and acted horizontally, we have next to consider the circumstances which modify the result, producing now one form of flexure or fracture, .now another. Such modifying circumstances are the depth to which a given stratum is buried, its thickness and rigidity, the character of the beds which are above and below it, and the intensity and rapidity with which the flexing force is applied. When in a mountain region one sees the manner in which vast masses of rigid strata are folded and crumpled like so many sheets of paper, one perceives the enormous power which is involved in these operations and the gradual, steady way in which that power must have been exerted. When strata are buried under a sufficient depth of overlying rock to crush them, they become virtually plastic and yield to the compressing force by bending. The movement would seem not to be a true molecular flow, but rather a gliding of the mineral particles one upon another. At such relatively great depths cavities cannot exist, and if the compressed rock should be broken by the compression, the particles are again welded together into a firm mass. We may accordingly distinguish a *shell of flowage*, in which the rocks all yield plastically, a more superficial *shell of fracture*, in which all but the softest rocks break on compression, and between the two a *shell of fracture and flowage*, in which some rocks break and others bend, according to their rigidity. The depth of the zone of flowage is estimated at 20,000 to 30,000 feet below the surface.

FIG. 187. — Folded and fractured iron ore and jaspilite, Lake Superior region. About ½ natural size

Strata which have not been buried to a sufficient depth to make them plastic, will yield to compression by breaking, though whether a given bed is faulted or flexed, will often depend upon whether the folding force is applied slowly or with comparative rapidity. A force long acting in a slow and steady fashion will

FIG. 188. — Plicated beds on unfolded ones; Mineral Ridge, Nevada.   (U. S. G. S.)

produce folds, when the same force applied more suddenly would shatter the beds. Near the surface, under light loads, rigid rocks will always break rather than bend, when compressed. Different stratified rocks differ much in their rigidity, and hence a load which is sufficient to cause one bed to bend and flow, when later-

ally compressed, will leave another unaffected, or cause it to break, if the compressing force overcomes its strength. In Bald Mountain, New York, the stiff limestones are left unchanged by a pressure which has crumpled and contorted the soft shales.

A certain amount of gentle folding may take place immediately at the surface and has actually been observed in process of formation, even in rigid rocks. In Wisconsin the limestone bed of the Fox River suddenly arched upward into a low anticline,

FIG. 189. — Plicated limestone, with sheet of igneous rock, near Rockland, Maine. (U. S. G. S.) The limestone has flowed under compression, and the igneous rock has fractured

crushing and bending the steel columns of a mill which had been built at that spot. The bed of the Chicago drainage canal, also in limestone, curved upward in similar fashion, when the excavation had removed the overlying load. Very many other surface folds are demonstrably of very recent origin.   Folds of this character are due to a gentle and gradual compression, to which the strata yield by a read-justment of the joint-blocks of which they are made up.

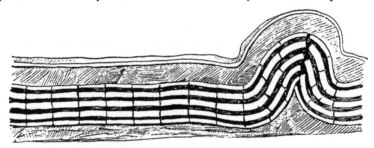

FIG. 190. — Model showing the slip of folded beds upon one another.   (Willis)

A factor of much importance in determining the character and position of folds is the mode in which the strata were originally laid down.   As we have already learned, the sheets of sediment which cover the sea-bottom are, on a large scale, nearly level, but they often show slight depar-tures from such horizontality along certain lines.   These *initial dips* often determine the place of flexures, because they divert the com-pression from its horizontal direc-tion.

The effects of lat-eral compression are shown in Figs. 190 and 191, taken

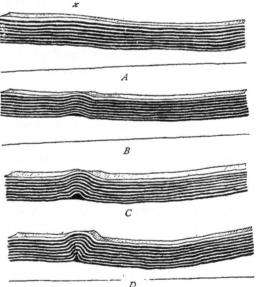

FIG. 191. — Model showing effects of lateral compres-sion.   *A*, before folding, with slight initial dip at *x*; *B*, *C*, *D*, in various stages of compression.   (Willis)

from the models experimented on by Mr. Willis, which, when strongly compressed, imitate with remarkable accuracy the structures which may be observed in folded rocks. Fig. 190 shows that in folding, the beds must slip upon each other, as is proved by the lines perpendicular to the bedding-planes, which were continuous before folding, but in the anticline are broken by the differential motion of the layers, each bed rising farther up the slope than the one beneath it. The same thing must occur in folded rocks, which sometimes show polished bedding-planes, due to the slipping of the beds upon one another. The series A to D (in Fig. 191) represents a model before and in various stages of lateral compression, and exhibits the effect of the slight initial dip at x in determining the position of the anticlinal fold, which is developed by compression. The formation of one fold assists in the development of another, for it both changes the direction of compression and redistributes the load of overlying strata. The arch of the anticline lifts the load and diminishes the weight upon the beds that lie beneath the flexure, but increases the weight upon the lines from which the arch springs.

There is much independent evidence to show that folding is a gradual process. The force exerted is enormous, but so is also the resistance to be overcome, and a steady or oft-renewed compression, acting upon strata under a great load of overlying masses, will produce regular flexures, where a sudden compression, however intense, could only shatter them.

Thrusts are likewise due to lateral compression, by which the rocks have been sheared and broken, and the beds on one side of the plane of fracture have been thrust up over those on the other. A plication or overturned fold may often be traced into a thrust, in a way that shows the direction of movement to have been the same in both fold and fracture. Numerous experiments also show that lateral compression will produce just such structures. A reduction of the overlying load, by diminishing the plasticity of the rocks, will occasion shearing and overthrusts, when, under a greater load, the same strata, exposed to an equal force of compression, will simply flex and bend.   As we have seen (see Fig. 186), an anticlinal fold whose load has been reduced by erosion, will, on renewed compression, fracture and develop a thrust.

While thrusts are associated with violent folding, overturning and plication of strata, faults occur, as a rule, in regions where folding is absent, or very subordinate, or, if in areas of folded rocks, the faults were, generally at least, formed at a period more or less subsequent to the period of folding.   The association of the differ-

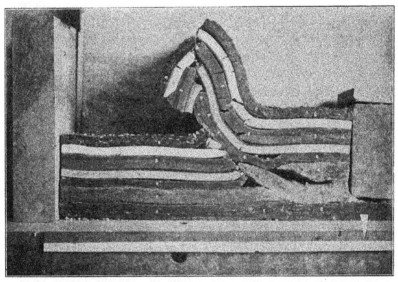

FIG. 192. — Model illustrating the development of a fold-thrust

ent classes of faults shows that locally tension and compression may be generated in the same area and probably simultaneously. Reversed and horizontal faults are due to compression, the force in the latter case acting parallel to the fault-plane and in the former case across it, while normal faults are the result of a local tension. It is still an open question how these local compressions and tensions are generated.

One explanation is that such phenomena are developed in regions that have been raised by upwarping above a position of adequate support, whence results a system of fractures and the settling and readjusting of the fault-blocks. " If we endeavour to restore a system of normal fault-blocks to the relations which they may have had before faulting, we must commonly construct a dome-shaped figure of some sort, whose surface occupies more space than the displaced blocks occupy. That is to say, in any cross-section an elongated arc has in consequence of faulting been brought into a shorter chord, commonly by bringing the narrower parts of wedges into juxtaposition. . . . The doming may produce elongation or stretching in superficial sections at least, and thus tend to provide the opportunity for the development of planes whose attitude is that of the normal fault-plane. In so far as the inadequacy of support gives rise to vertical displacements *pari passu* with the stretching, the blocks will adjust themselves with reference to each other by relative displacement in the direction of maximum stress and least resistance. . . . In this process elongation is the primary condition and a settling down of the blocks is a result. Through that settling a secondary effect of compression is set up. The large masses become wedged against one another, and as their magnitude is such that their own weight is sufficient to deform them, they suffer more or less folding and even reversed faulting as an after effect. . . . We may reasonably expect to find some reversed faulting in connection with normal faulting wherever the latter is developed on a truly large scale. The absence of folding or reversed faulting could only follow in case the blocks were free to move outward to the extent demanded by the elongation due to the attitudes of the normal fault-planes." (Willis.)

In some cases, normal faults are due to pressure acting along and parallel to the fault-plane and causing the strata to arch gently upward on the upthrow side, downward on the downthrow side. Faults of this class have been observed in central Pennsylvania, Tennessee, and Alabama. Though due thus to pressure, a tension is developed across the fault-plane.

Quite a different type of explanation seeks to account for the phenomena of faulting by the transfers of molten magmas deep

within the earth.   In certain regions, as in the Tonopah district
of Nevada, it has been made exceedingly probable that such transfers are the actual cause of the fracturing and dislocation of strata,
and some observers would give this principle a widespread, if not
a general, application.   " Not only are the violent migrations of
igneous material the cause of complex faulting, but also it is most
reasonable to conceive that the deeper and more gradual movements of the subcrust are the cause of the larger fault systems.
. . . Given this cause of faulting, the heretofore puzzling facts are
satisfactorily and easily explained.   Compression and tension still
remain true causes of faulting, but mainly as local and proximate
ones.   The common expression, *tilting* of fault-blocks, attains a
deeper significance, for this tilting may be more largely the result
of subcrust migrations than of the mere force of gravity.   Cases
of horizontal motion and pivotal motion become simple, for there
is no necessary unchangeable relation between the direction of the
force and the position of the fracture-plane."   (J. A. Reid.)

Even if it be granted that the effective forces which cause the
folding and dislocation of rocks are, in the last analysis, a horizontal or tangential compression, it still remains to inquire how this
great force was generated.   There is no general agreement concerning the solution of this problem.   For a long time it was supposed
that a satisfactory solution was given by the contraction of the earth
from cooling, and perhaps the majority of geologists still adhere
to this view, which may be briefly expressed as follows: The
earth's crust long ago reached a state of fairly constant temperature, but the highly heated interior is steadily cooling by radiation,
and consequently contracting.   As the crust cannot support itself,
it must follow the shrinking interior, and is thereby crowded into
a smaller space, thus setting up irresistible lateral stresses.   If the
earth were homogeneous, its surface would be wrinkled all over,
as is the skin of a withered apple, of which the pulp contracts from
loss of water, crowding the skin into a smaller space; but as the
crust is heterogeneous, with special lines of weakness, the compression results in the formation of long, narrow belts of folded

rocks, separated by broad areas of relatively little disturbed strata.

The contractional hypothesis has been attacked from many points of view, and very serious doubt has been thrown upon its adequacy to explain the facts, and even upon its reality. A modification of this hypothesis has been proposed by Professor Chamberlin, who regards the downward movement of segments of the earth's crust as primary and the horizontal movements as incidental to the former. The lithosphere is regarded as made up of a number of heavier and stronger segments, the surface of which forms the ocean basins, and of lighter and weaker segments which, on the surface, are the continental platforms. The general shrinkage of the earth causes the oceanic segments to descend, compressing the lighter continental segments and producing belts of folded rocks upon their borders.

The study of the radio-active substances and their distribution in the rocks of the earth's crust has led some observers to the conclusion that the earth's loss of heat is fully compensated by radio-activity and that since a very early period in the history of the globe, there has been no shrinkage at all, a standpoint which others have reached from entirely different lines of evidence and reasoning.

An elementary text-book is not the proper place for the discussion, or even the full statement, of all the different hypotheses which have been proposed in explanation of these most difficult problems. Suffice it to say that all the questions concerning the mechanics of the earth's interior are bound up together in an indivisible unity and that the full and satisfactory answer to any one question will involve the solution of all the cognate problems.

CHAPTER XIV

JOINTS. — STRUCTURES DUE TO EROSION

WITH the exception of loose, incoherent masses, such as soil, gravel, sand, etc., all rocks which are accessible to observation, are divided into blocks of greater or less size by systems of cracks and crevices, which are known as *joints*. These may be easily observed in any stone-quarry, where they are taken advantage of in getting out the stone.

In the igneous rocks all the division planes which separate the blocks are true joints, which vary greatly in their number and manner of intersection and in the consequent shape of the joint-blocks. Fine-grained basalts display a very general tendency to *columnar jointing*, forming more or less regularly prismatic columns, which are commonly hexagonal. Several modern lavas (see p. 76) display these hexagonal columns, as do the ancient basalts of very many regions. In certain cases, as in the famous Giant's Causeway of Ireland, the columns are divided transversely by concave joints, giving a ball-and-socket arrangement which has a curiously artificial appearance. Although the regular hexagonal columns are most frequent among the fine-grained basalts, they also occur in the coarser rocks of the gabbro family and in other families also. The acid glass of Obsidian Cliff (see Fig. 26) shows columnar jointing, and the phonolite of Mato Tepee in South Dakota is jointed in magnificent columns, and many other examples might be cited.

In many of the granites and other coarse-grained igneous rocks, the joints divide the mass into cubical blocks, or into long, rectangular prisms, or into broad, slab-like plates. In others, again, the blocks are of exceedingly irregular form and size.

FIG. 193. — Platy jointing in diabase; above, spheroidal weathering and transition to soil. Rocky Hill, N. J. (Photograph by Sinclair)

In sedimentary rocks the joints are ordinarily in only two planes, the third being given by the bedding-planes. In homogeneous,

FIG. 194. — Regular jointing in gneiss, near Washington. (U. S. G. S.)

heavily bedded sediments, such as limestones and massive sandstones, the joints are apt to form cubical or rectangular-prismatic blocks, making a weathered cliff look like a gigantic wall of regular masonry.   Other sedimentary rocks are, as a rule, more irregularly jointed.

FIG. 195. — Irregular jointing in gneiss, Little Falls, N.Y. (Photograph by van Ingen)

FIG. 196. — Jointing in shale, Cayuga Lake, N.Y. (U. S. G. S.)

Joints are of very different orders of importance: some, the *master joints*, traverse many strata and remain constant for long distances and considerable depths, while each layer usually has minor joints which are confined to that bed. One set of joints, the *strike joints*, run more or less parallel to the strike of the beds, while the second set, the *dip joints*, follow the dip; the former are usually the longer and more conspicuous. Oblique or diagonal joints intersect the other two systems, and many irregular cracks may occur. In general, the more disturbed the rocks have been, the more broken they are.

Cause of Joints. — With regard to the manner of their production, joints may be classified into two series: (1) those which are due to tension, the rock usually parting in planes normal to the directions of tension; (2) those which are due to compression, the cracks forming in the shearing-planes.

(1) *Tension Joints.* — In igneous rocks joints are caused by the cooling and consequent contraction of the highly heated mass. This shrinkage sets up tensile stresses in the mass to which the rock yields by cracking and parting, the shape of the blocks being largely controlled by the coarseness or fineness of the mass. Igneous rocks are subject to all the vicissitudes which affect other kinds of rocks; they are faulted, compressed, exposed to tension, etc. Hence, systems of joints may occur in them, which were formed subsequently to the shrinkage-joints due to the contraction of cooling. In some cases the jointing of sedimentary rocks may perhaps be caused by a shrinkage of the mass on drying, but this cannot be an important method of producing systems of joints.

The convex sides of anticlinal and synclinal folds are stretched, and (provided they are not too deeply buried) the stretching may result in a system of cracks radial to the curves which follow the strike of the beds. Folds are not horizontal, but pitch in the direction of their axes. This complex folding may produce two sets of tensile stresses perpendicular to each other, and thus cause two series of joints, one following the strike and the other the dip of the beds. Complex folding must produce a twisting and warping of the strata, and it has been experimentally shown that a brittle substance, when twisted, cracks in two sets of fractures which intersect nearly at right angles. How slight is the twisting and warping needful to produce joints is shown by the fact that strata which are perfectly horizontal, so far as can be detected, are jointed. The modern limestones which are formed in coral-reefs are jointed, even in cases where the movements resulting in fracture must have been minimal.

Tension joints produce either rough, or smooth and sharply cut surfaces, which is determined by the character of the rock. In sandstones which are weakly cemented the cracks pass be-

FIG. 197. — Jointing in limestone, Black Hills, South Dakota. (U. S. G. S.)

tween the grains, while in hard and firm rocks the fractures are clean.

(2) *Compression Joints* are caused when the rocks yield along the shearing-planes. In simply folded strata are produced two sets of strike joints which are inclined toward each other, but whether dip joints will be made by complex folding is not certain. In some conglomerates the joint planes pass through the hard quartz pebbles and leave a smooth, even, shining face. Tension would pull such

a pebble out of its socket and only by shearing could it be cleanly
cut.

The whole subject of joints in sedimentary rocks is a difficult
one and the explanations given of them are not altogether satisfactory, for several other agencies may be involved in their production.   It is, however, highly probable that the master joints which
roughly follow the strike and dip of the strata, have been caused
by the forces which produce folding.

FIG. 198. — Joints dying away downward, shown by pinching out of white calcite
veins.   (Photograph by van Ingen)

Joints cannot occur in the shell of flowage, and are best developed in the shell of fracture, being of less importance in the transition belt between the two.

STRUCTURES DUE TO EROSION

Unconformity. — We have hitherto considered the stratified
rocks as made up of beds which follow upon one another, in orderly

sequence, and as being affected alike by the elevation or depression, folding or dislocation, to which they may have been subjected. Strata which have thus been laid down in uninterrupted succession, with sensibly parallel bedding-planes, and which have been similarly affected by movements, are said to be *conformable*, and the structure is called *conformity*. In many places, however, the strata exposed in a section are very obviously divisible into two groups, each made up of a series of conformable beds, but the upper

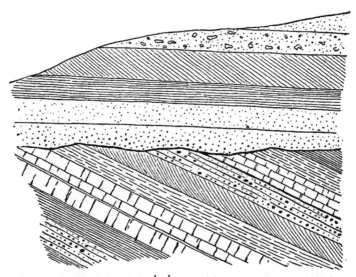

FIG. 199. — Unconformity with change of dip, or angular unconformity

group, as a whole, is not conformable with the lower, but rests upon its upturned edges, or its eroded surface. The two groups are said to be *unconformable* and the structure is named *unconformity*. The definition of unconformity here given includes certain not uncommon structures, which must be distinguished as having quite a different significance.

Unconformity is of two kinds: (1) There is a distinct difference in the dip of the two sets of strata, the upper beds lying across the upturned and truncated edges of the lower. This is the more

usual kind and is shown in Figs. 199–202. The structure implies that the lower series of beds was first laid down under water,
and that these beds were then upturned, tilted, or folded to form a
land surface. Erosion next truncated the folds, planing the edges
of the disturbed beds down to a more or less level surface. The
land surface was again depressed beneath the water, and the
second set of strata was deposited upon it. Finally, a renewed

FIG. 200. — Angular unconformity, Grand Cañon of the Colorado.
(Photograph by Sinclair) .

elevation, accompanied perhaps with folding or faulting, has
brought both series of strata above the sea-level.

While the older beds formed a land surface, they were eroded
and no deposition took place upon them. Consequently, between
the two sets of strata is a gap, unrecorded by sedimentation (at
that point), the length of which represents the time that the older
beds were above water. The processes involved in an unconform-

ity are of slow operation, so that the gap usually implies a very long lapse of time. In many cases whole geological ages, of incalculable duration, have intervened between the deposition of the two groups of strata.

FIG. 201. — Angular unconformity, old gravels on hard shale; Kingston, N.J. (Photograph by Sinclair.) Note the smooth joint-faces of the shale, in contrast to the rugged fracture-surfaces

(2) In the second kind of unconformity the two groups of strata have the same dip, the upper series resting upon the eroded surfaces of the lower. The processes involved in this kind of unconformity are nearly the same as in the first, so far, at least, as the alternation of land surface and sea-bottom, elevation and depression, are concerned. In this case, however, the first upheaval was not accompanied by any folding or fracturing of the beds. An

FIG. 202. — Angular unconformity, west of Altoona, Pa. (U. S. G. S.)

unconformity of the second class is sometimes exceedingly difficult to detect and then is called a *deceptive conformity*. Such a case arises when the surface of the ground is made by cutting down strata to the upper surface of a hard bed, which is then depressed beneath the water, as a flat pavement, upon which new material of a similar kind is laid down with hardly a perceptible break. In the Rocky Mountain region remarkable instances of this deceptive conformity occur, where, in the middle of a mass of limestone apparently formed without any interruption, there is, in reality, an enormous time-gap. Long and careful search has made clear the nature of the contact and exposed the deception. The existence of an unconformity, when none is apparent, may sometimes be detected by observing certain structural features which affect the lower and older beds, but not the upper. For example, the lower strata may be faulted, or intersected by a dyke of igneous rocks, the fault or dyke ending abruptly at a certain level and not continuing into the upper series.

The lowest member of the upper series of strata in an unconformity is very frequently a conglomerate or coarse sandstone, and

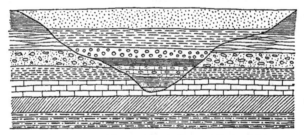

FIG. 203. — Unconformity without change of dip, and overlap

represents the beach formation of the sea advancing over the old land. These are called *basal conglomerates*. Such coarse beds are, however, not always present, and they may be only locally developed along a particular line.

Unconformities may be confined to relatively restricted regions, or they may extend over whole continents; they are very useful means of dividing the strata into natural chronological groups.

Overlap. — When a series of strata is deposited in a basin with sloping sides, or one sloping side, each bed will extend farther than the one upon which it lies, and thus in a thick mass of strata, if the shelving bottom be gently inclined, the upper beds will extend far beyond the lower ones, or *overlap* them (see Fig. 203). Overlap also occurs where the sea is advancing or transgressing slowly across a subsiding land surface, the rate of depression not much exceeding the rate of deposition. Here also each stratum extends farther across the old land surface than the one beneath it, and conceals the edges of the latter. The relation of overlap is between the successive layers of a *conformable* series.

Overlap may be a structure of much economic importance, if one of the lower strata, say a coal-bed, is mined. It is not safe to assume that wherever the upper beds of such a series are found, the lower will be found directly beneath them, an assumption which may result in costly failure.

Contemporaneous Erosion. — It was stated above that the definition of unconformity, as given, would include certain structures, which, nevertheless, must be distinguished from it: one of these is contemporaneous erosion. This structure is produced when a current of water excavates channels for itself in the still soft and submerged mass of sediment. After the current has ceased to flow, renewed deposition fills up the hollow with the same or a different kind of material as was thrown down before. This structure requires only a short pause in deposition, not a long, unrecorded break, and does not necessarily involve movements of elevation and depression. Furthermore, contemporaneous erosion is a local phenomenon, and though in a limited section it may not always be easy to distinguish it from an unconformity, the difference becomes apparent when a wider area is examined. If the structure be one of contemporaneous erosion, the two series of strata will be conformable except along the line of the channel or channels. Fig. 204 is an example of this structure and shows where a channel in an ancient sea-bottom was filled up by a later deposition of material.

The clay " horses " (as miners call them), which frequently interrupt coal beds, are the channels of streams which meandered through the ancient peat bog, and which were filled up with sediment when the swamp became submerged. The " horses " are usually of the same rock as that which forms the cap or roof of the coal seam.

Horizontal and Oblique Bedding. — Another kind of deceptive resemblance to unconformity is occasionally caused by the alternation of horizontal and oblique bedding, a horizontal bed resting upon a series of inclined layers. A conspicuous example of this is given by the Le Clair limestone of Iowa, which was at one time altogether misunderstood, but the deception is seldom one that a little care will not expose.

FIG. 204. — Contemporaneous erosion, channel in wall of Niagara Gorge. (U. S. G. S.)

Outliers. — An outlier is an isolated mass of strata, which is surrounded on all sides by beds older than itself. This definition does not imply that the older beds must actually rise to the level of the outlier and enclose it, but as *viewed on a map*, which brings all irregularities of surface down to one plane, the older beds appear to surround the outlier. An outlier has been cut off by denudation from its former connections, from which it is separated, in some cases by a few feet, in others by scores or even hundreds of miles. Outliers thus stand as monuments which show, partially at least, the former extension of strata long subject to denudation, though we never can be sure that the farthest outlier was at the actual original margin of the beds, and generally may be confident that it was not. Outliers are almost always composed of horizontal strata, or of isolated synclines.

If the outlier be brought to the surface by faulting, it is called a *faulted outlier*, in distinction from one which is entirely due to erosion. A faulted outlier may be found on the downthrow side of a fault-block, especially in a trough-fault, which is downcast with reference to the blocks on each side of it. In such a case the older beds actually surround and enclose the isolated mass of newer beds.

Inliers differ from outliers in not necessarily being *isolated masses* of rock, but merely isolated *outcrops* of older beds which are surrounded by newer strata, though underground they may be continuous with very extensive areas of beds. An inlier is thus a larger or smaller mass of rock surrounded by beds which are geologically younger than itself. The summit of an anticline or dome which has been truncated by denudation exposes older strata in the middle, newer ones on the sides. Inliers may also be due to faulting and occur on the upthrow side, as in a fault-block which is on the upthrow side with reference to the blocks on each side of it, or *Horst*.

Outliers may be converted into inliers by the deposition of newer beds around them. The isolated " stacks " and pillars on the sea-coast, as shown in Figs. 76 and 77, are outliers, but a movement of depression submerging them in the sea would eventually result in their being buried in newer deposits, thus changing them into inliers. There is abundant evidence that such changes have actually occurred in past times.

CHAPTER XV

UNSTRATIFIED OR MASSIVE ROCKS

THE unstratified or massive rocks have risen in a molten state from below toward the surface, though by no means always reaching it, and have forced, or perhaps have sometimes melted, their way through or between the stratified rocks. One of the most important points to determine with regard to a massive rock is its relation to the strata in which it occurs; for the earth's chronology is given by the stratified rocks. Considered only with reference to itself, an igneous mass gives no trustworthy evidence as to the time when it was formed. The term *eruptive* is frequently employed in the same sense as unstratified, because of the belief that most igneous masses have been connected with volcanoes; but as such a belief may not be well founded, it is better to use a non-committal term.

As in most departments of geology, there are unfortunately considerable differences in the meaning attached by various writers to the terms used in the description of the igneous, or massive, rocks. Since it is highly desirable that greater uniformity and exactness of nomenclature should be attained, the usage proposed by Professor Daly will be followed here, though his classification is more elaborate than is required in an elementary work.

We shall first take up the volcanic rocks, because modern volcanoes give us the key by which we may readily interpret them.

I. ANCIENT VOLCANOES AND THEIR ROCKS

Volcanic Necks. — Volcanoes, like all other mountains, are subject to the destructive effects of the atmosphere, rivers, and the

sea. In an active volcano the upbuilding by lava flows and fragmental ejections more than compensates for the loss by weathering, and the cone continues to grow in height and diameter.
When the volcano has become extinct, the destructive agencies
work unopposed. We find extinct volcanoes in all stages of
degradation, from those which look as though their activity
might be renewed at any moment, to those which require the

FIG. 205. — Volcanic neck, Colorado.   (U. S. G. S.)

careful examination of a skilled geologist to recognize them for
what they are.

In the Pacific States may be found admirable examples of
volcanic cones in various stages of erosion. In northern Arizona
the picturesque San Francisco mountains, themselves volcanic, are
surrounded by numerous small and very perfect cones, hardly
affected by weathering (Fig. 21). In northern California stands

the noble peak of Mt. Shasta (Fig. 29), which was active till a late geological date and still shows traces of activity in its hot vapours, but has begun to suffer notably from weathering. Still farther north, in the State of Washington, is Mt. Rainier, another volcanic cone, which has been longer exposed to the destructive agencies and has been worn into an exceedingly rugged peak.

FIG. 206. — Diamond mine, showing circular form of volcanic pipe in sandstone; Kimberley, South Africa. (Photograph by Hancox)

These mountains, however, merely exemplify the earliest stages of degradation; as time goes on, the loftiest cones will be worn away, and at last only the worn-down and hardly recognizable stump of the volcano remains, which is known as a volcanic neck. The neck consists of the funnel or vent filled up with the hardened lava of the last eruption, or, less commonly, with a mass of volcanic blocks. Associated with this plug of lava may be preserved the lowest lava flows or tuffs of which the cone was originally built up. If the land upon which the volcanic neck stands be covered

by the sea or other body of water, the remnant of the cone will
be buried beneath sediments, and a volcanic island may be similarly cut down and covered with sediments. Subsequent upheaval and denudation may at a long subsequent time once more
expose the buried cone to view. Several examples of this have been
found in Great Britain.

FIG. 207. — Irregularly and columnar-jointed lava flow on Sandstone; Island of
Staffa, Scotland

The diamond mines of South Africa are in almost cylindrical
pipes, which are cut through stratified rocks and are filled with an
irregular agglomerate. On the surface the pipes show no topographical indication of their presence, but are quite level with the
ground. The exact nature of these pipes is not well understood;
if they are truly volcanic, all traces of the cones and associated
ejected masses have been removed by denudation.

Lava Flows and Sheets which were poured out on the surface of the ground may be recognized by the aid of several criteria. In flows of only moderate antiquity, which have suffered little denudation, the nature of the mass may be determined at a glance, and traced to the vent whence it issued. Successive sheets, piled one over the other in a rude bedding, are also evidence that the rocks are surface lavas. Surface sheets may be overlaid by sedi-

FIG. 208.—Lava flow on Sandstone, Upper Montclair, N.J. (Photograph by van Ingen.) The white line shows the irregular contact

ments, which were deposited upon a submarine flow, or after depression of the land. Such a flow is then called a *contemporaneous* or *interbedded sheet*, and evidently its geological age follows the rule for strata; it is newer than the bed upon which it lies and older than the one which rests upon it.

Fragmental Products (Pyroclastic) are positive proof of volcanic action, for they cannot be formed underground. Coarse masses of agglomerate, blocks, and bombs show that the vent from which they issued was not far away, while beds of fine ashes and tuffs may be made at great distances from their source. All these varieties may be enclosed in true sediments, and may, in part, escape destruction long after the volcano which ejected them has been cut away. The fragmental products are always contemporaneous, and when interstratified with sediments are newer than the underlying, older than the overlying, stratum.

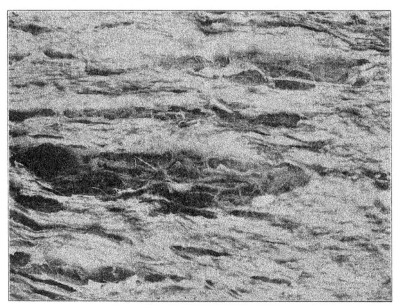

FIG. 209. — Pumice, natural Size

II. ROCKS SOLIDIFIED BELOW THE SURFACE (PLUTONIC)

We now come to a series of rocks which no one has ever observed in the course of formation, because they were solidified at greater or less depths underground. When such masses are exposed to view, it is not because they have been brought to the surface, but because the surface has been eroded down to them. Though these unstratified masses cannot be observed in the process of formation, as may the lavas and pyroclastic rocks, yet the nature of the rocks themselves, and their relations to the volcanic and stratified rocks, enable us to explain them satisfactorily. In whatever shape they occur, these masses are *intrusive*, and have forced or melted their way upward, filling fissures and cavities, or have thrust themselves between strata, following the path of least resistance. Intrusions are younger, it may be vastly so, than the strata which they penetrate and lie over or beneath; their géological date may be determined by a process of elimination, finding the newest strata which they have traversed and the oldest which they have not reached.

A primary division of the plutonic masses is into (1) *injected* and (2) *subjacent* bodies. "An injected body is one which is entirely enclosed within the invaded formations, except along the relatively narrow openings to the chamber where the latter has been in communication with the feeding reservoir." (Daly.) Subjacent bodies, on the other hand, have no floor upon which the intrusive mass rests, the communication with the earth's interior being by great openings which enlarge downward indefinitely within the limits of observation.

Both injected and subjacent bodies may be either *simple, i.e.* composed of material intruded at one period, *multiple, i.e.* composed of material of the same kind intruded at more than one period, or *composite, i.e.* made up of material derived from different kinds of magma intruded at more than one period of time.

I. *Injected Bodies*

These are of manifold variety of shapes and sizes and differ in their relations to the enclosing, or *country rock*, and different terms are accordingly used to describe them.

Dykes. — A dyke is a vertical or steeply inclined wall of igneous rock which was forced up into a fissure when molten and there consolidated. Dykes of a certain kind may actually be seen in the making, as when the lava column of a volcano bursts its way through fissures in the cone. The ordinary dyke is formed in fissures which traverse stratified rocks, breaking across the bedding-planes and usually approximating a vertical position, though sometimes it cuts through older and already consolidated igneous rocks. In thickness dykes vary from less than a foot to a hundred feet or more, and pursue nearly straight courses, it may be for many miles. The rock of a dyke has usually a compact texture, having

FIG. 210. — Parallel dykes, Cinnabar Mountain, Montana. (U. S. G. S.) The left-hand dyke forms the distant peak

cooled more slowly than the volcanic masses, though the edges, chilled by contact with the walls of the fissure, may be glassy. If the rock displays columnar jointing, the prisms are horizontal, normal to the cooling surfaces.

Dykes may be so numerous as to form a regular network of intersecting walls, just as we have found to be the case in faults and fissures.

The commonest rocks in dykes are basalt, quartz porphyry, andesite, and diabase.

When denudation has so far cut away the surface of the ground as to expose the dyke, the form which the latter takes will depend

FIG. 211.— Dyke trenched by weathering faster than country rock. (U. S. G. S.)

upon the relative destructibility of the igneous rock and the enclosing strata. If the latter wear away more rapidly, the dyke will be left standing above the surface like a wall (Fig. 210); but if the igneous mass be disintegrated more rapidly than the strata, a trench will mark the line of the dyke.

Dykes are common and conspicuous objects in the Connecticut valley and in the sandstone belt which runs, with interruptions, from the Hudson River to North Carolina.

Intrusive Veins are smaller and more irregular, frequently branching fissures which have been filled with an igneous magma; they may be only a few inches in thickness, and may often be traced to the mass which gave them off. The nature of the rock in a vein may be much modified by material derived from the walls. This

FIG. 212. — Veins of granite in cliff, near Gunnison, Col. (U. S. G. S.)

vein rock is often so coarsely crystalline, as in pegmatite veins, that it has been suggested that it could not have solidified from fusion, but was deposited from solution in superheated waters.

Sills or Intrusive Sheets. — These are horizontal or moderately inclined masses of igneous rock, which have small thickness as compared with their lateral extent. Sheets conform to the bedding-planes of the strata, often running long distances between the same two beds; but if they can be traced far enough, they may

generally be found cutting across the strata at one point or another.   In thickness they vary from a few feet to several hundreds
of feet.   The Palisades of the Hudson are formed by a sheet
of unusual thickness; its outcrop is 70 miles long from north
to south, and its thickness varies from 300 to 850 feet.

Sills are most commonly found in horizontal strata, which
offer less resistance to horizontal expansion than do the folded
beds; they are also very generally of the most fusible kind, the

FIG. 213. — Granite veins intrusive in diorite and both cut by a small dyke of aplite;
coast of Maine.   (U. S. G. S.)

gabbro family, because such magmas retain their fluidity and
flow for longer distances than do the highly siliceous rocks.   It is
probable that intrusive sheets can be formed at only moderate
depths, because the overlying strata must be lifted to an amount
equal to the thickness of the sheet, although certain cases are known
where the sill appears to have made its way by melting and incorporating some of the strata.   At great depths the weight to be
lifted is so enormous, that the easiest path of escape must be
by breaking through and across the strata.   If the beds are sub-

jected to compression after the intrusion of the igneous masses, the latter will be flexed or faulted like the stratified rocks.

In a limited exposure it is often difficult to distinguish at once between a sill and a contemporaneous sheet, but there are certain characteristic marks which enable the observer to decide. The presence of scoriæ shows that the sheet is contemporaneous. If, on the other hand, the *overlying* stratum be baked and altered

FIG. 214. — The Palisades, seen from Hastings, N.Y.   (Photograph by van Ingen)

by the heat, or if the sheet cuts across the bedding-planes at any point, or if it can be traced to a dyke which rises above it, or if it gives off tongues or veins, or if pieces of the overlying stratum be torn off and included in the sheet, it must be intrusive. The nature of the contact between the sheet and the stratum above it is also significant; if the former be contemporaneous, the cracks and fissures of its upper surface will be filled with the sedimen-

tary material.   Finally, the texture of the igneous mass gives valuable evidence; in the intrusive sheet the texture is compact
(without glassy ground mass) or even quite coarsely crystalline,
while the contemporaneous sheet will display the glassy or porphyritic texture of surface flows.

Laccoliths. — A laccolith (or laccolite) is a large, lenticular
mass of igneous rock, filling a chamber which it has made for

FIG. 215. — Contact of diabase sill with shales below. Base of Palisades, Wee-
hawken, N.J. (U. S. G. S.)

itself by lifting the overlying strata into a dome-like shape; the
magma was supplied from below through a relatively small pipe or
fissure.   The rock of which laccoliths are made is nearly always of
the highly siliceous and less fusible kinds, so that it can more easily
lift the strata than force its way between them.   Intrusive sheets
are, it is true, often given off from a laccolith, but these are of quite
subordinate importance, while dykes and irregular protrusions,

called *apophyses*, extend into the fissures of the surrounding and overlying strata. Subsequent erosion may remove the dome of

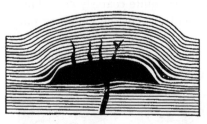

strata and cut deeply into the igneous mass beneath, leaving rugged mountains, the height of which depends upon the amount of original uplift and the subsequent denudation. Laccoliths in various stages of denudation occur in different parts of the West. Fig. 219 shows

FIG. 216. — Diagrammatic vertical section of a laccolith (Gilbert). The full black indicates igneous rock

Little Sun-Dance Hill in South Dakota, a small dome from which the overarching strata have not been removed and the igneous core has nowhere been exposed, yet there can be little doubt

FIG. 217. — Eroded laccolith, with many sills and apophyses; Colorado. (Holmes)

of its presence. Bear Butte (Fig. 220) represents a second stage of denudation; the strata have been removed, except those upturned around the foot of the butte, and the igneous core exposed, yet but little eroded. In the same region is Mato Tepee (also called the Devil's Tower), a magnificent shaft of columnar phonolite, which rises 700 feet above a platform of horizontal strata. This tower is the remnant of a laccolith from which the covering strata, and probably much of the igneous core, have been eroded away. In southern Utah the Henry Mountains are a group of laccoliths from which several thousand feet of overlying strata have been removed and the cores deeply dissected. In the Elk Mountains of Colorado are some enormous laccolithic masses.

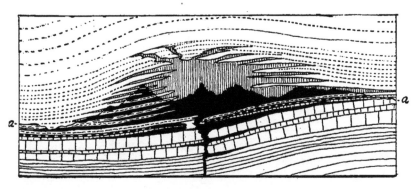

FIG. 218.— Vertical section through laccolith shown in Fig. 217 before denudation. *aa*, present surface: full black, remaining parts of intrusive body; vertical lines, portion of laccolith removed by denudation. (Holmes)

Chonoliths. — Sometimes the shape of an intruded igneous body is so irregular and its relations to the country rock are so complex that it cannot be referred to any of the preceding categories. For such irregular masses Daly has proposed the term *chonolith*, which he defines as follows: " an igneous body (*a*) injected into dislocated rock of any kind, stratified or not; (*b*) of shape and relations irregular in the sense that they are not those of a true dyke, vein, sheet, laccolith, bysmalith or neck; and (*c*) composed of magma either passively squeezed into a subterranean orogenic chamber, or actively forcing apart the country rocks." Chonoliths are probably much more numerous than true laccoliths.

2. *Subjacent Bodies*

The mode in which the plutonic masses of this group have reached their present position is highly problematical and still forms the subject of a lively discussion, to which attention has already been called in another connection (see p. 291). From the purely descriptive point of view, the special characteristic of these bodies is that their diameter increases downward to unknown depths and, consequently, that they do not rest upon a floor of country rock.

FIG. 219. — Little Sun-Dance Hill, South Dakota.   (U. S. G. S.)

Stocks or **Bosses** are rounded or irregular masses of intrusive rock, which vary in diameter from a few feet to several miles; they cut across the country rock, which they have sometimes pushed aside and sometimes cleanly perforated, and with which the contact is steeply inclined or vertical.   The structure of the country rocks, such as bedding-planes, has no effect upon the shape of the stock. From many stocks are given off tongues or apophyses, which penetrate the country rock as veins, dykes, sills, and various irregular protrusions.   Granite, diorite, and gabbro are especially common in stocks and the texture frequently becomes coarser from the

circumference to the centre of the mass. In many instances, perhaps generally, stocks themselves are but protrusions from larger
masses.

Batholiths are great masses of plutonic rock hundreds or even
thousands of miles in extent; in general characteristics they agree
with stocks, except for their very much greater size, yet small
batholiths and large stocks grade into one another, so that any line
of demarcation between them must be arbitrarily drawn; probably

FIG. 220. — Bear Butte, South Dakota. (U. S. G. S.)

all true stocks, could they be followed down, would prove to be protrusions from batholiths. Granite is the commonest batholithic
rock, and in such masses forms the core of many great mountain
ranges, like the Sierra Nevada and the Rocky Mountains.

THE MECHANICS OF INTRUSION

As in all questions which deal with the subterranean agencies,
the exact manner in which molten magmas make their way up

2 D

FIG. 221. — Mato Tepee, South Dakota, the core of a laccolith composed of columnar phonolite

through the overlying rocks is veiled in obscurity; in fact, it is the unsolved problem of the ascensive force of lava in another shape. The great variety of forms assumed by the intrusive bodies is due to the complex interaction of two main groups of factors — the ascensive force of the molten magma, however that may be generated, and the resistance to be overcome. With these are frequently associated factors of a third series, the orogenic compression of the rocks, which may squeeze a purely passive magma into the cavities and fissures made by compression. The description given in the preceding section of the various plutonic bodies left out of account the fact that the different kinds are connected by all sorts of transitions. Laccoliths grade into sills, on the one hand, and into chonoliths, on the other, and the same continuous body may be a dyke in part of its course, a sill in another, and so on. The character of the magma itself is also of importance in determining the result, whether the molten mass is thoroughly fluid or merely pasty, and how great the quantity of the imprisoned gases and vapours. In the complicated play of these different factors it is often extremely difficult to distinguish effect from cause, and it is this which gives rise to such radical divergences of opinion in interpreting the phenomena.

Igneous intrusions are most abundant in regions of disturbed rocks, and we find great areas of nearly horizontal strata, such as the Great Plains, in which intrusions are not known to occur. On the other hand, folded, even intensely compressed, strata may have no igneous rocks associated with them. The Appalachian Mountains, for example, are singularly free from intrusions. These associations have been differently interpreted. According to one view, the very general coincidence of extensive intrusions and orogenic compression implies that the magma is for the most part passive, and is squeezed by the compressing force into the actual or potential cavities which are generated by the compression. On the other hand, there is a growing tendency among many geologists to regard the deep-seated magmas as active and energetic agents of dislocation and to find in them the origin of the compressive force itself.   We have met with this tendency already in the discussion of earthquakes (p. 52), dislocations (p. 367), etc., and it must be reckoned with in all attempts to solve the great problem of subterranean activities.   As so often happens, it will probably be found that the truth lies between the extreme views.

In the chapter on the igneous rocks (p. 291) we learned that very different opinions were held concerning the proper answer to the question whether the magmas make their way entirely by mechani-

FIG. 222. — Inclusions (xenoliths) of schist in granite.   (U. S. G. S.)

cal means, taking advantage of fissures, cavities, and lines of weakness, and forcing the country rock aside, or whether they may make room for themselves by dissolving, fusing, and incorporating more or less of the rocks which formerly occupied the position now held by the plutonic bodies.   So far as the injected masses are concerned, it is seldom necessary to assume that they have done more than lift or push aside the enclosing rock, but the case is very different with the subjacent masses.   Frequently the contact between the country rock and a stock or batholith shows no evidence that the former has been compressed. or crowded to make room for the intrusion, and it seems impossible to account for the presence of the plutonic mass except on the assumption that space has been gained by fusing more or less of the enclosing country rock. It is not necessary to suppose that such fusion takes place only on the periphery of the intruding magma; on the contrary it seems more likely that the magma dislodges the joint-blocks which then sink in the highly heated mass and are gradually dissolved. It must be admitted that this hypothesis has not been established. Some of the highest authorities maintain that it is definitely disproved by the microscopic and chemical examination of the batholithic rocks, which are not affected by the character of the country rock through which they break. We have here a conflict of evidence which it remains for future studies to harmonize.

The energy of intrusion is eloquently displayed along the margins of many batholiths, where the country rock is shattered and great blocks are torn off and embedded in the plutonic mass. Such blocks are called *inclusions* or *xenoliths*, and, on a small scale, they occur in other plutonic bodies, such as sills and laccoliths. The existence of these blocks in their undissolved state has been made an argument against the assimilation hypothesis, but it must be remembered that the intact xenoliths are products of the dying energy of intrusion, when the magma was already so stiff that the blocks were no longer able to sink in it.

CHAPTER XVI

METAMORPHISM AND METAMORPHIC ROCKS

By the term *metamorphism* is meant the profound transformation of a rock from its original condition by means other than those of disintegration. The incipient changes of the latter class may very greatly modify a rock and its constituent minerals, but such changes are distinguished from metamorphism under the term *alteration*. Metamorphism usually implies an increase in hardness and in the degree of crystallization, and very frequently also the generation of an entirely new set of minerals, which take on a characteristic arrangement. The degree of metamorphism varies according to circumstances, and from the mere consolidation of loose sediments to the most radical reconstruction of the rock there is every possible transition. Fossils may be found in those metamorphic rocks of sedimentary origin which have not been completely changed. The more thorough the reconstruction of the rock, the more obscure do the fossils become, and in advanced stages nearly or quite all trace of them is obliterated.

It was long supposed that the metamorphic rocks were one and all transformed sediments, but later investigations have shown that many of them were originally igneous. Indeed, it is often quite impossible to decide whether a given metamorphic rock has been derived from a sedimentary or an igneous original. This is not surprising, for the ultimate chemical (not the mineralogical) composition of a basalt, a volcanic tuff, or a clay shale, may be the same, and the metamorphic processes may produce an identical rock from any one of these three as a starting-point. Much yet remains to be learned regarding the modes, causes, and results of metamorphism, and some of the most far-reaching problems of geology are bound up with these questions.

Metamorphism is of two quite distinct kinds: (1) contact or local, and (2) regional metamorphism.

I. CONTACT METAMORPHISM

This is the change effected in surrounding rocks by igneous magmas. There is a difference between the effects produced by a surface lava flow and those caused by a plutonic intrusive. In the former case the results are usually not very striking, because of the way in which a lava stream surrounds itself with non-conducting seoriæ, and are such as may be referred to the action of dry heat. Bituminous coal is changed into a natural coke by the removal of its volatile constituents; clay may be baked into a hard red rock, looking like earthenware, and limestone changed to quick lime, by driving off the CO_2. Plutonic intrusions, on the other hand, are more efficient agents of change, because they are presumably of a higher temperature and retain their heat longer, and because the vapours and gases which they contain cannot escape into the atmosphere, but strongly affect the invaded rocks. The rock invaded and metamorphosed may be either sedimentary, igneous, or already metamorphic, and the effects may be very marked, or surprisingly small; indeed, it is often quite impossible to say why the changes should be so insignificant. Magmas which contain an abundance of the mineralizing vapours (see p. 287) produce much more effect than those with only a small quantity of such vapours. For this reason acid magmas are more effective than basic. Much, too, depends upon the nature of the invaded rock; sediments which contain large percentages of alumina and lime are much more readily and profoundly changed than those which are made up almost entirely of silica. The distance to which the zone of change extends is wider when the intrusive mass cuts across the strata than when it follows the bedding-planes, so that a dyke or stock is more effective than a sill.

We may now consider some examples of contact metamorphism, and, for this purpose, shall select only the changes of sedimentary rocks; for those of the other classes require a treatment too
minute and refined for an elementary work. We may note, in
passing, however, that some of the veins given off from granite
stocks, which have invaded other igneous rocks, are probably of a
metamorphic nature and due to the penetration of vapours.

In a series of strata which have been invaded by an igneous
magma, we find a gradual change from the unmodified rock which
lies beyond the reach of the transforming agencies, to that at the
actual contact with the igneous mass. Along this line of contact
the strata are so thoroughly reconstructed that often only a microscopical examination will distinguish the changed sediment from
the igneous rock. A siliceous sandstone or conglomerate develops
no new minerals in the change, or only in insignificant quantity
from the impurities present. The bulk of the material simply
crystallizes and forms the white rock, *quartzite*. Clay rocks
undergo more radical change and are usually divisible into distinct
zones; the outermost zone is unchanged; in the intermediate
one the shale is changed to a dense *slate* spotted with biotite,
magnetite, or other dark minerals. The spotted slate passes
gradually into *mica schist*, a rock made up of flakes of mica,
with some quartz and felspar, arranged in rudely parallel planes.
At the contact the rock is converted into *hornfels*,[1] which is a very
dense substance, looking like trap, and filled with numerous silicated minerals, such as hornblende, felspar, and many others which
were not enumerated in the chapter on the rock-forming minerals.

Limestones are crystallized by the heat into marble, a dense aggregation of crystalline grains of calcite, usually with obliteration
of the bedding-planes and of any fossils which the rock may have
originally contained. Pure limestone gives rise to white marble,
but as most limestones contain impurities, they develop, when
metamorphosed, a large variety of minerals, such as biotite, gar-

[1] Also called *hornstone*, but as this term is used for flint, it is better to
retain it in the latter sense only.

net, amphiboles, pyroxenes, etc. Beds of bituminous coal are baked into a natural coke, as in Virginia and North Carolina, or changed to anthracite, as in Colorado, or even to graphite in the contact zone, and limonite is converted into magnetite.

Among investigators of the subject there is much difference of opinion as to how far there is an actual migration of material from the plutonic magma into the enclosing rock walls. When there is shattering along the contact, or fissures and crevices are opened in the country rock, material, both in solution and in a state of fusion, is introduced. *Cementation* is the deposition of mineral matters from solution in the interstices between the granules of the rock. Quartz, calcite, iron oxides, felspars, mica, augite, and other minerals may be thus introduced, and sometimes the quantity of new material brought into the rock is very large. *Injection* is the penetration of a rock by molten substances which may fill up all the minute crevices. The distinction between cementation and injection is not a very sharply marked one, because superheated water and molten magmas appear to mix in all proportions. The difference between the two processes seems thus to be largely a question of the quantity of water present. In some examples even into the unruptured walls fluorine and boron have penetrated, and probably the escaping hydro-fluosilicic acid has introduced silica and some bases for a short distance from the contact.

Contact metamorphism, as its name implies, is a local phenomenon, but a widely ramifying and complex system of igneous intrusions may change large areas of sedimentary rocks.

II. Regional or Dynamic Metamorphism

This term applies to the reconstruction of rocks upon a great scale, in areas covering, it may be, thousands of square miles, and evidently other processes in addition to those of contact metamorphism are needed to explain such widespread changes. Regionally metamorphic rocks are, with the exception of the slates, thoroughly crystalline and usually have lost all trace of whatever fossils and stratification planes they may originally have had.

The first step in metamorphism consists in a mere hardening of the rock, accompanied with the loss of water and other volatile substances. In the second stage the component minerals already present are crystallized, but new compounds are sparingly formed. This stage is frequently accompanied by cleavage, which, to distinguish it from that of minerals, is often called *slaty cleavage.*

FIG. 223. — Oblique synclinal fold in slate, showing cleavage planes at all angles to the bedding-planes. (U. S. G. S.)

Cleavage and Fissility. — *Cleavage* is " a capacity present in some rocks to break in certain directions more easily than in others," while *fissility* is a " structure in some rocks, by virtue of which they are already separated into parallel laminæ in a state of nature. The term *fissility* thus complements cleavage, and the two are included under cleavage as ordinarily defined." (Van Hise.)

Many unmodified igneous rocks have a marked cleavage, which is occasioned by the arrangement of the constituent mineral grains with their long axes parallel, or by a parallelism in the cleavage-planes of these minerals, or by both factors combined. In cleaved sedimentary rocks the cleavage-planes may coincide with the planes of stratification. Much more commonly, however, they intersect the latter at all possible angles, keeping a constant direction for long distances (parallel to the axes of the folds in which they

FIG. 224. — Fissile quartzite, California. (U. S. G. S.)

occur), while the bedding-planes change with the dip from point to point. Ordinary roofing slate is one of the best possible examples of a cleaved rock and in beds of slate interstratified with other rocks, the cleavage is usually quite perfect in the former, absent or but partially developed in the latter.

It is very generally agreed among geologists that slaty cleavage is a result of compression; for, disregarding certain igneous masses, it occurs only in rocks which show other evidences of having been subjected to compression. On the other hand, the mechanics of the problem are somewhat obscure and have given rise to differences of opinion. The most probable view seems to be that the cleavage-planes are developed at right angles to the compressing force, and are due to the arrangement of the constituent mineral particles of the rock with their longest diameters, their cleavage-planes, or both, in parallel directions. Further, that " this arrangement is caused, first and most important, by parallel development of new minerals ; second, by the flattening and parallel rotation of old and new mineral particles ; and third, and of least importance, by the rotation into approximately parallel positions of random original particles." (Van Hise.) Fissility is also due to compression, the rocks giving way along the *shearing-planes*, which are inclined to the direction of the pressure. Slaty cleavage is brought about in the softer rocks and fissility in the more rigid by similar compression.

A more advanced degree of metamorphism is characterized by the *schistosity* or *foliation* of the rocks, as is also true of contact metamorphism when such a rock as mica schist is formed. Schistosity or foliation is the arrangement of the component mineral particles of a rock into rudely parallel planes or undulating surfaces, in consequence of which the rock parts most readily along those planes or surfaces, and has a banded appearance. In the schistosity which is developed in the contact metamorphism of a sediment, the foliation appears to be determined by the stratification planes, but in regionally metamorphosed rocks this is generally not the case. Here the foliation, like cleavage and fissility, with which the former is connected by all grades of transition, as a rule, is independent of previous structures, and is determined by the direction of the compressing force. The intergradations between cleavage and fissility, on the one hand, and schistosity on the other, make it appear that all those structures are due to the same agency operating with different degrees of power under somewhat different circumstances.

The shearing and crushing of the rocks frequently change the component minerals into *paramorphic* forms, *i.e.* those which have the same chemical composition, but different crystal forms; for example, aragonite is thus converted into calcite and augite into hornblende. In the more complete stages of metamorphism an entire chemical reorganization is made, and new minerals are abundantly generated. Inasmuch as great areas of metamorphic rocks are almost invariably those which have been intensely and violently compressed, and moderately folded sedimentary rocks may sometimes be traced directly into intensely plicated metamorphic rocks, we are justified in concluding that the compression is the cause of the reconstruction, especially as the exceptions are more apparent than real. If this conclusion is well founded, it leads to the highly interesting and important generalization first clearly stated by President Van Hise, that *the structures impressed on the stratified rocks after their first formation, folds, faults, thrusts, joints, cleavage, fissility, and foliation are all due to lateral compression, acting with different degrees of intensity and at different depths, depth and overlying load being controlling factors of the first importance.*

There is some difference of opinion as to the relative importance of contact and dynamic metamorphism, though it is not disputed that large areas may be metamorphosed by frequent and extensive igneous intrusions, nor that such intrusions may aid very materially in the transformations made by intense compression.

Igneous masses, when subjected to the same processes, give rise to rocks entirely similar to those made by the metamorphism of sediments, so that it is sometimes impossible to distinguish between them. The compression may have been applied while the magmas were still pasty, or long after they had cooled and solidified. Certain rocks, like the gneiss of Manhattan Island, are believed to have been formed both from the metamorphism of sediments and the injection of igneous material, and thus to have had a highly complex origin.

The Causes of Metamorphism have already been suggested in the preceding paragraphs, but it will be well to summarize them, though it should be borne in mind that the metamorphic processes are by no means completely understood.

1. *Heat* is evidently a very important factor of change, as is shown by the phenomena of contact metamorphism and by numerous experiments by which the process has been imitated successfully. In contact metamorphism the heat is derived from the igneous magmas, and in dynamic it is in part mechanically generated, in part due to the interior heat of the earth invading deeply buried masses.

2. *Compression* is believed to be the great agent of dynamic metamorphism, and the amount of the change depends upon the intensity of compression and the depth at which it operates. Hence the varying results, ranging from gentle folding, at one end of the series, through violent folding to complete reconstruction, crystallization, and foliation, at the other.

3. *Moisture* is another potent agent of reconstruction. Superheated water under pressure is able to attack and dissolve the most refractory substances and to build them up into new combinations. Many minerals, such as the felspars, which have never been artificially crystallized by dry heat alone will crystallize readily in the presence of superheated water, and the water lowers the temperature necessary for metamorphism. Rocks which melt at 2500° F. dry heat, become pasty at 750° F. in the presence of water. In contact metamorphism, steam is a very important factor of change, but other vapours and gases play an efficient part.

4. *Pressure,* as distinguished from active compression, is a necessity for any extensive metamorphic action, whether contact or dynamic. It is the difference of pressure which is responsible for the different effects of surface flows of lava and of subterranean intrusions and which gives to depth its importance as a controlling factor. The dead-weight pressure of overlying rocks prevents the rapid escape of the mineralizing vapours and, when sufficiently great, causes the rock to shear and flow without fracture. Limestone heated at the pressure of the atmosphere, in a lime-kiln or an open vessel,

becomes quicklime (CaO) through the expulsion of CO_2, but heated under pressure, so that the gas cannot escape, it crystallizes into marble. Such pressure, also, is an essential factor in dynamic metamorphism as a precondition in enabling the rock to behave more or less plastically under active compression and without shattering. Dynamic metamorphism must therefore take place at considerable depths below the surface.

It is believed by many geologists that metamorphism may proceed so far as completely to melt a sedimentary rock, producing a magma which is indistinguishable from a typically igneous one. Such extreme metamorphism has not been demonstrated for any considerable body of rocks, but may be true, nevertheless, and if so, we should then have the cycle of rock transformation complete, from igneous rock, through sedimentary and metamorphic, back to igneous. Be this as it may, certain metamorphic rocks do undoubtedly form a common meeting place for the sedimentary and igneous classes.

THE METAMORPHIC ROCKS

In the scheme of classification it is not yet practicable to separate the metamorphic rocks of igneous origin from those which are transformed sediments, for it is often impossible to distinguish one from the other.

A. Non-foliated Rocks

These represent the less advanced stages of metamorphism, in which the forces of compression may have produced cleavage or fissility, but not foliation. The more important rocks of this class are of sedimentary origin, and it will be unnecessary for us to consider the igneous rocks which have been changed, though not to the extent of producing foliation.

Quartzite is derived from the metamorphosis of sandstone, and between the two kinds of rock are found such complete transitions, that the separation of them seems almost arbitrary. In a typical quartzite the rock is crystalline, and the quartz deposited around the sand-grains is in crystalline continuity with those grains, though the microscope still reveals the original fragmental nature of the rock. Quartzites also result from the metamorphism of conglomerates, and the pebbles are sometimes much flattened by compression. If the sandstone or conglomerate contained impurities, other minerals besides quartz are generated; if any considerable quantity of clay was present, mica will be produced and, it may be, in such abundance that the rock passes into mica schist (see below).

Quartzites are formed both in contact and regional metamorphism, but the change is principally due to cementation, large amounts of silica (estimated as one-sixth of the original quantity present in the sandstone) being brought in and deposited from solution, though this cementation may be effected by ordinary percolating waters bearing SiO_2 in solution, so that some quartzites should hardly be regarded as metamorphic. Many quartzites do not appear to have been subjected to great compression, while others are cleaved or fissile (Fig. 224).

Slate and Phyllite. — Slate is a fine-grained, dense, and hard rock, which, when metamorphosed by compression, is cleaved. It results from the transformation of clay shales, fine arkose, and sometimes of volcanic tuffs. Crushed fragments of felspar change into interlocking crystals of quartz and felspar, or quartz and mica. The mineral particles, both original and newly developed, have a parallel arrangement of their long axes and cleavage planes, which determines the cleavage of the rock. In colour, slates are usually drab, or dull dark blue, but they may be brick-red, green, or purple. When fine-grained and regularly cleaved, they are extensively quarried for roofing purposes. Great areas of them occur in Vermont, eastern Pennsylvania, Virginia, and Georgia, south of Lake Superior, and on the western flank of the Sierra Nevada.

Phyllite is slate in a more advanced stage of metamorphosis, in which the mica spangles are more abundant, and visible to the naked eye, giving lustrous surfaces to the cleavage-planes. Like micaceous quartzite, phyllite may often be traced into mica schist.

Marble is a metamorphic limestone, in which the fragments and particles of organic origin have been converted into crystalline calcite. Magnesian limestones yield crystalline dolomites, which are likewise included under marble. In the process of reconstruction, the fossils and even the bedding-planes of the original limestone are usually entirely obliterated. The grain of the rock varies much, from the fine, dense, loaf-sugar-like statuary marble to a very coarse texture of large crystals. Pure limestone gives rise to a white marble, but the presence of organic matter is betrayed by veins of graphite, which may indicate the lines of mashing and flow, along which the rock yielded to the compressing force. Iron and organic matters present in the limestone produce a great variety of coloured and variegated marbles, some of which are of extraordinary beauty. The sand and clay present in many limestones will, on metamorphosis, give rise to a variety of silicated minerals. Not all crystalline limestones are to be called marbles, for crystallization may be the work of surface waters at ordinary temperature, and even modern coral-rocks may be crystalline. Such non-metamorphic crystalline limestones differ from marbles in being less hard and in retaining the fossils and stratification planes which they originally had. Other crystalline limestones, like stalagmite and travertine (see p. 307), were deposited from solution.

Marble is an exceptional case of a completely crystalline rock derived from sediments by dynamic metamorphism, which is not foliated or schistose. This is believed to be due to the capacity of calcite to recrystallize freely after it has been subjected to compression and mashing.

The economic value of the marbles makes them largely sought after; in this country they are extensively developed along the Appalachian region, from Vermont to Georgia, in the Rocky Mountains, and the Sierra Nevada.

The **Ophicalcites** are crystalline magnesian limestones and dolo-

2 E

mites, with varying amounts of included serpentine, which gives ·
them a mottled appearance. They are not thoroughly understood,
and it appears that they may be formed in various ways. Some
ophicalcites are almost certainly marbles, in which inclusions of
olivine, pyroxene, or hornblende have been formed and afterward
altered into serpentine (see p. 19). Others would appear to be
broken and fissured serpentines, having the crevices filled up with
calcite deposited from solution.

Anthracite is usually regarded as a metamorphic form of coal,
and, as we have seen in a preceding paragraph of this chapter, it
is formed from bituminous coal by contact metamorphism. On a
large scale it occurs chiefly in areas of folded and disturbed rocks,
though not invariably so. A more intense metamorphism of carbonaceous material gives rise to *graphite* (or black lead), a semi-
crystalline form of carbon, which, however, is a mineral rather
than a rock.

B. FOLIATED ROCKS

The foliated or schistose rocks are those which are divided into
rudely parallel planes, with rough or undulating surfaces, due to
the flakes and spangles of some mineral. The planes of foliation
may coincide with the original bedding-planes or they may intersect the latter at any angle, just as do the planes of cleavage and
fissility. The foliated rocks represent the most advanced stage of
what we can confidently call metamorphism, and may be derived
from either sedimentary or igneous originals; it is not always possible to say which.

Gneiss is a term of wide significance, which includes a number
of rocks of different modes of origin and different mineralogical
composition. It is " a laminated metamorphic rock that usually
corresponds in mineralogy to some one of the plutonic types."
(Kemp.) The varieties of gneiss are ordinarily named in accordance with the most conspicuous dark silicate present, as *biotite
gneiss, hornblende gneiss*, etc.; but this system of nomenclature

gives an imperfect notion of the character of the rock. A better method has been suggested by Dr. C. H. Gordon and adopted by Professor Kemp, though the older scheme is still in general use. This is to name the varieties in accordance with the igneous rocks to which they correspond in mineralogical composition; as *granitic gneiss, syenitic gneiss, dioritic gneiss*, etc. The commonest variety is granitic gneiss, with mica or hornblende; the orthoclase and quartz are mingled together, with conspicuous laminæ and folia of the dark mineral.

FIG. 225. — Plicated gneiss, Montgomery County, Pa. (U. S. G. S.) This figure clearly displays the characteristic foliation

Most gneisses were generated by the dynamic metamorphism of granite, either before its consolidation or after it had cooled and hardened. Some authorities deny that gneiss has ever been formed from sedimentary rocks, but there is good reason to believe that it sometimes has such an origin, and in certain instances the crushed pebbles of the parent conglomerate are still distinctly visible, especially on a weathered surface. Still another series of these rocks are of complex origin, granitic magmas being injected along the foliation planes and into all the crevices of metamorphosed sediments.

Gneisses are widely spread in ancient formations, especially in the most ancient of all, and they cover vast areas in the northern part of North America.

The **Crystalline Schists** are more finely foliated than gneiss, into which they often grade imperceptibly, having very similar miner-

FIG. 226. — Boulder of gneiss, displaying its conglomeratic nature on weathered Surface. (International Boundary Survey)

alogical composition. They have very diverse modes of origin arising from both sedimentary and igneous rocks. Slates, impure sandstones and limestones, as well as felsites, andesites, diabases, tuffs, etc., may all give rise to crystalline schists by contact or dynamic metamorphism. The varieties are named from their most important ferro-magnesian mineral.

Quartz Schist is a foliated quartzite in which cleavage or fissility has developed into schistosity. The mashing and cementation of the original sandstone may take place at the same time, or the quartzite may be produced by the latter process and subsequently converted into schist by compression.

Mica Schist is principally composed of quartz, muscovite, and biotite, with more or less felspar. By an increase in the quantity of felspar present, and a coarser foliation, it grades into gneiss, and by an increase of quartz it may pass into quartzite and thence to sandstone. Through the phyllites mica schists are connected

FIG. 227. — Mica Schist with garnets. Nearly natural size

with the slates, and in another direction, by increase of lime they pass into argillaceous limestones. Mica schists are very largely exposed in New England and southward along the eastern flank of the Appalachian Mountain system.

Hornblende Schist is a foliated rock, consisting of hornblende with a varying proportion of felspar and less quartz. The hornblende schists are, for the most part, derived from the dynamic metamorphism of various basic igneous rocks, the augite being readily converted into hornblende by crushing, but in rare instances they are believed to have had a sedimentary origin. The hornblende schists occur as belts or bosses in metamorphic areas and are largely exposed around Lake Superior. The schists already described are much the most abundant members of the group, but there are several others. Thus, we have *talc* and *chlorite schists*, both of which are due to alteration, chiefly of hornblende schist, and *graphite schist*, which has quantities of that carbon mineral along its foliation planes.

CHAPTER XVII

MINERAL VEINS AND ORE DEPOSITS

MINERAL veins and ore deposits are of the greatest economic importance and have therefore received a great deal of attention, and a very extensive literature has grown up concerning them. Obviously, but a meagre outline of the subject can be attempted in this place, and the treatment of the much-disputed questions of the modes of formation cannot be given adequately or at length.

I. MINERAL VEINS

The crevices, fissures, and faults which traverse hard rocks generally remain open for a time and are then frequently filled up by the deposition of material which is quite different from the country rock of the walls. Fissures thus filled by crystalline deposits are mineral veins, and these vary greatly in dimensions, from a few inches to many miles in length. The minute veins are filled with material derived from the walls by solution and redeposited in the crevices, such as the veins of crystallized calcite in limestone. Great *fissure veins*, on the other hand, which may run unchanged for many miles and penetrate to depths beyond the reach of mining, are " characterized by regular, straight walls, by a fairly constant width, and by a definite direction of both strike and dip." (Spurr.) Such veins are usually very distinctly marked off from the wall of country rock, and may be either simple or banded, with the bands in general parallelism with the walls of the fissure. In a simple vein the mineral contents are deposited irregularly without any definite arrangement, or in a solid, homogeneous mass, while the banded structure is produced in several different ways. One of the commonest of these ways is by the deposition of minerals on the walls of an open fissure, for the more perfect ends of the crystals project from the walls toward the middle of the vein, and the bands are arranged usually in symmetrical pairs from the walls inward. In many instances the symmetrical arrangement is departed from, because a fissure once filled with crystalline minerals may be again opened by renewed diastrophic movements and a renewed deposition take place, the older vein forming the walls of the newer one. The parallel bands may be of the same mineral, or each pair may be of a different mineral. Banded structure may also be brought about by movements of the rock subsequent to the filling of the vein, and frequently both factors coöperate to produce the result in the same vein.

As we have already seen (p. 341), a fault zone is often a mass of shattered and sheared rock, consequently it is not surprising that many mineral veins should be highly complex, branching and anastomosing around the broken pieces of country rock. In such cases it is evident that the deposition of the minerals has taken place in a broader or narrower zone of fault rock. The nature of the country rock itself often determines whether a vein shall be simple or complex, and the same vein may be simple in one part of its course and complex in another, as the country rock changes from point to point, either vertically or horizontally. Before the deposition of the mineral contents, the fissure was open in part of its course, where the rocks yielded easily to tension, while in other parts the crack was represented by a mass of shattered rock, yet with abundant narrow openings, through which water could circulate.

A third class of mineral veins is composed of the *veins of replacement*, in which the circulating waters have not merely deposited minerals in an open fissure, but have gradually substituted one substance for another, by dissolving out the latter and replacing it with the former, it may be molecule by molecule, so that the replacing minerals are pseudomorphs after the older series (see p. 11), retaining the crystal form, sometimes the cleavage of the latter.

In this way fossils may be produced in newer minerals, even metals. A replacement vein represents a water channel of some kind, and so it has a more or less definite direction, but it seldom has sharply defined walls, for the new deposits impregnate the country rock and fade away into it. Sometimes, however, the replacement has been so complete that a vein results which is at first sight hardly distinguishable from a true fissure vein, and even a banded structure may occur in such veins, due to a previous banding in the rock which is replaced. This banding of the rock may be occasioned by a succession of shear-planes, along which the first deposition takes place, followed by the replacement of the rock included between the shear-planes, or by the occurrence of bands of more and less soluble material, replaced in the order of solubility.

Replacement veins are most commonly found in limestones, since those are the most readily soluble rocks, but they also occur in rocks which are relatively very insoluble, such as sandstones, and in igneous rocks like granite. The processes of molecular substitution, which are carried on very slowly, may take place where there is a very small amount of soluble material present.

Mineral veins are especially characteristic of disturbed, fractured, and dislocated rocks, and are practically absent from regions of undisturbed strata. This association is what we should expect to find, for deep fissures are to be found only among rocks which have been more or less violently shifted by diastrophism or by igneous intrusions. Such intrusions are very favourable to the formation of mineral veins, and many veins may be traced directly into plutonic bodies, and others are clearly results of contact metamorphism.

The substances which are found in mineral veins vary widely, in accordance with the mode of formation, and in the same vein may differ greatly from point to point. Sometimes, though not always, the character of the country rock exercises a controlling influence upon the contents of the vein, which change as the rock traversed changes. Among the commonest and most widely disseminated of vein minerals are quartz, calcite, and barite (heavy

spar, BaSO$_4$). Frequently the ores of the commercially valuable
metals are found in mineral veins, which then are called *metalliferous veins*. The minerals which fill up most of the vein, such
as quartz, calcite, etc., form what is called the *vein stuff*, or *gangue*,
and in the latter the ores may be disseminated in fine particles, or
gathered in threads, pockets, or nuggets, sometimes in the *native*,
or uncompounded state, but much more frequently as sulphides,

FIG. 228. — Dykes of sandstone in shales, Northern California. (U. S. G. S.)

oxides, carbonates, and other combinations. Mineral veins may
thus be regarded as a special case of ore deposits, and the mode of
their formation can most conveniently be discussed in connection
with the latter.

Sediment-filled Veins, though belonging in an entirely different
category from true mineral veins, may be briefly mentioned here.

Vertical fissures are sometimes filled up by sediment washed in from above, but more remarkable are the instances where the fissure was evidently filled from below with sediment different from the walls. In Fig. 229 is seen an example from northern California: the fissures which traverse the shale have been filled with sand, which has consolidated into firm sandstones and, as they resist weathering better than the enclosing soft shales, they·stand out in relief. These are called *sandstone dykes*, though they are not true dykes, which are always of igneous rocks. The explanation of these curious structures is given by many modern earthquakes, notably the great Indian quake of 1897 (see p. 42). It will be remembered that on that great disturbance the ground opened in innumerable fissures, through which " astounding quantities " of sand and water were discharged. Not all the fissures communicate with the surface, and if the superficial rocks rest upon unconsolidated beds of sand, the sand will be forced upward into any cracks that may be formed, as bore-holes are sometimes clogged at considerable depths with clay squeezed into them by the pressure of the overlying rock.

II. Ore Deposits

The term *ore* implies an economic conception and means a source of supply of a metal which can be profitably worked, hence the proportion of the metallic constituent which must be present for profitable working depends very largely upon the price of the metal. Iron ore, ready for the blast-furnace, must have at least 35 % of the metal, while a 3 % ore of copper may be employed. The table of the elements which chiefly make up the accessible parts of the earth's crust (see p. 6) shows that the only commercially important metals which are among the first eight elements are aluminium and iron, while the other metals form but an excessively small proportion of the crust. It has been estimated that lead, tin, and zinc form some hundred-thousandths of a per cent each, copper is in the hundred-thousandths or millionths of a per cent, silver a tenth or a hundredth as much as copper, and gold one tenth as much as silver. (Vogt.) Infinitesimal as these proportions seem, the metals are very widely disseminated in the rocks, and the processes of ore deposition are therefore, above all, *processes of concentration*, by which the scattered particles of the metallic compounds are brought together in relatively large quantity.

The variety of ore deposits, regarded from the standpoint either of their contents, their mode of formation, or the rocks in which they are found, is excessively great, and no classification of them is satisfactory. All that can be attempted here is a description of some of the commoner and more typical kinds of ore deposits, with a brief discussion of the problems concerning their origin, problems which are still very far from definitive solution.

Stratified Ore Deposits are usually of iron, or less commonly manganese, and occur in beds interstratified with other rocks. The ores themselves may be found in continuous sheets, thick beds, or scattered nodules, and were evidently deposited from solution in water, like the bog and lake ores which are now in process of formation. Very frequently bedded ores of iron are found among highly metamorphic rocks, especially the crystalline schists. *Placers* are river gravels which contain grains, or nuggets of heavy metals, such as gold, platinum, or tin oxide (stream tin). They are due to the concentration of the metallic particles, originally scattered through veins or rocks by erosion and stream transportation, and owing to their high specific gravity the metallic particles are thrown down where gravel is deposited. The stratified ore deposits thus offer no particular difficulty of explanation.

Ores due to Magmatic Segregation. — In our study of the igneous rocks, we learned that deep-seated molten masses in the slow process of cooling and consolidation frequently undergo differentiation, so that different parts of the same continuous magma consolidate into rocks of very different composition (see p. 290). Many basic rocks contain considerable quantities of metals, and there is good reason to believe that by segregation these metallic constituents may be so concentrated as to form ore bodies. The commonest ores which are referred to this mode of origin are magnetic iron oxide, generally containing titanium, such as those of the Adirondack Mountains, New York, and many other regions. Iron sulphides containing nickel in paying quantities occur in Pennsylvania and Canada, and nickeliferous olivines in Oregon, all of which are regarded as due to magmatic segregation. Chromite, the oxide of iron and chromium, also forms ore bodies of probably similar origin, and the great body of zinc ores at Franklin Furnace, New Jersey, has been referred to the same category.

Ores due to Contact Metamorphism. — When the country rock, which is invaded by a plutonic mass, is of a kind that permits extensive penetration by the magmatic vapours and gases, metamorphism may result for a considerable distance from the intruding igneous rocks. Among the new minerals which are generated along the contact zone, metallic ores may occur in sufficient quantity to be economically important. The minerals in question may be deposited in the interstices of the wall rock, or may replace that rock bodily for a greater or less distance from the actual contact. Ore bodies formed in this manner are usually characterized by the presence of garnets, oxides and sulphides of iron in association, and by fluorite and other minerals containing fluorine and boron. As we have seen (p. 409) there is a difference of opinion among geologists as to how far new mineral substances can be carried into the walls of country rock, but such a case as the Dolcoath mine in Montana, more than half a mile from the contact with the granite, which has been the chief agent in metamorphosing the district, is highy suggestive. "The ore-bearing stratum of the mine was originally a bed of impure limestone, which has been metamorphosed to garnet and pyroxene, with spots of calcite. Associated with these gangue minerals are sulphide and telluride of bismuth, containing gold." (Spurr.)

Metalliferous Veins (or Lodes). — These are particular varieties of mineral veins, the principal characters of which have been given in the preceding section. Metalliferous veins are no exception to the rule that subterranean activities are not well understood, and among students of the subject there are many and strong differences of opinion concerning the mode of formation of such veins. However, there is general agreement that the contents of veins, both gangue and ores, have been deposited from solution in thermal waters and vapours, just as certain existing hot springs are making similar deposits now (see p. 194). The first requisite for the formation of a lode is a water channel, because the metals are present in minute quantities, and immense quantities of water must pass before any considerable deposit can be accumulated. Hence, ore deposits are found in fissures, shattered rock-masses, in joints, in porous and soluble strata, where water may pass with comparative freedom, and further these waterways must, directly or indirectly, communicate with great depths, or with highly heated rocks, permitting supplies of hot water to reach them.

There is no general agreement as to the source of the waters that have filled the veins with gangue and ores. Perhaps the majority of geologists incline to the opinion that such waters are *meteoric, i.e.* of atmospheric origin, and that the waters descending through the rocks dissolve the metallic and other minerals and penetrate to great depths until they become highly heated and rise again through fissures. As the waters, thus charged with ore and gangue minerals in solution, ascend to the cooler layers nearer the surface, they are chilled and precipitate the greater part of the dissolved substances along the waterways. An alternative view, which seems to be better founded, is that the solvent hot waters are largely of *magmatic* origin; that is to say, that they are derived from the immense quantities of superheated steam which impregnate the igneous magmas. How vast is the amount of this water, is shown us by every great volcanic eruption, but the slowly solidifying plutonic bodies must give off their steam much more gradually. With the highly heated ascending magmatic waters are doubtless mingled a greater or less proportion of meteoric waters, varying in amount according to local circumstances.

The views held concerning the origin of the ore substances themselves are similarly divergent. The hypothesis that the solvent waters are mainly of meteoric origin seems to involve the conclusion that the metallic minerals are dissolved out of the rocks through which the waters descend, while the magmatic hypothesis finds the source of the metals in the plutonic masses. For lack of space, it is impracticable to present here the evidence for and against these conflicting opinions; it must suffice to point out that the exceptional occurrence of the metalliferous veins and the nearly or quite universal association of igneous rocks with such veins, seem, in the present state of knowledge, to lend greater probability to the magmatic hypothesis. Obviously, however, no definitive conclusion is yet possible.

Secondary Enrichment of Veins. — The outcrop of a mineral vein is much altered by weathering; the depth to which this alteration penetrates is determined by the level of the ground water. For example, in the deeper portion of many gold-bearing veins the gold is contained in crystals of pyrite, while above the ground-water level, in the shell of weathering, the gold is scattered in minute threads and grains of native metal through a mass of more or less shattered quartz, which is stained rusty red or brown, and the pyrite has disappeared. Pyrite, when exposed to air and water, is slowly converted into the soluble ferrous sulphate ($FeSO_4$), which in turn is oxidized into limonite, with liberation of sulphuric acid. Iron is an important constituent of most ore bodies, and its concentration and deposition below the surface as limonite or hæmatite forms the *gossan*, or *iron hat* of mining phraseology.

In many veins the process of weathering results in the formation of a zone of secondarily enriched sulphides. The unaltered ores in the depths of the vein are sulphides, but from the surface to the ground-water level they are oxidized, and below the zone of oxidation is found that of the secondary sulphides, which, when present, is apt to be much richer than the deeper portions of the vein, because it represents an additional stage of concentration. The metals are dissolved in the oxidized portion of the vein by percolating waters, carried downward and substituted for part of the iron in the original sulphides below.

Ore Deposits formed by Surface Waters. — The ore bodies of this class are formed by the concentration of the metals disseminated in the rocks, through solution and deposition by surface waters. Such deposits are made not far from the surface, to which they show a definite relation, and disappear downward. The most abundant ores of this class are those of iron and are exemplified in the famous Lake Superior region.

Many of the ores of lead and zinc, which also occur in veins, seem to be referable to this class, though the bodies have no definite relation in form to the surface. Such ores occur in limestones, in crevices, along joint or bedding-planes, in cavities, or by replacement of the country rock, and appear to have no connection with any fissures rising from great depths, nor with intrusive masses of igneous rock. The mode of formation of these ore bodies has been the subject of much discussion and is not yet entirely clear, but in the case of the upper Mississippi Valley, for example, it is very generally believed that the deposition has been accomplished by descending and circulating waters from the surface which have dissolved and concentrated the metallic sulphides originally disseminated thinly through the limestones. The disseminated sulphides are supposed to have been deposited in the limestones at the time the latter were accumulating in a great inland sea, being brought in solution from the land. There is ground for believing that lead is but one member of a series of radio-active elements, and, if this is true, we shall be unable to determine which of these elements was the one actually deposited in that ancient sea.

Summary of Structural Geology. — Structural geology brings vividly before us the innumerable changes through which the earth's surface has passed, and which are recorded in the rocks. The sedimentary rocks, originally laid down under water in approximately horizontal positions, have been upheaved into land surfaces, either without losing that horizontality, or being tilted, folded, compressed, or even violently overturned. Or, they may be fractured and dislocated in great faults and thrusts. These movements we have found to be due to enormous lateral compression set up within the crust of the earth, a compression generated in some manner not yet clearly understood. Whether folding or faulting shall result from a given compression depends upon the rigidity of the strata, upon the load which overlies them, and the sudden or gradual way in which compression is applied. The results of compression on a large scale are accompanied by certain minor changes not less characteristic. Compressed rocks are cleaved, fissile, or schistose, according to the intensity of the action, and whether the rocks affected are in the shell of flowage or of fracture. These changes may go so far as completely to reconstruct the minerals of the rocks, destroying the old, generating new, and obliterating the original character of the strata. *Thus, displacements, dislocations, cleavage, fissility, and dynamic metamorphism are but the varying results of lateral compression, acting under different conditions and at varying depths.*

Another class of rocks — the igneous, massive, or unstratified — we found to have penetrated and overflowed the strata, and to have consolidated in the fissures and cavities which they have made for themselves, or to have been poured out freely on the surface. According to the circumstances under which these masses have cooled, the resulting rock is of glassy, porphyritic, finely or coarsely crystalline texture. When solidified as sheets or dykes, the igneous rocks may be folded, faulted, cleaved, or metamorphosed like the strata, and when a region has been long and repeatedly subjected to compression, its structure may become excessively complex, and the metamorphosis of its rocks so complete that not even the most careful examination will suffice to distinguish those rocks which were originally sedimentary from those which were igneous.

Highly heated waters circulating through fissures and along the joint-planes of the rocks deposit the substances which form the mineral and metalliferous veins, though concerning the source of these substances and of the solvent waters there is much difference of opinion.

Our study has taught us that many of these processes go on

2 F

deep within the earth's crust, and hence cannot be directly observed, but must be inferred from their results. Encouraging progress has already been made in this work, but very much more remains to be done before our knowledge of structure and its full meaning shall be even approximately complete.

PART III

GEOMORPHOLOGY

CHAPTER XVIII

THE GEOGRAPHICAL CYCLE

GEOMORPHOLOGY, or physiography, is the study of the topographical features of the earth, and of the means by which, and the manner in which, they have been produced. In this country the term physiography, or physiographical geology, is firmly established and very widely used. This is unfortunate, because the term was originally proposed and still continues to be employed in a very different sense. It would be an advantage in clearness and precision of nomenclature, if *Geomorphogeny*, which is extensively made use of in Germany, could be substituted.

This subject is primarily a department of physical geography, but is of value to the geologist for the light which it throws upon the historical development of the land surfaces, and upon features of the past which are not recorded in the processes of sedimentation. The geographer endeavours to explain the topographical forms of the land, and, in order to do this, he must show how those forms have originated. The geologist, on the other hand, makes use of the topography to determine what changes have passed over the land, and in what order those changes have occurred. The old method of reading geological history concerned itself merely with the sedimentary accumulations and igneous intrusions. This method has the defect of leaving us without information regarding the changes of land surfaces (except where transgressions of the sea are recorded in unconformities) and the details of mountain-making. The physiographical method supplements this by adding, in part, the required information concerning the land surfaces. Each method is improved and strengthened when we use both of them together, and when we are able to correlate the accumulations of sediments with the denuding processes which furnished the material.

The topography of any land area may be considered as the outcome of a struggle between two opposing sets of agencies: (1) those which tend to upheave the region and thus increase its elevation; (2) those which tend to cut down the land in one place and build it up in another. The latter comprise the agencies of *degradation* and *aggradation* respectively, while the former are the *diastrophic agencies*.

The details of topography are, in large degree, controlled by still a third class of factors, which, however, are passive rather than active; namely, the character, arrangement, and attitude of the rock masses. A partially degraded region in which the rocks are homogeneous will have a very different kind of relief from one in which the rocks are heterogeneous and differ materially in their powers of resistance to the denuding agents. A region of horizontal strata will give rise to very different topographical forms from those which are developed in areas of folded or tilted strata. We must further distinguish between regions whose topography is, in the main, due to *constructive* processes and those in which denudation has prevailed. Examples of such constructive forms are volcanic mountains, and plains or plateaus formed by widely extended lava flows, plains newly deserted by the sea and due to sedimentation, alluvial plains of rivers, and the mounds, ridges, or sheets of drift spread out by the action of glaciers and of the waters derived from their melting. Still another important kind of topography is the *tectonic*, in which the main features have been determined by tectonic processes, more or less modified by subsequent denudation; the ridges are anticlines and the valleys synclines, while fault scarps may form long lines of cliff.

The topography of any region is, as we have seen, the resultant of the very complex interaction of many different kinds of factors, and is subject to continual change according to definite laws. Let us suppose, in the first instance, a region newly upheaved from beneath the sea into dry land. The topography of such an area will be *constructional*, due entirely to the processes of diastrophism and accumulation, and characterized by the absence of

FIG. 229. — Volcanic topography, northern Arizona. (U. S. G. S.)

a highly developed system of drainage by streams. The coastal plain of the middle and southern Atlantic States is an example of such topography but slightly modified.

Next, the processes of denudation begin their work upon the region. The sea attacks the coast-line by cutting it back in one place and building it out in another, until a condition of equilibrium is attained. Rivers are established, adjusting themselves to the structure of the underlying rocks, and cutting deep, trench-like valleys, while the atmospheric agencies widen out the valleys, slowly wearing down and washing away the sides and tops of the hills. This is the stage in which we find the greatest degree and variety of relief, and it may be called the stage of *maturity*, as contrasted with the first, which is a stage of *youth*. The continuance of the degrading operations will, if uninterrupted, eventually wear down the region to a nearly plane surface, through which

FIG. 230. — Glacial topography, eastern Washington. (U. S. G. S.)

sluggish streams meander, the featureless condition of *old age.* When the process is complete, the country is said to be *base-levelled.*

The conception of the *cycle of topographical development* is essential in geomorphological reasoning. Each cycle begins with the uplift of an area approximately at base level, the processes of denudation working with minimum efficiency and extreme slowness. The movement of upheaval revivifies the destructive agencies, and the work of carving out a surface of relief begins afresh, only to terminate, unless interrupted by renewed elevation, in once more base-levelling the region. A complete cycle is thus from base level back to base level, though, as it is a cycle, a beginning may be selected at any part of it.

The details of the cycle differ widely under different climatic conditions. If we take the successive stages as they are developed in a pluvial climate, with all basins filled, abundant rivers running to the sea, and all the snow of winter melting in summer, to constitute the *normal cycle*, we shall find that the *arid cycle* of desert climates deviates from the normal in very important ways.

The term *age* as applied to topographical features does not mean the length of time required for their formation, but merely the stage of development within the cycle which they have attained. The length of time required to reach a given stage of such development will vary greatly in different regions, in accordance with climatic conditions, the resistance of the rocks, their altitude above sea-level, and similar factors. An area of resistant rocks in an arid climate will be hardly at all affected in the time that a mass of soft rocks exposed to a heavy rainfall will be cut down to base-level.

It seldom, if ever, happens that the topographical development of a region proceeds uninterruptedly through the stages of youth, maturity, and old age. Oscillations of level introduce new condirions and cause the work of denudation to start afresh with renewed energy, or, if the movement be one of depression, it will check the work already in progress. The cycles of development are thus partial rather than complete, and a given region may display topographical forms dating from very different and widely separated cycles. The more resistant rocks retain the features acquired in an earlier cycle, while the weaker and more destructible rocks have already taken on the forms due to a later cycle. A landscape thus often includes features of different geological dates, and it is in the identification of these that the value of the physiographical method to historical geology consists.

In the production of new topographical forms, old ones are more or less completely destroyed, and thus, the farther back in time we go, the fewer subdivisions are recognizable, and only the outlines of the great cycles can be followed.   Very ancient features would be quite obliterated in the successive cycles of development, were they not sometimes buried under the sediments of an encroaching sea.   A subsequent reëlevation of the area into land, and a stripping away of the covering of newer sediments by the agencies of denudation, will again bring to light the ancient land surface which had been buried for ages.   An interesting example of this is presented by the Charnwood Forest in England, where an extremely ancient landscape is slowly coming to light, as the covering of soft rocks, which has so long preserved it, is removed by denudation.

In Part I we have already studied the agencies of denudation, but there we concerned ourselves principally with their modes of operation and their efficiency in destroying old rocks and in furnishing material for the construction of new.   We have now to consider these agencies from a somewhat different point of view; to determine the characteristic forms of land sculpture which they produce at the various stages of their work.

The Sea. — The work of the sea is confined to the coast-line, which it cuts back by the impact of its waves and currents.   Speaking broadly, the waves do but little effective work below the limits of low tide, and advance by undermining and cutting down the cliffs which form the coast.   The result of the work is to form a platform covered by shallow water, which is called a plain of marine denudation.   As observed in actual cases, these platforms are narrow; for so long as the sea-level remains constant with reference to the land, there is a limit to the effective assault of the waves upon the shore.   The water covering the platform is very shallow, and only in exceptional cases do the waves have sufficient power to overcome the friction of a wide platform.   The materials removed from the land are piled up at the seaward foot of the platform and extend it in that direction.

An example of a plain of marine denudation is found on the north coast of Spain, where there is a broad platform between the mountains and the sea, almost perfectly flat. This plain has been uplifted above the sea-level and has been but little dissected by the subaërial agents. Narrower platforms, still in process of extension, may be observed on most rocky and precipitous coasts, as those of Scotland, Ireland, and France. Along a slowly sinking coast the platforms may be cut back much farther, for the deepening water prevents the loss of wave power by the friction on a shoal bottom. If, on the other hand, the coast rises at intervals, a series of terrace-like platforms will be cut.

As we shall see in the following section, plains may be produced by the work of the subaërial agencies, and it is often important to distinguish between the plains of submarine and those of subaërial origin. This distinction cannot always be made with certainty, but not unfrequently the plain shows unmistakable signs of the manner in which it was made. In the plain of marine denudation the sediments formed from the waste of the land will be deposited upon the seaward portion of the platform, or upon a lower level of previous formation. Further, this sediment will show by its character that it actually was derived from the material cut away by smoothing the plain, and the whole of it, even its bottom layers, will be of marine origin. In such a plain the advancing sea must have obliterated the stream valleys which had been excavated when the region was land. This obliteration will be performed partly by shaving down the divides, or watersheds, between the streams and partly by filling up the valleys with sediment.

When the region is once more uplifted above the level of the sea, an entirely new system of drainage will be established upon it, determined by the slopes of the overlying cover of newly deposited sediments, and having no reference to the structure and arrangement of the underlying older rocks. These newly established streams may, if the upheaval of the country gives them sufficient fall, cut down through the newer sediments. Indeed,

the latter may eventually be swept away entirely by the various
subaërial agencies, but the stream courses, which were determined
originally by the slopes of that newer sediment, will show little or
no adjustment to the structure of the underlying older rocks.

These criteria are useful in identifying those plains which were
smoothed by the action of the sea; but when the processes of subaërial denudation have completely dissected the elevated area, all
such evidences may be removed and the origin of the plain may
become quite indeterminable.

The Subaërial Agents are those which operate over the entire
surface of the land. Their tendency is, in the first instance, to
carve out valleys and leave relative eminences standing, and thus
to increase the irregularity, or *relief*, of the land. This, however,
is merely a temporary stage, and if time enough be granted, these
agencies will sweep away the irregularities and plane the entire
region down to base-level.

Rivers cut down and deepen their channels so long as their
beds have sufficient slope and fall. The banks also are undermined, as the current swings from side to side, and frequently fall,
thus widening the channel. The sides of the trench, unless removed by other agencies, will be as steep as the nature of the
rock material will allow. Unassisted river action will, therefore,
cut nearly vertical trenches, which are continually deepened, until
the base-level is reached. Examples of such river-cut trenches
are the Au Sable Chasm (see Fig. 58, p. 142) and the inner gorge
of the Grand Cañon of the Colorado.

The trench-like valley, with nearly vertical sides, is, however,
not the usual form of river valley. The atmospheric agencies, the
undermining and sapping of springs, landslips, and the like, are
continually wearing away the sides of the excavation, the waste
thus produced being readily carried away by the stream. As the
upper part of each hillside and cliff is that which has been longest
exposed to the denuding agencies, the valley will be widened
at the top more than at the bottom, and will gradually become

widely open, unless the alternation of hard and soft strata be such as to favour the retention of the cliff-like form by undermining. A system of river valleys is normally accordant, the tributaries entering the main stream at grade, and each valley is winding, with projecting spurs from the sides, and of V-shaped cross-section.

The rapidity with which the deep and narrow trench is widened into the broad, gently sloping valley will depend upon two sets of conditions. (1) Upon the climate, which is as much as to say the intensity with which the denuding forces operate. Cañons and narrow gorges are much more frequent in arid regions than in those of abundant rainfall. (2) Upon the resistant power of the rocks. If the valley sides are composed of rocks which yield readily to weathering, the trench will be speedily broadened, while if the rocks offer great resistance to chemical and mechanical disintegration, the gorge-like form will be retained very much longer. This is illustrated by almost any considerable stream, such as the Delaware or the Potomac. In certain places the valley is widely open, while in other parts of the course are deep gorges, as at the Delaware Water Gap and Harper's Ferry. The gorges occur in the places where the stream cuts across hard, resistant rocks, and the open valleys are found where it intersects softer and more destructible rocks.

Rivers also produce changes in topography by constructional processes, as in their flood plains and terraces, processes which are most notable in the lower parts of the course, and which gain increased efficiency through a subsidence of the region.

Degradation is most rapid on the hillsides which border river valleys, because of the removal of waste by the rivers. Away from the streams the denudation of the country is much slower, because the waste is less readily removed. Those points will longest remain standing above the general level which are composed of the hardest rocks and are farthest removed from the principal lines of drainage.

A glaciated region has a topography marked by rounded,

flowing outlines, with smoothed, polished and striated rocks in the
central zone, where erosion was most active, and with lines of
moraine, sheets of drift and overwash plains, eskers and drumlins
in the peripheral zone, where denudation was feeblest and deposition more important. Glacially excavated valleys are over deepened, U-shaped in section, with the projecting spurs truncated,
or entirely removed, and the tributary valleys are not graded to
the main trunk, but left hanging on the retreat of the ice. Great
numbers of lakes are characteristic of such regions.

The subaërial agencies act with the greater efficiency the more
elevated the region upon which they operate. Consequently, so
long as the region be not again elevated, denudation operates at
a continually diminishing rate. The strong relief of hill and
valley is carved out with comparative rapidity, but the more
nearly the country is reduced to base-level, the more slowly does
degradation proceed, and the final stages of base-levelling must be
exceedingly slow. Nevertheless, if no renewed upheaval takes
place, the loftiest and most rugged land surface must be eventually cut down to that level. The universal and permanent base-
level is, of course, the sea; but other local and temporary base-
levels may for a time control the development of certain areas.
Tributaries cannot cut below the main stream into which they
flow; a lake forms the base-level for the streams which supply
it, until the lake is removed by draining away or being filled with
sediment. Regions, like the Great Basin, whose drainage finds
no outlet, may have base-levels either above or below the level of
the sea; e.g. the surface of the Dead Sea of Palestine is 1308 feet
below the Mediterranean.

It is perhaps a question whether any large region has ever
remained stationary for a sufficiently long time to be absolutely
base-levelled. On the other hand, there is abundant evidence
to show that such areas have been worn down to a low-lying,
featureless surface, with only occasional low protuberances rising
above the general level. Such a surface is called a *peneplain*,
and represents what is usually the final stage of a cycle of denuda-

tion. Here and there an isolated peak may remain high enough to deserve the name of mountain, which owes its preservation to the exceptionally resistant nature of the rocks of which it is composed, or to its exceptionally favourable position with reference to the drainage lines. A renewed upheaval of the peneplain will begin another cycle of denudation, revivifying and rejuvenating all the destructive agencies, and valleys and hills will be carved out of the approximately level surface. In a peneplain dissected by the revived streams the sky-line of the ridges is notably even, and all

FIG. 231. — Peneplain, with residual mountain, Southern California. (Photograph by H. W. Fairbanks)

the heights rise to nearly the same level. Differences of level are, however, frequently produced by a warping process, which may accompany the upheaval, raising some portions of the peneplain to greater heights than others. Excellent examples of reëlevated and subsequently dissected peneplains are the uplands of southern New England and the highlands of New Jersey.

In topography climatic differences are very obvious, because in each climate the dominant subaërial agents are characteristic, and the modifying effects of vegetation are likewise dependent upon climatic factors. In the polar lands destructive work is accomplished chiefly by the activities of frost and ice, while in temperate lands with normal rainfall rivers and rain are the principal agents. In arid regions changes of temperature and wind are the most active processes, though the scanty vegetation gives to the rare but violent rains an unusual effectiveness. All of these climatic differences are reflected in characteristic topographical forms.

The Arid Cycle

"The essential features of the arid climate . . . are: so small a rainfall that plant growth is scanty, that no basins of initial deformation are filled to overflowing, that no large trunk rivers are formed, and hence that the drainage does not reach the sea." (Davis.)

The peculiarities of erosion in arid climates have already been described; it remains to point out the characteristic features of the geographical cycle under arid conditions, as these are defined in the preceding paragraph. The successive steps of the cycle are much affected by the topography at the beginning, but it would lead us too far to take into consideration all the various cases, and only a general outline can be attempted. We need only assume the elevation of a large area, with more or less of deformation. The drainage will be consequent on the newly formed slopes, and the lowest part of each basin will form the local base-level, for the streams of each basin of deformation are confined to that basin and die away in the floor without uniting into a permanent trunk stream. Occasionally or periodically a playa lake may form, into which all the streams may flow, but as a rule they are disconnected fragments of a drainage system.

In the youth of the cycle the highlands are slowly eroded, and deposition takes place on the slopes and floor of each basin, diminishing the relief and raising the local base-level, a strong contrast to the corresponding stage of the normal cycle, in which

relief is increased by the excavation of stream valleys. Even
in arid regions, however, valleys are cut on the highland slopes,
while the basin floor is made nearly level by deposition. This
stage is exemplified by the Great Basin and its mountains. Water
is the chief agent of erosion and deposition during the period
of youth, but the wind is also important in eroding the bare
rocks and in distributing the finer waste, part of which it
carries outside of the arid region altogether. Extremely slow
as this process of complete removal of the finer débris by the

FIG. 232. — The Mohave Desert, California. (Photograph by H. W. Fairbanks)

wind undoubtedly is, yet it is the only agency which actually
lowers the *average* altitude of the region, for no water flows out
of the area we are considering.

Maturity of development is attained by the connection of
the separate initial basins into a continuous whole. Erosion
of the highlands and deposition on the basin floors may result
in the formation of a continous slope from a higher basin to a
lower one, so that, even in the absence of any permanent stream
connecting the two, the rain will wash material to the lower
basin and a new and lower base-level will be established for the

higher one, while the lower basin floor will be built up by the transfer to it of the waste accumulated in that which lies at the higher level. " As the coalescence of basins and the integration of stream systems progress, the changes of local base-levels will be fewer and slower, and the obliteration of the uplands, the development of graded piedmont slopes and the aggradation of the chief basins will be more and more extensive." (Davis.) In maturity the relative efficiency of the eroding agents is not the same as in the earlier stages. Large areas at the foot of the mountains and highlands have already been graded to an even slope by the torrents formed by the rare but violent rains. When vertical trenching can be carried no farther, lateral erosion removes the divides between the streams, which thus have no definite channels, and the water flows down the graded slope in a thin sheet. The wasting away of the highlands diminishes the scanty rainfall, and the quantity and efficiency of the water decrease with the diminution of relief, but the activity of the wind is not affected, and hence its relative importance is increased.

The continuance through vast periods of time of the slow processes above outlined will result in the destruction of the original relief, its place being taken by large plains of bare rock, sloping to plains of accumulation. If the strata have been disturbed by folding or tilting, the plain surface cuts across their structure to an even slope. The work is done without reference to the sea as a base-level and the altitude of the plain is determined by the loss of material through transportation by the wind.

The period of old age begins with the breaking up of the unified drainage system through the excavation of wind-made hollows in the softer rocks, a process unlike anything which we have found in pluvial climates, where a drainage system once established is not disintegrated during the progress of a normal cycle. If the desert were strictly rainless and erosion carried on by the wind alone, there would apparently be no limit to the excavation of such hollows, but no such instance is known, and the occasional heavy rains suffice to counteract it by filling up the hollows, but the drainage system is effectively disintegrated, and the waste, which in the stage of maturity gathered on the lower parts of the region, is washed about irregularly. The surface worn down by the wind may have no slope in any particular direction, for the wind has no base-level and is not affected by the inclination of the ground, but this surface may be reduced to an essentially plain-like character. Harder masses of rock, which have successfully resisted wear, may rise above the plain as residual mountains, just as they sometimes do in the peneplain of the normal cycle.

The deserts of North America have not reached the condition of old age, but it is realized in South Africa, where the desert plains have lately been studied by Dr. Passarge, whose conclusions are thus summarized by Professor Davis: " Passarge states that these desert plains are not undulating with low hills, but true plains of great extent, from which the isolated residual mountains rise like islands from the sea. The residuals may be low mounds, only a few metres high, or lofty mountain masses, rising several thousand metres above the plains. The plain surrounds the steep slope of the mountains with a table-like evenness; there is no transitional belt of piedmont hills, and no intermediate slope. The mountains consist of resistant rocks, such as granite, diorite, gabbro, quartzite, etc., granite being the most frequent; the plains are of more easily eroded rocks, such as gneiss, schists, slates, sandstones, and limestones. The bedding of the rocks is not flat, but disturbed; the plain therefore truncates the rock structures. . . . The products of weathering are usually spread as a thin veneer on the plain; the waste does not lie in place, on the rocks from which it was weathered, but has been drifted about by wind and flood and has gathered in slight depressions. . . . These rock-floored plains are not uplifted peneplains, but are the product of desert erosion unrelated to normal base-level, in which occasional water-action has co-operated with persistent wind-action."

We have definite evidence that the earth has undergone many

2 G

climatic changes and that among these changes the alternation of arid and pluvial conditions in the same region have not infrequently occurred.   Hence, in deciphering the history of any high-level plain the distinction between the normal and the arid cycle must always be borne in mind, and, if possible, evidence obtained which will enable the observer to determine whether the plateau is a reëlevated peneplain, or the product of an arid climate.

CHAPTER XIX

LAND SCULPTURE

WHILE the final effect of the subaërial denuding agencies is to sweep away all relief, and to cut the land surface down to low-lying base-levels or peneplains, yet in the process great irregularities are produced by the more rapid removal of some parts than of others. The topographical forms generated by this differential erosion vary much according to circumstances. We have already considered some of these differences with regard to the agencies which have produced them. Now we have to examine the differences with a view of learning how topographical forms are determined by the character and arrangement of the rocks which are undergoing degradation.

Forms in Horizontal Strata. — When a peneplain or plain of marine denudation is lifted high above sea-level, without folding or steep tilting of the strata, streams are soon established upon the new land, and proceed to cut deep trenches across the plateau, which are gradually widened out under the influence of weathering, and the arrangement of hard and soft rocks finds expression in the resulting forms. If the surface layers resist weathering, the plateau will be gradually dissected into flat-topped *mesas* and table-mountains, which in the progress of denudation are converted into pyramidal shapes; while if the whole mass of rocks be easily destructible, they weather down into dome-shaped and rounded hills, which are smallest at the top, the part longest exposed to weathering. The wild and grotesque scenery of the Western bad lands, with their chaos of peaks, ridges, mesas, and buttes, is merely the result of the differential weathering of horizontal strata, some beds and parts of beds yielding more readily than others. The bad lands are carved out of soft and scarcely indurated rocks, but firm rocks in climates of similar aridity give rise to many vertical-sided mesas (see Frontispiece), as is so notably the case in the sandstones of New Mexico. In pluvial climates the slopes are gentler.

If a series of more resistant beds underlies a mass of softer strata, a change in the topographical forms will occur when the underlying harder rocks are partially exposed. In the soft rocks the valley sides have gentle slopes, but when the harder mass is penetrated, the slopes become steep, or even vertical. When hard and soft strata alternate in a valley wall, the harder beds form cliffs. This is accomplished by cutting away the softer beds and thus undermining the harder ones, until the latter can no longer support their own weight, and masses fall from the face of the cliff, thus maintaining the verticality. The talus blocks form a slope, connecting the successive cliffs by gentler inclines. The Uinta Mountains in northern Utah are formed by a great anticlinal arch, so broad and gently curved that in a given section the strata appear almost horizontal. Out of these immensely thick and nearly level masses the subaërial denuding agencies have carved an infinite and most picturesque variety of peaks, pinnacles, columns, and amphitheatres, while the streams have cut profound and gloomy cañons. Vast talus slopes remain to indicate the amount of destruction.

Forms in Inclined Strata. — Inclined or tilted strata give rise to a different class of topographical forms. If, as is generally the case, harder and softer strata alternate, the latter will be swept away more rapidly than the former, which are left standing as ridges or cliffs, the height and steepness of which are determined by the thickness and inclination of the more resistant rocks. In case the strata are steeply inclined, a succession of hard beds alternating with soft will give rise to a series of ridges and valleys, the slopes of which depend upon the angle of dip. If the beds are standing in a vertical position, the two slopes of each ridge will be nearly equal, the hard strata forming the backbone of the ridge and the softer ones the sloping sides. Often narrow ridges, more or less closely resembling dykes of igneous rocks, are formed by the isolation of hard vertical strata, the softer beds on each side being removed by erosion. As the inclination departs from verticality, the more unequal do the two slopes of each ridge become, the longer and gentler one being in the direction of the dip. Ridges and valleys of this class are beautifully exemplified in the Appalachian Mountains. Figure 40 (P. 115) shows

FIG. 233. — Escarpments and dip slopes, Montana.   (U. S. G. S.)

Kittatinny Mountain, through which the Delaware River has cut the famous Water Gap; the crest of the ridge is formed by very hard and indestructible sandstones and conglomerates, while the broad valley above and below the gap is in slates or other destructible rock.

In gently inclined strata the abruptly truncated and cliff-like outcrops of the hard beds are called *escarpments*, and follow, with some irregularities and sinuosities, the line of strike.

Whether the general course of the escarpment shall be straight or curved will, therefore, be determined by the constancy or change in the direction of the dip; for, as we have already learned, the strike changes with the dip, always keeping at right angles to the latter. The upper surface of the gently inclined hard stratum may be completely exposed by the stripping away of the softer overlying mass, and then the slope of the ground is the same as that of the resistant stratum, and is called a *dip slope.* A series of gently inclined strata, made up of alternating harder and softer beds, will thus give rise to parallel ridges and valleys, or escarpments and dip slopes, according to the completeness with which the softer beds are removed and the harder ones exposed. A magnificent example of such escarpments and slopes is displayed in the high plateaus of Utah and Arizona, where the dip slopes are from 20 to 60 miles broad and the escarpments 1500 to 2000 feet high. The amount of denudation involved in the production of these vast amphitheatres staggers belief, though there is no escape from the enormous figures.

Under the influence of denudation escarpments are continually though slowly receding, being cut back in the direction of the dip. Rain and frost act directly upon the hard beds, but work more effectively by cutting away the softer beds below and thus undermining the hard strata, causing them to fall. The fallen masses are gradually disintegrated in their turn and washed away into the water-courses. The escarpments may follow a relatively straight or a very sinuous course. Sinuosities, when present, are commonly due to the action of springs, which undermine the escarpments and, by the recession of their heads, excavate the line of cliffs into bays and amphitheatres. A sinuous escarpment is more rapidly cut back than a straight one, because, in addition to the coöperation of the springs, it offers a larger surface to the attack of the destructive agencies. Every step in the recession of an escarpment lowers the ridge and brings it nearer to base-level, because the direction of retreat follows the line of dip, which carries the beds down to base-level with a rapidity determined by the angle of dip. A steeply inclined bed needs to be cut back only a short distance, when it will be reduced to base-level, whereas a bed dipping very gently remains above base-level for long distances. Of course the general elevation of the whole region above base-level is also an important factor in determining the amount of work to be done.

For reasons that will appear later, we assume that when denudation began its work upon a region of inclined strata, that region was a sloping plain, or peneplain, formed by the outcropping *edges* of the strata. The first lines of drainage established would necessarily follow this slope, and the first valley or valleys cut would be across the strike of the beds, trenching both hard and soft beds. Such valleys are called *transverse*, or *dip* valleys, and the streams which flow in them, *transverse streams*. Transverse streams cut steep-sided, cañon-like valleys, the rocks giving way along the joints and making the two valley-walls alike, as though the valley were cut in horizontal strata. A second series of valleys will be excavated along the strike of the softer beds, giving *longitudinal*, or *strike*, valleys and streams. In such a longitudinal valley, following the strike of a mass of soft strata, the stream which occupies it will tend to flow along the foot of the escarpment formed by the outcrop of hard strata, and to shift its course laterally in the direction of the dip, cutting away the soft beds in which it flows, and undermining the hard escarpment. Longitudinal or strike valleys tend to have one steep or vertical, and one gently sloping side. The strata dip across the stream and hence on one side are inclined toward the valley and on the other away from it. The former is the weaker structure, because the loosened joint-blocks glide into the stream, and the ground-water, following the stratification planes, forms springs on that side of the valley. The side on which the dip is away from the stream is attacked chiefly by the undermining action of the stream and thus kept vertical. Such a stream is a potent agent in causing the recession of the escarpment and may remove large areas of both hard and soft strata.

The steep ridges, or " hog-backs," which occur among the foothills of the Rocky Mountains, show interesting examples of streams flowing along the strike of inclined strata, though the ridges are themselves not formed quite in the way already described. They are composed of the steeply dipping limbs of *monoclinal* folds, of which the upper horizontal limbs have been removed by denudation (Fig. 234).

FIG. 234. — Hog-backs, east side of Laramie Mts., Wyo.   (U. S. G. S.)

Forms in Folded Strata. — A region of folded strata is, in the first instance, thrown into a series of ridges and valleys, the ridges formed by anticlines and the valleys by synclines; in other words, the topography is *tectonic* in character and determined by diastrophic movements.   If the folding be of moderate degree, so as to produce undulations of sweeping and gentle curves, the tendency of denudation is to reverse the original topography

and convert the anticlines into valleys and the synclines into
ridges. This apparently paradoxical result is found, when
examined, to be natural and simple enough. The crests of
newly formed anticlines have been subjected to tensile stresses
which open the joints in the strata and render them an easy
prey to the denuding agents. The surface of the synclines, on
the contrary, has been tightly compressed, and their joints are

FIG. 235. — Anticlinal ridge, Big Horn Mts., Wyo., hard beds in relief. (U. S. G. S.)

closed by crowding. Aside from this, another factor tends to
produce the same result. In a folded series of alternating harder
and softer beds denudation is most rapid on the exposed anticlines, and in them the hard strata are first reached and cut
through. When an underlying mass of soft strata is reached,
it is rapidly trenched into valleys which may soon be excavated
below the level of the synclinal troughs.

If the folds originally made by the force of lateral compression be steep and high, as in mountain ranges, the anticlines persist longer as ridges, but the wearing away of their summits gives rise to subordinate ridges and valleys within the limits of each anticlinal arch. Here also the ridges are the outcropping harder beds, and the valleys are cut in the softer ones. Even in mountain ranges denudation may reverse the original structural topography and give rise to anticlinal valleys and synclinal mountains.

FIG. 236. — Truncated anticlinal ridge, Montana.   (U. S. G. S.)

If a region of folded rocks has once been planed down to base-level or to a peneplain, and then reëlevated and subjected to denudation, the resulting topography will be determined by the same laws. Indeed, this is a frequent method in which regions of tilted or inclined strata are produced, for, as we saw in Chapter XII (p. 327), inclined beds are very often parts of truncated folds. In such regions drainage is first in accordance with the slopes of the planed and tilted surface, but as denudation proceeds, the structure and arrangement of the rocks make themselves felt, and bring about changes and adjustments of the drainage to the structure, as will be more fully explained in a subsequent chapter.

Forms in Volcanic Rocks. — Volcanic topography is primarily constructive. The cones vary in form and height in accordance with the amount and character of the material of which they are composed, and the nature of the eruptions. Thus, we find lofty, steep-sided cinder cones, like those of the Pacific coast, or very gently sloping lava cones, like those of the Sandwich Islands, or truncated cones and crater-rings, due to violent explosions and to remelting and engulfment of the upper part of the cone. Lava flows may take the form of long, narrow streams, or great floods poured out one upon another, until immense volcanic plateaus are built up, like those of Oregon and Washington.

The progress of denudation sculptures these volcanic masses in characteristic ways, already described in Chapter XV. Cones are first furrowed with ravines and valleys and then gradually degraded into necks, or into low hills of volcanic agglomerate. The infinite variety of combinations of soft tuffs, loose masses of scoriæ and ash, hard sheets, streams, pipes and dykes of lava, give rise to the manifold forms of volcanic mountains and islands in the course of denudation, the harder elements resisting longer and standing in relief. Lava flows are generally harder than the stratified rocks upon which they rest, and therefore, aside from their original form, their topographical effect is much the same as that of an exceptionally hard stratum among softer beds. A surface stream may be, indeed eventually must be, cut up by erosion into isolated masses, which protect the underlying softer rocks and thus form flat-topped table mountains and mesas, the lava cap with nearly vertical sides, the stratified rocks below with gentler slopes, especially in pluvial climates. A lava plateau is dissected by streams, first trenching steep cañons and then atmospheric erosion widens the cañons and narrows the divides, just as in ordinary plateaus of stratified rocks, but the great hardness of the volcanic masses renders the process very slow.

Forms in Plutonic Rocks. — It is exceptional that topographical features can be definitely referred to the constructive or tectonic effect of intrusive plutonic bodies, for the obvious reason that the presence of a plutonic mass at a given point can rarely be determined until the covering of overlying strata has been removed. An exception to this is given by many laccolithic hills and mountains, in which the covering of strata is more or less completely retained. In Little Sun-Dance Hill (Fig. 219) we have a dome-like hill, the strata of which are almost intact and the presence of the plutonic body is only inferred, not absolutely certain. A second stage of denudation is found in Bear Butte (Fig. 220) of the same region, the covering of strata being removed, except where they are upturned around the base of the butte, and finally, in Mato Tepee, we have only the central core of the laccolith preserved (Fig. 221). The Henry Mountains of southern Utah and several of the Colorado ranges display laccoliths in all stages of dissection.

The plutonic bodies are exposed by denudation, and since they are, as a rule, more resistant than the invaded rocks, they generally form prominences corresponding to the form of the intrusive mass. Sometimes, however, the intrusive body is less resistant than the enclosing rocks and then is marked by a depression. Dykes, when exposed by denudation, stand out in relief as long walls, the height of which is determined largely by the thickness of the dyke and by its resistance to destruction. In certain cases, as in North Carolina, the dyke-rock disintegrates more rapidly than the enclosing rock, and hence long trenches indicate the position of the dykes. Sills, so far as their effect upon topography is concerned, may be regarded simply as hard strata, but some sills are much thicker than strata often are.

Stocks, which increase in diameter downward, project as small hills when first exposed, but when they are slowly denuded and the country-rock is rapidly worn away, they become larger and *relatively* higher, as the surrounding area is lowered by denudation. If the region where the stock is found is sufficiently above base-level, a very high hill and even a mountain may thus be formed.

FIG. 237. — Palisade-sill, Fort Lee, N.J.   (Photograph by van Ingen)

Batholiths, like stocks, increase in size downward, lying upon no floor of country-rock. Hence, they give rise to great ridges, or irregular masses, often of enormous size, when laid bare by erosion. Many mountain ranges are composed of granite batholiths, from which the covering of strata has been stripped away and which are themselves deeply dissected into peaks and crags. On the other hand, such batholiths are no exception to the rule that plutonic bodies may sometimes wear away more rapidly than the rocks which enclose them. When this occurs, the batholith will be found as a plain, or depression, with the more resistant rocks rising above it.

In brief, *the controlling factor in a region of mature topography is the arrangement of the rock masses, prominences being due to the more resistant rocks, whatever their nature.*

CHAPTER XX

TOPOGRAPHY AS DETERMINED BY FAULTS AND JOINTS

NOT uncommonly faults have no direct effect upon topography, or whatever influence they may originally have exerted has been lost, denudation having·worn down the two sides to the same level or to a continuous slope, so that no evidence of the fault appears in the surface forms, but must be indirectly obtained. When newly formed, a fault is accompanied by the scarp, as a long line of cliff or bluff. As modern faults teach us, displacements of great throw are probably produced by repeated movements along the same plane of fracture, and thus require long periods for their completion; and during this process of dislocation erosion is actively at work, especially upon the scarp, which is most exposed to attack. The length of time during which the scarp persists, as in the case of topographical features generally, depends upon the activity of weathering and the resistant power of the rock. Well-preserved and lofty fault-scarps are therefore very much more frequent in arid than in pluvial climates. Long before its removal the scarp is deeply dissected and made rough and craggy, so as to resemble a mountain range. Examples of this kind of topography are to be seen on a grand scale in the arid parts of the West, their preservation being due in part to the geologically late date of the dislocations, and in part to the aridity of the climate which makes denudation very slow. The imposing lines of cliff which demarcate the plateau of the Colorado on the western side are fault scarps, and the plateau itself is crossed by many minor faults which still form prominent surface features, though much modified by erosion and some of them quite disguised.

The Great Basin, which is enclosed between the Wasatch Mountains on the east and the Sierra Nevada on the west, and comprises nearly all of Utah and Nevada, is bounded by the enormous fault-scarps of those ranges. The Sierra Nevada escarpment has a throw estimated at 15,000 feet, though this has been much reduced in height, and is deeply dissected by erosion, so that its true nature is not immediately apparent. On the eastern side the Wasatch escarpment is similarly dissected, but there have been additional very recent movements along the old fault-planes, and the lately formed scarps, though low, are remarkably fresh and unchanged

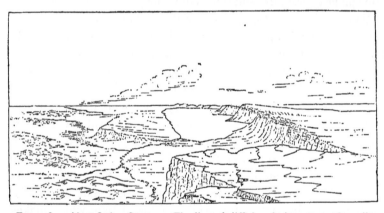

FIG. 238. — Abert Lake, Oregon. The line of cliffs is a fault-scarp. (Russell)

even in incoherent materials. The Basin itself is traversed by many north and south faults, and the blocks included between these parallel lines of dislocation are tilted and form great scarps, but with gently inclined top. Each block thus has an abrupt side, the rugged and worn scarp, and a long, gently sloping side, which gradually inclines to the foot of the next parallel scarp. Denudation has carved these tilted fault-blocks into ridges and peaks, giving them the appearance of an ordinary mountain range, when seen from the escarpment side, but formed in an entirely different manner from true mountains of folding, and known as *block mountains*. In southern Oregon the plateau of basalt has been fractured into a series of blocks, which are tilted, with very gentle ascent on one side, ending in an abrupt scarp on the other. These have been so little affected by erosion that their true character is immediately apparent.

The Adirondack Mountains display a topography which is strongly dominated by systems of faults; their general character is thus described by Professor Kemp: " The Adirondack region . . . is mountainous in its eastern half and has its highest peaks near its centre, but on the west the mountains disappear and the

FIG. 239. — Sierra Nevada fault-scarp, Mono Lake, Cal.   (U. S. G. S.)

area becomes a plateau ranging from 2000 feet above tide gradually downward to the west until it is but slightly higher than Lake Ontario and the St. Lawrence. . . .   The mountains are arranged in visible northeast and southwest lines, and are often very steep if not positively precipitous in the portions that look to the southeast or northwest. There are also other steep faces nearly at right angles with the above, but they are less pronounced." These features are due to three intersecting systems of faults, the principal one of which is northeast and southwest and determines the general trend of the mountains. The second system, which trends north-

2 H

east at right angles to the first, is less important, and acts chiefly in cutting the ridges into separate blocks. The third series, with north and south trends, is quite subordinate in its effects. The faults extend southward for many miles from the mountains into the plain, and some of them still retain their scarps.

In Chapter XIII (p. 347) we learned that a block isolated by faults hading *away* from it on all sides, so that the block is on the upthrow side with reference to the area all around it, is called a *Horst*, a name adopted from the German for lack of an English term. The Horst structure is thus much like that of the block mountains of the Great Basin and the Adirondacks, except for the direction of slope in the fault-planes and for the fact that there is usually no relation between the strike of the beds and that of the faults. Several examples of Horst mountains occur in Europe, such as the central plateau of France, the Vosges, the Black and Thuringian Forests, and the Hartz Mountains. These uplifts are composed chiefly of ancient, very strongly folded and hard rocks, from which most of the covering strata have been removed by denudation, and rise quite abruptly from the comparatively undisturbed strata of the surrounding lowlands.

Along the entire eastern coast of Asia the topography is controlled by gigantic systems of faults. The zigzag mountain ranges which bound the coastal plains on the west have a general trend approximately parallel to the coast, and mark a series of great faults with downthrows to the east. The coastal plain itself is divided into a series of more or less parallel blocks by step faults, also with downthrows to the eastward, resulting in a system of terraces, which, however, have not flat, but inclined tops, because of the tilting of the blocks, much as in the block mountains of southern Oregon, though on a vastly larger scale. An east-west section through northern China shows three such enormous blocks, the Mongolian block on the west, the Japanese block on the east, with the Manchurian between; the Japanese block is partly submerged, making the outer portion an island, as is true of the whole chain of islands which fringes the eastern Asiatic coast. These dislocations may be followed from the Tropic of Cancer to the Arctic Circle and through sixty degrees of longitude.

Not only are the grand features of topography in eastern Asia dominated by gigantic fault systems, but the minor features indicate similar control. The Korean peninsula, for example, has its topography determined by systems of intersecting faults, of which the principal series extends from N.N.W. to S.S.E., and the whole country has been compared to a chess-board of fault-blocks. Japan is divided into two parts by a great transverse fault, the *fossa magna*, north of which the island has a northerly trend and is dominated by fault lines, while south of the fossa the island turns westward and is controlled by the axes of folds.

Faults frequently determine the location of valleys, either by affording convenient courses which are taken possession of and excavated by streams, or directly by trough faulting, that is, parallel faults hading *toward* each other, including an elongate sunken block. A single fault may lead to the formation of a valley when followed by a stream. Fault valleys are very common in the Sierra Nevada and Great Basin; and in the Coast Range of California a series of valleys, together more than 400 miles long, has been excavated along the fault line which is the seat of the recent earthquake disturbances. The lower Hudson flows in a fault valley, of which the Palisades form the scarp, and in the Adirondacks the drainage is so largely along intersecting fault lines that its regularity is very striking and has occasioned the descriptive term of " lattice drainage " (Brigham); the zigzag course of the Au Sable Chasm is determined partly by faults and partly by joints. The valley of the Rhine above Strassburg is a great trough-fault included between the Black Forest highlands on the east and the Vosges Mountains on the west, both of which, as we have seen, are Horst mountains. The trough is very deep and is nearly filled with river deposits, which indicates the gradual character of the dislocating movements. Other European examples of such trough valleys are the basin of Lake Garda in northern Italy, the Christiania fjord in Norway, and the Gulf of Patras and Straits of Corinth in

In Chapter XIII (p. 347) we learned that a block isolated by faults hading away from it on all sides, so that the block is on the upthrow side with reference to the area all around it, is called a *Horst*, a name adopted from the German for lack of an English term. The Horst structure is thus much like that of the block mountains of the Great Basin and the Adirondacks, except for the direction of slope in the fault-planes and for the fact that there is usually no relation between the strike of the beds and that of the faults. Several examples of Horst mountains occur in Europe, such as the central plateau of France, the Vosges, the Black and Thuringian Forests, and the Hartz Mountains. These uplifts are composed chiefly of ancient, very strongly folded and hard rocks, from which most of the covering strata have been removed by denudation, and rise quite abruptly from the comparatively un-disturbed strata of the surrounding lowlands.

Along the entire eastern coast of Asia the topography is con-trolled by gigantic systems of faults. The zigzag mountain ranges which bound the coastal plains on the west have a general trend approximately parallel to the coast, and mark a series of great faults with downthrows to the east. The coastal plain itself is divided into a series of more or less parallel blocks by step faults, also with downthrows to the eastward, resulting in a system of terr which, however, have not flat, but inclined tops, becau tilting of the blocks, much as in the block mountain Oregon, though on a vastly larger scale. An through northern China shows three such e Mongolian block on the west, the Japanese the Manchurian between: the Japanese b making the outer portion   island, a of islands which fringes the eastern locations may be followed from the Tropic of Cancer tche Arctic Circle and through sixty degrees of longitude.

Not only are the grand features of topography in eaern Asia dominated by gigantic fault systems, but the minor feaures indicate similar control. The Korean peninsula, for examie, has its topography determined by systems of intersecting faultsof which the principal series extends from N.N.W. to S.S.E., andhe whole country has been compared to a chess-board of fat-blocks. Japan is divided into two parts by a great transverse ault, the *fossa magna*, north of which the island has a northerly end and is dominated by fault lines, while south of the fossa the islad turns westward and is controlled by the axes of folds.

Faults frequently determine the location of valleys, eher by affording convenient courses which are taken possessiorof and excavated by streams, or directly by trough faulting, hat is, parallel faults hading *toward* each other, including an longate sunken block. A single fault may lead to the formation of valley when followed by a stream. Fault valleys are very commo in the Sierra Nevada and Great Basin; and in the Coast Range f California a series of valleys, together more than 400 miles lcg, has been excavated the fault line which is the seat of theecent earthquake . The lower Hudson flows in fault valley, of ades form the scarp, and in the Airon- the argely ntersecting fault lis that casioned the descptive he zigzag coursef the lts and partly byints. is a great trougfault ds on the east a the est, h, as we have se are ugh nd is nearly fillewith ates racter of the discat- Curo fsuch trough veys Jard he Christnia Gul of Corin in

Greece. Such valleys are likewise extensively developed in Mongolia and southern Siberia; the basin of Lake Baikal is in a trough valley.

The most remarkable known instance of trough faulting is the

FIG. 240. — Normal fault, Au Sable Chasm, N.Y. . (Photograph by van Ingen.)
· At this point the stream leaves the line of fault to follow a joint-plane

great Rift Valley of eastern Africa, which begins about 15° S. lat., and after a short course sends off to the northwest the Central African branch, in which are situated lakes Tanganyika, Albert Edward, and Albert; both branches are also the seat of extensive volcanic action.   The main valley continues northward from Lake Nyassa and contains Lake Rudolph and several smaller lakes; passing between the plateaus of Abyssinia and Somaliland, it extends to the Red Sea, which is itself regarded as being but a very deep part of the same great downthrow inundated by the sea. At the northern end of the Red Sea, the Gulf of Akabah, on the east side of the Sinai Peninsula, is an extension of the trough which, growing shallower, rises to the land and keeps its northerly direction for a long distance.   Growing deeper again, it becomes the basin of the Dead Sea, the surface of which is 1300 feet below sea-level, and the valley of the Jordan, and north of that, the valley of the Orontes, reaching nearly to Antioch.   The valley of the Jordan and the Dead Sea is enclosed between parallel faults and monoclinal folds, the descent being very abrupt on the eastern side, while on the west it is more gradual and by means of several terraces and scarps.   In Africa the great Rift Valley is not bordered by mountain ranges, but cuts through high plateaus, and the form of the valley, with its constantly rising and falling bottom, shows that it is not a valley of erosion, but one of tectonic depression.

If this interpretation, which we owe to Suess, is correct, this vast dislocation extends through fifty degrees of latitude, and is one of the most striking structural features of the earth's surface.

Faults, especially those of large throw, frequently bring rocks of very different degrees of hardness into close juxtaposition.   If the harder rocks are on the upthrow side, then the scarp may persist for a very long period, because they are so much more slowly worn away than the softer rocks of the downthrow.   If the latter is composed of the harder rock, the scarp is first eroded away, and then continued denudation, removing the softer rocks more rapidly, forms a scarp on the downthrow side, a reversed scarp, so to speak, due entirely to differential erosion. In other words, the hard rocks will eventually be left standing at a higher level than the softer ones, whether they are on the upthrow or downthrow side, and thus conform to the general rule of topographic development, provided the region stand sufficiently high above base-level.

Under these circumstances, faults of very high geological antiquity may continue to dominate the topography of the faulted region. A classical example of this is given by the Central Lowlands of Scotland, which lie between the Highlands on the north and the southern Uplands, and consist chiefly of quite soft and easily destructible rocks, while the higher areas which bound them on each side are made up of much more resistant rocks. When the rocks on the two sides of the fault-plane differ notably in hardness, then the transition from higher to lower ground is well-defined and abrupt, but the transition is gradual and by gentle slopes when the difference in hardness is small. The great faults, with throws sometimes amounting to 6000 feet, which demarcate the Lowlands, are very ancient, but they still control the main features of topography.

When the faults pass through strata of approximately uniform durability, the scarps are removed with a rapidity which is chiefly dependent upon the climate, because the upthrow side is exposed to attack and the downthrow is an area of accumulation. In such rocks all surface evidences of the fault are removed, and the very existence of the dislocation is often very difficult to detect. Unquestionably, countless undetected faults remain to be discovered even in well-known regions.

Unless interrupted by diastrophic movements, the progress of denudation must eventually remove all surface indications of faults, however great the difference in hardness between the rocks of the upthrow and downthrow sides. When the region is base-levelled, or reduced to a peneplain, no fault-scarps will be apparent. On the other hand, the beginning of a new cycle of denudation, through the upheaval of the faulted area and the rejuvenation of the erosive forces, may result in the reappearance of the scarps, provided the rocks on the two sides of the fault differ notably in hardness. In this case the harder will stand above the level of the softer, whether they are on the upthrow or the downthrow side. If the dislocations have not brought together rocks of different resisting powers, then reëlevation, not accompanied by renewed displacements along the old fault-planes, will not bring the scarps again into prominence, though newly established streams may take advantage of these lines of weakness in trenching their channels.

FIG. 241. — Great thrust, near Highgate Springs, Vt. The upthrow side has been denuded away and the hammer spans the thrust-plane, connecting beds which are stratigraphically many thousands of feet apart. (U. S. G. S.)

From what·has been said it is sufficiently evident that faulting plays a very important part in the formation of topographical as well as of structural features, and to faults may be added thrusts, though the importance of the latter is relatively less, for thrusts occur among violently compressed and folded rocks, while faults are found chiefly in strata which otherwise are not greatly disturbed.

THE TOPOGRAPHICAL INFLUENCE OF JOINTS

As we have repeatedly had occasion to observe, all firm and coherent rocks are divided by sets of joints into blocks of varying
sizes and shapes, each kind of rock tending to display a certain
degree of characteristic uniformity in the shape and size of its joint-
blocks. There is, however, much latitude in this respect, for
jointing is largely determined by the stresses of compression,
tension, or torsion, to which the rock has been subjected after its
formation. In whatever manner they may have been formed,
joints exercise a very important control in the development of
topographical details. The reason for this becomes obvious when
we reflect that joints are lines of weakness, along which the rocks
are especially liable to attack, and that master joints are important
structural planes, akin to faults, which persist for long distances
and often penetrate to very considerable depths from the surface.

One of the most important kinds of control exerted by joints is
in conditioning the drainage lines of a given region. The general
direction in which the drainage of any region is carried is determined by the prevailing direction of slope existing when the river
system was established, but the flow of minor streams is very often
dependent upon the direction of the joints in the rocks through
which they flow. In several regions where the matter has been
carefully examined, as in France, Connecticut, and Wisconsin,
it is found that the network of streams closely coincides with
the network of dip, strike, and diagonal joints, which have
a constant trend over wide areas. The Au Sable Chasm, so
frequently referred to in preceding pages, pursues a zigzag course
through the eastern part of the Adirondack uplift; in part this
course is along a fault-plane, which is again forsaken to follow
a line of master-joint.

An especially notable case of joint-control is seen in the gorge
and Victoria Falls of the Zambesi in South Africa, cut through
an enormously thick volcanic plateau of basaltic lava, an account
of which is given on page 144 and need not be repeated here.

Suffice it to say that the course of the gorge has been determined by the master-joints, the river endeavouring to keep a southerly course, but repeatedly deflected by the joint-planes. The great chasm is cut along a line of fault, which yielded readily to abrasion, and may be traced into the sloping cliff at each end.   At the present time the western cataract of the Victoria Falls, called Leaping Water, has discovered another line of weakness and has cut its bed fifteen feet below the level of the other falls, manifestly preparing the way for the excavation of a new zigzag.

FIG. 242. — Gorge of the Zambesi below Victoria Falls, South Africa

Another respect in which joints are important is in controlling the details of topography, for the obvious reason that it is along the joint-planes that rocks yield to the attack of the denuding agencies, most of which loosen and detach the joint-blocks.   Evidently, the form of surface left after the detachment of the blocks is conditioned by the joint-planes.   In the plutonic igneous rocks there is great variety of form and size in the joint-blocks. Granite is often divided by three sets of nearly rectangular joints, one of which is horizontal or gently inclined, and widely spaced so as to form large rectangular blocks. When this is the case, weathered granite cliffs look like old masonry and rise in step-like terraces. When the joints are very close together, the rock breaks up into small angular blocks. In other examples, only the vertical

FIG. 243. — The Chasm, Victoria Falls of the Zambesi

joints are strongly pronounced, and then wild and craggy cliffs and needle-like summits result from weathering. The characteristic granite domes so frequently found, to which reference has already been made, are due to exfoliation of granites in which the joints are few and widely spaced.

Columnar jointing in sheets is chiefly vertical, and hence produces vertical faces, which change and follow any curvature that may develop in the columns. Most lavas are closely and very irregularly jointed, so that they are rapidly broken up by frost. Such rocks, when exposed in mountain tops, give rise to sharp ridges, jagged, irregular peaks, and very rough slopes, portions of the rock where the joints are more widely spaced and the blocks larger, yielding less rapidly to destruction, and projecting as ridges and buttresses. Aside from rectangularly jointed granite, with well-defined horizontal division planes, masses of igneous rock

FIG. 244. — Granite dome, Yosemite Valley, Cal.   (Photograph by Sinclair)

yield rough, irregular, and craggy surfaces.   On the other hand, the direct influence of joints is often masked by the effects of chemical decomposition produced by water descending along the joint-planes, in consequence of which a freshly exposed surface is often already rotted and ready to yield to the rain and wind.   Thus, with general characteristic features, igneous rocks give an infinite variety of details in form.

In stratified rocks the joints are usually vertical to the bedding-planes, and when the strata are hard and do not crumble easily on weathering, the surfaces are vertical in horizontal strata, which thus give rise to flat-topped, vertical-sided mesas, and pyramidal mountains. Inclination of the strata changes the position of the joint-planes, resulting in escarpment, dip slopes, and the other classes of forms already described. The fact to be emphasized here is the share which the joints have in producing these forms.

FIG. 245. — Limestone cliffs, Black Hills, S. D.   (U. S. G. S.) ·

The crystalline schists are, for the most part, jointed in a highly irregular manner, though gneiss sometimes has rectangular blocks, like those of granite. As a result of this confused jointing, the schists give rise to forms which display a maximum of irregularity.

Under given climatic conditions, each kind of rock weathers in a characteristic manner, and this manner is, in large degree, determined by the joint-blocks into which it is divided.

CHAPTER XXI

ADJUSTMENT OF RIVERS

RIVERS are among the most powerful of the agents of topographical development, and it is important to understand something of their modes of change and adjustment. These changes
are sometimes exceedingly complex and puzzling, for rivers do the

FIG. 246. — Stream cutting through a ridge, Middle Park, Col. (U. S. G. S.)

most unexpected things in what seems an utterly capricious and
whimsical way. We often see rivers breaching hills and even vast
mountain ranges, cutting their way through enormous obstacles,
which a slight deviation from their course would, seemingly, have
enabled them to avoid. They apparently choose the difficult and

477

shun the easy path. The general explanation of these paradoxical results is, that the river began its flow when the topography was entirely different from its present state of development. It is this fact which renders the rivers such valuable aids to the geologist in his attempts to reconstruct the past, for the apparent whims and caprices are really the necessary results of law.

Consequent Rivers. — A river has its stages of development, youth, maturity, and old age, just as has a land surface, each stage displaying its characteristic marks. When an entirely new land surface is upheaved from the sea, it has no rivers, and its drainage must consist merely of the surface rain wash following the initial slopes of the new land. No instance of any considerable area of newly uplifted land has ever been observed, but the sequence of events may be readily inferred from known facts. Since the slopes cannot be absolutely plane nor the material entirely homogeneous, there must be slight depressions along which the rain water will gather into rills, and these will wear out little trenches. The more favourably situated trenches will receive more water and be more rapidly deepened and enlarged into ravines. In this early stage of drainage development there will be many ravines, more or less parallel, which are dry except after rains. Those ravines which are most rapidly deepened will be cut down to the level of the ground water and will there be fed by springs and become permanent streams when a level is reached below which the ground water does not sink in the driest seasons. If the new land is not simply sloping, but folded, valleys for drainage are afforded by the synclines. The principal valleys are thus longitudinal, the main streams flowing in the synclinal troughs and passing from one syncline to another at the points where the anticlines are lowest, owing to the descending pitch of the folds. Such a drainage system is exemplified in the Jura Mountains of Switzerland. Thus, In a newly upheaved or newly folded land the streams are determined entirely by the slopes of the new surface and are called *consequent streams*. In its earliest stages a river can drain its territory or basin in only imperfect fashion, and

FIG. 247. — Two very young gulches, Colorado. (U. S. G. S.)

whatever depressions exist in the surface of the new land are filled up with water and form lakes. Tributaries are much fewer than in later stages of development; the divides between the tributaries are obscurely marked, and in plains these divides are broad areas, not lines. The Red River of the North is an example of a stream in a very youthful stage, which flows across the level floor of an abandoned lake. In this plain the divides between the streams are so wide and flat that water gathers on them after heavy rains, having no reason to flow in one direction rather than another. In northern Minnesota is the watershed or divide between the Mississippi, St. Lawrence, and Hudson's Bay drainage systems, which is hardly visible, the sluggish streams wandering over an almost flat surface, which has countless marshes and lakes.

As the river system becomes somewhat older, the stream channels are deepened, the larger ones being cut down to base-level, and if the region be one of considerable elevation, deep gorges and cañons are excavated. If the streams flow across strata of different hardness, waterfalls result where a hard ridge crosses them, but in the main stream these cascades and rapids are ephemeral and soon removed by the stream's wearing down the obstacle. On the head-waters of streams, however, waterfalls may persist for a long period. The river valleys are widened out by atmospheric denudation, and channels are formed on their sloping sides, which gradually grow into side valleys. The lakes are for the most part drained or silted up, only the more important and deeper ones remaining, while the system of tributary streams and rills is greatly expanded. A mature river system is characterized by the complete development of its tributaries and drainage, so that every part of its basin is reached by the ramifying channels, and rivers of the same grade tend to be separated by nearly equal interspaces. The waterfalls have disappeared, except near the stream-heads, and the stream-channels have sought out and utilized every weakness in the strata, adjusting themselves to the structure of the rocks and the alternations of hard and soft beds.

Valley floors are broadened and deposition begins upon them, and the streams, reaching a condition of equilibrium between erosion and deposition, are said to be *graded*. In graded streams the slope attained varies greatly; a small stream or one loaded with sediment requires a steeper slope than a large one, or one carrying but a small load. Thus, the lower Mississippi and its tributaries are graded, but while the great river flows in a valley with hardly any slope, the valleys of the smaller streams are still quite steep. In the process of development the stream gradients are continually readjusted, with the general result of diminishing the slope. When the stream has reached base-level and no longer erodes vertically, save in seasons of flood, it continues to cut laterally and to receive and transport the material washed into it by rains, and thus the divides are worn away.

The complete network of streams has enlarged the valley surfaces, which increases the rate of destruction and brings to the river a greater load of sediment to carry. In maturity the river receives its maximum load, sometimes so great that the lower reaches of the main stream are unable to transport it all, and spread the excess out over the flood-plain. The channel of an overloaded stream may be so raised and banked in by its own deposits that some of the tributaries are deflected and made to run for some distance parallel to the main stream, perhaps even reaching the sea independently. An example of this is the Loup Fork of the Platte in Nebraska. "The Platte flows there upon a ridge of its own creation. The Loup comes down into its valley and flows parallel with it for many miles." (Gannett.)

The final stages of river development are reached when the base-level is attained, and the drainage basin reduced to a peneplain by the combined action of the streams and weathering. The flood-plain deposits may now be partially or completely removed, for the main trunk no longer receives an excessive load, and hence it is able to carry away some of that sediment which it had previously deposited. With its drainage basin smoothed down into a peneplain, the river's work is done; it has reached old age.

The course of river evolution above described is the ideal cycle

21

of development, which, however, may be and generally is inter-rupted by diastrophic movements. An elevation of the region may simply rejuvenate the streams and start them afresh upon a career of wearing down the land. But if accompanied by extensive warping or folding of the rocks, the drainage system of the entire region may be revolutionized. A depression of the region will have the contrary effect, checking or stopping the work in which the streams were engaged, drowning their lower reaches and con-verting them into estuaries. A lowered land surface has less material to lose before it is reduced to base-level, but the work of denudation is accomplished more slowly.

When it was first suggested that rivers had cut their own valleys and had not merely taken possession of ready-made trenches, it was objected that such an explanation required many streams to begin their course by flowing uphill. It is very common to find a stream flowing across a region, cutting its way through ridge after ridge, instead of following the easy path of the longitudinal valleys. This is just what the principal streams of the northern Appalachians, such as the Delaware, the Susquehanna and the Potomac, have done, and at first sight, their course is very difficult to explain. Without going very far back into the history of these mountains, we may simply state that the ridges through which the rivers named have cut are the remnants of a reëlevated and dis-sected peneplain, across which the streams flowed to the sea, cut-ting transverse valleys that were rapidly deepened into gorges. On the soft strata longitudinal valleys were opened out which, however, were formed after the transverse streams and could not be deepened faster than they, because the main stream flowing in each transverse valley gave a temporary base-level for the tribu-taries flowing in the longitudinal valleys. The hard beds were sawed through by the descending streams, but elsewhere these beds stood up as ridges, and thus the ridges are also younger than the streams. The mystery disappears at once, if we simply remem-ber that the transverse streams began their flow upon a sloping plain above which the present ridges did not project.

Antecedent Rivers. — Another way in which rivers have been enabled to cut their way through opposing ranges of hills and even mountains is by occupying the district before the hills or mountains were made. Such streams are called *antecedent* and are defined as " those that during and for a time after a disturbance of their drainage area maintain the courses that they had taken before the disturbance." (Davis.) In this manner a stream originally consequent may become antecedent. The simplest case of antecedent drainage is where an area is uplifted without deformation and without changing the direction of the slopes. Under such circumstances all the streams retain their old channels, and simply gain renewed power to cut them into deeper trenches, down to the new base-level. Such streams are said to be *revived*. Revived streams which had begun to meander may be held in these windings and trench them into deep gorges. Even if the upheaval be accompanied by folding or deformation, one or more of the streams may persist in its ancient course, provided the folding be very slow and gradual, so that the river is able to cut down through the obstacles which are raised athwart its course. A revolving saw cuts its way through a log which is pushed against it, so the river cuts its way through the rising barrier. If the latter be raised faster than the river can cut, then the stream will be dammed back into a lake, or will be diverted to a new course. Naturally, the great trunk rivers are more likely to hold their previous courses than the smaller streams.

A fine example of an antecedent river is the Columbia, of Washington and Oregon, which is deflected to the westward by the volcanic plateau of central Washington as far as the foot of the Cascade Mountains, where it turns southward, following the mountains for some distance, then it once more turns to the westward and cuts through the Cascades in a great cañon. This course the river has maintained despite a differential uplift of thousands of feet, and probably also the rising of the Cascades athwart its course. The Snake River, a tributary of the Columbia, has cut a cañon 6000 feet deep through lava and granite, through a slowly rising upwarp.

Several rivers in the Alps and Himalayas, which rise in the inner part of the ranges and cut their way out through deep chasms, are believed to be antecedent.

Superimposed Rivers. —·An old land-surface with well-defined topography may be deeply buried under newer accumulations, as of lava floods, great bodies of volcanic ash and tuff, sheets of glacial drift, lake deposits, or, after depression beneath the sea, by marine deposits. In each of these cases, the new surface has no reference whatever to the old; the more ancient and buried rocks may and generally do have an entirely different character, arrangement, and structure from those which overlie them. The drainage system established upon the new surface is consequent upon the initial slopes of the latter, and when the streams have cut through the mantle of newer rocks and reach the ancient surface below, they are entirely out of adjustment with that surface and its rocks. As the streams cut their trenches through the overlying mantle of newer strata, they encounter the older rocks below, first laying bare the higher ridges of the latter, which will cause waterfalls and rapids. The upper Mississippi has in many places excavated its channel through the surface sheet of glacial drift and is now engaged in eroding the ancient crystalline rocks which the drift had covered. When the stream has everywhere cut through the newer rocks, its course will be seen to have no relation to the structure of the older rocks which it is now trenching. If, as frequently has happened, denudation has stripped away almost all the newer strata, the drainage of the country seems to be quite inexplicable and to be arranged without any reference to the structure of the rocks across which the streams flow. Such a system of drainage is said to be *superimposed, inherited*, or *epigenetic*.

Examples of superimposed streams may be found in great numbers in the United States. Almost all of the minor streams in that part of the country which is covered with glacial drift belong in this class, as do also the rivers, like the Columbia, the Snake, and the Des Chutes, which trench the great volcanic plateau of Idaho, Oregon, and Washington. An especially curious and interesting case is found in western Colorado in the valleys of the Uncompahgre and Gunnison. When these streams were first established, the region was a plateau with westerly slope largely built up of volcanic ash, beneath which an old topography was buried. The Gunnison flowed over a concealed mountain of granite, and when its valley had been cut down to the granite, it was compelled to hold the same course, and has trenched a cañon 2000 feet deep in that rock. The Uncompahgre, which, though a tributary of the Gunnison, flows parallel with it for a considerable distance, followed a course which took it over an old valley buried under soft materials which were rapidly removed. At the present time the course of the Gunnison seems to be quite paradoxical, though it is easily explained by its history.

Subsequent Streams. — As a river system approaches maturity, and as the drainage of the area becomes more complete, it will increase the number of its branches. Those branches which were not at all represented in the youthful stages of the system, and are opened out along lines of yielding rocks, are called *subsequent*, and all streams will develop more or fewer of such branches as they advance to maturity.

The foregoing classification of streams does not involve categories which are entirely exclusive of one another. Any stream, whatever its mode of origin, may become antecedent through diastrophic movements. Most superimposed streams are also consequent, but by no means are all consequent streams superimposed. The Columbia and Snake are both superimposed and antecedent.

Adjustment of Rivers. — However the streams of a district may have been established in the first instance, whether they were consequent, antecedent, or superimposed, they are liable to changes more or less profound and far-reaching. These changes, which belong to the normal development of the drainage system and are not dependent upon diastrophism, are due to adjustment of the streams to the rock structure of the district, the streams searching out the lines of weakness and least resistance, and everywhere taking the easiest path to their destination. The upstream extension of branches and the shifting of divides result in the *capture* of streams, or parts of such, by others more favourably situated, one master stream gradually absorbing many smaller ones which had originally been independent.

A divide, or water-parting, between two streams is gradually shifted by the lengthening of the more favourably situated stream, or of one of its subsequent branches. This more favourable situation may be because it has a shorter course and greater fall, giving a swifter flow, or because it flows at a lower level, giving greater fall to its tributaries, or because its course is through soft and easily eroded rocks, while its rival is embarrassed by hard rocks and ledges. Another favourable circumstance which may decide between streams otherwise equal is given by the attitude of the strata. In regions of inclined strata, as we have already learned, the escarpments formed by outcropping ledges of harder rocks tend to migrate in the direction of the dip. As such escarpments frequently form divides between minor streams, the stream toward which the escarpment migrates will be at a disadvantage. This shifting of divides is a very slow process, but after a long time of insidious advance the actual capture and diversion of part of a stream may be quite suddenly effected.

Stream capture may be effected in a great variety of ways, but a few examples must suffice. We may, in the first place, suppose two neighbouring streams following roughly parallel courses, but owing to the original conformation of the region, flowing at different levels. The stream that flows at the lower level will allow greater fall to its tributaries, which will thus work upward more rapidly. One of these tributaries will eventually work its way through the divide and tap the rival stream, all of whose waters above the point of tapping will be diverted to the main stream which flows at the lower level.

The same mode of capture may be effected in the case of two streams which head on opposite sides of the same divide, one of which has a much steeper grade than the other. The stream which has the steeper slope will work headward more rapidly, and will eventually tap the head-waters of the opposite stream. This method is illustrated in the Catskill Mountains, which are carved out of a table-land sloping gently westward, but having a steep escarpment on the eastern face, with recesses in which easterly flowing streams were established. At the same time westwardly flowing streams formed on the gently sloping summit level. The eastern streams, especially the Kaaters Kill and Plaaters Kill, have a steep descent and swift flow and have thus been able to extend their heads rapidly and to capture and divert many tributaries of Schoharie Creek, which runs westward. These captured tributaries keep their original course in the direction of Schoharie Creek, but make sharp turns and reverses to join the capturing streams. The Yellowstone Lake once discharged southwestward into the Snake River. The Yellowstone River was then a small stream, but was favourably situated for establishing a steep gradient as it flowed over a plateau of comparatively soft lava. Extending upward, the stream tapped the lake and reversed the direction of its discharge.

Another method of stream capture is well illustrated by the Delaware, the Potomac, and other transverse rivers which have cut deep gorges through the Appalachian ridges. Suppose two parallel transverse streams flowing across a gently sloping peneplain which is composed of tilted rocks of different degrees of hardness. In the manner already explained (p. 482) these streams cut gorges through the ridges of hard rock, while longitudinal valleys are worn out along the strike of softer strata, which valleys are occupied by tributaries of the transverse streams. If one of the two transverse streams be considerably larger than the other, it will saw its way through the hard ridges at a correspondingly faster rate and establish a lower base-level for its tributaries. One of the tributaries with its more rapid fall will be thus enabled to shift its divide at the expense of a branch of the rival transverse stream, capture it, and by reversing the direction of its flow draw off the waters of the smaller main stream above the point where its captured tributary entered it. Or a tributary of the larger main stream may push its way up a longitudinal valley until it taps and diverts the smaller transverse stream without the intermediation of any tributary of the latter. Examples of both of these varieties of capture may be found among the Appalachian rivers; an excellent illustration of the latter method is given by the Potomac and Shenandoah.

When the Potomac was beginning to cut its gap through the Blue Ridge at Harper's Ferry, a smaller stream, Beaverdam Creek, was cutting a similar gorge through the same ridge a few miles to the south. The Shenandoah was then a young and short tributary of the Potomac, which it entered from the south, flowing through the longitudinal valley which was opening along the strike of the softer strata to the west of the Blue Ridge. As the Potomac is much larger than Beaverdam Creek, it cut its gap much more rapidly, thus giving a steep and swift course to the Shenandoah. The latter pushed its way up the longitudinal valley until it tapped Beaverdam Creek and captured its upper course, diverting its waters to the Potomac. Beaverdam Creek no longer flowed through the gorge which it had cut in the Blue Ridge and which was thus abandoned and became a " wind-gap," the beheaded Beaverdam now rising to the eastward of the abandoned gorge. This gorge is known as Snickers Gap. The great number of wind-gaps in the Appalachian ridges show how frequently the capture and diversion of smaller streams by larger ones has been accomplished among those mountains.

Figures 248 and 249 show two stages in the evolution of a river system. Figure 248 represents the first stage, in which several transverse streams, a, c, e, f, g, are breaching the escarpments indicated by shaded lines. Of these streams, c carries the most water, and will therefore deepen its gorges through the hard ridges more rapidly than the others, and give its tributaries the advantage of a greater fall. In the second stage (Fig. 249), c has captured the upper courses of all the other streams except g, which has not yet been reached. The branch l has captured a, beheading it, *diverting* the portion a'' and *reversing* the portion a'. Similarly, m has captured and divided e, n has done the same with b, and p with d, while g must eventually suffer the same fate. Wind-gaps will be left in the ridges where the captured streams once crossed them.

In regions of folded rocks thrown into a series of parallel anticlines and synclines, the process of adjustment may become exceedingly complicated. Suppose an original consequent stream flowing in a syncline of hard rock considerably above base-level, whose subsequent branches have opened out valleys in softer rocks along the crests of the anticlines, where the harder surface

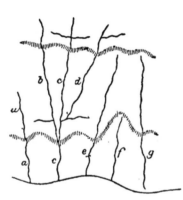

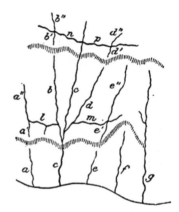

FIG. 248. — Evolution of a river system, first stage. The shaded lines represent escarpments of hard rock. (De Lapparent)

FIG. 249. — Evolution of a river system, second stage. (De Lapparent)

stratum is first cut through. The extension and junction of these subsequent branches may offer a more advantageous course than the hard syncline, and cause the latter to be wholly or partially deserted. The streams originally flowing in the synclinal troughs may gradually be shifted to the degraded anticlines which, as we have seen, are wasted away more rapidly and this transfer is facilitated by the fact that the synclinal troughs have very gentle slopes longitudinally, giving but small velocity to the rivers which flow in them.

A thoroughly mature drainage system is characterized by a complete adjustment of its streams to the structure of the rocks. The rivers as finally established are thus apt to be a patchwork of streams captured and diverted, and the result of adjustment is the production of a system often radically different from the original one. Even after a river system has become maturely adjusted, a reëlevation of the country may produce a new and .entirely different adjustment, by changing the relation of the folds and outcrops of hard and soft strata to the base-level. A region of great antiquity which has repeatedly been worn down and reëlevated will have experienced many revolutions of its drainage systems.

Warping of the surface nearly always produces extensive changes in the drainage systems affected by diverting the course of many streams, though the master streams may excavate with sufficient rapidity to hold their channels as antecedent rivers. As will be shown in a subsequent chapter, the Appalachian Mountains had, by the close of the Cretaceous period, been worn away to a peneplain, across which the Delaware, Susquehanna, and Potomac flowed in transverse valleys to the Atlantic, while the New River in Virginia and the French Broad in North Carolina flowed westward to the Mississippi. The southern part of the peneplain was drained by a longitudinal river called the Appalachian River, and smaller streams running westward and southwestward to an extension of the Gulf of Mexico drained the southwestern side of the peneplain. Next followed an upwarping of the peneplain along a north-south axis, through which the northern rivers continued in their old courses, but one of the southwestern streams extended headward and captured the headwaters of the Appalachian River. Still another upwarp succeeded, this time on an axis running nearly east and west through northern Alabama and Mississippi, in consequence of which a tributary of the Ohio extended itself southward and captured the southwestern stream which had before beheaded the Appalachian River. Thus arose the Tennessee, which enters the

Ohio after such a curious course and which is made up of parts of three originally independent river systems, and in its changes records the history of the region through which it flows.

Accidents to Rivers. — This term is employed to express the interruptions which hinder or prevent the normal development of a river system. The diastrophic changes and their effects we have already considered, but there are others which should be mentioned. A change of climate from moist to arid greatly interferes with the development and adjustment of a river system. Many stream channels are abandoned and others are occupied only after rains, while the reduced flow in the permanent streams diminishes their erosive powers. Large areas, like the Great Basin region, may have no outlet to the sea, because the mountain streams all lose themselves in the desert sands. Lake Bonneville (see p. 219) had an outlet until the increasing dryness of the climate so lowered its waters that the outlet could no longer be reached, evaporation exceeding influx. Great lava flows may obliterate the drainage system of a region and compel the establishment of an entirely new one, as has happened in southern Idaho and southeastern Oregon, a region of exceedingly immature topography and drainage. Extensive ice-sheets, by spreading a thick mantle of drift which fills up the valleys, may produce the same effects as lava flows, except that the drift is more easily removed. In the northeastern United States many streams have been displaced by the sheets of glacial drift, and forced to seek new channels at a comparatively recent date; they still preserve all the signs of youth, such as deep, trench-like gorges (see Fig. 58), waterfalls, and rapids. The larger rivers have, for the most part, been able to reoccupy their old valleys, but the smaller streams have generally been compelled to excavate new channels.

CHAPTER XXII

SEA–COASTS

THE sea-coast is not merely a line, but a zone of varying breadth, sloping toward the sea, and with a subaërial and a submarine portion. The submarine portion of the coast frequently continues the slope of the subaërial portion, with an interruption formed by the actual beach, upon which the surf breaks. On flat, gently sloping coasts, the beach is generally broad, especially if the range of the tides is great, and a beach wall is present (see p. 247), frequently with a belt of sand-dunes behind it, while on steep, rocky coasts the beach is narrow and may occasionally be absent altogether.

The coasts of the different parts of the world display a great variety of form and structure, but they may all be included in a small number of classes. An obvious primary division is into (1) regular, (2) irregular, and (3) lobate coast-lines.

1. **Regular Coasts** continue " for great distances without notable indentations and, for the most part, in gentle curves, convex toward the land, which are connected by curved lines or meet at obtuse angles. . . . The flatter the coast, the more perfectly is this type developed, and the coast-line runs for many kilometers in the same curve. With a steep slope the course is regular only in general; in detail it seems as though drawn by a trembling hand, with numerous little prominences, which project but a few hundred metres beyond the general coast-line and separated from one another by shallow, curved indentations." (Penek.) Flat coasts in most cases border coastal plains and are very generally regular, while the regular steep coasts are marked by lines of cliff, which abruptly break the slope of the land toward

the sea. Certain regular coasts, like that of eastern North America, for example, are uniformly flat or steep for long distances, others are alternately flat and steep, and are then said to be *adjusted*. Regular coasts have few islands.

Coastal plains are absent from desert regions and are built up by the activity of rivers; they may be coalesced deltas, or of littoral origin, or submarine with the river sediments distributed along the shore by currents; subsequent elevation has converted the sea-bottom into land. The longest known coastal plain, that of the middle and south Atlantic States, is chiefly of this submarine origin. Other coastal plains, like that of Holland, are of mixed origin, subaërial, littoral and marine deposits occurring near and upon one another. Diastrophic movements have but little effect in changing the character of a regular coast bordering a coastal plain; elevation merely makes a new coastline along what was before the flat sea-bottom, and depression causes the sea to advance over the very gently sloping plain, in either case without changing the regularity of the coast.

Along the steep, regular coasts the line of cliffs, though pursuing a very uniform course, is broken by small bays, giving a serrate coast-line. It is on such coasts that the destructive work of the waves is most advantageously seen (Figs. 74–77, pp. 168–171); sea-caves, isolated pillars and stacks, and lines of rocky ledges and shoals abound. The submarine part of the slope is usually gentler than the subaërial, and descends gradually to depths of ten to twenty fathoms. Above water the height of the sea-cliff is moderate, seldom more than 300–400 feet and, consequently, this kind of coast is most typically developed in rather low lands, mountainous coasts giving rise to other forms. The line of cliffs intersects hill and valley and interrupts the system of connected valleys, making it evident that land has been lost along that line. Such a coast is obviously the work of wave destruction. The serrations are due to differences in the hardness of the rocks, the softer rocks being cut into bays and the more resistant ones standing out as headlands. The bays, however, remain small,

because in them the power of the waves is diminished, and soon
a point is reached where no further retreat of the land is possible
until the headlands have been cut back.

In brief, the low-lying flat coasts which border coastal plains
are areas of accumulation, where the land is still, or has lately
been, gaining at the expense of the sea, while steep, rocky coasts,
bordered by lines of cliff, are areas along which the sea is eating
away the land.

Diastrophic movements have a much greater power in changing
the character of the cliff coasts than of those which are low-
lying and flat. If the land is sinking, its valleys become submerged and converted into bays, thus forming an irregular coast-
line, while, if the land is rising, the wave-cut platform forms a
plain at the foot of the cliffs, which are now inland and beyond
the reach of the surf. Only in the very rare instances of the
cutting action of the sea keeping exact pace with the movement
of elevation or depression, will the steep and regular character
of the coast be maintained. Hence, such coasts are restricted
to regions which are stationary or in extremely slow movement.

Geographical cycles are seldom so clearly distinguishable along
the sea-coast as in the interior of the continents, chiefly because
coast topography is determined more by diastrophic movements
than by marine erosion, which works very slowly on account of
its limited sphere of action. Subaërial denudation also, as will
be shown in the sequel, is an extremely important factor in controlling the character of the coast. Nevertheless, indications
of the cycle may not infrequently be found. A newly upheaved
coast tends to be regular and straight, because the sea-bottom
is nearly flat and, when elevated, the sea-level marks a straight
line upon it. As is true of the subaërial agencies, the first effect
of wave erosion is to produce irregularities, cutting out bays along
the softer rocks and leaving the more resistant ones to stand out
as headlands. But, as already explained, the depth to which
these bays can invade the land is very limited, since in them
the power of the waves is greatly reduced, and hence the surf

is unable to produce an irregular coast in the full sense of that term. However, such a coast may be formed by the combined work of the surf and of depression upon a land of strong relief. The submerged valleys of subaërial origin become bays and estuaries that run far into the land, like Delaware and Chesapeake Bays, and the coast-line becomes highly irregular.

In either case, whether the irregularities are relatively insignificant and formed by surf erosion, or of great prominence and due to depression, the tendency of marine action is to straighten the coast-line and remove the irregularities and thus to produce an *adjusted coast*. This adjustment is brought about by the combined work of erosion and deposition. The bays are places of sedimentary accumulation, especially if rivers enter them; the mouth of the bay is first partially or completely closed by a barrier deposited by the shore-current, behind which the lagoon is silted up and converted into a marsh and eventually into a plain. Meantime the headlands are slowly worn back, until the power of the waves is insufficient to cut them further. Provided the coast remains stationary for a long period, adjustment follows, and the ancient coast becomes regular, as the youthful one was, but with a difference of structure, for the adjusted coast is made up of bold, truncated headlands, alternating with low-lying plains and marshes. The Italian coast, with its broad, gently curving gulfs, is an example of an adjusted coast, and in the province of Tuscany these changes have largely taken place within historic times: "The bays of Piombino and Grosseto were cut off from the sea by bars, the lagoons thus formed were transformed into swamp, the dreaded maremmas, which in their turn were filled in. The former island of Monte Argentario was connected with the mainland by two bars." (Penck.)

2. **Irregular Coasts** display a great variety of forms due to the manner of their origin, and several subdivisions are employed to express this diversity of origin. All the forms, however, have this in common, that they are produced by the depression and submergence of land-surfaces, and it is the variety of the latter

which causes the manifold differences of the irregular coasts. On such coasts the numerous bays penetrate far into the land, sometimes diminishing regularly in width, but frequently of varying width, now expanding into lake-like form, now contracting to a strait, often winding and even branching, and always ending in a land valley. Islands are numerous and are in line with the land between the bays. The land may be high or low, gently or steeply sloping, and the subaërial slope is continued beneath the water without change, until the flat sea-bottom is reached. The type of an irregular coast is given by its bays, which vary from short, funnel-shaped indentations, to long, narrow, winding, and branching channels. The subdivisions of the class are made in accordance with these variations.

a. *Fjord Coasts* are found in the high latitudes of both hemispheres and in regions which have undergone intense glaciation; in the northern hemisphere they are limited by the 49th parallel, and in the southern hemisphere by the 41st. Alaska, British Columbia, Greenland, Scotland, Norway, the southern end of South America, and New Zealand, are typical examples of fjord coasts. The fjords are long, narrow, frequently branched, and usually very deep; the bottom is divided into several basins and the fjords are generally much deeper in the middle of their course than at the seaward end, though sometimes they are continued across the sea-floor as submarine valleys. The ridges of land which separate adjoining fjords are frequently notched by low passes, which seaward become straits, connecting the fjords and cutting up the ridges into islands, which are always very numerous along coasts of this class. The famous " inside passage " from Puget Sound to Sitka, Alaska, is a network of deep waterways among countless islands.

Fjords are not confined to any particular type of land topography, nor to any single kind of structure. In Norway, western North America and southern Chili, they pierce lofty, mountainous coasts; in Scotland the coasts are of low mountains, while in southern Sweden and Finland the fjord coasts are flat. Similarly, they occur on coasts where the lines are determined by great fault-scarps, as well as on those where the control is due to folding. The one indispensable condition is former or present glaciation, and in Norway, Greenland, and Alaska the landward extensions of many fjords are still occupied by glaciers.

Fjords are clearly glaciated valleys; whether they have been merely remodelled by glaciers, or whether they are entirely due

FIG. 250. — Branching fjord, Lynn Canal, Alaska. (U. S. G. S.)

to glacial excavation, they bear all the characteristic marks of ice-action, as these have been elsewhere enumerated (see p. 162). Glaciers have the power of overdeepening their valleys and of excavating them below sea-level, but it is not yet definitely known just how far this overdeepening may proceed. At all events, the known fjord coasts show other evidence of being much de-

2 K

pressed and invaded by the sea, and thus a fjord coast results from the partial submergence of a glacially modelled region. The great depths of the fjords and their freedom from sedimentary

FIG. 251. — Fjord, Wrangel, Alaska. (U. S. G. S.)

deposits are explained by the fact that at the time of their submergence these valleys were occupied by the ice, which thus prevented the accumulation of sediments. Had rivers been flowing in them when the depression occurred, their mouths would first have been drowned, checking the current and causing a deposition proportionate to the load. This has happened in the case of the Hudson River, which is a drowned cañon of great depth in which are accumulated immense thicknesses of river mud, even above the Highlands. That a fjord coast is due to depression is not contradicted by the fact that many such coasts are now rising, like that of Scandinavia; the elevation is still far from compensating for the depression.

b. Rias Coasts. — This term is derived from northwestern Spain, where the Ria de Vigo, de la Coruña, del Ferrol, and several others, form long, fjord-like bays, though branching little, which extend far into the land. Coasts of this type have frequently been regarded as fjord coasts, but there are essential differences; the bays are shorter, more funnel-shaped, broadening and deepening seaward, and are not nearly so deep as fjords. The excavation was not glacial, and hence the rias are not confined to high latitudes, but occur abundantly in the temperate and tropical regions, as in Brittany, Cornwall, Ireland, the east shore of the Adriatic, Brazil, southern China, and eastern Australia.

Rias coasts are most frequent at the margin of low mountains and lands of moderate height, but they may occur along areas of any elevation. Nor are they associated with any special type of geological structure; in Spain, Brittany, Cornwall, and other regions the rock is a granite without recognizable structure. In other instances, structure has exerted an evident control, as in the southwest of Ireland, where the bays follow the strike of the rocks and thus occupy longitudinal valleys, which are cut out along the soft sandstones, while the intervening ridges are made up of more resistant limestones. In still other cases the rias are found in valleys of folding, where the coast-line intersects the line of strike. An unusual case is the Bay of San Francisco,

which is T-shaped and fills a longitudinal strike-valley parallel
to the coast, with a transverse connection with the Pacific by way
of the Golden Gate.

The mode of origin of .rias coasts is by the depression and
partial submergence of short slopes cut by deep valleys of subaërial origin, whether excavated by rivers or formed by tectonic
processes. The rivers were short and carried no great load of
sediment, hence the bays were not filled up with silt and mud
during the slow submergence. In some examples, as in those of
Brittany, the valleys are continued for some distance across the
sea-floor, a circumstance which in itself is an evidence of depression. The characteristic difference between fjord and rias
coasts is that the former are due to glaciation and the latter are
not.

c. Calas Coasts are typically displayed in the Balearic Islands
and are marked by numerous short, semicircular, and rather
shallow bays, separated by narrow peninsulas. On the coasts
of the Red Sea the bays have a more or less rectangular outline
not narrowing inland. Obviously, coasts of this class differ but
little from the serrate regular coasts into which they grade; their
mode of origin, however, renders it important to make the
distinction. Calas coasts owe their irregularities not to wave
erosion, but to the submergence of land valleys; those of the
typical kind are due to the depression of mountain slopes, furrowed by numerous short ravines. The coasts of the Red Sea
type arise on the depression of desert mountains, in which valleys
are few and small.

The irregular coasts are thus in all cases due to the submergence
of land, and their characteristic features are to be explained by
the differences of the land-surfaces before submergence, which in
turn are determined *chiefly by the subaërial agents.*

While the classification of coastal forms serves a useful purpose,
it must not be supposed that it is always easy to refer a given coast
to a definite type. In travelling along a coast, one type is fre-

quently found to change and give way to another, as in eastern North America, for example. This entire coast has recently been depressed, and from the end of Florida to 45° N. lat. the subsidence is still in progress, but the effects of the submergence are very different in accordance with the former land-surfaces. The depressed margin of the coastal plain is a regular coast, with very few islands, while that of Maine, Nova Scotia, and New Brunswick is highly irregular and has numerous rocky islands, while the Hudson River, Delaware and Chesapeake Bays are drowned valleys. Between the two types the transition is gradual. Here one kind of coast succeeds another, but two or more may occur together; calas are frequent on rias coasts, and rias are found among fjords. In such cases the general character of the coast determines its reference.

3. **Lobate Coasts.** — In the preceding classes of irregular coasts the bays are, after all, of comparatively small dimensions, and the general trend of the coast is not greatly affected by them, but there are other coasts, where the land is invaded by very broad, deep gulfs, which are not mere indentations, and the interlocking gulfs and peninsulas are of the same order of magnitude. Greece is a typical instance of the lobate coast, and the islands of Celebes, Japan, and Haiti are other examples of the same type. The origin of the gulfs, which are usually very deep, is probably to be ascribed to faulting.

Professor Penck has calculated that, in round numbers, 37% of the sea-coasts of the earth belong to one or other of the irregular types; and of this amount nearly one-third is of the fjord coast class and rather less than half of the rias coast class. Of the continental coasts 43% are regular, and two-thirds of these are low, flat coasts, and nearly one-third adjusted. As the flat coasts are due to accumulation and many of them to upheaval, but a small amount remains as directly formed by wave erosion. From these figures it appears that diastrophism is of more importance than marine denudation in determining the character of the coast-lines.

Relations of Sea-coasts to Structure. — As has been repeatedly mentioned in the foregoing paragraphs, the details of coastal topography often show no obvious relations to the geological structure of the rocks. On a grand scale, however, the position and trend of coast-lines are controlled by tectonic features, and frequently the details are similarly determined. Many coasts are due to great systems of faults, which may, as in the Atlantic shores of Europe, intersect the prevailing lines of strike, or may run approximately parallel with those lines, as in eastern Asia, where a series of tilted fault-blocks form the plains, the coast-lines, and the lines of fringing islands. Other coasts are coincident with long lines of folding, as is illustrated by the entire west coast of North and South America, the foreland of the great mountain chains being more or less completely submerged. The Dalmatian coast on the east side of the Adriatic is a half-submerged belt of folding and the coast-line obliquely truncates the strike. Fjord, rias and calas coasts may occur in association with any type of structure; they are determined by the dominant class of subaërial denuding agents.

CHAPTER XXIII

MOUNTAIN RANGES

THE term *mountain* is somewhat loosely employed for any lofty eminence, and the distinction between mountains and hills, as ordinarily made, is principally a question of height. Some so-called mountain peaks and ridges are merely the portions of dissected plateaus left standing, such as Lookout Mountain and Missionary Ridge in Tennessee, and the Alleghany Front in Pennsylvania. Such mountains usually have flat tops (table mountains), are composed of strata which are nearly or quite horizontal, and owe their existence either to their being composed of more resistant rocks than the denuded parts of the plateau, or to their favourable situation with reference to the drainage lines. Another type of mountain is the volcanic, which is usually an isolated cone and may be built up to great heights; it is simply the accumulation of volcanic material which has been piled up around the vent. To the same general class, as due to igneous rocks, might be referred the *laccolithic mountains*, in which an intrusive magma has pushed up the overlying strata into a dome. Such mountains may stand isolated (Fig. 220), or several may be grouped together (Henry Mountains of southern Utah), or they may form extensive parts of true ranges (Elk Mountains, Colorado). Block mountains, which are tilted and eroded fault-blocks, form a third class; these may be single or in groups, or lineally extended as a range. Typical mountain ranges and chains are mountains of *folding*, and differ materially from any of these classes, both in their structure and their mode of origin. Before proceeding to discuss the origin and history of mountains, it will be necessary to define the terms to be used.

503

A **Mountain Range** is made up of a series of more or less parallel ridges, all of which were formed within a single geosyncline (p. 330) or on its borders. The ridges are separated from one another by longitudinal valleys, and may be formed either by the successive folds or by denudation within the limits of the folds. In the latter case the outcropping harder strata make the ridges. A mountain range is always very long in proportion to its width, and its ridges have a persistent trend. These features distinguish a true range from the ridges cut out of a plateau by denudation. The Appalachian range, the Wasatch, the Coast Range, are examples of typical mountain ranges.

A **Mountain System** is made up of a number of parallel or consecutive ranges, formed in separate geosynclines, but of approximately similar dates of upheaval. The Appalachian system comprises the Appalachian range, running from New York to Georgia, the Acadian range in Nova Scotia and New Brunswick, and the Ouachita range in Arkansas. Each of these ranges was formed in a different geosynclinal, but at the same geological date, and they are consecutive, having a common direction.

A **Mountain Chain** comprises two or more systems in the same general region of elevation, but of different dates of origin. The Appalachian chain includes the Appalachian system, the Blue Ridge, the Highlands of New Jersey and the Hudson, a system of different date, and the Taconic system of western New England, which was not formed at the same time as either of the others.

A **Cordillera** consists of several mountain chains in the same part of the continent. Thus, the chains of the Rocky Mountains, Sierra Nevada, Coast Range, and their prolongations in Canada, together make up the Rocky Mountain or Western Cordillera.

From these definitions it will appear that the mountain range has a unity of structure and origin which fits it especially for study. If the history of the ranges be understood, the systems and chains will offer little additional difficulty.

A mountain range (disregarding, for the present, certain exceptional cases) consists of a very thick mass of strata, which are much thicker in the mountains than the same strata in the adjoining plains. In the Appalachian range, for example, the stratified rocks are more than 25,000 feet thick, but on tracing the same series of beds westward into the Mississippi Valley, they are found to become very much thinner, hardly exceeding one-tenth of the thickness in the mountains. This immense thickness of the component strata is not peculiar to the Appalachians, but reappears in the typical mountain ranges everywhere; the Wasatch range has 31,000 feet of strata, the Coast Range 30,000 feet, the Alps 50,000 feet, etc. The thick series of strata which make up a mountain range are usually conformable throughout, though this conformity may in some cases be deceptive and due to the obliteration of unconformities by folding. Deposition usually appears to have been without conspicuous breaks, and there was little or no loss from denudation, though in some cases the region which subsequently was upheaved into the range had its oscillations of level, recorded now in unconformities. This may be seen, for example, in the Ouachita range of Arkansas.

Another well-nigh universal fact concerning the structure of mountain ranges is the intense folding or plication of their strata, often accompanied by great thrusts. The degree of plication varies much in different ranges. The Uinta Mountains are formed by a single great and gently swelling arch of strata, faulted along its northern slope. So gentle is the curvature of the beds that in a single view they often seem to be quite horizontal. The Black Hills, South Dakota, form a great dome, with somewhat oval ground-plan. Much more commonly the strata are thrown into a series of parallel folds, which sometimes are open, upright, and symmetrical, as in some of the ridges of the Jura Mountains of Switzerland, in which the folds are so symmetrical and regular that a section across the parallel ridges looks like a diagram. This comparatively gentle folding is, however, not the rule, but rather an intense compression and plication. The Appalachians are thrown into closed, asymmetrical, and overturned folds, with frequent great thrusts. The Sierra Nevada is so intensely plicated that the thickness of its strata has not yet been estimated. The Alps have undergone such enormous compression that many of the ridges are in the form of fan folds (*i.e.* the anticlines are broader at the crest than at the base), while others have been pushed over to an inverted position. The combination of this violent contortion with faults and thrusts often results in an indescribable confusion and chaos of forms, which it is exceedingly difficult to comprehend.

In folded mountain ranges three zones may be distinguished: (1) A rigid, unyielding mass which is not folded, (2) the zone of folding, (3) the zone of diminishing action, where the folding gradually dies away or ends in a fault. Many, perhaps most, ranges are bounded by faults on one or more sides, as is true of the Sierra Nevada, Wasatch and Uinta Mountains, the Alps, etc. The side of the range toward which the overturned folds incline is called the *foreland*, and may be either the unfolded mass or the zone of diminishing action; the former arrangement occurs in the Alps, the latter in the Appalachians.

The two main characteristic features of mountain ranges are, then, the immense thickness of the strata of which they are made, and the compression and folding or thrusting which they have undergone. Certain minor structures which accompany these more striking features should, however, not be overlooked. In the first place, the folded strata of mountain ranges are very generally cleaved, or fissile, or both, the planes of cleavage or fissility running parallel with the axes of the folds. (2) The major folds are themselves composed of successive series of minor folds in descending order of magnitude, the smallest of them being visible only with the microscope. (3) Dynamic metamorphism is an almost universal feature of mountain ranges, the transformation of the rocks being in proportion to the intensity of the plication. The microscope gives eloquent testimony to the enormous forces which have been at work, by showing how the minerals have been mashed and flattened, rendered plastic and flowing like wax in a hydraulic press. (4) Masses of igneous rocks are very often, though not always, associated with mountain ranges, and many such ranges have a core of igneous rock, often granite, with strata flanking it on both sides.

Origin of Mountain Ranges

The manner in which mountain ranges have been formed must be deduced from a careful study of their structure, for no one has ever witnessed the process of that formation. Mountain building may be going on at the present time; indeed, there is no reason to suppose that it is not, but so slowly is the work carried on that it withdraws itself entirely from observation. Nevertheless, the general course of events may be inferred with much confidence from the structure of the range.

The first step in the formation of a mountain range must evidently be the accumulation of an immensely thick body of strata. This, of course, must have taken place chiefly under water, and the only body of water large enough is the sea. Furthermore, our studies of modern marine deposits have taught us that thick strata can be accumulated only in rather shallow water and parallel with shore-lines. This shoal-water origin of their strata is confirmed by the examination of actual mountain ranges, where we find great masses of conglomerates, ripple-marked and sun-cracked sandstones and shales, and abundant other testimony of deposition in shallow water, in deltas and on flood plains of rivers. To accumulate thick strata in shoal water, the bottom must subside as the sediments are piled upon it, else the water would be filled up and deposition cease. Such a sinking trough is a geosyncline, and in geosyclines filled with sediments is the cradle of the mountains. The area of the trough varies from time to time, as do also the position of the line of maximum subsidence and the relative rate of depression and sedimentation, so that the depth of water varies. We saw above that the strata of mountain ranges are very much thicker than the same strata in the adjoining plains, which means that the ranges have been formed along the lines of maximum sedimentation.

The second stage in the building of a range is the upheaval of the thick mass of strata into a series of anticlinal and synclinal folds, which may be upright, open, and symmetrical, or closed, asymmetrical, inclined, or inverted. This, as we have already learned, can be produced only by *lateral compression*, a conclusion which is sustained not only by the mechanics of folding and faulting, but also by the less obvious structures, such as cleavage and fissility, metamorphism, the microscopic crumplings and plications, and the crushing and flowage of the mineral particles. The compressing force does not raise anticlines with great cavities beneath them, for such arches could not well be self-supporting, but mashes together the whole mass of strata, raising them into folds and wrinkles, crowding the beds into a greatly reduced breadth; or, when they are not sufficiently loaded to be plastic, breaking and dislocating them in great thrusts. It is not necessary to suppose that a mountain range was thrown up by one steady movement. On the contrary, there is good reason to believe that repeated movements, separated, it may be, by long intervals of time, have been engaged in the work.

The great forces of compression which have upheaved mountain ranges have manifested themselves recurrently from the earliest to the latest recorded periods of the earth's history, and from these recurrences form conspicuous landmarks in the chronological scheme.

There are certain mountain ranges which have a different structure and must have had a correspondingly different mode of origin. As already pointed out, in the Great Basin, which lies between the Sierra Nevada and the Wasatch Mountains, are a number of parallel mountain ranges with a prevalent north and south trend, which are collectively called the Basin Ranges. These mountains are not folds of very thick strata, but tilted *fault-blocks*, which have been made by normal faults, each upthrow side standing as a great escarpment, but with a tilted top that gradually slopes back to the foot of the next block, to which it stands as the *downthrow* side. The processes of denudation have carved these tilted blocks into peaks and ridges of the ordinary kind. The boundary ranges, the Sierra Nevada and the Wasatch, although mountains of folding, have themselves been modified by the same process, for each of these ranges has a great fault along its base, the Great Basin being on the downthrow side with reference to each of them.

The **Date of Mountain Ranges** means the geological period in which they were first upheaved above the sea. This date is subsequent to the newest strata which are involved in the movement, and earlier than that of the oldest strata which did not take part in it, but must have done so, had they been present. Strata which rest unconformably against the flanks of a range must have been deposited after the folding movement was accomplished. If the newest folded strata and the oldest unmoved strata be of successive geological periods, the date of the upheaval is placed between those two periods and said to close the older one for the particular region involved. The subsequent history of a mountain range after its final upheaval above the sea must be read in its denudation and in the evolution of its topography and drainage.

DENUDATION OF MOUNTAINS

Mountains as we see them are never in the shape which they would present were the forces of compression and upheaval alone concerned in their formation. Every mountain range has been profoundly affected by the agencies of denudation, and their ridges and peaks, their cliffs and valleys, have been carved out of swelling folds and domes, or angular, tilted fault-blocks. As upheaval is a slow process, denudation must have begun its work as soon as the crests of the folds made their appearance above the sea, or above the level of the ground, so that probably no range ever had the full height which the strata, if free from denudation, would have given to it. Upheaval, though sometimes slow enough to allow rivers to keep their channels open, is yet too rapid to be kept in check by the processes of general atmospheric weathering, and so the ranges grew into great uplifts.   But as soon as the movement of elevation ceased, denudation began to get the upper hand, for as we have learned, mountains are the scene of rapid erosion.

Lofty, or *alpine*, ranges are subject to peculiarly rapid and effective denudation, quite different in character from that which operates in the lowlands.   Above the limit of the growth of trees (tree- or timber-line) rock destruction goes on with great rapidity, as is indicated by the wild and chaotic confusion of rock-pieces. In German this is called *Felsenmeer* (sea of rock), but there is no English term for it, and it is preëminently characteristic of lofty mountain slopes.   These masses of shattered rock are not only evidences of rapid disintegration by frost, but afford an immensely increased surface to destructive weathering.   The wind, which blows with great violence, is an important agent of destruction; avalanches carry down great quantities of rock, and the combined agencies of frost and gravity produce the vast talus slopes of all high mountains.

Another very effective agent among alpine summits is the glacier, which, by widening and cutting back the cirque at its head, eats rapidly into the mountain mass.   When many glaciers rise on the different sides of a mountain, the recession of the cirques will develop sharp crests and knife-edges.   To this cause has been chiefly attributed the extreme ruggedness of the high Alps, Sierra Nevada, and other ranges.

For a long period the effect of denudation is greatly to increase the ruggedness of the mountains, carving folds into ridges and cliffs, and ridges into bold and inaccessible peaks, but sooner or later the mountains are worn down lower and lower, and are eventually levelled with the plains from which they spring.   In the process of degradation, the synclines often resist wear better than the anticlines, and standing up above the level, form the synclinal mountains of many ancient ranges.

From the geological point of view mountains must be regarded as short-lived and ephemeral; low-lying plains persist for a longer time than do lofty ranges, as rivers may outlast many generations of lakes. Consequently, among the mountain chains of the globe, we everywhere find that the lofty ranges are those of comparatively recent date, while ancient mountains have been worn down into mere stumps. The Appalachians have been reduced nearly to base-level, and their present condition is that of a reëlevated and dissected peneplain, the ridges and valleys of which are determined by the position, attitude, and alternation of harder and softer strata. In its pristine state this very ancient range may have been as lofty as the Alps or Andes. Of course, there is no mathematical ratio between the youth of a range and its height, for moderately folded strata of moderate thickness never could have formed very high mountains, but in a general way it is true, that very high ranges are youthful, and that very old ranges are low. The process of degradation may go so far as to sweep away a mountain range to its very roots, leaving only the intensely plicated strata of the plain as evidence that mountains ever existed there. Of such a nature is the upland of southern New England and the great metamorphic area of Canada, both of which probably once carried ranges of high mountains.

APPALACHIAN CYCLES

We have seen that any region, however lofty and rugged, must eventually be worn down to base-level, provided only that the country remain stationary with reference to the sea until the process of degradation is complete. It is doubtful, however, whether any extensive region of hard rocks has ever been absolutely reduced to base-level: the usual result is the formation of a peneplain, a low-lying, featureless surface of gentle slopes and with only occasional eminences rising above the general level. Reëlevation of such a peneplain at once revivifies the streams and gives all the destructive agencies new powers. The peneplain is attacked and carved into valleys and hills, the valleys being rapidly cut down to base-level, while the divides and hills are much more slowly removed. If time enough be granted, the rugged country formed from a dissected peneplain is in its turn worn down to a second peneplain at a lower base-level. This alternate upheaval and wearing down together constitute a *cycle of denudation*, from base-level back to base-level. A complete cycle is one in which the whole region is reduced to a peneplain before the reëlevation occurs, and a partial or incomplete cycle is one which is interrupted by upheaval before the region is cut down, and only small and local peneplains have been formed. From a study of an old region several cycles of denudation may frequently be made out, represented by the remnants of dissected peneplains at different levels preserved in the harder rocks. The successive adjustments of the drainage system are a valuable auxiliary in working out the history of the cycles.

As an excellent example of these cycles of denudation whose marks are preserved in the structure of the rocks, we may take the Appalachian Mountains, which have been studied with great care. The cycles have been worked out elaborately, but only an outline of the more striking events can be given here.

These mountains began as a great geosyncline in which throughout the vast lengths of the Palæozoic era were accumulated enormously thick masses of shoal-water sediments. Toward the close of that era a number of crustal movements set in, crushing the sides of the geosynclinal trough, and crumpling the mass of strata contained in it into a series of roughly parallel, closed, inclined, or overturned folds, forming doubtless a very lofty range of mountains. During the long ages of the Mesozoic era the mountains were attacked and worn down by the destructive agencies; and by the time the Cretaceous period was reached the range had been reduced to a peneplain, with only a few hills rising above its almost featureless level, — hills which are now the peaks of western North Carolina, the highest points of the range at present. The present height of these peaks is due to subsequent reëlevation. This plain is called the Kittatinny peneplain, because the ridge

of that name in Pennsylvania and New Jersey is one of the remnants of it. To the observer who can overlook the billowy ridges
of the present range their even sky-line is very striking, and these
ridges are all composed of the hardest rocks, which all rise to
nearly the same level. To reproduce the plain it would be necessary to fill the valleys between the Blue Ridge on the east and
the plateau on the west up to the level attained by the hard ridges,
and this would give a gently arched surface, sloping very gradually to the Mississippi Valley and the Atlantic. On this peneplain
were already established the great streams which flow to the
ocean, such as the Susquehanna and the Potomac.

Next the peneplain was raised very gradually to a height of
1400 feet in Virginia, diminishing in both directions from this
point, and the denuding forces once more attacked and dissected
the plateau, the larger streams holding their transverse courses and
sawing through the hard strata, which were left standing as ridges
by the cutting of the longitudinal valleys along the more destructible beds. Denudation had cut down the softer beds to one general level, called the Shenandoah peneplain, the period of rest
being long enough to bring all the areas of soft and soluble beds
to this level, but not materially to lower the ridges of the more
resistant strata.

"The swelling of the Appalachian dome began again. It rose
200 feet in New Jersey, 600 feet in Pennsylvania, 1700 feet in
southern Virginia, and thence southward sloped to the Gulf of
Mexico. . . . In consequence of the renewed elevation, the
streams were revived. Once more falling swiftly, they have
sawed, and are sawing, their channels down, and are preparing
for the development of a future base-level." (Willis.)

Accordance of Alpine Summits. — It is perhaps generally true
of very high ranges that their highest peaks and ridges are arranged so as to be in accordance. If we imagine a surface which
shall everywhere touch these summit-levels, it will be found to
form a gently arching dome, with major axis coinciding with the
general trend of the mountains and highest in the interior of the

2 I.

range. This accordance has been observed in many alpine ranges, such as the Pyrenees, Alps, and Caucasus, the Sierra Nevada and Cascades, the Selkirks and Coast Range of British Columbia, and the mountains of Alaska.

In attempting to explain this remarkable coincidence, the hypothesis of peneplanation and subsequent upwarping, which is so satisfactory and so widely accepted as applied to the Appalachians and similar cases in Europe, has also been employed for alpine ranges of comparatively recent geological date. This explanation has not, however, found so general an acceptance for the high mountains, and other hypotheses should be considered as possible alternatives in each case. These rival hypotheses may be grouped into two classes: (1) those which regard the accordance as due to original structure in the processes of upheaval, and (2) those which refer the phenomenon to the spontaneous action of the denuding forces.

(1) There is no reason to believe that mountains could, or ever did, rise to indefinite heights; on the contrary, their height must be limited by the ability of their foundations to sustain them, and the principle of isostatic adjustment might well operate to produce some rough accordance of height, while denudation during the slow upheaval would tend to remove the higher summits faster than the lower ones. "The downcrushing of higher, heavier blocks with the simultaneous rise of their lower, lighter neighbors, coupled with the likewise simultaneous, especially rapid loss of substance on the higher summits, form a compound process leading toward a single, relatively simple result." (Daly.)

(2) The core of many high ranges is a granite batholith, from which are carved the higher peaks and ridges, the original covering of strata having been swept away by denudation. The form of the uneroded batholith was that of a gently arching dome, such as might be expected to give rise to accordant summits in the peaks carved from it. In ranges, like the Alps, which have no batholithic core, there is yet almost always great metamorphism due to the compression of the rocks and the weight of overlying masses.

The surface of the metamorphic core was probably quite regular before the covering strata were removed, and would tend to result in accordant summits among the peaks which are sculptured out of it. The thoroughly metamorphosed rocks, such as gneisses, schists, quartzites, marbles, etc., are not very different from one another in their resistance to weathering, and hence denudation would tend, in a general way, to maintain the accordance. Indeed, many investigators of the Alps ascribe the accordant summits to denudation alone. "The longer a mountainous region is exposed to denudation, the more completely do the indications of original inequalities in the relative heights of its peaks disappear, till finally the summits are determined entirely by the character of the rocks, the most resistant rocks forming the highest peaks." (Penck.)

It is not necessary to assume that any of these alternative hypotheses is true to the exclusion of the others. All the factors mentioned may coöperate to a common end, and for every mountain range the problem should be regarded as an individual one, without taking for granted that one explanation will cover all cases.

PART IV

HISTORICAL GEOLOGY

CHAPTER XXIV

FOSSILS — GEOLOGICAL CHRONOLOGY

A fossil is the impression or remains of an animal or plant preserved in the rocks.

A knowledge of fossils is indispensable to the geologist because they give him the means of establishing a consecutive chronology of the earth, and teach him much concerning the changes of land and sea, of climate, and of the distribution of living things upon the globe.

I. How Fossils were embedded in the Rocks

The conditions of the preservation of fossils are much more favourable to some kinds of organisms than to others. It is only under the rarest circumstances that soft, gelatinous animals, which (like jelly-fish) have no hard parts, can leave traces in the rocks. The vast majority of fossilized animals are those which have hard shells, scales, teeth, or bones; and of plants, those which contain a sufficient amount of woody tissue.

Again, the conditions under which organisms live have a great influence upon the chances of their preservation as fossils. Land animals and plants are much less favourably situated than are aquatic forms, and since the greater number of sedimentary

516

rocks were laid down in the sea, marine organisms are much more common as fossils than are those of fresh water.

On land, fossils have been preserved, sometimes in astonishing numbers, under wind-made accumulations of sand, dust, or volcanic ash, and in the flood-plain deposits of rivers. Peat-bogs are excellent places for fossilization, and the coal seams have yielded great numbers of fossils, principally of plants. The remains of land animals and plants, especially of the latter, are sometimes swept out to sea, sink to the bottom, and are there covered up and preserved in the deposits; but such occurrences are relatively uncommon. Small lakes offer more favourable conditions for the preservation of terrestrial organisms. Surrounding trees drop their leaves, flowers, and fruit upon the mud-flats, insects fall into the quiet waters, while quadrupeds are mired in mud or quicksand and soon buried out of sight. Flooded streams bring in quantities of vegetable débris, together with the carcasses of land animals, drowned by the sudden rise of the flood.

The great series of fresh-water and volcanic-ash deposits, which for long ages were formed in various parts of our West, have proved to be a marvellous museum of the land and fresh-water life of that region. On the fine-grained shales are preserved innumerable insects and fishes, with multitudes of leaves, many fruits, and occasionally flowers, while in the sands, clays, and tuffs, are entombed the bones of the reptiles, mammals, and, more rarely, birds of the land, mingled with those of the crocodiles, turtles, and fishes that lived in the water. Similar deposits are known in other continents.

It is on the sea-bed that the conditions are most favourable to the preservation of the greatest number and variety of fossils. Among the littoral deposits ground by the ceaseless action of the surf, fossils are not likely to be abundant or well-preserved, but in quieter and deeper waters vast numbers of dead shells and the like accumulate and are buried in sediments. The fossils are not, however, uniformly distributed over the sea-bottom; in some places they are crowded together in multitudes, while large areas will be almost devoid of them. The differences are due to varia-tions in temperature, in the character of the bottom, in food supply, and other conditions. Even under the most favourable circum-stances, the fossils can never represent more than a fraction of the life of their times. Indeed, the wonder is that so much of the life systems of past ages has been preserved, rather than that so large a part has been irretrievably lost.

The ways in which fossils are preserved vary much, but three groups include all the principal kinds.

(1) Preservation of more or less of the original substance. In certain rare instances an organism may be preserved intact,

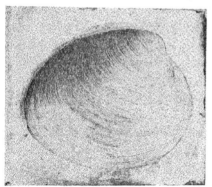 

FIG. 252. — Artificial external mould of clam shell (on left) and original shell (on right)

as have been the carcasses of the extinct species of elephant and rhinoceros in the frozen gravels of Siberia. Much more common is the decomposition of the soft structures and the preservation of the hard parts, — bones, shells, etc. Most of the shells and bones found in the rocks of later geological date are composed of the material originally belonging to them, though they have suffered much loss of substance.

(2) Entire loss of substance and retention of form. In this class of fossils all the original material of the organism has been lost, and no trace of its internal structure is retained, but only the external form has been reproduced in some different material.

Under this class we may distinguish two principal varieties: (a) _Moulds_ and (b) _Casts._ A _mould_ is formed when the fossil is embedded in sediments, which accurately reproduce its external form, and harden so as not to collapse when the fossil is removed. Percolating waters then dissolve away the organism entirely, leaving only a cavity, which is the mould.   Impressions of footprints, which may be placed in the same category as moulds, have already been explained (see pp. 206–7).

Casts are formed when the mould is filled by some solid substance deposited from percolating waters, thus reproducing the

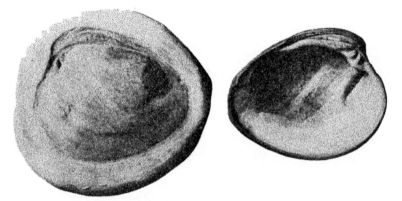

FIG. 253. — Artificial internal cast of clam shell (on left) and inner view of original shell (on right)

form of the fossil, as is done artificially with plaster or gutta-percha. If the fossil were hollow, like a shell, we frequently find a combination of internal cast with an external mould in the same specimen. At the time the fossil is embedded its interior is filled with the same sediment which hardens and forms an internal cast, exactly reproducing the form of the interior.   The shell itself is then dissolved away, leaving a space between the outer mould and the inner cast. Moulds and casts are commonest in rocks which permit percolating waters to traverse them freely, such as sandstones and coarse-grained limestones.

(3) Loss of substance with reproduction of form and structure. Fossils of this class are also called petrifactions and pseudomorphs (the latter a term borrowed from mineralogy). Here the original material of the organism has been more or less completely removed, and other material substituted for it; but the substitution has been so gradual, molecule by molecule, that not only the external form but also the microscopic structure has been perfectly reproduced. Several scantily soluble substances act as petrifying

FIG. 254. — Petrified logs, exposed by weathering of tuffs, Arizona. (Photograph by Sinclair)

materials, the most perfect results being given by silica. A silicified bone, or tooth, or bit of wood, differs from the original only in weight, colour, and hardness, and when a thin section is examined under the microscope, the minutest details of structure may be made out as perfectly as from the unaltered original. $CaCO_3$ is a very common petrifying agent, but it often obliterates structure by crystallizing after deposition; less usual are pyrite and siderite.

II. What may be learned from Fossils

The principal value which fossils possess for the geologist lies in the assistance which they give him in reconstructing the history of the globe. This they do in several ways.

(1) **In determining Geological Chronology.** — The most obvious way in which to make out the relative ages of a series of stratified rocks is to determine their order of superposition, for the oldest will be at the bottom and the newest at the top (see p. 321). But this method is of only local application and will not carry us far in an endeavour to compile a history of the whole earth. It cannot enable us to compare even the rocks of different parts of the same continent, for any exposed section is but a small fraction of the whole series of strata. More embarrassing still, strata change their character from point to point, limestone being laid down in one place while sandstone is accumulating in another. Still less can the order of superposition help to determine the relative ages of rocks in different continents, for this order in North America can be no guide to the succession in Africa or Australia. This conclusion does not imply that order of superposition may be safely neglected; on the contrary, it is of fundamental importance, but it must be studied in connection with the fossils.

Life, since its first introduction on the globe, has gone on advancing, diversifying, and continually rising to higher and higher planes. We need not stop to inquire how this progression has been effected; for our present purpose it is sufficient to know that progress and change have been unceasing and gradual, though not necessarily occurring at a uniform rate. Accepting, then, the undoubted fact of the universal change in the character of the organic beings which have successively lived on the earth, it follows that rocks which have been formed in widely separated periods of time will contain markedly different fossils, while those which were laid down more or less contemporaneously will have similar fossils. This principle enables us to compare and correlate rocks from all

the continents and, in a general way, to arrange the great events
of the earth's history in chronological order.

The general principle that similarity of fossils indicates the approximate contemporaneity of the rocks in which they are found
must not be taken too literally, for it is subject to certain limitations and exceptions.

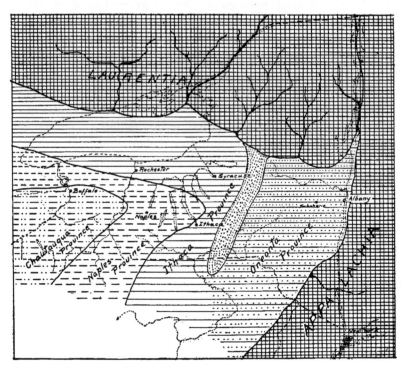

FIG. 255. — Geological map of Central New York, showing the faunal provinces of
the Upper Devonian (Portage stage).   (Clarke)

(a) Exact contemporaneity is not meant, for the progress of
life is very slow, and rocks formed thousands of years apart may
yet contain precisely similar fossils.

(b) Animals and plants differ in different parts of the world, so
that contemporaneous rocks formed in widely separated regions

will always show a certain amount of difference in their contained fossils. In comparing the rocks of two continents, it is often exceedingly difficult to decide just how much of a given difference in the fossils is to be ascribed to a difference in the time of rock formation, and how much to mere geographical separation. There is a great difference in the value of the various classes of organisms for chronological purposes, the most useful being those which have the most efficient means of very wide simultaneous dispersal. Such are the *pelagic* organisms, which live at the surface of the open sea and, either alive or dead, are carried for vast distances by the ocean currents.

A geographical distinction which should be emphasized is that of *facies*, by which is meant the sum total of environmental con-

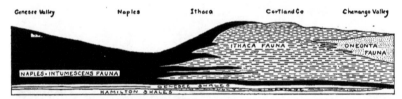

FIG. 256. — East-west section through area given in Fig. 255, showing changes of facies. (Clarke)

ditions, organic and inorganic. Deep or shallow water, salt water or fresh, muddy or sandy, or rocky bottom, are all differences of facies, and quite closely adjoining, strictly contemporaneous deposits may display them. Different assemblages of contemporaneous fossils also constitute facial differences; thus we speak of the graptolitic, or the cephalopod facies of a geological division.

A possible source of error in the inferences drawn from fossils lies in the incorporation of more ancient remains washed out of an older rock and embedded in a newer one. Figure 257 is a photograph taken on the York River, Virginia, which shows fossil shells washed out of the bluffs and lying on the beach, where they are mingled with modern organic remains.

Despite these limitations we find that, speaking broadly, the

order of succession in the appearance and extinction of the great
groups of fossils is much the same for all parts of the earth, and we
may confidently assume that the grander divisions of geological
time are of world-wide significance.

It will now be easy to understand why the fossils in two groups
of unconformable strata are generally so radically different.   It is

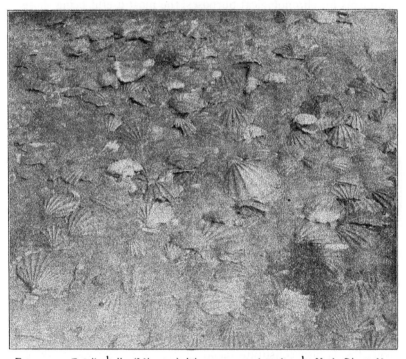

FIG. 257. — Fossil shells (Miocene) lying on a modern beach, York River, Va.
(Photograph by van Ingen)

because of the long lapse of unrecorded time at that point, during
which organic progress continued; when deposition was resumed,
the animals and plants were all new, and so the change is abrupt.
If one is reading a book from which a dozen chapters have been
torn out, the change of subject will appear violently abrupt; to
bridge over the gap one must find another copy of the book.

Likewise, to fill up the gap of a great unconformity, we must go to some region where deposition went on uninterruptedly, and there we may trace the gradual and steady change in the fossils.

A geological chronology is constructed by carefully determining, first of all, the order of superposition of the stratified rocks, and next by learning the fossils characteristic of each group of strata. To many it has seemed that this is reasoning in a circle, but that is because the argument is not fully stated, some of its steps being omitted. *The order of succession among the fossils is determined from the order of superposition of the strata in which they occur.* When that succession has been thus established, it may be employed as a general standard.

Great physical events, such as the upheaval of mountain ranges, widespread transgressions of the sea and changes of climate, often give us a means of correlating the strata of different continents with greater precision than can be done with the aid of fossils only. The latter are, however, indispensable means of first determining which of these events are comparable in different regions. The history is recorded partly in the nature and structure of the rocks, partly in the fossils, and partly in the topographical forms of the land and the courses of the streams. By combining these different lines of evidence, local histories are constructed for each region, until from these the story of the whole continent may be compiled. The comparative study of the fossils then gives the clew for uniting the history of the different continents into the history of the earth. Much remains to be done before this great task can be accomplished, but already we have an outline of the scheme which future investigations may fill up.

It necessarily follows from the way in which sedimentary rocks are formed, and the local nature of upheavals and depressions of land, that in no single locality can the entire series of strata be observed, and that each region can display but a certain proportion of the whole record. The different parts of our continent are of vastly different geological dates, and even the same area may have been many times a land-surface, and as often a sea-bottom. Unconformities, more or less widespread, offer a natural and convenient mode of dividing the strata into groups, but the difficulty with this method is that the dates of elevation and depression so seldom correspond in different regions, that divisions thus made are apt to be of more or less local validity. The only standard yet devised which is applicable to all the world is that founded upon the progress of life.

The comparison between human history and geological history is one that has very often been made, but trite and hackneyed as it is, it is none the less instructive. The history of civilized nations is the record of continuous development, not without retrogressions and periods of comparative stagnation, but having no actual gaps in it. For the sake of convenience, history is divided into certain periods in accordance with the predominance of certain great ideas and principles, and these periods are real, representing the salient facts in the progress of development. Each period is, however, but the outcome of the antecedent periods, and the ideas and principles which characterize it were slowly maturing, it may be through centuries, and even after other ideas have risen to predominance, older ones continue to live and influence the world. For example, when we speak of the age of the French Revolution, we refer to a time when a certain set of political ideas and principles were the most striking and influential factors in the development of the civilized world, beginning with the visible changes of 1789 and ending with the restoration of the Bourbons in 1815. But the tremendous outbreak was slowly preparing throughout the eighteenth century; the conflagration was proportionate to the materials that had been gathered for it. Nor, on the other hand, could the effects of the great movement be undone by the return of the exiled king. To this day the whole civilized world feels the effects of the convulsion, and the entire course of the nineteenth century would have been different but for the French Revolution.

Historians are careful to distinguish between events and the record of them. Events are continuous and bound up into a chain of consequences, every one of which is dependent upon others, while the records may be scanty, interrupted, confused, unintelligible, even misleading and falsified, so that it is no easy task to write history accurately and without attributing undue importance to this or that principle or policy.

These considerations fully apply to geological history; its divisions are founded upon the rise and culmination of great groups of animals and plants, which one after another have risen to predominance and then declined, their place being taken by others better fitted for the new conditions. These successive culminations are not sudden, but gradual and continuous, and the beginnings of each group are to be found in times long before the period of its predominance. Nor is decline immediately followed by extinction; one group slowly gives way to another, but long after the first has ceased to be the principal fact in the world's life, it may linger on in diminished importance until, perhaps, it finally disappears. The geological periods, therefore, like historical periods, had not definite beginnings and endings, for one slowly fades into another, but they are none the less actual because the lines of separation between them must often be somewhat arbitrarily drawn, and they cannot always be made to correspond in different regions.

In geology, as in history, we must distinguish between the events and the records of them. The more complete the records, the more obviously continuous and gradual was the course of events; only imperfect records can make the history seem broken and disjointed. As our science was first developed in western Europe, where the great groups of strata are mostly separated by unconformities, with abrupt changes in the fossils, the older geologists very naturally concluded that the great divisions of geological time were marked by frightful catastrophes which devastated the earth, destroying every living thing upon it. Each group of rocks was looked upon as the product of a long and tranquil period, and its fossils were believed to represent an entirely new creation. Though opposed by some far-seeing minds, the doctrine of *Catastrophism,* as it was called, long held sway, but was shown to be erroneous, when the study of geology was carried to other parts of the world. Then it appeared that the supposed catastrophes, if they occurred at all, were not general, but local, and that records missing in Europe had been preserved, partially at least, in other continents. Enough of these missing records has been recovered to show that the earth's progress was not by a series of abrupt and sudden changes, but by a continuous, orderly development.

Major divisions of *geological time* are founded upon the more striking changes in the animals and plants, while for minor divisions the more detailed differences in the organisms are employed. Parallel with the divisions of *time* run the groups or systems of the *strata*, for characterizing which both the physical nature and structure of the rocks and the fossils are employed. In the very difficult and complicated task of compiling the earth's history, no kind of evidence can be ignored, and wide knowledge and sound judgment are needed in the work, so that no particular class of records shall be either over- or undervalued.

Though the goal of geological inquiries is to construct the history of the earth as a unit, this goal can be reached only by the minute and exhaustive study of the local histories. Each of the latter has certain peculiarities of its own which must be determined, and hence arises the multiplicity of local names for groups of strata, so confusing to the student. Local names are useful, because they avoid the necessity of premature correlations, which may lead to the direst mistakes.

(2) **As Evidence of Geographical Changes.** — We have seen that from the composition and structure of the stratified rocks themselves much may be learned concerning the geographical conditions under which they were formed, and of the subsequent geographical changes of the region in which they occur. Fossils supplement this information regarding the body of water in which the rocks were laid down, whether fresh or salt, deep or shallow, near or far from land, in an open sea or a closed basin, and whether such a closed basin had occasional or constant communication

with the ocean. Most of the stratified rocks which now form part of the land-surface give us information only concerning the former extension of bodies of water over what is now the land, but they can tell us nothing of the land areas which have disappeared beneath the sea. In this connection fossils are of great assistance, for, in certain instances, the distribution of marine fossils points to the presence of land barriers to migration which no longer exist, or to a continuity of coast-lines which are now broken up, while the fossils of land animals may demonstrate the former existence of land bridges between regions which have long been separated by water. Thus it may be shown that North America was frequently and for long periods of time connected with Asia across Bering Sea, and that its union with South America is of geologically late date.

(3) **As Evidence of Climatic Changes.** — The remarkable climatic changes through which various parts of the earth have passed are indicated by fossils. Indeed, with the exception of glacial marks and ice-formed deposits, fossils offer almost the only trustworthy evidence available as to changes of temperature. Thus, when we find in the rocks of Greenland the remains of extensive forests of such trees as now grow in temperate latitudes, the only possible inference is that Greenland now has a far colder climate than when those forests existed. The same conclusion follows from the presence in the rocks of Wyoming and Idaho of great palm leaves and other subtropical plants associated with the bones of crocodiles and other reptiles, such as live only in warm regions. In deposits of a far later date occur bones of the reindeer in southern New England and in the south of France, walrus bones in the sands of New Jersey, and those of the musk-ox in Arkansas; all of which shows that at one time these regions had a much colder climate than at present.

The evidence as to climatic changes which is presented by fossils must, however, be treated with great caution, because even nearly allied species often have entirely different habits, and flourish in quite different climates. Most fossils belong to extinct

2 M

species, as to whose climatic relations we have no knowledge.
Before any conclusion concerning changes of climate can be regarded as established, we should have the testimony of species
still living, or, if that is not.possible, the evidence must be drawn
,from large assemblages of different kinds of animals and plants.
Such an extreme case as the fossil plants of Greenland is sufficient
evidence without further corroboration.

III. CLASSIFICATION OF GEOLOGICAL TIME

The method of making the divisions and subdivisions of geological time is not . yet a fixed one, and there is much difference
in the usage of various writers. The names of the divisions also
have been given at various times and in many lands, according to
no particular system. Most of these names have been taken from
the locality or district where the rocks in question were first
studied or are most typically displayed ; as Devonian from Devon-
shire, Jurassic from the Jura Mountains. Some are named from
a characteristic or prevalent kind of rock, such as Cretaceous
(Latin *creta*, chalk) and Carboniferous. Of late there has been a
tendency toward a more uniform method of nomenclature, and
to the use of one set of terms for the divisions of time, and another and corresponding set for the divisions of the strata. The
grander divisions of time are called eras, and in descending order
we have periods, epochs, and ages. The following table represents
the divisions in the scale of time and the scale of rocks which have
been adopted by the International Geological Congress.

TIME SCALE	ROCK SCALE
Era	Group
Period	System
Epoch	Series
Age	Stage
	Substage
	Zone

It will be observed that the subdivision is carried farther in the scale of rocks than in that of time, because of the generally local character of these minor subdivisions. The names employed are, as yet, the same for both scales, and we speak of the Palæozoic Era or Group, and of the Silurian Period or System. It has been proposed to give separate names to the divisions of the two scales, and this would be an improvement in some respects.

TABLE OF MAJOR GEOLOGICAL DIVISIONS

Cenozoic Era	Quaternary Period
	Tertiary Period
Mesozoic Era	Cretaceous Period
	Jurassic Period
	Triassic Period
Palæozoic Era	Permian Period
	Carboniferous Period
	Devonian Period
	Silurian Period
	Ordovician Period
	Cambrian Period
Pre-Cambrian Eras	Algonkian Period
	Archæan Period

CHAPTER XXV·

ORIGINAL CONDITION OF THE EARTH — PRE-CAMBRIAN PERIODS

As we trace the history of mankind back to very ancient times, we find that the records become more and more scanty and less intelligible, until history fades into myth and tradition. Of a still earlier age we have not even a tradition; it is *prehistoric.* Similarly, among the geological records the earliest are in a state of such excessive confusion that they are exceedingly difficult to understand, and between different observers there are radical differences of opinion both as to the facts and as to their interpretation. Furthermore, there must have been an inconceivably long time earlier than the most ancient recorded periods, as to which conjecture and inference are the only resource. In these difficult straits astronomy offers valuable assistance to the baffled geologist. The *Nebular Hypothesis* is a scheme of the development of the solar system which is very generally accepted by astronomers, in some form, as essentially true.

The term *nebular hypothesis* is usually, though not with exactness, limited to one particular form, according to which the place of the present solar system was originally· occupied by a vast rotating nebula, a mass of intensely heated vapour, or possibly clouds of meteorites, extending beyond the orbit of the outermost planet. As the nebula cooled by radiation, it contracted, leaving behind it successive rings, like those of the planet Saturn, but on a vastly larger scale. The rings kept up the rotation imparted by the nebula, and all of them lay in nearly the same plane. Unequal

532

contraction in various parts of each revolving ring caused it to break up and gather by mutual attraction into masses. If these rings were composed of relatively small solid masses, like meteorites, or if they had solidified by condensation of the vapours, the heat generated by the collisions, as the broken ring was gathered into a mass, would suffice to raise the temperature and liquefy or vapourize the mass. By revolution the nebulous masses would assume a spheroidal shape and become planets. The central mass of the original nebula forms the sun, which is still in an intensely heated, incandescent state.

Another form of the nebular hypothesis, called for the sake of distinction the *Planetesimal Hypothesis*, has recently been proposed by Professor Chamberlin. This postulates, as the beginning of the solar system, a spiral nebula, " and that the matter of this parent nebula was in a finely divided solid or liquid state before aggregation. . . . It regards the knots of the nebula as the nuclei of the future planets, and the nebulous haze as matter to be added to these nuclei to form the planets. It assumes that both the knots and the particles of the nebulous haze moved about the central mass in elliptical orbits of considerable, but not excessive, eccentricity. . . . It deduces a relatively slow growth of the earth, with a rising internal temperature developed in the central parts and creeping outward." (Chamberlin and Salisbury.)

This is not the place to discuss the evidence for an astronomical speculation, but it is clear that the hypothesis regarding the development of the solar system which we adopt must condition our views as to the early unrecorded stages of the earth's history. From the strictly geological standpoint the most important difference between the Nebular and the Planetesimal Hypotheses is that, according to the former, the earth has passed through a gaseous and a molten stage, and therefore must have formed a crust by solidification, while the latter leads to the conclusion that the earth has been solid from the beginning, and consequently never formed a crust of solidification.

THE PRE–CAMBRIAN PERIODS — I. ARCHÆAN

It is unfortunate that an account of historical geology should necessarily begin with the most difficult and obscure part of the whole subject, but the treatment must be in accordance with the chronological order, and the oldest rocks are the least intelligible. The ordinary criteria of the historical method, namely, the stratigraphical succession and the comparison of fossils, fail us here almost entirely, and the only way of correlating the rocks of different regions and continents is by means of the characters of the rocks themselves. In the present state of knowledge, " lithological similarity " is not a safe guide. So many metamorphic rocks, once referred to the Archæan, have proved to be of much later date, that some cautious geologists, who have no confidence in " lithological similarities," prefer not to use the term *Archæan* at all, but to employ local terms for the oldest crystalline rocks exposed in a given district.

The Archæan includes the most ancient rocks, often spoken of as the " basement, or basal complex." Its antiquity is best assured in regions where it is separated by thick series of sedimentary or metamorphic rocks from the Lower Cambrian, which can be certainly identified by its fossils. The character and relations of the pre-Cambrian rocks differ so much in different areas that it will be best to describe them in two or three typical regions. In the Canadian provinces of Quebec and eastern Ontario and in the Adirondack Mountains, the oldest rocks are a series of intensely metamorphosed sediments, including great bodies of limestones, quartzites, schists, etc. In the Adirondacks, especially, this series is invaded by enormous bodies of intrusives, which preceded and were involved in a great period of metamorphism. In eastern Ontario the thickness of the metamorphosed sedimentaries is exceedingly great: " Along its whole northern border this sedimentary series is torn to pieces by an enormous volume of gneissic granite of igneous origin which rises from beneath it, and which along its margin also wells up through it in the form of great intrusive bathyliths." Though underlying the metamorphosed sediments of the *Grenville series*, the gneissic granite, or *Laurentian*, is the younger, as the contact is an intrusive one.

In the region around Lake Superior the pre-Cambrian rocks are displayed in enormous thickness and are divided into groups by four great unconformities. Of these the most ancient is the *Keewatin*, intensely metamorphosed rocks derived from the transformation of lava flows, tuffs, and other volcanic rocks, with some of sedimentary origin, and forming a great series of schists, which are underlaid and penetrated by the Laurentian gneissic granites. It immediately suggests itself that the Grenville series of the east is the equivalent of the Lake Superior Keewatin, but the committee of the Canadian and United States Geological Surveys, which has investigated these problems, report that they consider it "inadvisable in the present state of our knowledge to attempt any correlation of the Grenville series with the Huronian or Keewatin." The classification proposed by the committee is contained in the subjoined table, though emphasis must again be laid upon the fact that no equivalence between the Grenville and the Keewatin is asserted. They may be separated by vast periods of time, and yet both must be older than the intrusion of the Laurentian granites.

PRE–CAMBRIAN ROCKS

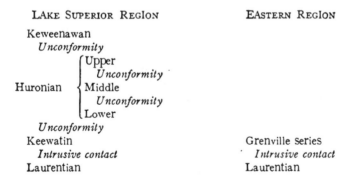

LAKE SUPERIOR REGION	EASTERN REGION
Keweenawan	
Unconformity	
Huronian { Upper *Unconformity* Middle *Unconformity* Lower	
Unconformity	
Keewatin	Grenville series
Intrusive contact	*Intrusive contact*
Laurentian	Laurentian

N.B. — In the classification adopted in this book the Archæan comprises the Keewatin and Laurentian and probably also the Grenville series.

The Archæan, then, is composed of completely crystalline rocks of various types. Massive rocks, such as granite and basic eruptives, and foliated rocks, like gneissoid granite, gneiss, many varieties of schists, are intermingled in the most intricate way, a characteristic well expressed in the oft-used phrase of the basal or fundamental complex for the Archæan. The component mineral particles show plainly the intense dynamic metamorphism to which they have been subjected, in their extremely complex arrangement and in their laminated and crushed condition. The rocks thus referred to the Archæan are not necessarily all of the same age, but they are all of vast antiquity and older than any other known series. They are of very great but unknown thickness, for the bottom of them is nowhere to be seen, and even when thrown up into mountain ranges, erosion has in no case cut so deeply into these rocks as to expose anything different below them.

The reason for uniting these rocks into one group is not merely their likeness in composition, which is not a sufficient criterion, but because of their unique and uniformly complex structure, their resemblance to one another and difference from any other group of rocks, and their invariably fundamental position.

The Distribution of the Archæan Rocks. — At the outset of our historical studies it is essential to understand clearly just what is meant by the term *distribution of a given formation.* It means: (1) that the given rock is at the surface over a certain area, disregarding the covering of soil, drift, or other loose materials; (2) that the concealed extension of the formation beneath newer rocks may be confidently inferred from surface observations. So far as the Archæan rocks are concerned, their surface distribution can at present be stated only with much reserve, for they often grade into crystalline schists of demonstrably later date, and much that once was referred to the Archæan is now known to be far more recent. Accurately to determine the distribution of the basal complex will require the most extensive, minute, and laborious

FIG. 258.—Map of known pre-Cambrian surface exposures in North America.
The black areas are outcrops

investigation. The northern part of North America, from the Arctic Ocean to the Great Lakes, probably including Greenland, is made up of an immense area of schistose rocks, estimated at more than 2,000,000 square miles in extent. Over this vast region occur numerous areas of Archæan rocks, but it is not yet possible to say how much of it belongs in that group and how much is newer.

Beside this principal region are several other minor ones. A narrow band of schistose rocks extends, with some interruptions, from Newfoundland to Alabama, with shorter parallel belts in eastern Canada and New England. Another great axis is on the site of the Rocky Mountain chain, with several shorter and generally parallel belts from Mexico to Alaska. Isolated areas occur in Missouri, central Texas, New Mexico, and Arizona. In all of these regions are found rocks like the typical Archæan, which stand in the same relation to the newer groups, but how much should be referred to the oldest series is still a question.

In the other continents occur great areas of very ancient gneisses and crystalline schists, but even less than in North America has the distinction been made between the fundamental complex and newer groups. In the following statements no attempt is made to determine how much of the areas mentioned is properly Archæan.

In *Europe* the principal area lies to the north, covering parts of Ireland and the Highlands of Scotland, with which was probably once connected the great continuous mass of Scandinavia, Finland, and Lapland. Considerable areas also occur in central and southern Europe, as the central plateau of France, parts of Germany and Bohemia, and long, narrow belts in the Pyrenees, Alps, and Balkans. In *Asia* these ancient crystalline rocks are found in the great mountain ranges, such as the Himalayas, Altai, etc. They make up a large part of the Indian peninsula, and are extensively displayed in China, Japan, and the islands of the Malay Archipelago. The vast central plateau which occupies so much of *Africa* is principally composed of these rocks, which are also largely exposed in *Australia*. In *South America* similar rocks appear in the highlands of Brazil and in the Andes.

It is estimated that the Archæan rocks form somewhat more than one-fifth of the land-surface of the earth, and there is reason to believe that they are actually universal, and that a boring made at any point, if sufficiently deep, would encounter them. They are found at the bottom of many deep cañons, and borings frequently penetrate them at points where there are no surface indications of their presence. If these rocks are really distributed over the entire globe, they are the only formation of which this is true.

Origin of the Archæan Rocks. — This is a problem which has given rise to a great deal of discussion, but a solution appears to be near. Independently, in many countries, observers have reached the conclusion that these rocks are divisible into two great series, a schist series composed chiefly of highly metamorphosed sedimentary and volcanic rocks, and a gneissoid granite series, which is intrusive and later than the former.

Assuming that this conclusion is true, at least as a working hypothesis, it involves certain curious consequences. Surface lava flows and volcanic tuffs, and still more, sedimentary rocks, necessarily imply a solid floor upon which they were laid down, but of this floor not a trace has anywhere been found. The question immediately arises, what has become of it? No answer to this question can yet be given, but apparently the most likely suggestion is that the ascending floods of molten magma, which gave rise to the gneissoid granites, must have melted and assimilated it. If this were only a local phenomenon, there would be nothing very surprising about it, but it would seem to be true of the entire globe, and this is a startling conclusion. We are then to suppose that a solid crust, however formed, was for a very long time sufficiently rigid and stable to allow a great thickness of sedimentary and volcanic rocks to be accumulated upon it and then was ingulfed and destroyed by a *universally* ascending magma, though it is not necessary to suppose that this took place simultaneously over the whole earth, or even within a relatively short period of time; it may have required ages in the accomplishment. Furthermore, it must not be forgotten that remnants of the floor may yet be discovered in little-known regions. If this complete and universal assimilation actually took place, it is an absolutely unique phenomenon in the recorded history of the earth, though something more or less similar may have happened many times before that record began.

Many other hypotheses have been propounded to account for the origin of the Archæan rocks, but as they are not supported by any strong evidence, it is not worth while to consider them here; several of them have been formally abandoned by their authors.

That the oldest known rocks were not the first to be formed is manifest from the derivative nature of many of them, for sediments necessarily imply some preëxisting rock to furnish the materials, and volcanic outbursts involve a solid surface through which they break.

From the extreme degree of dynamic and thermal metamorphism which the Archæan rocks have undergone, we should not expect to find any recognizable fossils in them. On the other hand, there are indirect evidences that life was already present on the earth at that period. The limestones, iron ores, and graphite found in these rocks appear to have been organically accumulated, but it is possible that they were chemically formed, and so the evidence, while probable, is not altogether conclusive.

II. ALGONKIAN

This is the name proposed by the United States Geological Survey for the great series of sedimentary and metamorphic rocks which lie between the basal Archæan complex and the oldest Palæozoic strata; it is but little used outside of this country and is not universally employed even here, but it is beginning to make its way in Europe, and serves a useful, though possibly a temporary, purpose. While it is possible, though not very likely, that more advanced knowledge may lead us to distribute these rocks partly into the Archæan and partly into the Palæozoic, yet for the present, at least, it is better to form a separate grand division for them.

The Algonkian rocks, which are widely distributed in North America, form an immensely thick mass of strata and of metamorphic rocks which are believed to represent those strata in other regions. These metamorphic rocks were long generally referred to as the *Huronian*, which was regarded as the upper portion of the Archæan, but, so far as can be learned, they occupy the same stratigraphical position as certain little changed sediments, between the fundamental complex below and the Cambrian above. At the base of the magnificent section exposed in the Grand Cañon of the Colorado is a very thick mass of strata, separated by great unconformities from the Archæan gneiss below and from the overlying Cambrian. This mass is again subdivided by minor unconformities into three series. The lower series, at least 1000 feet thick, and perhaps more, is made up of stratified quartzites and semi-crystalline schists, cut by intrusive granite. Above this come nearly 7000 feet of sandstones, with included lava sheets, and at the top more than 5000 feet of shales and limestones, in which a few fossils have been found. The two upper series are not at all metamorphic. All these strata are steeply inclined, and upon their truncated edges rests the sandstone referred by Mr. Walcott to the Middle Cambrian.

In central Montana is a very extensive exposure of Algonkian rocks, 12,000 feet thick, composed of immense bodies of sandstones, quartzites, limestones, and hard arenaceous shales. These beds, called the Belt series, are upturned in the Belt Mountains, and unconformably overlaid by a Middle Cambrian sandstone. " In late Algonkian times an orographic movement raised the indurated sediments of the Belt terrane above sea-level, . . . [and] folding of the Belt rocks formed ridges of considerable elevation, and areal [aërial] erosion and the Cambrian sea cut away in places from 3000 to 4000 feet of the upper formations of the Belt terrane before the sands that now form the Middle Cambrian sandstones were deposited." (Walcott.)

A very similar succession of rocks to that of the Grand Cañon is found in the Lake Superior region, intervening between the Archæan complex and the Upper Cambrian, from both of which they are separated by great unconformities. As in the Grand Cañon section, these rocks are divisible into four series by unconformities. The lower two series, with a maximum thickness probably exceeding 5000 feet, are much crumpled, metamorphosed, and semi-crystalline. They comprise limestones, quartzites, mica schists, etc., cut by igneous dykes, also much volcanic tuff and agglomerate. Next follows a series of 12,000 feet of less intensely folded but still metamorphic rocks, quartzites, shales, slates, mica schists, with dykes and interbedded sheets of diorite. A few fossils have been found in the quartzites of this series. The fourth series has a maximum thickness of 50,000 feet, though usually much less. The lower part of this series is formed by thick lava sheets, interbedded with sandstone and conglomerate, and above is a mass of sedimentary rocks largely derived from the volcanic materials. This uppermost series is by some authorities referred to the Cambrian, but, in the absence of fossils, there seems to be no way of definitely deciding the question.

Over the great Archæan area of Canada occur many districts of metamorphic rocks which are plainly of sedimentary origin, such as crystalline limestones, schistose conglomerates, as well as volcanic tuffs and agglomerates. In the Archæan region of Canada and in New England the Algonkian metamorphics seem to grade into the Archæan complex without unconformity. This apparent conformity may, however, very well be due to subsequent dynamic metamorphism, which, as has been proved, may obliterate nearly all traces of a great unconformity. Through the Rocky Mountain region and the Pacific coast mountains, the Archæan is in very many places overlaid by great thicknesses of metamorphic Algonkian rocks, such as quartzites, sandstones, and schists, which are sometimes as much as 12,000 feet thick, as in the Wasatch and Uinta mountains. Other isolated areas are found, as in the Black Hills, where a great mass of schists, slates, and quartzites is separated by a very marked unconformity from the overlying Cambrian; also in Missouri and Texas. The Algonkian rocks of the West have not been subjected to such extreme folding as have most of those of the East, and hence their distinctness from the Archæan is more clearly marked. In the southern Appalachians are some little changed strata which are referred to the Algonkian.

In other continents the distinction between the Archæan and the Algonkian is beginning to demand recognition. In Great Britain, for instance, are found very interesting parallels with the Algonkian of this country. In Scotland the Torridon sandstones, 8000 to 10,000 feet thick, which are nearly horizontal and almost unchanged, lie unconformably between the oldest Cambrian and the basal Archæan; and in other areas, metamorphic rocks of sedimentary origin occupy a similar position. In Finland and Sweden 10,000 feet of sedimentary and igneous rocks and schists occur between the Archæan and the Cambrian. Many of the crystalline schists of the European pre-Cambrian areas appear to correspond in character and position to the metamorphic Algonkian.

Lately the surprising announcement has been made of extensive glaciation in the early Algonkian of North America The conglomerate at the base of the Lower Huronian of Canada is regarded as of glacial origin, " since it contains angular and subangular boulders of all sizes up to cubic yards, enclosed in an unstratified matrix. These boulders are often miles from any possible source. Recently, striated stones have been broken out of their matrix in the Lower Huronian of the Cobalt Silver region, giving still stronger proofs that the formation is ancient boulder clay." (Coleman.)

In South Africa we find clear evidence of three distinct ice ages, before the close of the Palæozoic Era, and the most ancient of these, which cannot be definitely dated, may correspond in time to that above-mentioned in Canada.

Life in the Algonkian. — In the Grand Cañon and Montana determinable fossils have been found in the less changed sediments, but they are too few and scanty to tell us much of the life of the times.   Evidences of life are not wanting in the metamorphic rocks of the eastern and northern regions, but they are indirect.   The strata of crystallized limestone are indications of the presence of animal life in the Algonkian seas.   The great quantities of graphite diffused through many of the schists, and the beds of iron ore likewise tend to show the existence of plants at the same time.   More conclusive are the determinable fossils obtained in the Belt series of Montana and in the Grand Cañon series, which include the tracks of worms, brachiopods, and fragments of large Crustacea referable to the Eurypterida.   Such remains imply a long antecedent history of life, the records of which remain to be discovered.

The pre-Cambrian rocks are remarkable for their wealth of valuable minerals.   Immense accumulations of iron ore occur in Canada, New York, New Jersey, along the Appalachians from Virginia to Georgia, in Michigan, the Lake Superior region, Missouri, and the Southwest.   The great copper mines of Lake Superior are associated with igneous rocks which intersect sandstones referred to the Algonkian.

It will be obvious to the student how very little is really known regarding the most ancient rocks of the earth's crust, the Archæan. They are enormously thick metamorphic masses of vast geographical extent.   In all the continents they form the foundation upon which the oldest fossiliferous sediments were laid down, and, in brief, they are the oldest, the thickest, the most widely distributed and the most important of all the accessible constituents of the earth's crust.   Their uniform character, wherever found, the extreme plication and metamorphism which they have undergone, and their world-wide distribution, are all extremely remarkable features, such as recur in rocks of no other age.   The Algonkian sedimentary rocks present the earliest chapters in the recorded history of life.   The pre-Cambrian rocks indicate that vast periods of time had elapsed before the clearly recorded part of the earth's history began, a time probably longer than all subsequent periods.

CHAPTER XXVI

PALÆOZOIC ERA — CAMBRIAN PERIOD

THE Palæozoic is the oldest of the three main groups into which the normal fossiliferous strata are divided; it forms the first legible volume of the earth's history, and in interpreting it speculation and hypothesis play a much less prominent part than in the pre-Cambrian volume. The Palæozoic rocks are conglomerates, sandstones, shales, and limestones, with quite extensive areas of metamorphic rocks, and associated igneous masses, both volcanic and plutonic. The thickness of these rocks is very great, estimated in Europe at a maximum of 100,000 feet. This does not imply that such a thickness is found in any one place, but that if the maximum thicknesses of each of the subordinate divisions be added together, they will amount to that sum. In this country more than 25,000 feet of Palæozoic strata are exposed in the much folded and profoundly denuded Appalachian Mountains, but in the Mississippi valley they attain only a fraction of that thickness. These rocks are, in the vast majority of cases, of marine origin, but some fresh-water beds are known, and very extensive swamp and river deposits have preserved a record of much of the land life of the era, especially of its later portions. That there must have been land-surfaces is abundantly shown by the immense thickness and extent of the strata, all of which were derived from the waste of the land. Both in Europe and in North America, the land areas were prevailingly toward the north, and are doubtless indicated, in part, by the great regions of the pre-Cambrian metamorphic rocks. The general character of the Palæozoic beds shows that they were, in large measure, laid down in shallow water in the neighbourhood of land. Their great thickness indicates,

2 N   545

further, the enormous denudation which the land areas underwent. The calculation has not been made for this country, but
for Great Britain Geikie states that the lower half of the Palæozoic group represents the waste of a plateau larger than Spain and
5000 feet high, cut down to base-level.

Very widespread disturbances of the earth's crust before the
beginning of the Palæozoic era and at its close have produced well-
nigh universal unconformities with both the underlying pre-
Cambrian and the overlying Mesozoic rocks; at only a few points
are transitional series found.

Early in Palæozoic time were established the main geographical
outlines which dominated the growth of the North American
continent, — a growth which was, for the most part, steady and
tranquil. These conditions may be briefly stated as the formation of a great interior continental sea, divided from the Atlantic
and the Pacific by more or less extensive and variable land areas.
There are thus three principal regions of continental development: those of the Atlantic and Pacific borders and the interior.
In addition, the eastern border is subdivided by pre-Cambrian
ridges into subordinate areas of deposition. At the present time
the surface rocks over the eastern half of the continent are prevailingly Palæozoic, extending chiefly southward and southeastward from the great pre-Cambrian mass of the north.

Palæozoic time was of vast length, perhaps exceeding that of
the combined Mesozoic and Cenozoic eras.

The subdivisions of the Palæozoic are very clearly marked,
locally often by unconformities, but on a wide scale by the changes
in the character of the fossils. There is some difference in the
practice concerning these divisions, not as to their limits or order
of succession, but merely as to their rank, whether certain ones
should be called systems (periods) or series (epochs). This is
a difference more about names than facts. The successive steps
of organic and geographical development are best displayed by
dividing the group into six systems, or periods, which are as follows, beginning with the oldest: 1. Cambrian; 2. Ordovician;

3. Silurian; 4. Devonian; 5. Carboniferous; 6. Permian. By many geologists the Ordovician and Silurian are comprised in one system, and the Carboniferous and Permian in another; but the present tendency is in favour of maintaining all six as equal in rank. It must not be supposed that these systems represent equal spaces of time as measured by the thickness of rocks, or equal geographical extent; on the contrary, they are very unequal in both these respects. The classification means that the six systems, or periods, stand for approximately equivalent changes in the character of the animals and plants.

Palæozoic Life possesses an individuality not less distinctly marked than that of the group of strata, which demarcates it very sharply from the life of succeeding periods, and gives a certain unity of character to the successive assemblages of plants (*floras*) and of animals (*faunas*). The era is remarkable both for what it possesses and what it lacks. Among plants, the vegetation is made up principally of Cryptogams, seaweeds, ferns, club-mosses, and horsetails. Especially characteristic are the gigantic, tree-like club-mosses and horsetails, which are now represented only by very small, herbaceous forms. The only flowering plants known are the Gymnosperms, the Cycads, and their allies; no Angiosperms have been discovered. Palæozoic forests must have been singularly gloomy and monotonous, lacking entirely the bright flowers and changing foliage of later periods.

The Palæozoic fauna is largely made up of marine invertebrates, in the earlier periods entirely so; *i.e.* so far as we have yet learned, though land life surely began before the oldest records of it yet discovered. Graptolites and Hydroid Corals, true Corals, Echinoderms (especially Crinoids, Cystideans, and Blastoids), long-hinged and hingeless Brachiopods, Mollusca (particularly the Nautiloid Cephalopods), and the crustacean groups of Trilobites and Eurypterida are the most abundant and characteristic types of animal life. Insects, centipedes, and spiders were common toward the end of the era. Cambrian rocks contain no fossil vertebrates, but they make their appearance in

the Ordovician. For long ages the only vertebrates were
fishes and certain low types allied to the fishes, but at the end
of the Devonian and in the Carboniferous appeared the Amphibia, followed in the Permian by true Reptiles. Teleosts,
such as make up by far the largest part of the modern fish-fauna,
both marine and fresh-water, as well as birds and mammals, are
entirely absent from the Palæozoic.

The overwhelming majority of Palæozoic species, and even
genera, fail to pass over into the Mesozoic, and even in the larger
groups which continued to flourish almost always a more or less
complete change of structure occurs, so that Palæozoic corals,
Echinoderms, and fishes, for example, are very markedly distinct
from those which succeeded them. The difference is generally
in the direction of greater primitiveness of structure in the older
forms, Palæozoic types standing in somewhat the same relation
to subsequent types as the embryo does to the adult.

In the vast periods of time included in the Palæozoic era
occurred some remarkable climatic vicissitudes, which will be
more fully described in the succeeding sections. Times of widespread glaciation occurred in the Lower Cambrian of Norway
and China, probably of Australia, and perhaps also of South
Africa; in the Devonian of South Africa, and in the Permian of
the latter region, India, Australia, and South America, perhaps
also in Europe and North America.

For most of the era, however, the climate appears to have
been mild and equable on the whole, very much the same kinds
of animals and plants occurring in high as in low latitudes. In
short, we can detect no evidence of climatic zones as being distinctly marked in those periods.

THE CAMBRIAN PERIOD

The rocks older than the coal measures were for a long time
heaped indiscriminately together, under the name of Greywacke,
or Transition Rocks, and were little regarded by geologists.

About 1831, the problem of these ancients rocks was attacked by two eminent English geologists, Sedgwick and Murchison, who soon brought order out of the chaos. There was much discussion and dispute as to the limits of the systems into which the Greywacke should be divided, and as to the names which should be given to them. The oldest fossiliferous strata were by Sedgwick called *Cambrian* (from the Latin name for Wales), but were included by Murchison in his Lower Silurian. The latter example was long followed by most geologists, but the advance of knowledge has fully vindicated the claim of the Cambrian to rank as a distinct system. The divisions of the American Cambrian are as follows: —

3. Upper Cambrian, Saratogan Epoch, *Dikellocephalus* Fauna.

2. Middle Cambrian, Acadian Epoch, *Paradoxides* Fauna.

1. Lower Cambrian, Georgian Epoch, *Olenellus* Fauna.

American. — In North America, Cambrian rocks are not exposed at the surface over large areas, being, for the most part, deeply buried under later sediments; their maximum thickness, so far as known, does not exceed 12,000 feet. While not forming extensive areas of the present surface, Cambrian strata are very widely distributed over the continent, usually resting uncomformably upon the plicated and metamorphosed rocks of the Archæan and Algonkian. These strata are found in the pre-Cambrian depressions, from the Adirondacks to Newfoundland, and along the flank of the Appalachian uplift, from the St. Lawrence to Alabama. They also fringe Archæan or Algonkian areas in other regions, as in Wisconsin, Missouri, Texas, in the Rocky Mountain chain, from Colorado to British Columbia, and in the mountains of Nevada. Cambrian beds are exposed in the Colorado Cañon, and doubtless would be found throughout the larger part of the continent were the overlying beds stripped away.

So far as they are accessible to observation, the Cambrian rocks are chiefly such as are laid down in shallow water near shore,

conglomerates, sandstones, shales, which are ripple-marked in
a way that betrays their shoal-water origin. There are also some
areas of deeper water accumulations, found in the thick limestones of western Vermont, the Appalachian Mountains, Nevada,
and British Columbia. Very little igneous rock is found in the
Cambrian of North America. Small intrusions occur in Newfoundland and New England, and quite considerable ones in
British Columbia, but some of these may be long post-Cambrian
in date.

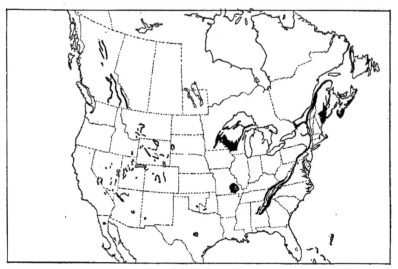

FIG. 259. — Map of known Cambrian outcrops in the United States and Canada

As is indicated by the geographical distribution of the fossils,
North America was, in the Cambrian period, divided into two
provinces of very unequal size and faunally very different. The
Atlantic province, comprising Newfoundland, New Brunswick,
Nova Scotia, and New England, shows so close a connection
with Europe as to justify the inference that in high latitudes a
land bridge spanned the Atlantic, or at least that a chain of
islands and shoals permitted the migration of shore-loving marine
animals from one continent to the other. All the rest of North

America belongs to the Pacific province, though there are many local faunas within that vast area.

During Cambrian times the sea was slowly advancing over the land in North America, and the geography of the continent was very different at the close of the period from what it had been at the beginning. In the Lower Cambrian the land areas are inferred to have been somewhat as follows: First, there was the great northern mass of crystalline Archæan and Algonkian rocks, but this was probably much more extensive than the present exposures of pre-Cambrian rocks would indicate. It probably covered the whole Mississippi valley down to 30° N. lat. and extended westward beyond the Rocky Mountains. Long, narrow strips of land, alternating with narrow sounds, occupied part of New England and the maritime provinces of Canada, while an Appalachian land, whose western line is marked by the present Blue Ridge, extended eastward an unknown distance into the Atlantic. On the *western* shore of the Appalachian land was a narrow arm of the sea, which opened south and nearly separated this land area from the great mass of the continent. During the Lower and Middle Cambrian, this long and narrow bay or sound must have been closed, or only occasionally and partially opened, at the northern end. In later Cambrian times it was perhaps open. The site of the Sierra Nevada was occupied by a long, narrow land, running from Puget Sound to Mexico, and another such area was found in eastern British Columbia. The Great Basin region was under water. Around these shores were laid down the coarser deposits of the Lower Cambrian, with great masses of shales and thick limestones in deeper water.

Middle Cambrian sediments have quite a similar distribution to those of the Lower, but the sea was slowly advancing over the continent from the south. Nothing is known of the condition of Central America and Mexico at this time, but from Arizona eastward to Alabama the land was submerged. This transgression of the sea continued, and reached its maximum in the Upper Cambrian. Toward the close of the period a large part of the continent had been submerged and, in particular, a vast interior sea had been established over the Mississippi Valley.

The Cambrian of Other Continents. — In *Europe* the Cambrian is very extensively developed, with remarkable differences of thickness in different regions. The maximum thickness occurs on the western side of the continent in Wales and Spain. In Wales are 20,000 feet of conglomerates, sandstones, shales, slates, and volcanic rocks, while in southern Sweden and northwestern Russia the entire period is represented by only 400 feet of beds. Much of the Welsh Cambrian is regarded by Professor Penck as being of continental origin, with occasional incursions of the sea. The Lower Cambrian appears to be limited to the north of Europe, while the Middle Cambrian witnessed the widest transgression of the sea, beds of this date occurring in France, Germany, Bohemia, Spain, and Sardinia. The Middle Cambrian is characterized by the Trilobite genus *Paradoxides* (Plate III, Fig. 6) which is very common in Europe and in the Atlantic province of North America, but is not found in other parts of the latter continent. The Upper division, like the Lower, is restricted to northern Europe, so that there was extensive submergence in the Middle Cambrian, but a withdrawal of the sea before the beginning of the later portion of the period. This is in decided contrast to the geographical changes of North America, where the most widespread extension of the sea took place in the Upper Cambrian. In Russia the Cambrian sediments are remarkable for their unconsolidated condition; some of them look as though just abandoned by the sea.

In Norway, 70° N. lat., have been found glacial deposits which are either basal Cambrian or late Algonkian. "The conglomerate in some places is seen to be formed of old moraines. . . . The stones have not the habitus of water-worn rolling stones, but of ice-worn stones. On some of the dolomite ones clear glacial striæ were observed. . . . At one place plain glacial striæ have been found upon the surface of the hard sandstone under a mass of conglomerate." (Reusch.)

Cambrian rocks cover great areas in eastern *Asia*, northern Siberia, Korea, and China. In China these strata, which are but little disturbed, attain the great thickness of 20,000 feet, and consist predominantly of sandstones and limestones. In the lower part are 170 feet of boulder clay of evidently glacial origin. "On the Yangtse River, in 31° lat., *i.e.* as far south as New Orleans, not high above sea-level, a large body of glacial material was discovered. . . . It demonstrates the existence of glacial conditions in a very low latitude in the early Palæozoic." (Willis.)

Cambrian also occurs in northern India, but none has yet been identified in Africa. It is found in the south of *Australia* and in Tasmania, apparently belonging to all three of the divisions. Evidences of glacial action have been observed in the Australian Cambrian, showing that this climatic change was not local but very widespread, especially as the earliest of the South African ice ages, mentioned under the Algonkian, may have been early Cambrian in date.

In *South America*, Cambrian has as yet been found only in the northern part of Argentina; it is apparently referable to the Middle division.

CAMBRIAN LIFE

The Cambrian fauna is of extraordinary interest, because it is the most ancient that we know with any fulness, though, of course, it does not represent the beginnings of life. Almost all the great types of invertebrates are already present and very definitely characterized, indicating that life had been differentiating for a vast period before the lowest Cambrian rocks had been laid down. As compared with the faunas of other Palæozoic periods, that of the Cambrian is very scanty, but our knowledge of it has been greatly increased of late and may be expected to increase in the future.

Though the successive Cambrian faunas have a very uniform distribution over wide areas, there are already indications of

local differences which mark out faunal provinces; thus, the
Lower and Middle Cambrian fossils of Newfoundland are more
similar to those of Europe than to those of the Appalachian and

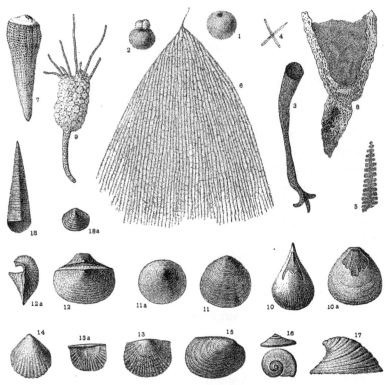

PLATE I.—CAMBRIAN FOSSILS

Fig. 1, *Orbulina universa* Lam., × 12, L. C.   2, *Globigerina cambrica* Matthew, × 5,
L. C.   3, *Leptomitus zitteli* Walcott, × ½, L. C.   4, Spicule of *Protospongia*, × ½, L. C.
5, *Climacograptus emmonsi* Walc., × ½, L. C.   6, *Dictyonema flabelliforme* Eichwald,
× ½, U. C.   7, *Archæocyathellus rensselaricus* Ford, × ½, L. C.   8, *Archæocyathus
profundus* Billings, × ½, L. C.   9, *Eocystités longidactylus* Walc., M. C.   10, *Lingulepis
pinniformis* Owen, × 1, U. C.   11, *Linarssonia taconica* Walc., × 4, L. C.   12, *Kutorgina
cingulata* Billings, × ½, L. C.   13, *Protorthis billingsi* Hartt, × 1, M. C.   14, *Camer-
ella antiquata* Bill., × 1, L. C.   15, *Fordilla troyensis* Barrande, × ⁶/₂, L. C.   16, *Rhaphi-
stoma attleborensis* S. and F. × ⁶/₂.   17, *Stenotheca rugosa* Hall, × 2, L. C.   18, *Hyolithes princeps* Bill., × ½, L. C.

interior regions of America.   A fauna of similar date but different
facies occurs in Alabama, and farther north in the Appalachians.

Of **Plants** nothing is surely known; certain marks on the bedding-planes of strata have been regarded as seaweeds, but they are too obscure for determination, and many are worm tracks.

The fauna is principally made up of Brachiopods and Trilobites, but many other types are represented also.

Foraminifera very like those of the modern seas (Plate I, Figs. 1, 2) are found even in the Lower Cambrian.

Spongida. — Siliceous Sponges are not uncommon.

Cœlenterata. — The Hydrozoa are represented by the *Graptolites*, a series of forms which are confined to the older Palæozoic rocks. *Dictyonema* (I, 6) is a complex Graptolite, found abundantly in a thin band of shale near the top of the Upper Cambrian, which is of nearly world-wide distribution. It shows the great value of the organisms which live at the surface of the open sea (pelagic fauna) in fixing contemporaneous deposits over enormous areas.

Other Hydrozoa are the jellyfish, of which recognizable casts have been found in large numbers. *Stromatopora* formed reefs in some of the Cambrian limestones.

It is still a question whether *Corals* were present in the Cambrian; certain fossils (*Archæocyathus*, I, 7, 8,) which by some authorities are called corals, are by others regarded as sponges. Though sufficiently abundant in some parts of the West to form reefs, the genus has only a few species, and, except locally, they are not conspicuous elements in the fauna.

Echinodermata. — The Echinoderms are rare, and belong to the Cystoids, a very primitive grade of the type.

Worms. — The presence of marine worms is abundantly indicated by tubes, tracks, and borings in the sands which have now consolidated into hard rocks. *Hyolithes*, a worm-tube (I, 18), is very common.

Arthropoda. — The only known Cambrian Arthropods are the *Crustacea*, and of these much the most abundant group is that of the *Trilobita*, which are altogether confined to the Palæozoic rocks, and are by far the most important of Cambrian fossils. It

is only within recent years that the systematic position of the
Trilobites has been established through the fortunate discovery
of specimens with their appendages attached (see Pl. VII, Figs. 1,
1 a). Trilobites have a more or less distinctly three-lobed body, at
one end of which is the head-shield, usually with a pair of fixed
compound eyes; at the other end is the tail-shield, and between the
two shields is a ringed or jointed body made up of a variable
number of movable segments. The Trilobites display an ex-

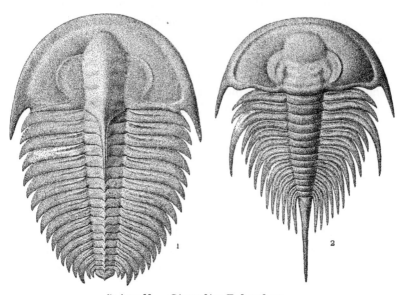

PLATE II. — CAMBRIAN TRILOBITES

Fig. 1. *Holmia bröggeri* Walc., x ⅓, L. C. 2, *Olenellus thompsoni* Hall, x ½ L. C. (Walcott)

traordinary variety in form and size, in the proportions of the
head- and tail-shields, in the number of free segments, and in the
development of spines. Already in the Cambrian this wealth of
forms is notable, though far less so than it became in the Ordovician. As compared with those of later times, the Cambrian Trilobites are marked by the (usually) very small size of
the tail-shield, the large number of free segments, and their in-

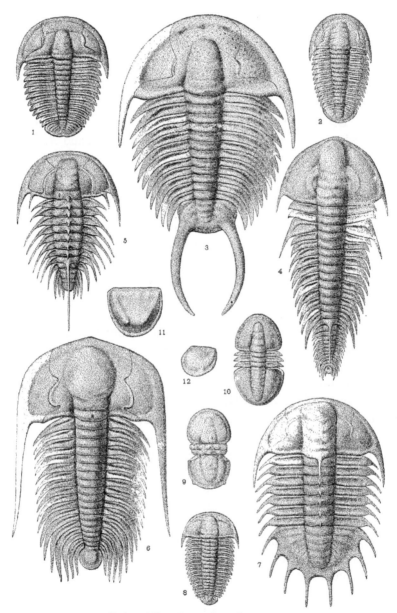

PLATE III. — CAMBRIAN CRUSTACEA

Fig. 1, *Ptychoparia kingi* Meek, × ½ M. C.   2, *P. antiquitata* Salter, × ½, M. C.   3, *Crepicephalus texanus* Shumard, × ½, M. C.   4, *Mesonacis vermontana* Walc., × ½, L. C.   5, *Zacanthoides typicalis* Walc., × 1, M. C.   6, *Paradoxides harlani* Green, × ¼, M. C.   7, *Dorypyge curticei* Walc., × ½, M. C.   8, *Atops trilineatus* Emmons, × ½, L. C.   9, *Agnostus interstrictus* White, ×³/₂, M. C.   10, *Microdiscus speciosus* Ford, × 1, L. C.   11, *Hipponicharion eos* Matthew, × 4, L. C.   12, *Aristozoe rotundata* Walc., × ²/₃, L. C.

ability to roll themselves up. The large Trilobites with long eye-lobes are very distinctively Cambrian. Some of them, like *Paradoxides* (III, 6), are very large (from 10 inches to 2 feet in length). *Olenellus* (II, 2) and *Holmia* (II, 1) also have large species, while *Agnostus* (III, 9), *Microdiscus* (III, 10), and *Atops* (III, 8) are small and without eyes.

The great importance of the Trilobites for Cambrian stratigraphy is indicated by the fact that the three divisions of the system are named for the three dominant genera of these crustaceans, *Olenellus*, *Paradoxides*, and *Dikellocephalus*.

Two other divisions of the Crustacea are found in the Cambrian: the *Ostracoda*, little bivalve forms, whose shells look deceptively like those of molluscs; and the *Phyllocarida*, which have a large shield on the head and thorax, and a many-jointed abdomen, with terminal spine.

Brachiopoda. — These are among the most abundant of Cambrian fossils; most of them belong to the lower order of the class (*Inarticulata*), in which the shells are mostly horny and the two valves are not articulated together by a hinge. The horny-shelled types, *Linnarssonia* (I, 11), *Lingulepis* (I, 10), and *Lingulella* are of great interest, as they differ but little from certain brachiopods which still exist. The second order of Brachiopods, the *Articulata*, which have calcareous shells connected by an elaborate hinge, were more common in the Upper Cambrian. In subsequent periods they became vastly more numerous than the Inarticulata, and throughout the post-Cambrian divisions of the Palæozoic their shells are found in incalculable numbers.

The **Mollusca** are already represented by their principal divisions. The *Pelecypoda*, or Bivalves (I, 15), are of very small size and found very scantily; their variety and relative importance have gone on increasing ever since Cambrian times. *Gastropoda* (I, 16, 17) occur in small numbers, especially in the Upper Cambrian. Fossils formerly referred to the *Pteropoda*, but now, generally regarded as worm-tubes, are among the most frequent of shells found in these rocks, but display no great variety. The

Cephalopoda, which are the highest group of molluscs, are perhaps represented in the Cambrian by shells which are rare and minute in size, and almost confined to the uppermost part of the system; that is, assuming that the tiny *Volborthella* found in Russia and New Brunswick is really a cephalopod, but this is not certain.

The Cambrian fauna displays steady progress, being distinctly more advanced in the upper than in the lower division.

CHAPTER XXVII

THE ORDOVICIAN (OR LOWER SILURIAN) PERIOD

ORDOVICIAN SYSTEM

WALES	NEW YORK	
Bala	Cincinnatian Series	Richmond Stage (Medina) Lorraine Stage Utica Stage
Llandeilo Llanvirn	Mohawkian Series	Trenton Stage Black River Stage Lowville Stage
Arenig Tremadoc	Canadian Series	Chazy Stage Beekmantown Stage ,

SIR RODERICK MURCHISON divided his great Silurian system primarily into two parts, Upper and Lower. This. method of classification is generally followed even at the present day, although it is widely recognized that the most decided break in the entire Palæozoic group is the one between these divisions. In 1879 Professor Lapworth proposed to give due emphasis to this distinction by erecting the Lower Silurian into a separate system, the *Ordovician*. The name is taken from the *Ordovici*, an ancient British tribe which dwelt in Wales during Roman times. Lapworth's example is now largely followed in England and the United States, but on the continent of Europe the name *Silurian* is still retained for both systems.

The classification and subdivision of the American Ordovician were first worked out in the State of New York, and consequently the New York scale serves as a standard of reference for the rest of the continent.

In the preceding table the classification lately issued by the New York Survey is given in comparison with that of the Welsh Ordovician. It is not to be supposed, however, that the subdivisions in the two continents are exactly equivalent, but merely that they correspond to one another in a general way.

DISTRIBUTION OF ORDOVICIAN ROCKS

American. — The passage from Cambrian to Ordovician was gradual, without any marked physical break. Only where the Upper Cambrian is sandy, as in New York, is there a decided change in the character of sedimentation. In the latter part of the Cambrian a great inland sea had been established over what is now the Mississippi valley and, with frequent fluctuations in depth and modifications in form, it was to persist for long periods as one of the salient features of Palæozoic geography. This sea was separated from the Atlantic by the land mass called Appalachia, and on the western side it was demarcated from the Pacific by islands of undetermined size. A generalized representation of the arrangement of land and water in Ordovician North America is presented in the map, Fig. 260. Such a map, which can only be a rude approximation to the truth, is constructed by marking as water all those areas where Ordovician rocks are known, or confidently inferred to be present, even though concealed by overlying, newer strata, and as land those areas where the Ordovician is wanting. Frequently, however, it is impossible to determine whether the absence of the strata is due to their never having been present at the given point, or to their removal by denudation. On the other hand, the deep-lying and buried extensions of the strata may be subject to many interruptions, of which there is no surface indication, for ancient islands and peninsulas may be covered over and concealed by newer strata. Hence, the shore-lines appear unduly simple, for the

2 O

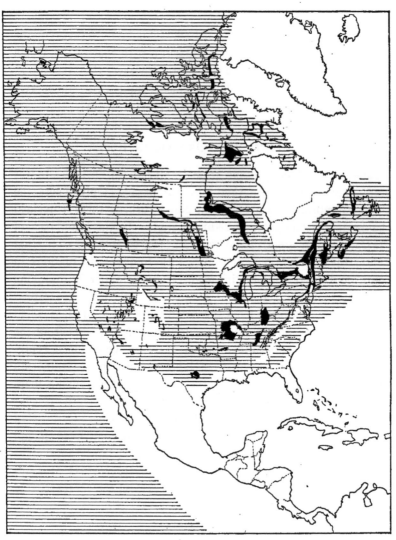

FIG. 260. — Generalized map of North America in the Ordovician period. Known
exposures of Ordovician rocks in black. Horizontally lined area = sea; white
= land. The condition of Mexico and Central America is not known

details have been destroyed by erosion, or hidden by deposition. Most attempts to reconstruct on a map the long-vanished geography of some ancient period probably err in the direction of not making sufficient allowance for the removal of the strata by denudation. An example will make this clear. At Elmhurst, fifteen miles west of Chicago, from which the nearest known exposure of Devonian rocks is distant eighty miles, and where the surface is made by a Silurian limestone (Niagara), was found a fissure containing Devonian fossils, brachiopods, and fish teeth. This proves that the Devonian once covered that whole region, but has been entirely swept away, leaving hardly a trace behind. "The presence of this Upper Devonian fauna at Elmhurst, buried as it is deep down in the Niagara limestone, indicates with certainty that during the greater part of Devonian time, the region now known as northern Illinois was above sea-level. . . . At a later period, near the close of the Devonian, the sea again occupied the region, sand was sifted down into these open joints, and with it the teeth of fishes." (Weller.)

The Cambrian subsidence continued into the earliest Ordovician, "when more of the continent was under water and the sea probably deeper than at any subsequent period." (Ulrich and Schuchert.) At the end of the Beekmantown stage, extensive geographical changes occurred; two parallel folds, extending from Alabama to Newfoundland and following the trend of the future Appalachian Mountains and the Cambrian trough, were raised as low barriers, which played a very important part in separating the Atlantic from the Interior Sea. Inclosed between these parallel folds was a narrow basin, which was frequently invaded by the sea, though rarely completely submerged from end to end. At the same time there was a widespread elevation, which greatly reduced the area and depth of the Interior Sea, but comparatively soon there followed a less extensive submergence (*Chazy*) both at the south and in the north. In the latter region a long, narrow gulf extended up the valley of the St. Lawrence, dividing at the Adirondack Mountains and sending one arm where now is the valley of the Ottawa, and the other over the present site of Lake Champlain. Another narrow body of water, called the Levis channel, and believed to be separate from the Chazy bay, though very near to it, extended from Newfoundland into the northeastern corner of New York. From the south, the invasion extended into Kentucky and Tennessee and persisted for a longer time.

. In the upper Mississippi valley, the lower Ordovician limestones are followed by a very extensively developed sandstone (the *St. Peter*), which occurs, at the surface or underground, over nearly the whole of Illinois, Iowa, and Missouri, much of Wisconsin, Minnesota, and Michigan and Indiana, and smaller parts of other states from Kansas to Oklahoma.

The Middle Ordovician (Mohawkian series) was a time of lime-stone-making on an extraordinarily wide scale, which implies that the Interior Sea received less terrigenous sediment than before, and this, in turn, probably indicates that the surrounding lands were low and flat, and that the sluggish streams carried but small loads of fine silt. During the Chazy age the Interior Sea had been reduced in size, and nearly all of New York was above water, but a renewed submergence again extended the sea over most of New York and reopened the northeastern connection with the Atlantic. The Mohawkian limestones, especially the *Trenton*, occur in New Brunswick, New York, Canada, over the upper Mississippi valley, in the Black Hills, Bighorn, Rocky, Wasatch, and Uinta Mountains, and the Great Basin. In Kentucky and Tennessee unconformities point to oscillations of level, emergence and submergence alternating.

The Upper Ordovician consists largely of a thick mass of shales (*Utica and Lorraine*) formed from the terrigenous silts spread widely over the sea-bottom, due perhaps to an elevation of the land, which, rejuvenating the streams, increased the loads carried to the sea, and perhaps also to a concomitant shoaling of the sea. These shales and slates are thickest toward the east, and extend along the Appalachian line from the St. Lawrence to Tennessee and westward into Indiana. The northeastern passage to the Atlantic was more freely opened, and a fauna with strong European affinities invaded the Interior Sea, though not for long. The immensely thick mass of shales and slates, which occurs in the valley of the Hudson near Albany, and follows the trend of the Appalachian Mountains to Tennessee, was once regarded as a distinct series (*Hudson River*) and placed at the top of the Ordovician. It is now known, however, to be a separate facies, representing at least the Trenton, Utica, and Lorraine stages, and possibly the whole of the Ordovician.

The uppermost of the Ordovician stages (*Richmond*) is most typically developed in Ohio and Indiana, with a littoral, and perhaps partly continental, facies in central New York, and along the Appalachians, the *Oneida* conglomerate and *Medina* sandstone, which, until very lately, have been regarded as the base of the Silurian. The "eastern Oneida," or Shawangunk grit, has recently been proved to belong to the upper part of the Silurian, and is therefore of much later date than the Oneida-Medina of central New York. "Ulrich has reëxamined the Medina deposits of the Appalachian region, more especially in Pennsylvania, Virginia, and Tennessee, and has concluded that they are the eastern shore deposits equivalent to the Richmond series of the Ohio and Mississippi valleys. This result therefore forces stratigraphers to place the line separating the Siluric from the Ordovicic, not at the base of the Medina formation of the New York standard section, but at its uppermost limit and beneath the Clinton." (Schuchert.)

In the western portion of the continent the Ordovician is not so well known as in the eastern, because it is so generally buried under newer strata, and, over great areas where it probably once existed, it has been removed by denudation. In the Lower Ordovician great areas of the northwest were land, but these were very extensively submerged in the Middle Ordovician, when the sea probably covered most of the Great Plains and much of the plateau region. The Upper Ordovician is much more restricted, and in many places lies unconformably upon the Middle, pointing to emergences and prolonged denudation. The Great Basin, or that portion of it known as the Nevada trough, seems to have been submerged throughout the Ordovician, as it had been in the Cambrian, and, indeed, during nearly the whole Palæozoic era. Ordovician rocks fringe the western side of the great northern pre-Cambrian area and occur in the islands of the Arctic Sea.

Aside from the slow and gentle oscillations of level above mentioned, the Ordovician was a period of tranquillity, generally speaking, without violent diastrophic movements, nor have any signs of volcanoes of that date been discovered in North America. Igneous intrusions are rare, though they have been found in New York and in the Wichita Mountains of Oklahoma, where the deposition of Trenton beds was followed by the upheaval of the mountains and the intrusion of a great mass of granite, which has metamorphosed the overlying sedimentaries.   .

Foreign. — In *Europe* the Ordovician rocks appear to have been laid down in two distinct seas separated by a ridge of land. The northern area extends from Ireland far into Russia, while the southern is represented by numerous scattered masses. These rocks cover a much wider surface than do the Cambrian. In Great Britain, especially in Wales, they form very thick masses of shales and slates, with but little limestone, intercalated with much volcanic lava and tuff, the volcanic activity being in very marked contrast to the quiet of North America. In Scandinavia these rocks are nearly horizontal limestones and shales, and in Russia they cover very large areas and are so perfectly undisturbed that many are still incoherent sediments. In the southern sea were laid down the Ordovician strata of Bohemia, Germany, northwestern and central France, Spain, Portugal, Sardinia, and Morocco.

The very marked difference between the fossils of the two great European areas, and the fact that the Ordovician fossils of other continents agree with those of northern Europe, while those of the southern district are peculiar, indicate that the latter region was a partially closed sea, which occupied the Mediterranean basin, though extending far beyond its present limits.

Asia was largely above water in Ordovician times, but a broad Indo-Chinese sea covered much of the eastern coast, and in northern Siberia are great areas of Ordovician strata, the upper members of which are red sandstones with gypsum and salt. This points to an arid climate. Marine rocks of Ordovician date are found in north *Africa*, but the equatorial and southern regions are highly peculiar among the continents in the very subordinate part taken by marine rocks of any period, the land being built up almost entirely of continental rocks. In South Africa a thick series of barren sandstones underlies marine Devonian, and probably some of these are referable to the Ordovician. Ordovician rocks are found in *New Zealand*, *Tasmania*, and the southern part of *Australia*. In *South America* they are not extensively developed, but have been found in Argentina and Peru.

The **Climate** of the Ordovician, so far as at present known, was uniformly mild and equable, as appears from the fossils of the Arctic lands. No glacial deposits have yet been discovered, though arid conditions obtained in northern Asia.

Close of Ordovician. — At the end of the period came a time of widespread disturbance, upheaval, and mountain-making, the traces of which are still plain in North America and Europe, especially along the Atlantic slope of each continent. In Nova Scotia and New Brunswick the Silurian strata lie unconformably upon the upturned Ordovician. Along the line between New York and New England the Taconic range was upheaved, its rocks greatly compressed, plicated, faulted, and metamorphosed. Many of the crystalline schists of this region, it has been proved, were derived from the metamorphosis of Cambrian and Ordovician sedimentary rocks. Evidences of this disturbance have been traced as far south as Virginia. The effects of the upheaval were not felt in the northern part of the Gulf of St. Lawrence, for on Anticosti Island the great limestone, which was begun in Ordovician times, continued without a break into the Silurian. The disturbance was along a line of especially thick accumulations, as appears from the comparative measurements of the same strata in different areas.   The Interior Sea appears to have been entirely drained;   at all events no deposits transitional to the Silurian are known from that region.   In the West and Northwest large areas remained land for long periods, but the Interior Sea was soon reëstablished in the Mississippi valley.   Some narrow strips of land were added to the margin of the Cambrian coasts, and on a line running through southern Ohio, Kentucky, and Tennessee a low, broad arch, the formation of which appears to have begun early in the Ordovician, was forced up by lateral compression. This is called the " Cincinnati anticline or axis."

In Europe the disturbances which brought the Ordovician to a close produced their maximum effects in England, Wales, and the Highlands of Scotland, where the thickness of the sediments is greatest.   In these regions the Ordovician beds are folded and often greatly metamorphosed, the Silurian strata lying upon their upturned edges.

THE LIFE OF THE ORDOVICIAN

Ordovician life displays a notable advance over that of the Cambrian, becoming not only very much more varied and luxuriant, but also of a distinctly higher grade.   During the long ages of the period also very decided progress was made, and when the Ordovician came to its close, all of the great types of marine invertebrates and most of their more important subdivisions had come into existence.   In a general way the life of the Ordovician is an expansion of that of the Cambrian, though but little direct connection between the two can yet be traced, and evidently there were great migrations of marine animals from some region which cannot yet be identified.   Several groups of invertebrates attained their culmination and began to decline in the Ordovician, becoming much less important in subsequent periods.   Thus the Graptolites, the Cystoidean order of Echinoderms, the straight-shelled

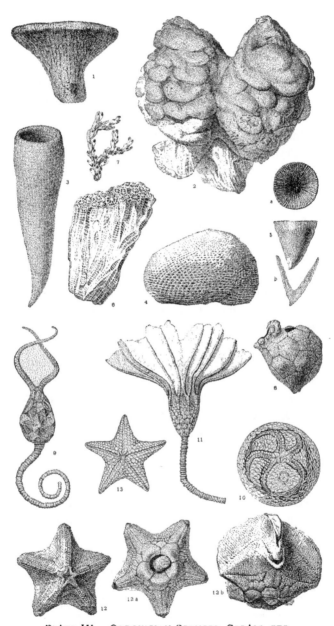

PLATE IV.—ORDOVICIAN SPONGES, CORALS, ETC.

Fig. 1, *Zittelella typicalis* Ulrich and Everett, x ½, Trenton. 2, *Strobilospongia tuberosa* Beecher, x ½, Trenton. 3, *Cyathophycus reticulatus* Walc., x ½, Utica. 4, *Receptaculites fungosus* Hall, x ½, Trenton. 5, *Petraia profunda* Conrad, x ½, Trenton. 5*a*, *The same*, top view. 5*b*, *The same*, vertical section. 6, *Columnaria stellata* Hall, x ½, Trenton. 7, *Romingeria trentonensis* Weller, x ½, Trenton. 8, *Malocystites emmonsi* Hudson, x 2, Chazy. 9, *Pleurocystites filitextus* Bill., x 1, Trenton. 10, *Lepidodiscus cincinnatiensis* Hall, x ½, Richmond. 11, *Glyptocrinus dyeri* Meek, x ½, Richmond. 12, *Blastoidocrinus carchariædens*, Bill, x ¾, Chazy. 12*a*, *The same*, basal view. 12*b*, *The same*, side view. 13, *Palæasterina stellata* Bill., x ½, Trenton.

Cephalopods (orthoceratites) among Molluscs, and the Trilobites, were never so abundant and so varied as during this period.

Plants. — In America no plants above the grade of seaweeds and coralline Algæ have been discovered, but in Europe a few of the higher Cryptogams are doubtfully reported. The flora of the Devonian, however, renders it highly probable that land plants were already well advanced in the Ordovician, and their remains may be discovered at any time. This must remain a matter of accident, for the known Ordovician rocks are almost all marine, which is not a favourable circumstance for the preservation of land plants. Such discoveries have, indeed, already been reported, but the evidence for them is not satisfactory.

Foraminifera and Radiolaria have been found in a few regions in great numbers, sufficient to prove that they were abundant in the Ordovician seas.

Spongida. — Sponges are much more numerous and varied than in the Cambrian. Of course it is only those sponges with skeletons of lime or flint which can be well preserved in the fossil state, and of these the Ordovician has many (Pl. IV; Figs. 1–4). The horny sponges, of which the common bath sponge is a familiar example, are necessarily much rarer and less satisfactory as fossils.

Cœlenterata. — The *Graptolites* are very numerous and varied, wherever conditions are favourable to their preservation, as in fine-grained rocks with smooth bedding-planes. The Ordovician is the time of their culmination and is especially characterized by the double forms, with rows of cells on both sides of the stem

Beekmantown. 6, *Constellaria polystomella*, Whitfield, × ½, Richmond. 7, *Stictoporella cribrosa* Ulrich, × ½. Trenton. 7a, *The same*, a fragment, × 9. 8. *Stomatopora inflata* Hall, on a brachiopod shell, × ½, Richmond. 8a, *The same*, a portion × 4. 9, *Trematis ottawaensis* Bill., + ½, Trenton. 10, *Schizambon canadensis* Ami., × 1, Utica. 11, *Orthis tricenaria* Conrad, × ½, Trenton. 11a, *The same*, lateral view. 12, *Platystrophia lynx* Eichw., × ½, Trenton. 13. *Hebertella sinuata* Hall, × ½, Richmond. 13a, *The same*, inner side of ventral valve. 13b, *The same*, inner side of dorsal valve. 14. *Dalmanella testudinaria* Dalman, × ½, Trenton. 14a, *The same*, inner side of ventral valve. 14b, *The same*, inner side of dorsal valve. 15, *Plectambonites sericeus* Sowerby, × 1, Richmond. 16, *Rafinesquina alternata* Conrad × ½, Richmond. 16a, *The same*, longitudinal section. 17, *Strophomena planumbonum* Hall, × ½, Richmond, 17a, *The same*, longitudinal section. 18, *Zygospira modesta* Say, × 1, Richmond. 18a, *The same*, from the side. 10, *Rhynchotrema capax* Conrad × ½, Richmond. 19a, *The same*, anterior view. 20, *Triplecia extans* Emmons, × ½, Trenton. 20a, *The same*, anterior view.

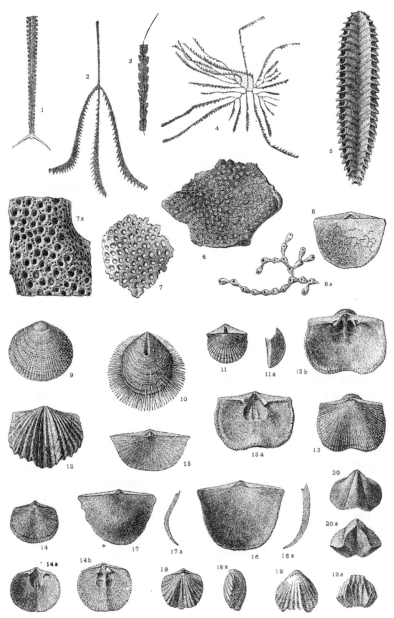

PLATE V. — ORDOVICIAN GRAPTOLITES, BRACHIOPODS, ETC.

Fig. 1, *Climacograptus bicornis* Hall, × 1, Utica. 2, *Tetragraptus fruticosus* Hall, × ½, Beekmantown. 3, *Climacograptus pungens* Ruedemann, × ³/₂, Chazy. 4, *Goniograptus postremus* Rued., × ½, Beekmantown. 5, *Phyllograptus augustifolius* Hall, × 2, (see Pl. V, Figs. 1–5). So abundant are the Graptolites that in many parts of the system they are almost the only fossils, and are employed to divide the substages into zones. Graptolite zones, with the same or closely similar species, and in the same order of succession, are found in Great Britain, the St. Lawrence and Champlain valleys, and in Australia. Hydroid corals, *Stromatocerium*, are abundant, and form reefs in the Chazy and Black River limestones. The few and doubtful Cambrian *Corals* are succeeded by a considerable number of Ordovician genera and species. Like other Palæozoic Corals, these are characteristically different from the reef-builders of the present day in showing a marked bilateral symmetry and having the septa arranged in multiples of four (*Tetracoralla*). Solitary cup-corals, like *Streptelasma* and *Petraia* (IV, 5), and compound colonies, like *Columnaria* (IV, 6), are examples of the range of differences among these early corals.

The **Echinodermata** have greatly increased in importance, and all the main subdivisions of the group are represented, all of which, except the Cystoids, first appear in the Ordovician. The *Cystoidea*, which we have already found in the Cambrian, attain their greatest development in the Ordovician. In these curious animals the body is either irregularly shaped, or symmetrical, with a short, tapering stem, by which the animal was attached to the sea-floor, and weakly developed arms. The body, or calyx, is made up of a number of calcareous plates; when these plates are very numerous, they are of irregular size and arrangement (IV, 8, 10), while the forms with few plates have them of a definite number, size, and shape (IV, 9). Some of the more regular Cystoidea have much resemblance to the true *Crinoids*. The latter are not rare, though less abundant than they afterward became. These animals (IV, 11) have a symmetrical calyx, with long, branching arms; the number and arrangement of the component plates are definite and characteristic for each genus. Most, but not all, of the Crinoids have a long, jointed stem, by which they are attached to the sea-bottom. The earliest Blastoidea, represented by *Blastoidocrinus* (IV, 12), appear in the Chazy. This genus, a very primitive form which retains notable cystoidean characters, is the first member of a group which was to become important in a long subsequent period, the Carboniferous. *Asteroids* (starfishes) and *Ophiuroids* (brittle-stars) are found, but cannot be called common, or abundant. The *Echinoidea*, or sea-urchins, are represented by very primitive forms.

Arthropoda. — The Trilobites (Pl. VII) increase very greatly in the number of genera and species, and most of the Cambrian genera are replaced by new ones. This is the period in which the group of Trilobites attains its highest development, gradually declining afterward and becoming extinct with the close of the Palæozoic. The most characteristic and widely spread genera of Ordovician Trilobites are: *Asaphus, Isotelus* (VII, 7), *Bumastus* (VII, 2), *Triarthrus* (VII, 1), *Calymmene* (VII, 8), *Ceraurus* (VII, 6), *Trinucleus* (VII, 4), *Acidaspis* (VII, 3), *Bronteus* (VII, 5), *Pterygometopus*, etc. These genera differ in aspect from those of the Cambrian in their much larger tail-shields, in their ability to roll themselves up (see VII, 2 a, 8 a), and in their rounder and better-developed, faceted eyes.

Other Crustacea mark advances in the Ordovician. Thus, the *Eurypterida*, a group which dates from the Algonkian, and was destined to a remarkable development in the Silurian and Devonian, is represented, though not abundantly. *Ostracoda* and *Phyllocarida* undergo no marked change. That terrestrial animal life had already begun is demonstrated by the occurrence of an *Insect, Protocimex*, in Scandinavia. From this we may be assured that terrestial vegetation was already established and that the atmosphere was fitted for the existence of air-breathers.

Brachiopoda. — These shells increase very largely in abundance and variety, the genera with hinged calcareous shells (Articulata) now gaining the upper hand and reducing the horny-shelled kinds to comparative insignificance. The most important genera are: *Orthis* (V, 11); *Platystrophia* (V, 12); *Dalmanella* (V, 14); *Plectambonites* (V, 15); *Rafinesquina* (V, 16); *Leptæna, Stro-*

PLATE VI. — ORDOVICIAN MOLLUSCA

Fig. 1, *Byssonychia radiata* Hall. x ½, left valve, Trenton. 2, *Ambonychia planistriata* Hall, x ½, left valve. 3, *Opisthoptera fissicosta* Meek, x ½, right valve, Richmond. 4, *Pterinea demissa* Conrad, x ½, left valve, Trenton. 5, *Cyrtodonta huronensis* Bill., x ½, right valve, Lowville. 6, *Cymatonota attenuata* Ulrich, x ½, right valve, Richmond. 7, *Cyclonema humerosum* Ulrich, x ½, Lorraine. 8, *Eotomaria supracingulata* Bill., x ½. 9, *Trochonema umbilicatum* Hall, x ½, Trenton 10. *Hormotoma gracilis* Hall, x ½, Trenton 11, *Cyrtolites ornatus* Conrad, x ½, Lorraine. 12, *Protowarthia cancellata* Hall, x ½, Black River. 13, *Maclurea logani* Salter, x ½, Trenton. 14, *Ophileta compacta* Salter, x ½, Beekmantown 15, *Conularia trentonensis* Hall, x ½, Trenton. 16, *Orthoceras multicameratum* Hall, x ⅛, Lowville. 17, *Cyrtoceras juvenalis* Bill., x ½, Trenton. 18, *Eurystomites occidentalis* Hall, x ¼. 19, *Schræderoceras eatoni* Whitfield, x ½.

phomena (V, 17), and *Rhynshotrema* (V, 19). Spine-bearing shells begin in *Zygospira* (V, 18).

Bryozoa. — This is a group which has yet yielded no representatives from the Cambrian, but appears abundantly in the Ordovician (V, 6–8). The genera differ little from those which live in the modern seas.

Mollusca. — One of the most striking differences between the Cambrian and the Ordovician is the great advance made by the Molluscs in the latter period. The Bivalves (*Pelecypoda*) are larger, more numerous, and more like modern forms (see Pl. VI, Figs. 1–6). The *Gastropoda* likewise increase notably in size and in numbers, especially the spirally coiled shells. Important genera are: *Eotomaria* (VI, 8); *Hormotoma* (VI, 10); *Protowarthia* (VI, 12); *Trochonema* (VI, 9); *Maclurea* (VI, 13). Neither Bivalves nor Gastropods had anything like the relative importance which they possess in modern times; the latter all had the mouth of the shell forming a complete ring (holostomate).

Much the most significant change in the Mollusca, however, is the great expansion of the *Cephalopoda*, a few of which perhaps appear in the uppermost Cambrian, but in the Ordovician have become one of the predominant elements in the marine life of the times. The Cephalopods, which are the highest group of molluscs, are in modern times represented by two suborders; in one, the squids and cuttlefishes (*Dibranchiata*), the shell is rudimentary and internal; while in the other (*Tetrabranchiata*) the shell is external. Such an external shell is divided by transverse *septa* into chambers, which are connected by means of a tube, the *siphuncle*, the animal living only in the terminal chamber at the mouth of the shell, the remainder of which is empty. The only existing representative of the Tetrabranchiata is the Pearly Nautilus, but throughout the Mesozoic and most of the Palæozoic eras there was an extraordinary variety of these chambered shells. In the Ordovician the Cephalopods were all *Nautiloids*, most nearly allied to the modern Pearly Nautilus, with chambered

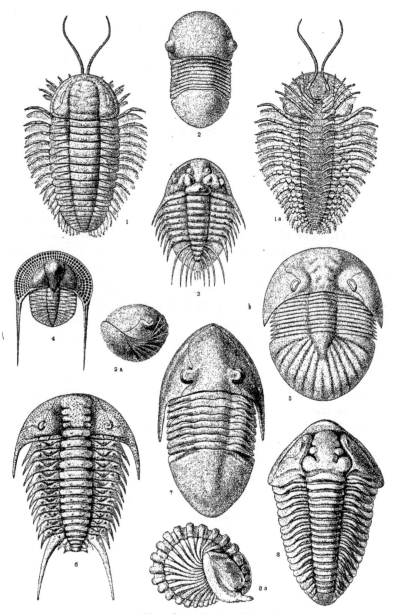

PLATE VII. — ORDOVICIAN TRILOBITES

Fig. 1, 1a, *Triarthrus becki* Green, × ³/₂, Utica. Restoration by Beecher of dorsal and ventral sides. 2, *Bumastus trentonensis* Emmons, × ½, Trenton. 2a, *The same*, from the side, rolled up. 3, *Acidaspis crosotus* Locke, × 4, Richmond. 4, *Trinucleus concentricus* Eaton, × 1, Trenton. 5, *Brontens lunatus* Bill., × 1, Trenton. 6, *Ceraurus pleurexanthmus* Green, × 1, Trenton. 7, *Isotelus maximus* Locke, × 1, Trenton. 8, *Calymmene callicephala* Green, × 1, Richmond. 8a, *The same*, rolled up, from the side.

shells, divided internally by simple septa. The commonest shell of this type is *Orthoceras*, which is a straight and very elongate cone (VI, 16) and sometimes attains a length of 10 feet; the genus persists throughout the Palæozoic and into the Mesozoic. *Endoceras*, which likewise has a straight shell, with a curiously complex siphuncle, is confined to the Ordovician. Besides these straight forms we find curved shells like *Cyrtoceras* (VI, 17), shells like *Eurystomites* (VI, 18), and *Schræderoceras* (VI, 19), which have the young shell coiled and the portion formed in old age straight, resembling an Orthoceras with its smaller end rolled up into a coil. Others again, like *Trocholites*, have the shell coiled in a close, flat spiral.

A peculiar shell, *Conularia* (VI, 15), which has a four-sided, pyramidal shape, with four triangular pieces to close the mouth, is a genus referred to the Pteropoda.

Vertebrata. — The curious, mail-clad *Ostracoderms*, primitive vertebrates which somewhat resemble the fishes in appearance, have been found in the Ordovician sandstones of Colorado and Wyoming. As' these remains are imperfect, description of the Ostracoderms will be deferred till a later chapter.

CHAPTER XXVIII

THE SILURIAN (UPPER SILURIAN) PERIOD

THE name *Silurian*, like Cambrian and Ordovician, refers to Wales. The term was proposed by Murchison in 1835 for a great system of strata older than the Devonian, and was taken from the *Silures*, another ancient tribe of Britons which inhabited part of Wales. Murchison gave great extension to his Silurian system, including in it most of Sedgwick's Cambrian, but, as already pointed out, the present tendency is to divide this vast succession of rocks into three systems of equivalent rank. It is unfortunate, and even unjust, that Murchison's term should not have been retained for the more important and widely developed lower division, now called the Ordovician, rather than for the upper division.

As in the Ordovician and Devonian, the New York classification, given in tabular form below in comparison with that of Wales, is the standard of reference for the American Silurian: —

SILURIAN SYSTEM

WALES	NEW YORK	
Ludlow, or Clunian	Cayugan Series	Manlius Stage
		Rondout Stage
		Cobleskill Stage
		Salina Stage
Wenlock, or Salopian	Niagaran Series	Guelph Stage
		Lockport Stage
May Hill, or Valentian		Rochester Stage
		Clinton Stage

N.B. Until very lately a basal series, the Oswegan, with the Oneida conglomerate and Medina sandstone, has been regarded as the lowest member of the Silurian, but this is now referred to the Ordovician. (See p. 565.)

Distribution of the Silurian Rocks

American. — The general disturbance which closed the Ordovician period appears to have greatly increased the extent of the continent. A relatively narrow strip of coast lands had been added to the northern pre-Cambrian area, converting much of Minnesota, Wisconsin, and the province of Ontario, northern New York and New Jersey, and western New England into land. Southern Ohio and central Kentucky and Tennessee had been raised into the Cincinnati anticline, but it is doubtful whether they remained as islands in the Silurian sea. Much of the Interior Sea had withdrawn, but the emergence was not long, geologically speaking, and the sea was soon reëstablished, but with entirely different boundaries and connections. What changes affected the land masses of the West and Southwest cannot yet be definitely determined, but the absence of the Silurian from extensive areas where the Ordovician is found indicates that these masses were greatly enlarged. How much of this enlargement came at a later date and how far the absence of the Silurian is the result of denudation, there is no present means of finding out.

The Silurian rocks are far thicker in the East, especially along the Appalachian range, than in the interior or western regions, where they thin out and are wanting over large areas.

An important feature in the Silurian geography of eastern North America was the establishment of the *Cumberland Basin,* or "Appalachian Mediterranean," as it has been called. This large sea lay to the eastward of the Interior Sea, from which it would seem to have been either completely separated, or so nearly so that the species of marine animals inhabiting the two bodies of water were very different. The Cumberland Basin was east of the Catskill-Helderberg line in New York, and its western shore crossed New Jersey and curved westward beyond the centre of Pennsylvania, whence it ran southwest more or less parallel with the Appalachian line, toward which it curved

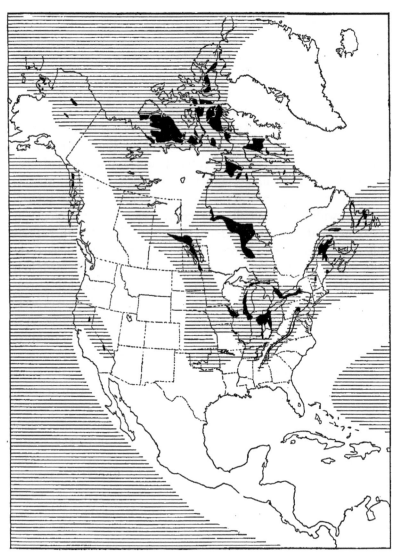

FIG. 261. — Generalized map of North America in the Silurian. Black areas = known exposures. Lined areas = sea. White areas = land, or unknown

eastward in southern West Virginia. This basin began apparently in western Maryland and adjoining areas very early in Silurian times, and continued to grow larger and deeper until the Devonian was well advanced. The Interior Sea underwent a succession of oscillations much like those which had affected it during the Ordovician; it was apparently closed at the south, but extended northwestward to the Arctic Sea, while its east-west diameter had been greatly reduced from that of the Ordovician.

The submergence which inaugurated the Silurian period (*i.e.* on the assumption that the Medina is properly referable to the Ordovician) brought the Interior Sea up to the narrow barrier which separated it from the Cumberland Basin, and in it were laid down the sediments of the *Clinton* stage, shales in the east passing westward into limestones, which extend through New York to Indiana, and perhaps also through Illinois to Missouri. In the Cumberland Basin the Clinton shales followed the trend of the Appalachians to Alabama. It must be remembered, however, that the Appalachian Mountains were not then in existence, as such, but they were foreshadowed by structural lines of depression and low folding which exerted a definite control of the coast-lines and basins through most of the Palæozoic era. Northeastward, the Clinton recurs in Nova Scotia and at other points in eastern Canada, but is not always readily identifiable. In many places interstratified concretionary hæmatites are found in the Clinton, especially along the Appalachian line, but also in Wisconsin, New York, and Nova Scotia.

A time of limestone-making on a great scale (the *Lockport* and *Guelph* stages), preceded in New York by the *Rochester* shales, followed the Clinton. In the East this great limestone has but a limited extension southward, but southwestward it stretches for nearly 1000 miles, to Wisconsin and thence across Illinois, Iowa, Missouri, and western Tennessee. Many scattered outliers in Manitoba and the region west of Hudson's Bay indicate the probable former extension of the limestone unbrokenly to the Arctic shores and islands. Rocks of corresponding date, laid down in the Cumberland Basin, but with marked faunal differences in the fossils from those of the Interior Sea, are found in western Maryland and Virginia, New England, New Brunswick, and Nova Scotia. East Tennessee, on the other hand, was elevated at the end of the Clinton stage and remained as a land-surface till the middle of the Devonian.

Little is known of the Silurian of the West, for, as already pointed out, there is reason to believe that nearly all of that region was then land. However, the Nevada trough continued to be submerged, presumably forming a gulf from the Pacific, and here the Niagara series is represented by the upper part of a thick mass of limestone which extends upward unbrokenly from the Trenton, the great limestone of the Ordovician.

The limestone of the Niagara epoch (Lockport) is very largely made up of corals, and in some places, as in eastern Wisconsin and areas to the south, distinct coral reefs may be observed, the most ancient which have as yet been found. As we have seen, corals flourished abundantly in the Ordovician, but, so far, no definite reefs have been noted in the rocks of that period.

The next change (*Salina* stage) was the separation, along the northern part of the northeastern arm of the Interior Sea, of a series of salt lagoons, in which were deposited red marls and shales, interstratified with gypsum and rock-salt, from which are obtained the brines of New York, Ontario, and Ohio. In part contemporaneous with these is a hydraulic limestone, called the Water-lime, which has much the same distribution as the salt-bearing beds, but is thickest where they are thin. The Water-lime indicates the freshening of the Salina lagoons and has preserved a remarkable assemblage of Crustacean fossils, belonging to the Eurypterida. The rocks of the Salina stage are thickest in New York and Pennsylvania, thinning to the westward. In the Shawangunk Mountains of eastern New York the Salina is represented by thick conglomerates of quartz pebbles, which were formerly referred to the Oneida, but are now known to contain, in interstratified shales, the remarkable Eurypterida of the Water-lime. This formation extends along the Appalachian line to Tennessee as a very thick mass of sandstones and conglomerates. The beds of salt and gypsum are strong evidence that the climate of Salina times, at least in the northeastern part of the continent, was arid, but how far this aridity was local, cannot be determined.

Throughout the Salina age the Interior Sea had been growing shallower, and shortly after the close of that stage the whole interior of the continent became land and remained so for a considerable period, but the Cumberland Basin persisted, and in it the limestones of the closing part of the Cayugan epoch and Silurian period were accumulated, the *Cobleskill, Rondout,* and *Manlius,* which are variously distributed in different parts of the basin, due probably to differential movements of the sea-floor, but all occur in succession in New York and in western Maryland, and for some distance north and south of the latter.

The Silurian rocks of North America have yielded no sign of volcanic material; in a few places they are traversed by igneous intrusions, most of which may, however, be of much later date. In Maine are some igneous rocks which decidedly appear to be Silurian, and the same may be true of certain areas in New Brunswick and Nova Scotia.

Foreign. — The division of Europe into northern and southern areas which we found in the Ordovician was maintained in Silurian times, and the southern sea was as peculiar in its animal life as it had been before, the northern being the typical Silurian which is found in the other continents. In the west of Ireland, Wales, northern England, and Scotland, Silurian beds accompany and overlie the Ordovician, but the much greater development of limestone points to a deepening of the water in those seas or a lowering of the surrounding lands. The volcanic materials, so conspicuous in the Ordovician, are no longer found. The Wenlock limestone of Great Britain, which corresponds to the American Niagara (Lockport-Guelph) is, like the latter, largely coralline. In Scandinavia also there is a great development of Silurian limestones, which extend far into Russia. In the latter. country the sea had retreated much from its extension in the Ordovician, except toward the southeast, where it was carried into Bessarabia. Most of the Russian Silurian was formed in an interior sea, connected with that of southern Europe. In the southern European countries, which display the Bohemian *facies*, the Silurian rocks have nearly the same general distribution as the Ordovician. The two systems are also associated in the Arctic islands, in China, north Africa, South America, and Australia. In South Africa the Silurian is probably represented in some of the barren Table Mountain sandstones. In all of these areas, as also in North America, the fossils resemble those of the northern European region, rather than those of the southern. In general the Silurian rocks are less extensively exposed at the surface than the Ordovician.

Climate. — The Silurian climate seems to have been like that of the Ordovician, very uniform over the earth and with no indication of climatic zones. The aridity of the New York region in the Salina age, which has already been mentioned, corresponds to the similar conditions in the Ordovician of Siberia. Both were probably local.

Close of the Silurian. — In parts of North America the Silurian passed so gradually and gently into the Devonian, that it is difficult to draw the line between the two systems. Some disturbances, however, took place in Ireland, Wales, and the north of England, for in these localities the Devonian lies unconformably upon the Silurian. In other parts of Europe the transition was gradual.

The Life of the Silurian

Silurian life is the continuation and advance of the same organic system as flourished in the Ordovician, certain groups diminishing, others expanding; and some new groups now make their first appearance

---

PLATE VIII. — (*Continued*)

sonoceras americanum Foord, × ⅜, Guelph. 17, *Trochoceras desplainese* McChesney, × ½, Guelph. 19, *Lichas breviceps* Hall, × ½, Lockport. 20, *Staurocephalus murchisoni*, Barrande, × 1, Lockport. 21, *Deiphon forbesi* Barrande, × 1, Niagaran of Arkansas.

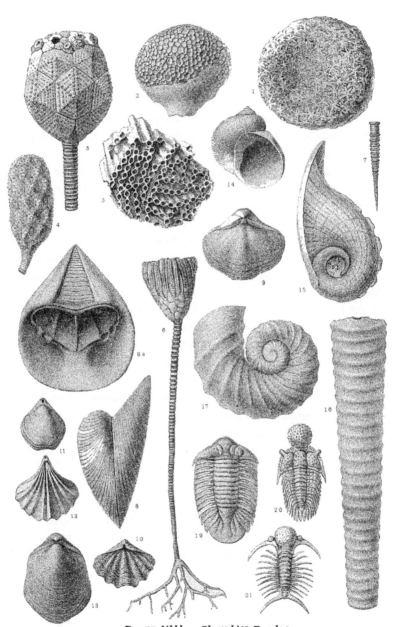

PLATE VIII. — SILURIAN ·FOSSILS

Fig 1, *Astræospongia meniscus* Roemer, × 1, Lockport. 2, *Favosites occidentalis* Hall, × 1, Lockport. 3, *Halysites catenulatus* Linn., × 1, Lockport. 4, *Holocystites cylindricus* Hall, × ⅓, Guelph. 5, *Caryocrinus ornatus* Say, × 1, Lockport. 6, *Eucalyptocrinus crassus* Hall, × ½, Lockport. 7, *Tentaculites gyracanthus*, Eaton, × ⁵⁄₂, Manlius. 8, *Monomerella noveboracum* Clarke and Rued., × ½, Guelph. 8a, *The same*, interior of ventral valve. 9, *Spirifer radiatus* Sowerby, × ½, Lockport. 10, *S crispus* Hisinger, × 1, Lockport. 11, *Meristina nitida* Hall × ½, Lockport. 12, *Rhynchotreta americana* Hall, × 1, Lockport. 13, *Pentamerus oblongus* Sowerby, × ½, Clinton. 14, *Strophostylus cyclostomus* Hall, × 1½, Lockport. 15, *Trematonotus alpheus* Hall, × ⅝, Guelph. 16, *Daw⁻*

Plants. — Our knowledge concerning the land vegetation of the Silurian is not much more definite than concerning that of the Ordovician. Most of the remains referred to land plants are of disputable character; the best authenticated is a fern (*Neuropteris*) from the Silurian of France.

Spongida are still common. Well-known forms are *Astylospongia* and *Astræospongia* (Pl. VIII, Fig. 1); both are almost restricted to Tennessee.

Cœlenterata. — The Graptolites have greatly diminished, especially the branching forms and those with two or more rows of cells. Those that persist are, for the most part, straight and simple. The Hydroid Corals, on the other hand, such as *Heliolites*, *Stromatopora*, etc., become important elements of marine life and in the formation of the reefs. The true *Corals* likewise increase largely, and play a more important rôle than in the preceding period. The increase is principally in the enlarged number of species belonging to much the same genera. *Favosites* (VIII, 2) is a characteristic new genus, and *Halysites* (VIII, 3), the chain coral, *Syringopora*, and others are much commoner than before.

Echinodermata. — In this group we observe a great expansion of the *Cystoidea*, which are very abundant in the Chicago limestone (Niagara) and the Manlius of Maryland and West Virginia. *Holocystites* and *Caryocrinus* (VIII, 4, 5) are typical of the fauna. There is also a marked increase of the *Crinoids; Eucalyptocrinus* (VIII, 6) is a good example. Starfishes also have grown more abundant. The *Blastoidea*, which originated in the Ordovician and became extinct at the end of the Palæozoic era, remain rare in the Silurian and Devonian, first becoming important in the Carboniferous. The Echinoids, or sea-urchins, which were commoner than before, have no arms, but a closed spheroidal or discoidal test, made up of calcareous plates, which in all the modern sea-urchins are arranged in just twenty vertical rows, and are closely fitted together by their edges, like a mosaic pavement. In the Palæozoic sea-urchins the number of rows of plates is either more or less than twenty; in some of the Silurian genera the plates are loosely fitted, and slightly overlapping, like fish-scales.

Arthropoda. — Among the Crustacea the Trilobites are still numerous, though decidedly less so than they were in the Ordovician; they represent, for the most part, new species of genera which have survived from the preceding period. The commonest genera are *Calymmene, Illænus, Dalmanites Lichas* (VIII, 19), *Phacops, Proetus, Encrinurus,* etc. Especially characteristic are the genera *Staurocephalus* (VIII, 20) and *Deiphon* (VIII, 21) and the large number of spiny forms. *Eurypterids* continue to increase in numbers and size, though not reaching their maximum in either respect. In these extraordinary Crustacea the head is quadrate and is followed by a long, tapering body, composed of thirteen movable segments; the last segment is either a pointed spine, as in *Eurypterus* (see Fig. 262), or a broad tail-fin, as in *Pterygotus.* Five pairs of appendages are attached to the head, the bases of four of which, on each side of the mouth, form the jaws, as in the existing horseshoe crab. The first pair

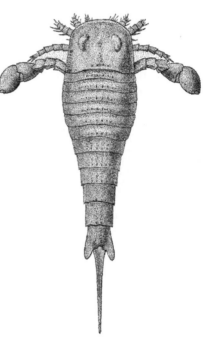

FIG. 262. — *Eurypterus fischeri* Eichw., Island of OeSel, Russia. (Schmidt)

of appendages are either short and simple (*Eurypterus, Stylonurus*), or are much elongated, and armed with pincers (*Pterygotus*). The fifth pair are either very long, or enlarged to serve as swimming paddles. The first body-segment carries a pair of apron-like appendages, with a narrow median extension, but the other segments have no appendages. The horse-shoe crabs find their most ancient representative in the genus *Hemiaspis* of the European Silurian. The Eurypterida would appear not to have been marine animals but to have lived in the freshening lagoons of the upper Salina and to have been entombed in the shoal-water calcareous muds of the Water-lime. Other Crustacea are much as in the preceding period.

Scorpions have been found in the Silurian of Europé and America, and some remains of *Insects* in the former continent. These animals prove the existence of a contemporaneous land vegetation, and confirm the doubtful evidence of the Ordovician and Silurian plants.

Bryozoa decline, but are still quite abundant, and contribute in an important way to the growth of the coral reefs as well as forming reefs by themselves in the Clinton and Lockport limestones, but they are less important reef-builders than they had been in the Ordovician, chiefly because the massive kinds are so largely replaced by delicate, lace-like colonies.

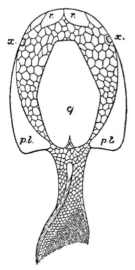

FIG. 263. — *Lanarkia sp.;* a primitive Ostracoderm. (Traquair.) Restored

Brachiopoda continue to be present in multitudes, but with a distinct change in dominant genera from those which were commonest in the Ordovician. Especially characteristic is the increase in the families of the *Spiriferidæ* and *Pentameridæ*, both of which continue prominent in the Devonian. The most important genera are *Atrypa, Spirifer* (VIII, 9, 10), *Pentamerus* (VIII, 13), *Rhynchotreta* (VIII, 12), *Leptæna, Streptis,* and *Whitfieldella.* A curious family of Inarticulate Brachiopods is the Trimerellidæ, several forms of which are found in the coral reef dolomites of the Guelph. *Monomerella,* a genus of this family, is figured (VIII, 8).

Mollusca. — The Bivalves show no very significant changes
from the Ordovician, but the Gastropods, especially such forms
as *Capulus*, increase decidedly; other well-represented genera of
these shells are *Platyostoma, Cyclonema, Strophostylus* (VIII, 14),
and *Trematonotus* (VIII, 15). *Pteropods* are smaller and less
numerous than before; a very common form is the little nail-
shaped shell, *Tentaculites* (VIII, 7), which is doubtfully referred
to this group, but may belong to the Worms. Among the Cephalopods Orthoceratites are less frequent than in the Ordovician,
but are more commonly ornamented by rings or longitudinal
ridges (VIII, 16), and *Endoceras* has disappeared, while low tur-
reted shells, like *Trochoceras* (VIII, 17), are characteristic. The
shells with curiously contracted mouth openings, like *Phragmoceras*, are more commonly found than in the Ordovician.

Vertebrata. — The remains of Ostracoderms of a primitive
kind have been found in the uppermost Silurian of Scotland;
they are of small size and very peculiar appearance, but are related to the genera which were to attain such prominence in the
Devonian. Sharks, doubtless of extremely primitive character,
existed in the Silurian seas, but very little is known about them.

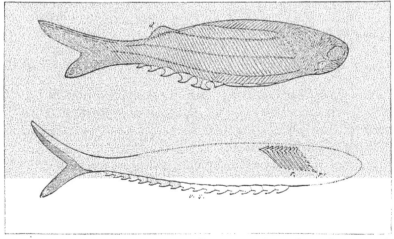

FIG. 264. — Upper figure, *Birkenia elegans* Traq., × 3/2. Lower figure, *Lasanius prob-
lematicus* Traq., enlarged. Both figures restored. (Traquair)

THE DEVONIAN PERIOD

DEVONIAN SYSTEM

RHINE-BELGIUM SECTION		NEW YORK	
Cypridina Slates	Upper Devonian	Chautauquan Series	Chemung and Catskill Stage
Clymenia Limestone		Senecan Series	Portage Stage / Genesee Stage / Tully Stage
Cuboides Zone			
Stringocephalus Limestone	Middle Devonian	Erian Series	Hamilton Stage / Marcellus Stage
Calceola Beds		Ulsterian Series	Onondaga Stage / Schoharie Stage / Esopus Stage
Coblenzian Series	Lower Devonian	Oriskanian Series	Oriskany Stage
Gedinnian Series		Helderbergian Series	Port Ewen Stage / Becraft Stage / New Scotland Stage / Coeymans Stage

THE name *Devonian*, taken from the English county Devon-shire, was proposed by Sedgwick and Murchison in 1839; it has found universal acceptance and has passed into the geological literature of all languages.

As in the Ordovician and Silurian, and for the same reason, the divisions of the New York Devonian are taken as the standard of

reference for North America, but the general standard is no longer to be found in Great Britain, in spite of the first description of these rocks in Devonshire. This is because these rocks in England are so much disturbed, metamorphosed, and faulted, that the order of succession among them could not be determined until the Devonian of the Rhine and Belgium, the largest and best-developed Devonian area of western Europe, had first been made out. Hence, the Rhenish section is widely employed as a standard for the different continents.

DISTRIBUTION OF THE DEVONIAN

American. — In certain areas, notably in that of the Cumberland Basin, the transition from Silurian to Devonian is so gradual that the boundary between them remained long in doubt and has been shifted more than once. The Helderbergian series has, until quite recently, been very generally referred to the Silurian, and at one time even the Oriskanian was included in the same system.

At the beginning of the Devonian (*Helderbergian* epoch) most of the continent west of the Cumberland Basin was land, but the Basin itself continued to deepen and enlarge, and subsidence in the South brought the sea in over western Tennessee into Missouri and southern Illinois, reaching into Oklahoma. The limestones of the Helderbergian series, which were laid down in the Cumberland Basin, extend, with some interruptions, from southwestern Virginia along the line of the Appalachians to Albany, N.Y. The Gaspé Peninsula, which forms part of the west coast of the Gulf of St. Lawrence, was submerged at this time, for here we find 1500 feet of limestone representing the Helderbergian and Oriskanian series. "The St. Lawrence tidal waters of this period must have extended westward to the border of Vermont and Montreal and southward along the Connecticut valley." (Dana.) Northern and southern New Brunswick, northern Nova Scotia, northern Maine and part of its coast were under the Helder-

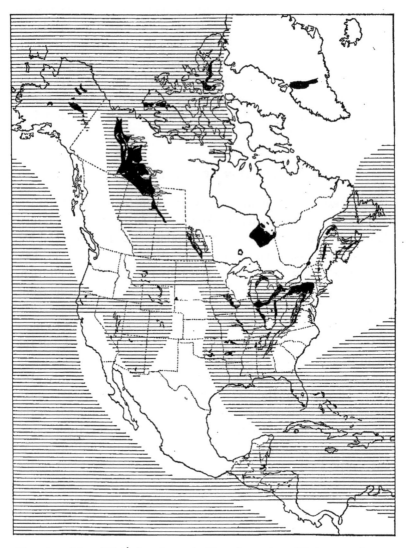

FIG. 265. — Map of North America in the Devonian period. Black areas = known
exposures; white areas = land, or unknown; horizontal lines = sea

bergian sea, and the New England troughs, the Connecticut valley, and the Gaspé-Worcester trough were submerged. In the West the Helderbergian has been identified only in the Nevada trough, where marine sedimentation continued uninterruptedly from the Silurian, while in the far North on the shore of Kennedy Channel, 80° N. lat., Helderbergian fossils have been found.

The *Oriskanian* epoch witnessed some geographical changes. The rocks of this series, which are prevailingly sandstones laid down in the Cumberland Basin, are thickest in western Maryland, and thin away both north and south from that region; to the south the beds are chiefly lower Oriskany and to the north mostly upper Oriskany, indicating oscillations of the Basin floor. In central New York the Oriskanian waters extended themselves westward across the state, breaking through the barrier in the southeast. In the Mississippi valley, the Oriskany has about the same limits as the Helderbergian, but it has not been found in Oklahoma, while, on the other hand, the sea transgressed over northern Georgia and Alabama, where the Oriskany sandstone, only 20 feet thick, successively overlaps the older formations from the Middle Cambrian to the late Ordovician. In addition to the Oriskanian limestones of the Gaspé Peninsula, rocks of this stage are found in Nova Scotia and New Brunswick and northern Maine, in which State they are enormously thick, 5000 feet.

The Lower Devonian of the Maritime Provinces, and especially of Maine, shows clear indications of an invasion of the Coblenzian fauna of Europe, from which may be inferred the existence of a land bridge across the North Atlantic, affording the necessary conditions for the migration of the shoal-water animals. "The evidence then is fairly conclusive that during the period represented by the Coblenzian-Oriskany, the arenaceous epicontinental sediments was the ground traversed by the Coblenz fauna westward along the North Atlantic continert." (Clarke.)

Very extensive changes characterized the Middle Devonian; the Cumberland Basin was elevated into land at the end of the Oriskany, and in eastern New York we have only the coarse sands

2 Q

and grits of the *Esopus* (which may be a phase of the Oriskany) and the *Schoharie*, but the Interior Sea was once more established, with a restricted area, and not improbably connecting with the Gulf of St. Lawrence by way of the Connecticut valley trough. In it was accumulated the great *Onondaga* limestone, which extends from the Hudson River across New York into Michigan, and around what may have been the islands of the Cincinnati anticline into Indiana, Illinois, and Kentucky. Interpretations differ concerning the form and size of this Mississippi valley sea; according to one view, it extended southward to the Gulf of Mexico, while another makes it completely enclosed on the south and postulates a northern extension across Hudson's Bay to the Arctic Sea. Another great invasion, called the Dakota Sea, which may have had no direct communication with the former, opened from the Arctic Ocean, where now is the mouth of the Mackenzie River, and crossed British America, the Dakotas, Nebraska, and Kansas, sending a gulf into Iowa, Minnesota, and northern Illinois and southern Wisconsin, and reaching or perhaps passing through Texas. "This channel was bounded on the west by the extensive Archæan islands or edges of land constituting the eastern axis of the present Rocky Mountains." (Williams.) However, another interpretation separates the Dakota Sea from the Mississippian by the width of the Great Plains, with a possible cross-connection in southern Canada, and includes all the eastern Onondaga in the Interior Sea. A third sea, the Cordilleran, covered much of the Great Basin, probably extending to the Pacific and connecting on the north with the Dakota Sea. In the northeastern part of the continent the Gaspé area was converted into a coastal lagoon, into which great masses of sand were swept by rapid streams; these sands contain numerous land plants. A remnant of one of the old channels is marked by a coralline limestone in northern Vermont.

The Onondaga limestone is largely a coral formation, and in some places the reefs are still beautifully preserved, as in the famous example at the Falls of the Ohio, above Louisville, Ky.

Limestone formation on such a scale implies the very general absence of terrigenous sediments from the epicontinental seas, which may be interpreted as due in part to the base-levelling of the surrounding lands.

A change of conditions in the northeastern bay of the Interior Sea, probably the elevation of the land and resulting rejuvenation of the streams, checked the accumulation of limestone and brought in great quantities of mud and silt (*Marcellus* and *Hamilton*), though in the Mississippi valley limestone-making still went on, and even in New York thin limestones occur here and there in the great mass of the Hamilton shales. At Gaspé continental sedimentation continued, and there the Erian series is represented by a very thick mass of sandstones, bearing fossils of land plants, and showing occasional brief incursions of the sea. In New Brunswick and Nova Scotia are sandstones and shales. Western Maryland and the adjoining parts of the Virginias were again submerged, but this time apparently by an expansion of the Interior Sea, that seems to have covered part of the old Cumberland Basin, which had emerged at the end of the Oriskanian. In the West the seas remained much as they had been during the Onondaga; indeed, the difference in interpretations of the latter stage, referred to on the preceding page, is chiefly a question of distinguishing Hamilton from Onondaga. A depression submerged the coast of British Columbia up to Alaska.

The Upper Devonian has much the same distribution as the Middle, but the Interior Sea appears to have lost its connection with the Gulf of Mexico and to have become joined with the Dakota Sea, while here and there in the Mississippi valley, the Upper Devonian overlaps older rocks where the Lower and Middle are absent. The *Tully* stage is a locally developed limestone in New York, which is of interest as corresponding to the *Cuboides* zone in the Rhenish section; both contain the very widespread brachiopod species, *Hypothyris cuboides*, a very useful guide or index fossil, as marking the base of the Upper Devonian. The *Genesee* is a mass of black shales, which increases in thickness from Lake Erie to central Pennsylvania, where it is 300 feet thick, and is succeeded by the *Portage*, which is largely arenaceous. The Portage exceeds 1000 feet in thickness. In western New York both the Genesee and the Portage have embedded in them a limestone which carries a highly interesting assemblage of animals called the "Naples Fauna." This has very little in common with that of the Hamilton, but is well marked in many parts of the world. (Figs. 255–6, pp. 522–3.) "The migration path of this pelagic fauna has been traced toward the Northwest through Manitoba into Siberia, thence through Russia into Westphalia. . . . In New York, where its fauna became extensive, it was alien and short-lived." (Clarke.)

The *Chemung* is a great mass of sandstones and conglomerates, which reaches its maximum thickness (8000 feet) in Pennsylvania, thinning greatly to the westward. Indeed, the stages of the Upper Devonian given in the New York scale, can seldom be recognized except in that State and in Pennsylvania. In Ohio the whole Upper Devonian is represented by the *Ohio Shale*, which varies from 300 to 2600 feet in thickness. The *Catskill*, which was originally regarded as a distinct series, is a very thick mass of sandstones, representing a facies of the Upper Devonian extending through the Senecan and Chautauquan epochs and, in Pennsylvania, into the earliest Carboniferous. These beds are believed to have been accumulated in a long and narrow estuary, running from eastern New York into Pennsylvania, where the beds reach a thickness of 7500 feet, and containing fresh water in part of its course. Areas of similar rocks, with fresh or brackish water fossils, occur in the Portage and Genesee and represent deposition in coastal lagoons.

The Western Devonian indicates such a different succession of physical events, that its subdivisions can seldom be correlated with those of the East. Devonian strata are not known to underlie the Great Plains, and the Front Range of Colorado appears to have been a land area, for there Carboniferous strata overlap and rest upon Cambrian and Ordovician. On the other hand, in the plateau region, from Arizona to Montana and along the Canadian Rockies are many Devonian outcrops, indicating that much or all of this region was submerged at one epoch or another of the period. Oscillations of level are also shown, as in the Grand Cañon region, where thin and worn patches of Devonian, evidently remnants of a much thicker and more widespread mass, lie unconformably upon Cambrian and below Carboniferous beds.   In the Wasatch Mountains of eastern Utah the Devonian is represented by 2400 feet of quartzites and limestones, and in the Nevada trough, where deposition seems to have been unbroken, 2000 feet of shale and 6000 feet of limestone are assigned to this system. Though the faunas differ notably from those of the East and have more affinity with those of Europe and Asia, correspondences with the Helderberg, Onondaga, and Hamilton have been observed. Affinities with the Old World are also shown by the Devonian of northern and southern California, British Columbia, and Alaska.

Like the rocks of the other Palæozoic systems in North America, those of the Devonian are quite free from igneous intrusions, except in several widely separated localities. Contemporaneous lava flows are interbedded with the Lower Devonian shales of northern New Brunswick, pointing to volcanic eruptions in that region, and the same is true regarding the Devonian of Nova Scotia, and that of northern California has beds of tuff and lava sheets. In central New York and various places in the West the Devonian strata are cut by intrusions, but these may be post-Devonian in date.

Comparing the rocks of the Ordovician, Silurian, and Devonian, as these are developed in the Appalachian and adjoining regions, a certain rhythmic or periodic recurrence of events may be discovered among them. Each system is characterized by a great and very widespread limestone formation, the Trenton, Lockport-Guelph, and Onondaga, respectively, and in each the limestone is succeeded by shales or other clastic rocks, the Utica, Salina, and Portage, due to an increase of terrigenous material, and each was closed by a more or less widespread emergence of the sea-bottom. Each began with a subsidence which gradually extended to a maximum at the time when the great limestone was formed. The parallelisms are not exact, but they are certainly suggestive.

Foreign. — The *European* Devonian appears in three different *facies;* one of these is the " Old Red Sandstone," which is largely of continental origin, and lies to the north. The second facies is of marine, shoal-water deposits and runs from Devonshire, through Belgium, the northern part of the lower Rhenish and the Hartz Mountains, to Poland; and the third, extending from northwestern France, through Germany to Bohemia, was laid down in deeper water. On the other hand, great changes took place in the extent and depth of the Devonian seas, those of the Lower Devonian being far less extensive than those of the Middle and Upper parts of the period, as is also true of North America.

The period began in Europe with an advance of the sea over the land in many places, reaching its maximum extension in the latter part of the period, but beginning to retire before the opening of the Carboniferous. This subsidence removed the barrier which in Ordovician and Silurian times had separated the northern and southern seas, but was accompanied by the formation of closed basins farther to the north. Europe then was largely an open sea with many islands, and where the waters were sufficiently clear and free from terrigenous sediment, coral reefs were extensively formed.

The marine Devonian occurs in the southwest of England, over large areas of Germany, in northwestern and southern France, and on an enormous scale in Russia. During the Silurian the sea had withdrawn almost entirely from Russia west of the Ural Mountains. In the Lower Devonian the sea broke in from the north over Siberia, reaching far into central Asia. In the Middle Devonian a great basin was formed by the depression of central Russia, the sea advancing from the north and the east. Devonian limestones and great coral reefs occur in the Alps, as do limestones, shales, and sandstones of the same period in Spain and Portugal.

The "Old Red Sandstone" is of particular interest, because, owing to the peculiar circumstances of its formation, it has preserved a record of Devonian land life, which, though fragmentary, is far more complete than anything we possess from the more ancient periods. These strata were laid down in closed basins (sometimes, perhaps, in fresh-water lakes), which had only a restricted communication with the sea, and it may be that these accumulations were partly made by the wind, though there is no gypsum or salt in the beds to indicate the prevalence of desert conditions. The Old Red is found in south Wales and the adjoining part of England, and on a much larger scale in Scotland; also in the Baltic provinces of Russia, where it is interstratified with beds of the marine Devonian; in Spitzbergen and Greenland the same formation recurs. These sandstones are said to be 10,000 feet thick, but according to some authorities, the lowermost part of them is Silurian. The Catskill of New York is very like the Old Red, and contains similar fossils, and the Old Red facies is found in northern New Brunswick on Chaleur Bay.

The European Devonian is full of the evidences of volcanic activity, in the shape of great lava-flows and tuffs. In central Scotland the volcanic accumulations exceed 6000 feet in thickness.

Besides the Devonian areas already mentioned in *Asia*, rocks of this system are found in China, the Altai, and in Asia Minor. They recur in northern *Africa*. The Bokkeveld beds of South Africa are among the rare marine formations of that region, and these, which are Lower Devonian, thin away northward and die out within a hundred miles of the coast. Below the Bokkeveld, in the upper part of the Table Mountain sandstone, is a boulder clay with striated boulders of evidently glacial origin, pointing to the establishment of rigorous climatic conditions in South Africa in the earliest Devonian, or perhaps late Silurian. In *South America* occurred a great transgression of the sea, and Devonian strata form larger areas of the surface than those of any other Palæozoic system. Shallow-water deposits are found in Peru, Bolivia, over large parts of Brazil, especially the basin of the Amazon, and in the Falkland Islands. The Bolivian Devonian, which belongs to the lower and middle parts of the system, contains a very similar fauna to that of North America, and connects the latter with Brazil, the Falkland Islands, and South Africa.

Kayser distinguishes two great marine provinces during the Devonian: (1) the Eurasian, which extended from Europe eastward over northern and central Asia to Manitoba, Canada, and (2) the American, which reached from the United States to South America and thence to South Africa.

Climate. — With the exception of South Africa, the distribution of Devonian fossils leads us to infer that the climate of the period was, like that of the Ordovician and Silurian, generally uniform over the earth and without distinction of zones.

DEVONIAN LIFE

The life of the Devonian is, in its larger outlines, very like that of the Silurian, but with many significant differences, which are due, on the one hand, to the dying out of several of the older groups of animals, and on the other, to the great expansion of forms which in the Silurian had played but a subordinate rôle.

Plants. — The fossils show that in Devonian times the land was already clothed with a varied, rich, and luxuriant vegetation of the same general type as that whose scanty traces are found in Silurian strata. All the higher Cryptogams are represented, and by large, tree-like forms, as well as by small herbaceous plants. The bulk of the flora is composed of *Ferns*, *Lycopods*, or Club-Mosses (especially the great tree-like *Lepidodendrids*), and *Equisetales*, or Horsetails. The highly important extinct groups, the *Sphenophyllales* and *Cycadofilices*, which were already in existence, will be described from the more complete Carboniferous fossils. Besides these Cryptogams, we find representatives of the lower

---

Helderberg. 10, *Tropidoleptus carinatus* Conr., × ½, Hamilton. 11, *Chonetes coronatus* Conr., × ½, Hamilton. 12, *Stropheodonta demissa* Conr., × ½, Hamilton. 13, *Pterinea flabellum* Conr., × ½, Hamilton. 14, *Platyceras dumosum* Conr., × ½, Onondaga. 15, *Manticoceras oxy* Clarke, × ⅙, Portage. 16, *Phacops rana* Green, × ⅛, Hamilton. 17, *Odontocephalus selenurus* Eaton, × ¼, Onondaga. 18, *Terataspis grandis* Hall, × ¹/₁₂, Schoharie. 19, *Echinocaris punctatus* Hall, × ½, Hamilton.

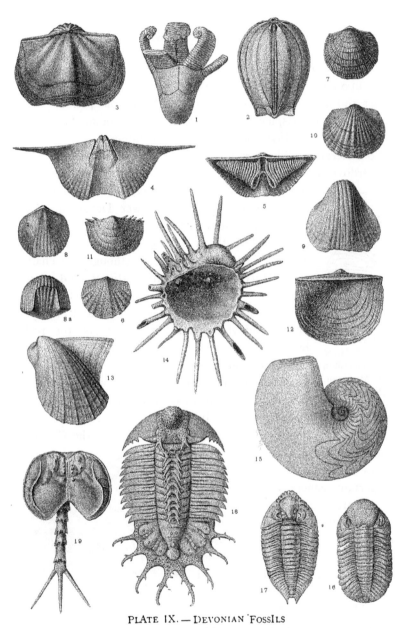

PLATE IX. — DEVONIAN FOSSILS.

Fig. 1, *Edriocrinus sacculus* Hall, × ½, Oriskany. 2, *Nucleocrinus verneuli* Troost.
× 1, Onondaga. 3, *Spirifer macropleurus* Conr., × ½, Helderberg. 4, *S. disjunctus*
Sowerby, × ½; internal cast of ventral valve, Chemung. 5, *S. mucronatus* Conr., × ½;
inner view of dorsal valve, showing arm-supports, Hamilton. 6, *Vitulina pustulosa* Hall,
Atrypa reticularis Linn., × ½, Hamilton. 8, *Hypothyris venustula*
Hall, × ⅚, Tully. 8a, *The same*, anterior view. 9, *Cyrtidula galeata* Dalman, × ½,
× ⅜, Hamilton.

kinds of flowering plants in the *Gymnosperms*, which presumably grew upon the higher lands. We shall meet this same flora in richer and more varied display in the Carboniferous period.

Sponges are conspicuous elements of the Devonian fauna, as, for example, the *Dictyospongidæ* in the Chemung.

Cœlenterata. — The Graptolites, which were so abundant in the Ordovician and had become much less common in the Silu-

rian, are now almost extinct, only a few simple species occurring in the Lower Devonian. The *Corals*, on the contrary, expand and multiply enormously both in numbers and in size. Most of the Silurian genera persist (though the chain-coral *Halysites* has become extinct), and many new forms are added. *Heliophyllum* (Fig. 266) is an example of the solitary corals, and *Phillips-*

FIG. 266. — *Heliophyllum halli* E. & H., × ½. Hamilton of Michigan. (Rominger)

astræa and *Acervularia* (Fig. 267) of the reef-builders.

Echinodermata. — The Cystoids have become much rarer than before, and are on the point of extinction; the Blastoids (Pl. IX, Fig. 2) are still in the background, though locally abundant in a few places, and the Echinoids have not yet become common; but the Crinoids and Star-fishes have greatly increased in number and variety. Important genera of the former group are *Cupressocrinus, Platycrinus, Actinocrinus, Dolatocrinus,* etc. *Edriocrinus* (IX, 1), a free-swimming crinoid, without a stem, occurs in the Helderbergian and Oriskany. The multitude of the crinoids contributed largely to the building up of the calcareous sea-bottom on which they flourished.

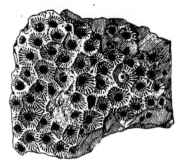

FIG. 267. — *Acervularia davidsoni* E. & H., × ½. Hamilton of Michigan. (Rominger)

Arthropoda. — The Trilobites had already begun to decline ın the Silurian, while in the Devonian the decline becomes much more marked, though they are still far from rare. New species of Silurian genera, like *Phacops* (IX, 16), *Homalonotus*, *Lichas*, *Acidaspis*, *Odontocephalus* (IX, 17), etc., are the commonest. *Terataspis* (IX, 18), is one of the largest and most curious of Trilobites. A characteristic of the Devonian Trilobites is the ornamentation of the margin of the head, as well as the extraordinary development of spines which many display on the head-and tail-shields. (IX, 17, 18.)

The other Crustacea make notable progress in this period. The *Phyllocarida* are abundant in the Upper and Middle Devonian. The first of the *Isopoda* and of the long-tailed *Schizopoda* make their appearance in the Devonian. The *Eurypterids* now attain their culmination in size, being actually gigantic for Crustacea, and some of them are as much as six feet long. The genera (*Eurypterus*, *Stylonurus*, and *Pterygotus*) are the same as in the Silurian. *Insects*, though still rare as fossils, are very much commoner than in the Silurian; they represent both Orthopters and Neuropters, which are among the primitive groups.

Brachiopoda. — As in the Silurian, Brachiopods continue to be the most abundant fossils, both in species and individuals; indeed, the Devonian is the culminating period of brachiopod abundance and variety; in North America at least it has many more genera than any other. Many Silurian genera have died out, but of others, like *Orthis*, *Stropheodonta* (IX, 12), *Orthothetes*, *Atrypa* (IX, 7), *Chonetes* (IX, 11), and *Productella*, the species are more numerous. The most characteristic shells are those belonging to the genera *Spirifer*, especially the very broad " winged " species (IX, 4, 5), *Hypothyris* (IX, 8), *Athyris*, *Gypidula* (IX, 9), *Tropidoleptus* (IX, 10), *Vitulina* (IX, 6), and those belonging to the still existing family *Terebratulidæ*, of which *Renssellaeria* and *Stringocephalus* are Devonian genera.

Mollusca. — Bivalves and Gastropods are much as in the Silurian: examples of the former are *Pterinea* (IX, 13), *Conocardium*, and *Grammysia*, while larger species of the Gastropod

Euomphalus, and spiny *Platycerids* (IX, 14), are characteristic. A minute Pteropod, *Styliolina*, is extraordinarily abundant in the Upper Devonian, forming limestone masses. The Cephalopods have been revolutionized; the wealth of Nautiloid shells which we found in the Silurian has been much diminished, though *Orthoceras, Phragmoceras, Gomphoceras,* and *Cyrtoceras* still persist abundantly in the Lower Devonian, while many other genera have disappeared. More significant is the first appearance of the Ammonoid division of the Tetrabranchiate Cephalopods, a group of shells which was destined to attain extraordinary development in the Mesozoic era. The Ammonoids are distinguished by the complexity of the " sutures," or lines made by the junction of the septa with the outer wall of the shell. In the Devonian Ammonoids, of which the Goniatites (IX, 15) are the common forms, the sutures are much less complex than in the Mesozoic shells. Another member of the group which is far more abundant in Europe than in America is *Clymenia*, the only Ammonoid in which the siphuncle is on the inner side of the spiral. *Bactrites* has a straight shell, like that of *Orthoceras*, but with the complex sutures which show it to be an Ammonoid.

Vertebrata. — One of the most characteristic features of Devonian life is the great development of the aquatic Vertebrates, which is so striking that the period is often called the " Age of Fishes." So numerous and so finely preserved are these fossils that a satisfactory account may be given of the structure and systematic position of many of the genera. This great assemblage of fishes and fish-like forms, it should be remembered, was not something entirely new in the earth's history, but was rather the wonderful expansion of types which during the Ordovician and Silurian had remained rare and obscure.

Of the Devonian Vertebrates none are more peculiar and characteristic than the *Ostracoderms*, which, though generally called fishes, really belong to a type much below the true fishes, being devoid of true jaws and of paired fins. The head and more or less of the body are sheathed in heavy plates of bone, and the remainder of the body and the tail are covered with scales. No trace of the internal skeleton is preserved, and it evidently was not ossified. The genus *Cephalaspis* of this group is curiously like a Trilobite in appearance, though, of course, the resemblance is entirely superficial. The head-shield is formed of a single great plate of bone, shaped like a saddler's knife, with rounded front edge and with the hinder angles drawn out into spines; the eyes are on the top of the head and very close together. The body is covered with large, angular plates of bone, arranged in rows; a small median dorsal fin and a larger triangular tail-fin make up the locomotor apparatus.

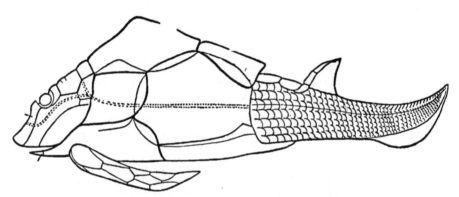

FIG. 268. — *Pterichthys testudinarius.* Restored. Old Red Sandstone. (Dean after Smith Woodward)

Pteraspis has a bony plate over the snout, a large shield on the back, and another on the belly, with rhomboidal scales covering the rest of the body.

A most extraordinary-looking creature is *Pterichthys* (Fig. 268), in which the head and most of the body are encased in heavy plates, the remainder in overlapping scale-like bones; the eyes are even closer together than in *Cephalaspis.* Dorsal and tail-fins are present and what appear to be pectoral fins. The pair of appendages referred to doubtless acted as fins, but they are not comparable to the paired fins of the true fishes, being merely jointed extensions of the head-shield. These three genera, *Cephalaspis*, *Pteraspis*, and *Pterichthys*, have been selected as types of the Ostracoderms, each one of which has several allies, differing from it in one or other particular.

Of the true *Fishes* there is great variety in the Devonian. The *Selachians* are well represented, one of which is *Cladoselache*

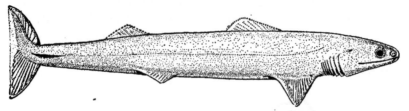

FIG. 269.— *Cladoselache newberryi.* Restored, × ¹/₅. Ohio shale. (Dean)

(Fig. 269), a small shark, from two to six feet in length, and the most primitive known member of the group. The *Dipnoi*, or Lung-fishes, were important elements of the Devonian fish fauna. *Dipterus* (Fig. 270), an example of this group, is very like the modern lung-fishes, which have dwindled to three genera, one

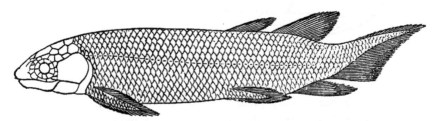

FIG. 270. — *Dipterus valenciennesi* Sedg. and Murch. Restored. Old Red Sandstone. (Smith Woodward after Traquair)

in South America, one in Africa, and one in Australia. A remarkable series of fishes, the *Arthrodira*, is very characteristically Devonian. One of the best-known members of this group is the European genus *Coccosteus* (Fig. 271), in which the head, back, and belly are covered with bony plates, but the rest of the body is naked. This bony armour gives the fish something of the appearance of the Ostracoderms, with which group it is often, though erroneously, classified. The backbone is represented by an unsegmented rod (the notochord, N, Fig. 271), to which arches of bone are attached (N, H, Fig. 271). Paired ventral fins were present, but pectorals have not been found. The jaws were pro-

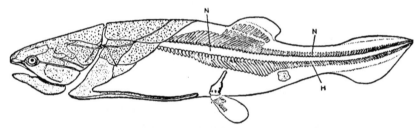

FIG. 271. — *Coccosteus decipiens* Ag. Restored. Old Red Sandstone. (Dean, after Smith Woodward)

vided with teeth, which fuse into plates. In the uppermost Devonian of Ohio are found some huge fishes allied to *Coccosteus*, but much larger and more formidable. The most important of these are *Dinichthys* and *Titanichthys*, the latter attaining a length of 25 feet.

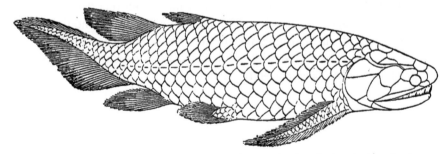

FIG. 272. — *Holoptychius nobilissimus* Ag. Restored. Old Red. (Smith Woodward after Traquair.) The ornamentation of the scales is not shown

A higher type of Devonian fish is that of the *Crossopterygii*, an ancient group of which but two representatives remain at present, both of them African. These fishes, like the Dipnoans, have " lobate " paired fins (see Fig. 272), *i.e.* the part of the fin belonging to the internal skeleton is large and covered with scales, form-ing a lobe to which the fin-rays are attached. Most of the Devo-nian members of the group have massive rhomboidal scales, but in others, like *Holoptychius*, the scales are thinner, rounded, and overlapping.

The most advanced fishes of the period are the Ganoid mem-bers of the Actinopteri, which from the Devonian until nearly the end of the Mesozoic era continue to be the dominant type of fishes. Nearly all of them have thick, shining scales of rhom-boidal shape.

The Devonian fish fauna (using that term in a very compre-hensive sense) is thus seen to be a rich and varied one, includ-ing Ostracoderms, Sharks, Lung-fishes, Arthrodirans, Crossop-terygians and Actinopteri, each with many representatives and mostly of very curious and bizarre forms. While thus varied and plentiful, this assemblage differs from the modern fish fauna in the primitive character of the groups which are represented, and in the entire absence of the Bony Fishes (*Teleosts*), which now make up the vast majority of fishes, both fresh-water and marine.

Amphibia. — Certain footprints which have been reported from the Upper Devonian of Pennsylvania, show that the Amphibia, the lowest of air-breathing vertebrates, had already begun their career; that is, if the correlation of the rocks in which these foot-prints occur has been correctly made.

CHAPTER XXX

THE CARBONIFEROUS PERIOD

THE name *Carboniferous* was given in the early part of the last century, when it was supposed that every geological system was characterized by the presence of some peculiar kind of rock. We now know that this conception is erroneous, and that workable coal seams have been formed in all the periods since the Carboniferous. It still remains true, however, that the latter contains much the most important share of the world's supply of mineral fuel, upon which the whole fabric of modern industrial civilization is founded. The great economic importance of the coal measures has caused them to be most carefully surveyed in all civilized lands, a process greatly assisted by the innumerable shafts and borings which penetrate these rocks. One result of this gigantic work is, that the history and life of the Carboniferous are better known than those of any other Palæozoic period, though our knowledge is still very far from complete.

The Carboniferous rocks are displayed in very different aspects or facies in the various parts of the continent and even in contiguous regions. New York no longer gives the standard scale, for that state has very little that is newer than the Devonian. For the eastern part of the country the sequence of strata in Pennsylvania serves as the scale of reference, while a very different one is needed for the Mississippi valley. In the Rocky Mountain region, again, the character of deposition deviated markedly from what occurred in the East, and all over the far West the Carboniferous is almost entirely marine, without coal. Even in this region, however, the distinction between the Lower and Upper Carboniferous may be drawn. The following table gives the succession in England, Pennsylvania, and the middle West, Illinois, Missouri, Iowa, etc.

CARBONIFEROUS SYSTEM

ENGLAND		PENNSYLVANIA		MISSISSIPPI VALLEY
Ardwick Series Middle Coal Gannister Series Millstone Grit	Upper Carbonif.	Monongahela Stage Conemaugh Stage Allegheny Stage Pottsville Stage	Pennsyl- vanian	Coal Measures Millstone Grit
Yoredale Series Scaur Limestone Series Limestone Shales	Lower Carbonif.	Mauch Chunk Stage (Greenbrier) Pocono Stage	Missis- sippian	Kaskaskia Stage St. Louis Stage Osage Stage { Warsaw / Keokuk / Burlington } Kinderhook Stage

DISTRIBUTION OF THE CARBONIFEROUS ROCKS

American. — The Carboniferous is divisible into two sharply marked portions, the Lower, or Mississippian, and Upper, or Pennsylvanian, a distinction which is applicable in all the continents in which the strata of this period have been carefully studied.

In most parts of North America the Devonian passed so gradually into the Carboniferous that there is no definite line of division between them, but at Gaspé, and in Nova Scotia, New Brunswick and Maine, there was a time of upheaval and erosion toward the end of the Devonian, followed by a depression, in consequence of which there is an unconformity between the two systems. When the Carboniferous period began, most of New York, New England, and eastern Canada were above sea-level, but the Gulf of St. Lawrence covered western Newfoundland, most of New Brunswick, and part of Nova Scotia.

The Interior Sea expanded widely, probably covering nearly the whole of the Great Plains, and most of the old land areas of the West and Southwest, which had persisted through more or less of the Silurian and Devonian, were extensively submerged,

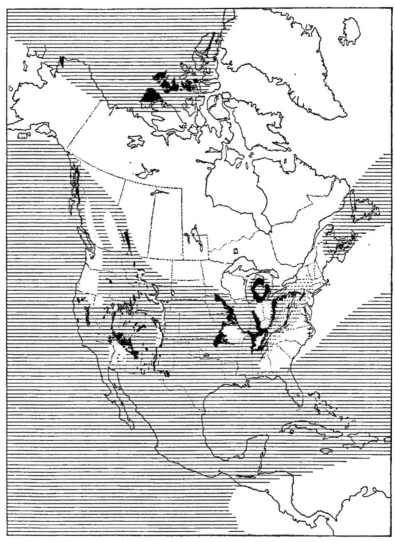

FIG. 273. — Map of North America in the Lower Carboniferous. Black areas = known exposures; white areas = land, or unknown; horizontal lines = sea

probably including all of Mexico and the northern part of Central America. West of the Rocky Mountains the Carboniferous is much the most widely extended of any of the Palæozoic systems, the sea reaching through British Columbia, on both sides of the Gold Range into southeastern Alaska.

In consequence of this great transgression, the Carboniferous strata rest unconformably, or sometimes in apparent conformity, upon all the older systems from the Pre-Cambrian to the Devonian. The Arctic coast of Alaska was submerged and several islands of the Arctic Sea, but the main portion of Alaska, which has great areas of Carboniferous rocks, appears not to have been invaded by the sea till Upper Carboniferous, or Pennsylvanian times. According to Girty none of the known Alaskan faunas, except from the north coast, " can be confidently referred to the Lower Carboniferous. The typical Mississippian is certainly absent as far as evidence has come to hand, and but one occurrence of a fauna definitely related to the Lower Carboniferous of California (Baird) has been found." (Girty.) The fossils of the *Baird* stage in California are entirely different from those of the Interior Sea, and if actually contemporaneous with the latter, must have been separated from them by some barrier. The Great Basin sea appears also to have been separated for a time from the Interior Sea, but communication was established before the end of the Lower Carboniferous. Over the Central States and the West, the Mississippian rocks are almost uniformly limestones, showing that this vast sea was clear and free from terrigenous sediments, but probably of moderate depth.

The northeastern portion of the Interior Sea was divided by the islands of the Cincinnati anticline into two bays, the eastern one of which covered most of Ohio, western Pennsylvania and Maryland; and the western bay occupied the southern peninsula of Michigan and had but a narrow communication with the first. The Appalachian valley trough has in its middle third a nearly or quite complete succession of the Devonian, but lacks the earliest Mississippian, indicating a brief elevation of this region. In the

South the eastern edge of the Interior Sea followed the line of the Appalachian fold probably into Virginia and there broke across the barrier and sent off some narrow sounds southward into the Appalachian Valley.

In the Acadian province, comprising Nova Scotia, New Brunswick and Newfoundland, the Lower Carboniferous is remarkably like that of Great Britain. In Nova Scotia, the Horton sandstone corresponds to the Calciferous of Scotland, and contains thin seams of coal; it is followed by the Windsor stage of marine limestone (= the British Scaur, or Mountain, Limestone) which contains beds of gypsum. The presence of gypsum shows not only that occasionally bodies of sea-water were shut off in closed lagoons, but also that an arid climate prevailed in the region.

In eastern Pennsylvania the Lower Carboniferous has a maximum thickness of 4000 feet, but thins rapidly southward and westward. The Pocono is a thick, hard sandstone, which caps many of the mountain ridges; it follows down the Appalachian line, thinning as it goes. The area of maximum sedimentation may have received largely continental deposits. Under different local names, the Pocono extends to eastern Kentucky. The Mauch Chunk shales form the remainder of the Lower Carboniferous in northeastern Pennsylvania, where the thickest portion not improbably represents a great delta, and the prevalence of sun cracks is indicative of an arid climate, such as probably prevailed in Nova Scotia. In Maryland, West Virginia, and Kentucky, part of this series is composed of a marine limestone (Greenbrier, Newman). The Lower Carboniferous of Virginia contains some workable beds of coal, the " false Coal Measures." Ohio was a region of slow sedimentation; here the Lower Carboniferous is formed by the Waverly series, which is divisible into seven stages, some of them carrying marine faunas, and is only 700 feet thick. In the Michigan bay were deposited the sandstones, grits, and shales of the Marshall series, followed by shales with some limestone and gypsum, whence we may infer that the bay was for a time converted into a salt lake and that the climate was dry, as in Nova Scotia and Pennsylvania. The bay was soon again invaded by the sea, for a marine limestone overlies the gypsiferous beds.

Southwest of these more or less completely enclosed bays, the Interior Sea was clear and free from terrigenous material, so that in it were deposited great masses of limestone (1500 feet maximum thickness) formed from a most luxuriant growth of corals, brachiopods and crinoids. In the Central States many different stages and substages may be distinguished in these limestones, and evidences are not wanting of fluctuating shore-lines. The *Kinderhook* extends farther north than the *Osage*, while the *St. Louis* sea again extended northward. The Osage series is remarkable for the extraordinary abundance and variety of its crinoids, unequalled, perhaps, in the world. This produces a peculiar facies which is nearly confined to the Mississippi valley, but was extended to the southwest, into New Mexico. " But this condition appears not to have invaded other western parts of the Mississippian sea, where I believe, under uniform conditions, the Kinderhook faunas persisted through Burlington and Keokuk [*i.e.* Osage] time without feeling, save in a subordinate degree, the influences which helped to differentiate the early Mississippian faunas of the Mississippi Valley." (Girty.)

The Lower Carboniferous was brought to a close by a very widespread upheaval, which removed nearly the whole Interior Sea and resulted in a very general unconformity between the Mississippian and the overlying Pennsylvanian. In only a few areas does there appear to be a continuity of sedimentation between them and in some of these there is reason to think that the conformity is apparent, not real. The Kaskaskia faunas are entirely wanting in the western portion of the Mississippian sea, but this should probably be interpreted to mean that the uppermost beds of the Lower Carboniferous were stripped away by denudation during the interval between the two formations, when so much of the continent was land. If this was not the case, the upheaval must have affected the western portion of the Interior

Sea considerably before it drained the central and eastern portions. Some folding accompanied the upheaval in certain areas, as in Iowa and northeastern Pennsylvania.

After a time of quite prolonged erosion over a great part of the continent, a new series of events was inaugurated. In Pennsylvania an orogenic movement took place, raising the low-lying mud flats of the Mauch Chunk, but down-warping their eastern border into a long and narrow trough which extended to Alabama, and in this trough a rapid sedimentation occurred, forming great bodies of gravel and sand, the *Pottsville* stage. In southwestern Virginia there is, apparently at least, continuity of deposition from the Mauch Chunk into the Pottsville. During the latter epoch the trough continued to subside under its accumulating load of sediment and, from time to time, to transgress westward, in which direction the higher members of the series extend. The subsidence of the trough was intermittent, and fresh-water peatbogs were established upon its surface, resulting in the formation of coal-beds, especially in the southern Appalachians, where the Pottsville is 6000 feet thick. The water which filled the trough varied in character; in the middle of the epoch marine faunas extend as far north as central West Virginia, but the northern portion appears to have been an estuary. Over the Mississippi valley the Millstone Grit and various sandstones with local names represent the Pottsville, but the evidence of the fossil plants shows that the Millstone Grit is not a single bed, but that in different places it corresponds to different levels of the Pottsville. At the end of the age the sea covered much of the Mississippi valley, perhaps connecting with the Michigan basin, though this is doubtful, but it spread over western Pennsylvania and a great part of Ohio. "It is highly probable that by this time the Pottsville sea swept across the Cincinnati arch in southern Kentucky and Tennessee so as to connect with the interior region." (D. White.)

In Arkansas the Pottsville, as shown by the fossils, is represented by a series of limestones, shales, and sandstones, which have till very lately been placed in the Lower Carboniferous, and so many

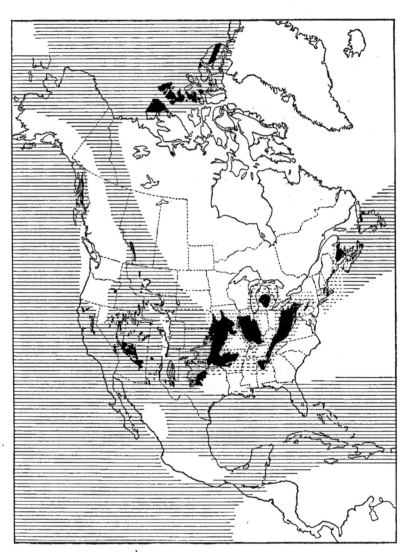

FIG. 274. — Map of North America in the Upper Carboniferous. Black areas = known exposures; white areas = land, or unknown; horizontal lines = sea; dotted area = Permian of Kansas and Texas

localities of Pottsville marine animals have been found over the West as to make it probable that the sea extended to Nevada (Girty), while in that state the same stage appears to be represented by beds which contain a fauna transitional between the Lower and Upper Carboniferous.

Though coal accumulation had begun in the Pottsville and even in the Mississippian, the time of its formation on the greatest scale was in the second half of the Upper Carboniferous. Vast areas of low-lying swamps bordered the Interior Sea, and in these vegetation flourished most luxuriantly. A very slow subsidence, often intermittent, allowed great thicknesses of vegetable material to accumulate, but frequently a more rapid sinking brought in the sea, or bodies of fresh water over the bogs, killing the trees which grew there. We cannot yet determine how far the different coal regions represent separate basins, and how far their separation is due to the subsequent removal of connecting strata, but even in connected areas we find great differences in the nature and thickness of the beds. This indicates that oscillations of level of different amounts took place in particular parts of the same basin. Thus, in one portion may occur a coal seam of great thickness, divided into two or more layers by exceedingly thin " partings " of shale. As we trace the coal seam in the proper direction, the partings gradually grow thicker, until, perhaps, they become strata, that intervene between very distinct and quite widely separated coal seams, each of which is continuous with the corresponding portion of the thick seam. The meaning of such a structure is, that while one part of the bog subsided very slowly, permitting the almost uninterrupted accumulation of vegetable matter, other portions sank more rapidly and were inundated with water, which deposited mechanical sediments on the surface of the submerged bog.

Hardly more than 2% of the thickness of the coal measures consists of workable coal. The strata are mostly sandstones, shales, clays, and in some regions limestones, interstratified with numerous seams of coal of very varied thicknesses. This alternation of coal with mechanical deposits does not necessarily, or even probably, imply oft-repeated oscillations of level, but may be explained by assuming a general, slow, but intermittent subsidence. After each submergence, we may suppose, the movement was nearly or quite arrested, and the shallow water was filled up with sediment, until a bog could again be formed. Doubtless, movements of elevation also occurred at times, but the general movement was downward. In the Nova Scotia field are 76 distinct coal seams, each of which implies the formation of a separate bog. Beneath most coal seams occurs what miners call the " seatstone " or " underclay," which is ordinarily a fire-clay, or it may be siliceous, but is always evidently an ancient soil. The underclay is filled with fossil roots, from which often rise the stumps of trees that penetrate the coal seam, or may even extend many feet above it. The rock which lies on a coal seam is usually a shale, stained. black by organic matter, but may be a sandstone or even a limestone, according to the depth of water over the submerged bog.

That coal is of vegetable origin is no longer questioned. Such a mode of origin is directly proved by microscopical examination, which shows that even the hardest anthracite is a mass of carbonized but determinable vegetable fibres. On the other hand, there has been much difference of opinion concerning the way in which such immense masses of vegetable matter were brought together. Much the most probable view is, that the coal was formed in position in great peat-bogs, added to, no doubt, by more or less drifted material. The evidence for this view is to be found: (1) In the great extent and uniform thickness and purity of many coal seams, which we cannot account for in any other way. Had the vegetable matter been largely drifted together, it must have been contaminated with sediment and could not have been spread out so evenly over great areas. This objection to the " driftwood theory " becomes all the stronger when it is remembered that the process of converting vegetable matter into coal greatly reduces its bulk, a given thickness of coal representing only about 7% of the original thickness of vegetable substance. Thus a 20-foot seam of coal implies the accumulation of nearly 300 feet of plants, and it is highly improbable that such a mass could have been evenly spread as drift over hundreds (or even thousands) of square miles, without a large admixture of mud or sand. (2) The very general presence of the underclay beneath coal seams points to the same conclusion. An underclay, as we have seen, is an ancient soil, and is of just the same character as that which we find under such modern peat-bogs as the Great Dismal Swamp.

The subsidence of the bogs and the deposition of sediments upon them gradually built up the great series of strata which are called the coal measures. The peat was thus subjected to the steadily increasing pressure of the overlying masses, which greatly aided in the transformation of the vegetable accumulations into coal. Where the coal measures have been folded, the still greater pressure, aided by heat, and perhaps by steam, has resulted in the formation of anthracite. The greater number of the Carboniferous bogs appear to have been covered by fresh water, though some were coast swamps, extending out into brackish or even salt water.

On the other hand, it must be admitted that there are not a few cases to which the peat-bog theory does not apply, for example, to the small coal basins in the central plateau of France. The famous basin of Commentry is, according to Fayol, explained without difficulty by regarding it as a delta formation in a large lake. The coarse gravel was deposited in inclined beds immediately at the mouth of a swift stream, the finer sediment was carried farther out and the floating masses of vegetation still farther. The vegetable matter became water-logged and sank to the lake-bottom, where it was free from sediment. Such a case, however, has little bearing upon the great coal-fields.

The workable coal-fields of North America, belonging to the Carboniferous system, are found in several distinct areas some of which were doubtless separate basins of accumulation, while others have become disconnected by denudation.

(1) In the Acadian province the coal measures occur in the island of Cape Breton, Nova Scotia, and New Brunswick; in Nova Scotia they are of immense thickness, 7000 feet, with 6000 feet of underlying conglomerate. The Coal Measures of Nova Scotia, the Upper Carboniferous, strongly resemble the type of development in central England. The immensely thick basal conglomerate is the Millstone Grit. A second basin of this province is near Worcester (Mass.), and a third extends through Rhode Island into southeastern Massachusetts The latter basins are metamorphic and yield a very hard anthracite.

(2) The great Appalachian field has an area of more than 50,000 square miles. It covers most of central and western Pennsylvania, eastern Ohio, western Maryland and Virginia, and West Virginia, eastern Kentucky and Tennessee, to northern Alabama. In this field the measures are thinner than in Nova Scotia; the beds are thickest along the Appalachian shore-line, about 4000 feet in western Pennsylvania and 6000 in Alabama, thinning much to the westward.

(3) In Michigan the measures are only about 300 feet thick, and were doubtless laid down in an isolated basin.

(4) The Indiana-Illinois field, which extends into Kentucky, is from 600 to 1000 feet thick.

(5) The Iowa-Missouri field extends southward around the Palæozoic island of southern Missouri into Arkansas and Texas. In Arkansas the Carboniferous system attains a greater thickness than anywhere else in North America, and all but an insignificant amount of this is Pennsylvanian.

The two latter fields are separated by a very narrow interval, and almost certainly were once continuous; the Indiana-Illinois field was probably also connected with the Appalachian area across western Kentucky and Tennessee.

As the coal measures are traced westward into Kansas, Nebraska, and adjoining states, we find them dipping beneath strata of a very much later date. When they once more return to the surface, as in the Rocky Mountain region, they appear under an entirely new aspect, being here altogether marine and containing no coal.

After the Pottsville age and during the formation of the Coal Measures, the Interior Sea was greatly restricted in the Mississippi valley by the broad, surrounding fringe of swamps and bogs, which the sea periodically invaded. The same succession of great swamps followed the Appalachian line into northeastern Pennsylvania and probably into southern New York. In eastern Pennsylvania the sea rarely came in during the formation of the coal measures, but once, at least, penetrated to Wilkes-Barré. Westward the Interior Sea probably did not extend to Nevada, as that of Pottsville time had done, but ended farther east along a line not yet determined. A shore-line in Colorado is indicated by the generally sandy and conglomeratic character of the Pennsylvanian rocks in that state, which have an eastern type of fauna. On that account, Girty has " tentatively assumed that the line of division between the Eastern and Western provinces passes through western Texas, central or eastern New Mexico, western Colorado, and so on upward, in a northwesterly direction, following nearly the trend of the Rocky Mountains."

The northwestern arm of the Interior Sea was shifted eastward from the position it had occupied in the Lower Carboniferous, and apparently joined the Arctic Ocean instead of the Pacific, submerging nearly the whole of Alaska, except a broad belt on the Pacific side, this belt of land continuing southward through British Columbia to Oregon. The area of the sea was somewhat diminished in southern and western Mexico and in Central America.

The Carboniferous period in North America was, on the whole, a time of tranquillity, with oscillations of level and shifting of the boundaries of land and sea from time to time, such as have been described in the foregoing pages. A very general upheaval of the continent brought the Mississippian to a close and the succeeding time of erosion was followed by some folding in the Appalachian region, the formation of the Pottsville trough and the renewed transgression of the sea. Volcanic action appears to have been confined to the Pacific coast region; the Lower Carboniferous of British Columbia is largely made up of volcanic material, and vulcanism was manifested in the Upper Carboniferous of the coast from Alaska to California. In northwestern Kentucky and southern Illinois the Carboniferous rocks are cut by dykes, but these may have been formed at a long-subsequent time.

Foreign. — In *Europe* the Carboniferous system is developed in a very interesting way. In the western and central parts of the continent (and in Great Britain) the succession of strata is very similar to that of the eastern half of North America, while in Russia it has more analogy with the western half of our continent. The changes of level which opened the period converted much of the Devonian sea-bed into land, but at the same time the sea broke in over many of the closed basins in which the Old Red Sandstone had been laid down. From the west of Ireland to central Germany, a distance of 750 miles, stretched a clear sea, free from terrigenous sediments, in which flourished an incredible number of corals, crinoids, and other calcareous organisms. From their remains was constructed an immense mass of limestone, having a thickness of 6000 feet in the northwest of England and of 2500 feet in Belgium. Above this great " Mountain Limestone," as it is called in England, come the coal measures. In Scotland the limestone is replaced by shore and shallow-water formations, such as sandstones, with some coal. In the southwest of England and east of the Rhine in Germany, the Lower Carboniferous is represented, not by a limestone, but by a series of sandstones and slates, called the *Culm*, with the coal measures above. In Russia the order of succession is reversed, the productive coal beds being below and the great bulk of the limestone above, but there is some productive coal interstratified in the marine limestones of the Donjetz basin in the south. This younger Carboniferous limestone is principally composed of shells of Foraminifera. Great areas of southern and eastern Asia are covered by this limestone, which is also largely developed in western North

America, extending as far east as Illinois. In southern Europe, Spain, the south of France, the Alps, and the Balkan peninsula, the Lower Carboniferous is partly limestone and partly Culm, while the Upper is largely made up of the foraminiferal limestone associated with clastic rocks. In the Arctic Sea, Nova Zembla, Bear Island, Spitzbergen, and Greenland have Upper Carboniferous limestones.

The following table, from Kayser, displays the relations of the Carboniferous beds in eastern and western Europe: —

	LITTORAL AND LACUSTRINE FACIES	MARINE FACIES	
Upper Carboniferous.	Productive Coal Measures (Western Europe).	Younger Carboniferous or Fusulina Limestone (Russia, etc.).	
Lower Carboniferous.	Productive Coal Measures (Russia, etc.).	Lower Carboniferous Limestone (Western Europe).	Culm (Germany).

In western Europe the Carboniferous period did not run such a tranquil course as in North America, but was broken by disturbances, of which the greatest were at the close of the Lower Carboniferous epoch, when the rocks were folded and upturned over extensive regions. These movements were accompanied and followed by volcanic outbursts, especially in Scotland, France, and Germany, and great eruptions occurred in China at the end of the period.

In *Asia* are large areas of Lower Carboniferous limestone and Culm, and of the Upper Carboniferous both the foraminiferal limestone and productive coal measures. China is one of the richest countries in the world in supplies of coal. Foraminiferal limestones of the Upper Carboniferous are found in Japan, Borneo, and Sumatra.

Africa. — Carboniferous limestones are found in Morocco and the Sahara and Egypt. In the southern part of the continent no marine rocks of the period are known. In Cape Colony the Witteberg quartzites, which overlie the Devonian, have Lower Carboniferous plants, and in the Zambesi district near the east coast is a coal basin of Upper Carboniferous age, which has a flora like that of the higher Coal Measures of Europe.

According to Frech, the Carboniferous of *Australia* is confined to the Lower division and appears in the eastern half of the continent, and in Tasmania. In the second half of the period was a time of elevation and erosion.

In *South America* the Carboniferous is not nearly so extensive as the Devonian; the Lower Carboniferous is principally composed of sandstones, which in Argentina contain plants so similar to those of South Africa and Australia as to indicate the probability of a land connection between these continents; limestones of this date have been reported from Chili. The Upper division, largely of limestones, has been found in Peru, Bolivia, and Brazil; in the latter it has a great extension in the Amazon valley and belongs to the uppermost part of the system.

Climate. — The striking uniformity of the climate during the Carboniferous is indicated by the distribution of the fossils, more especially of the plants, which are almost the same in the Arctic and Tropical regions. The formation of coal in vast peat-bogs does not imply a tropical climate, but rather conditions of moisture and moderate temperature. In the Lower Carboniferous, aridity prevailed in northeastern America, and in the Upper, gypsum and rock-salt were formed east of the Ural Mountains. There is some evidence of an American ice-epoch at the end of the Lower Carboniferous.

CARBONIFEROUS LIFE

The life of this period is thoroughly Palæozoic and continues along the lines already marked out in the Devonian, but there are some notable changes and advances which look toward the Mesozoic order of things.

Plants. — The Carboniferous vegetation is of very much the same character as that of the Devonian, but owing to the peculiar physical geography of the times, the plants were preserved as fossils in a much more complete state and in vastly larger numbers. The flora is composed entirely of the higher Cryptogams and the Gymnosperms, no plant with conspicuous flowers having come into existence, so far as we yet know. By far the most abundant of Carboniferous plants are the *Ferns* (Filicales) which flourished in multitudes of species and individuals, both as tall trees and as lowly, herbaceous plants. Many of these ferns cannot yet be compared with modern ones, because the organs necessary for trustworthy classification have not been recovered, and such are named in accordance with the venation of the leaves. In other cases the comparison with existing ferns may be definitely made, and these remains show that many of the modern families (*Marattiaceæ, Ophioglossaceæ*, etc.) had representatives in the Carboniferous forests and swamps.

Even more conspicuous, though much less varied, were the *Lycopods* (Lycopodiales) the remarkable character of which has been elucidated by the long-continued and laborious efforts of many investigators. While the Ferns have remained an important group of plants to the present time, the Lycopods have dwindled to a few insignificant herbaceous forms, but in Carboniferous times they were the abundant and conspicuous forest trees, at least of the swampy lowlands. One of the most characteristic of these trees was *Lepidodendron* (Pl. X), of which many species have been found in the coal measures. These great club-mosses had trunks of 2 or 3 feet in diameter and 50 to 75 feet high, which possessed the remarkable quality, for a Cryptogam, of an annual growth in thickness. At a considerable height above the ground the trunk divides into two main branches, each of these again into two, and so ón (dichotomous division). The younger parts of the tree are covered with long, narrow, stiff, and pointed leaves, while the older parts are without leaves, which have dropped off, making conspicuous scars, arranged in spiral lines around the

2 S

PLATE X. — CARBONIFEROUS VEGETATION

Lepidodendron, central tree with cones. *Sigillaria*, each side of middle, with leafy trunks. *Calamites*, right side. *Cordaites*, left side on raised ground. *Cycadofilices*, fern-like growth in foreground.

stem. At the ends of the twigs in some species, or on the sides of the trunk and larger branches, in others, are found the spore-bearing bodies, which have much the appearance of pine-cones. The stem was, to a large extent, filled with loose tissue and had only a relatively small amount of wood.

Another very characteristic and abundant tree is *Sigillaria* (Pl. X); it is closely allied to Lepidodendron, but has a very different appearance. The trunk is quite short and thick, rarely branching, and with a pointed or rounded tip, much as in the great Cactus; the leaves are similar to those of Lepidodendron, but are arranged between vertical ridges. Sigillaria also possessed the power of annual increase in diameter. Both Lepidodendron and Sigillaria are provided with branching rhizomes, or underground stems, which carry finger-like appendages inserted into pits. Before the nature of these rhizomes was understood, they were regarded as distinct plants and named *Stigmaria*.

A third group of Cryptogams, the *Equisetales*, or Horsetails, were of great importance in the Carboniferous forests. The *Calamites* were decidedly superior to the existing horsetails, not only in size, but in many features of organization as well. These plants had tall, slender stems divided by transverse joints, with a soft inner pith, surrounded by a ring of woody tissue, which grew annually in thickness. The shape and arrangement of the leaves differ much in the various genera, and even in different parts of the same plant; for example, they are needle-like in *Astrophyllites*, while in *Annularia* they are broad and at the base united into a ring around the stem, but some species of Annularia, at least, are probably merely the branches of larger calamites. The shape, size, and position of the spore-bearing organs likewise differ in the different genera, but often resemble those of the modern horsetails. The base of the stem tapers abruptly, and is either connected with a horizontal rhizome or gives off a bundle of roots. Fragments of calamite stems are among the commonest fossils of the coal measures.

The three preceding groups of Cryptogams all have representatives in the modern world, and one of them, the Ferns, is still abundant and varied. In addition to these, Carboniferous vegetation had two other cryptogamic classes of great interest, which are now extinct, and are not known to have passed beyond the Palæozoic era. Of these, the first is the class *Sphenophyllales*, a group of very slender, probably climbing and trailing plants, with small leaves varying in shape in different plants and different parts of the same plant. Some of the leaves, which are always small, are wedge-shaped, others are divided and others again are narrow and simple. The great interest of the class lies in the fact that it is intermediate between the horsetails and club-mosses, and doubtless its Carboniferous representatives were the survivors of an ancient group which was ancestral to both club-mosses and horsetails.

Even more remarkable is the class *Cycadofilices*, which was extremely abundant in the Carboniferous forests and swamps, and which affords the long-sought transition between the flowerless and the flowering plants, connecting, as its name implies, the ferns and cycads. In external appearance of stem and foliage these plants most resembled tree-ferns.

The *Flowering Plants* are still represented only by the Gymnosperms, of which the dominant group is the *Cordaiteæ* (see Pl. X), which were slender, very tall trees, " with trunks rising to a great height before branching, and bearing at the top a dense crown, composed of branches of various orders, on which simple leaves of large size were produced in great abundance." (D. H. Scott.) The centre of the trunk was occupied by a large soft pith, and the leaves, with their parallel venation resembling those of lilies and grasses, were long, broad in most species, narrow in others, and either sharply pointed or bluntly rounded. The Cordaiteæ had affinities with each of the three existing orders of Gymnosperms, the Cycads, Conifers, and Gingkos, but is not referable to any of them. The three orders named may all have existed in the Carboniferous, but this is not definitely known.

The Carboniferous flora is merely the Devonian flora somewhat advanced and diversified, and the forests were of the same gloomy, monotonous character as before. The wide distribution and uniform character of this flora are very remarkable; we find the same or nearly allied species of plants spread over North America, Europe (even in the polar lands, like Spitzbergen and Nova Zembla), Siberia, China, the Sinai peninsula, Brazil, Australia, and Tasmania.

Foraminifera. — For the first time these animals assume considerable importance in the earth's economy. Many genera which are still living had representatives in the Carboniferous seas, but the most conspicuous and abundant is the extinct *Fusulina* (XII, 1, 1 a), a very large kind, with shells resembling grains of wheat in size and shape. This genus is especially developed in the Upper Carboniferous, while *Schwagerina* characterizes the uppermost part of the system. In the Salem limestone of Indiana, a well-known building stone, of the Mississippian series, *Endothyra* (XI, 1) is abundant.

Sponges are common, though rarely found in good preservation.

Cœlenterata. — Corals were abundant, and contributed largely to the limestones; the genus *Lithostrotion* (XI, 2), which is peculiar to this period, plays a very prominent part. *Lophophyllum* is found in the Upper Carboniferous.

Echinodermata make up an exceedingly important part of the Carboniferous marine fauna. The *Cystoids* have disappeared, but the *Blastoids* have developed in great numbers, and are highly characteristic of the Carboniferous limestones. As the group is entirely extinct and does not pass beyond the Carboniferous system, its structure has much that is problematical about it. The delicate, symmetrical body, or *calyx*, which is carried on a short stem, is composed of a small, definite number of plates, and has five " pseudo-ambulacral " areas, which look much like the ambulacra of a sea-urchin. In exceptionally well-preserved specimens numbers of delicate pinnules are attached to these areas. The must abundant genera are *Pentremites* (XI, 4–5) and *Granatocrinus*.

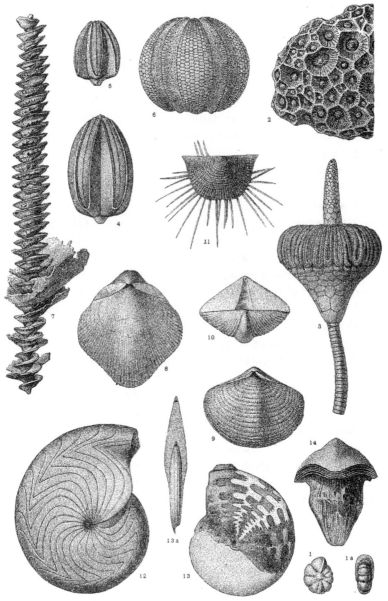

PLATE XI. — LOWER CARBONIFEROUS FOSSILS

Figs. 1, 1a, *Endothyra baileyi* Hall, × 9, St. Louis, side and end views. 2, *Lithostro-tion canadense* Castelnau, × ½, St. Louis. 3, *Eutrochocrinus christyi* Shumard, × ½, Burlington. 4, *Pentremites elongatus* Shumard, × 1, Burlington. 5, *P. conoideus* Hall, × 1, Keokuk. 6, *Melonites multiporus* Norwood and Owen, × ½, St. Louis. 7, *Archi-*

All other Echinoderms of the Carboniferous seas were utterly insignificant as compared with the *Crinoids*, which reach their culmination of development in this period: more than 600 species have been described from the Carboniferous limestones of North America alone. Certain localities, such as Burlington (Ia.) and Crawfordsville (Ind.), are famous for the vast numbers and exquisite preservation of their fossil sea-lilies. The crinoid remains occur in such multitudes that in many places the limestones are principally composed of them; in such places they must have covered the sea-bottom like miniature forests. But this extraordinary abundance is not general over North America, but characterizes the Central States only and Mississippian time, especially the Osage. All the Carboniferous Crinoids, like those of the earlier periods, belong to the extinct order *Camerata*, none of which passed over into the Mesozoic era. Of the long list of Crinoids found in the rocks of this system may be mentioned *Actinocrinus, Platycrinus, Rhodocrinus, Eutrochocrinus* (XI, 3), *Onychocrinus, Æsiocrinus,* and *Eupachycrinus* (XII, 2).

The *Echinoids*, or sea-urchins, are still far less abundant than the Crinoids, but they are much more numerous and varied, and of larger size than they had been before; some, indeed, are as large as any sea-urchins that are known from any period. The Carboniferous sea-urchins are, like those of the preceding periods, members of the ancient and now extinct subclass, *Palæechinoidea*, and the commonest genera are *Melonites* (XI, 6), *Oligoporus*, and *Archæocidaris*. In addition to these should be noted the presence of the modern subclass, *Euechinoidea*, as the ancestor of the still existing genus *Cidaris* is reported from the Carboniferous.

The first known *Holothuroidea*, or sea-cucumbers, date from this period.

Arthropoda. — The Trilobites have become rare and are soon to die out altogether; most of the species belong to the peculiarly

medes wortheni Hall, + ½, Keokuk. 8, *Spirifer grimesi* Hall, × ¼, Burlington. 9, *Reticularia pseudolineata* Hall, × ½, Keokuk. 10, *Syringothyris subcuspidatus* Hall, Keokuk. 11, *Productus magnus* Meek & Worthen, × ¼, Keokuk. 12, *Brancoceras ixion* Hall, × ½, Kinderhook. 13, 13*a, Prodromites præmaturus* Smith and Weller, × ½, side and end views. 14, *Petalodus linguifer* Newb. and Worth., × ½, Kaskaskia.

Carboniferous genera *Phillipsia* (XII, 21) and *Griffithides*, but the
Devonian *Proetus* still persists. The *Eurypterids* continue, even
into the coal measures, where they lived in the fresh-water swamps,
but they cannot compare in size or numbers with the great Devonian forms. The horse-shoe crabs are represented by *Prestwichia*.
Phyllopods and *Ostracods* are abundant, and in the coal measures
are found crustaceans formerly incorrectly referred to the Decapods,
which they resemble; of these *Anthracopalæmon* is the best-known
genus.

Centipedes and *Scorpions* are much commoner than in the
Devonian, and the first of the true *Spiders* are found here. Insects
likewise show a great increase in numbers, though the Orthopters
and Neuropters are still the principal orders represented. Many
of the Carboniferous insects are remarkable for their great size,
some of them measuring 30 inches across the extended wings,
and more remarkable is the fact that several insects of this period
had three pairs of wings, corresponding to the number of legs.
The character of the vegetation has a very direct influence upon
insect life, and the monotonous, flowerless Carboniferous forests
could not have supported butterflies, bees, wasps, ants, or flies.
No insects of these groups have been found in the rocks of that
system, and it is not yet certain whether even beetles were then
in existence.

The land life of the Carboniferous seems to be very much more
varied and luxuriant than that of the Devonian, and it probably
was so in reality. It must be remembered, however, that the immense development of fresh-water and marshy deposits in the
Carboniferous was much more favourable to the preservation of
such fossils than any conditions that the Devonian had to offer.
Part, at least, of the striking difference in the terrestrial fossils of
the two periods is to be accounted for in this way.

---

subquadrata Shumard, × ½. 13, *Aviculopecten occidentalis* Shumard, × ½. 14, *Bakevellia parva* Meek and Hayden, × 2, PERMIAN. 15, *Pleurophorus subcuneatus* M. and
H., × 1, PERMIAN. 16, *Pseudomonotis hawni* M. and H., × ½, PERMIAN. 17, *Bellerophon
percarinatus* Conrad. × 1. 18, *Pleurotomaria sphærulata* Conr., × ½. 19, 19a, *Strapa-
rollus pronodosus* M. and W , × ½. 20, *Waangenoceras cumminsi* White, × ⁴/₅. 20a, *The
same*, a suture line, PERMIAN. 21, *Phillipsia major* Shumard, × ½.

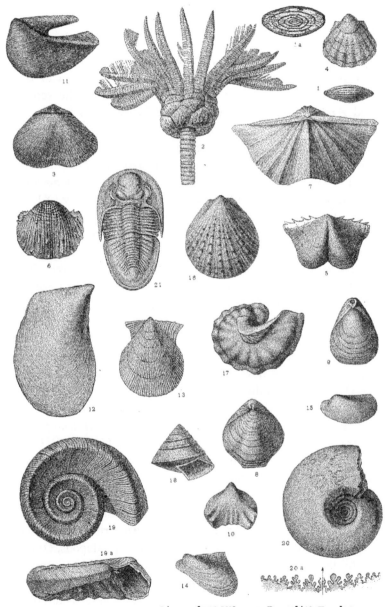

PLATE XII. — UPPER CARBONIFEROUS AND PERMIAN FOSSILS

UPPER CARBONIFEROUS. Fig 1, *Fusulina cylindrica* Fischer, × ⁵/₄. 1a, *F. secalica* Say, × ⁵/₂, longitudinal section. 2, *Eupachycrinus verrucosus* White and St. John, × ½. 3, *Derbya biloba* Hall, × ½. 4, *Meekella striatocostata* Cox, × ½. 5, *Chonetes verneuiliana* Norwood and Pratten, × 2. 6, *Productus costatus* Sowerby, × ½. 7, *Spirifer cameratus* Morton, × ½. 8, *Seminula argentea* Shephard, × ½. 9, *Dielasma bovidens* Morton, × ½. 10, *Pugnax uta* Marcou, × 2. 11, *Monopteria longispina* Cox, × ½. 12, *Myalina*

The **Bryozoa** become much more important than they had been before, and contribute materially to the formation of the limestones. A characteristic Carboniferous genus is the screw-shaped *Archimedes* (XI, 7), while *Fenestella* continues to be very abundant.

The **Brachiopoda** have undergone a marked diminution, as compared with those of the Devonian, though they are still very common.   Genera of long standing, like *Atrypa* and *Pentamerus*, have died out, but others, like *Chonetes* (XII, 5), *Spirifer* (XI, 8; XII, 7), and *Rhynchonella*, are still represented, but most important of all the Carboniferous genera is *Productus* (XI, 11; XII, 6), which has a very large number of species, among them *P. giganteus*, the largest known brachiopod. *Syringothyris* (XI, 10) and *Reticularia* (XI, 9) are allies of Spirifer; while *Meekella* (XII, 4) and *Derbya* (XII, 3) are extreme developments of the Strophomenoid stock, of Ordovician origin. The genus *Terebratula*, which became exceedingly abundant in the Mesozoic periods, has its beginning in the Carboniferous genus *Dielasma* (XII, 9), though we have already found the family represented in the Devonian and Silurian.

Mollusca. — The Bivalves are somewhat more abundant than in the earlier periods. Examples of these are *Aviculopecten* (XII, 13), *Monopteria* (XII, 11), and *Myalina* (XII, 12).   Of Gastropods, the same genera that occur in the Silurian and Devonian are continued into the Carboniferous, such as *Bellerophon* (XII, 17), *Euomphalus*, *Pleurotomaria* (XII, 18), *Loxonema*, *Platyceras*, with the interesting addition of the most ancient landshells yet discovered.   The genus *Conularia*, referred to the Pteropods, is common.   Among the Nautiloid Cephalopods, *Orthoceras* still persists, but this group reaches its acme in the number and variety of the coiled shells, many of which represent new genera, such as *Cycloceras*, *Trigonoceras*, etc. These Nautiloids have shells ornamented with prominent ridges or tubercles. The Ammonoids continue to be represented by Goniatites, but the Carboniferous forms of this group, such as *Brancoceras* (XI, 12) and *Prodromites* (XI, 13), display an advance over those of the Devonian in the greater complexity of their sutures, looking forward to the remarkable condition attained in Mesozoic times.

Vertebrata. — It is in this group that the most marked advances of Carboniferous life are to be observed, and the incipient stages of Mesozoic development are clearly shown. The extraordinary and bizarre Ostracoderms have become extinct, though the Arthrodirans continue into the coal measures.

The *Selachians* are numerous and varied, having developed so enormously that they give the Carboniferous fish-fauna a very different aspect from that of the Devonian. *Acanthodes* is a small shark covered with a dense armour of exceedingly minute square scales, and the fins are supported by heavy spines along their anterior borders. Another remarkable shark is *Pleuracanthus* (a Permian species is shown in Fig. 279), which has many features in common with the Dipnoi, such as the shape of the tail, the character of the pectoral fins, and the bones which form the roof of the skull, while the skin is naked. Isolated fin-spines and teeth show that many other kinds of sharks existed in the Carboniferous, in some of which the teeth were converted into a crushing pavement, adapted for a diet of shell-fish. (See Pl. XI, Fig. 14.)

The *Dipnoi* continue, though in diminished numbers, and their most prominent representative is the genus *Ctenodus*.

The *Crossopterygians* are much less abundant than in the Devonian; the commonest American genus is *Cœlacanthus*, which, though unmistakably a member of this group, has assumed the form of a bony fish, and looks much like a chub.

The *Actinopteri* are still represented only by the Ganoid cohort; these hold their own and even increase their numbers, many new genera replacing those of Devonian times. *Eurylepis, Palæoniscus, Eurynotus,* and *Cheirodus* are the best-known genera; they are all of moderate size and in appearance are not strikingly different from modern fishes.

The *Amphibians*, which we have seen some reason to believe existed in the Devonian, are of greatly increased importance in the Carboniferous. At the present time the Amphibia are represented by the dwarfed and specialized frogs and toads, newts and salamanders, which give but an imperfect notion of the structure of the extinct members of the class. The Carboniferous Amphibia all belong to the extinct order *Stegocephalia*, in which the skull is well covered with a roof of sculptured bones, and which are of moderate or small size, not exceeding seven or eight feet in length and mostly much smaller. The backbone is not ossified, the limbs are weak, the tail short and broad, and in many forms the belly is protected by an armour of bony scutes. An extraordinary number of genera of Carboniferous *Stegocephalia* are known, most of them like the Salamanders in shape, but some are elongate, slender, and of snake-like form. Examples are *Archegosaurus, Branchiosaurus, Dendrerpeton, Ptyonius,* and many others.

CHAPTER XXXI

THE PERMIAN PERIOD

THE name *Permian* was given by Sir Roderick Murchison in 1841 to a series of rocks which is very extensively developed in the province of Perm in Russia. In North America the Permian followed upon the Carboniferous with hardly a break, so that the distinction between the two systems must be made entirely upon the fossils, which change very gradually, by drawing a somewhat arbitrary line of demarcation. In various countries there is no general agreement regarding the upper boundary of the Carboniferous, and there are very great differences of opinion as to the correlation of the rocks of different continents, due chiefly to the want of an unbroken succession of fossils in any single area. The regions where the Permian beds are best known, Germany, England, and Russia, are not those which yield the normal facies of marine life, and hence any correlation with the latter is full of difficulties, and in the following table of the American Permian, no comparison is attempted with that of other countries. In consequence of these uncertainties, many geologists, especially in this country and in France, regard the Permian as a mere subdivision of the Carboniferous. Its relations with the overlying Triassic system are, however, nearly as close, and by some authorities it has been referred to the latter. The Permian is, on the whole, distinctively Palæozoic, but it has several features which mark it out as transitional to the Mesozoic.

PERMIAN SYSTEM

	W. TEXAS	E. TEXAS	PENNSYLVANIA
Guadalupian Series {	Capitan Stage	Double Mt. Stage	
	Delaware Mt. Stage	Clear Fork Stage	
		Wichita Stage	Dunkard Stage

Distribution of Permian Rocks

American. — Orogenic movements in the Appalachians had probably begun in the middle Carboniferous, as was seen in the folding which inaugurated the Pottsville trough, and toward the end of the Carboniferous there was in the low-lying Appalachian coal-field a slowly progressive movement of elevation, resulting in the draining and drying up of most of the region over which the peat-bogs had been extended. The movement spread east, north, and south, leaving in the middle of the region a smaller area in which the conditions of the coal measures continued very much as before. In the northern part of the Acadian province Permian beds overlie the coal measures in Prince Edward Island, Nova Scotia, and New Brunswick. These beds are soft red shales and sandstones, which were laid down in closed basins, not in the sea. In Pennsylvania, Ohio, West Virginia, and Maryland, the Permian beds follow directly and without any break upon the Monongahela stage of the coal measures: they were formerly called the Upper Barren Measures, and consist of 1000 feet of sandstones and shales with some limestone and a few seams of coal. The character of these beds is entirely like that of the coal measures, to which they were once referred, and their reference to the Permian is due to the marked change which had come over the vegetation. South of West Virginia no Permian beds have been found in the Appalachian area, owing to the elevation of this part of the region at the close of the Carboniferous, but the Permian occurs in Illinois, in what appears to be a stream-channel cut in the coal measures.

As we proceed westward and southward through Missouri into Nebraska, Kansas, and Texas, we find the Permian assuming much greater importance, and becoming more and more prominently developed in extent and thickness. A study of this region reveals the fact that only a part — the lower — of the Permian is developed in the Acadian and Appalachian areas. At the end of the Lower Permian the entire series of the coal measures east of the Mississippi River was elevated and the deposition of strata apparently ended, though there is no way of determining exactly when this elevation took place, nor how great a thickness of beds has been removed by denudation since the upheaval. In the region beyond the Mississippi the Permian beds thicken southward, attaining in southern Kansas a thickness of 2000 feet, and in Texas of more than 5000 feet. The mountains of Oklahoma, which may have been raised late in the Carboniferous or early in the Permian, separate the Texas and Kansas areas.

During the greater part of the Permian period the geographical state of North America was somewhat as follows. Except for the coastal plain on the Atlantic and Gulf of Mexico, the eastern portion of the continent had very much its present limits, though the position of the eastern and southern coasts cannot be determined. The coastal plain is deeply buried under deposits which are much younger than the Permian, and the continent may have extended farther into the Atlantic than at present, or the ocean may have extended more over the land. The Interior Sea was greatly changed both in extent and character from what it had been in the Upper Carboniferous, as is shown by the nature and distribution of its sediments. From most of the Mississippi valley the sea had withdrawn entirely, but still extended, as shallow and shifting waters, into southeastern Nebraska and Kansas, possibly into Iowa, and from eastern Kansas the line ran westward and southwestward across Oklahoma far into Texas. In the latter part of the period, lagoons were cut off from the sea and converted into salt and bitter lakes in which the salt and gypsum of Kansas and the gypsum of Oklahoma and Texas were precipitated. Occasionally the sea broke into these lakes, bringing a marine fauna with it for a short time.

The *Wichita* beds of Texas have two very distinct facies; in the north they are made up chiefly of fine red clays, with some beds of sandstone, conglomerate and impure limestone. The clays are principally river deposits made in a delta, or along a very flat coast, but with marine conditions at intervals. Passing southward, these beds gradually merge into marine limestones which were originally named the *Albany* stage and placed at the top of the Coal Measures. The succeeding *Clear Fork* beds, which are chiefly clays like those of the Wichita, but cut by many channels filled with cross-bedded sandstones, extend southward over the marine limestones of the Albany facies, but even in the north thin layers of limestone containing marine fossils are indicative of transgressions of the southern sea. The *Double Mountain* beds are largely the deposits of a salt lake and contain much gypsum; without any marine fossils, though in Oklahoma beds of a corresponding horizon have a scanty fauna between and above the gypsum layers.

Westward from Texas, the inland Permian sea extended over northern Arizona into southern Utah, where the beds are sandy shales, with much gypsum. The sea continued northward through eastern Utah, western Colorado, and probably east of the Rocky Mountains also, to the Black Hills and central Wyoming, forming an island in central Colorado. This immense body of water, or perhaps series of smaller bodies, was land-locked and salt, and in it were formed the characteristic " Red Beds " so widely distributed over the region mentioned, pointing to an arid climate. The Red Beds are not all Permian, however, and the rarity of fossils in them makes it often impossible to decide whether a given area of these beds should be referred to the Permian, to the subsequent Triassic, or to both. In southern Wyoming thin bands of sandstone and limestone in the Red Beds carry fossils very like those of the Kansas and Nebraska Permian, but the course of this marine invasion cannot yet be made out. Whether the Permian has been removed from the Great Basin by denudation, or never deposited over the greater part of it, is uncertain; but when this is determined, it will give the date of the upheaval of a land much of which had been submerged throughout the Palæozoic era.

Another and altogether different facies of the North American Permian is the purely marine development found in the mountains of western Texas and on the Pacific coast, especially in Alaska, where more than 6000 feet of marine Permian have been found in the region of the Copper River. The Pacific coast fossils differ strongly from those of the more eastern regions, and the fauna of western Texas shows affinity with the Mediterranean and Indian.

Foreign. — In *Europe*, as in North America, the Permian is developed in two very distinct facies. Southern Europe, Sicily, and the Alps have almost purely marine rocks and faunas, which resemble those of the Texas Guadalupian series. In central and western Europe the disturbances which, in many places, occurred at the end of the Carboniferous and, in those areas, produced a marked unconformity between the Carboniferous and Permian, resulted in the formation of a great inland sea, extending from Ireland to central Germany. In this great salt lake were deposited masses of red sandstones, shales, and marls, a predominant colour which strongly suggests desert conditions, though the coal-beds of France and Bohemia and central Germany might seem to contradict this. Occasionally the ocean broke into this closed basin, but the invading marine faunas soon perished. In Germany the Permian is in two very strongly marked divisions, the *Rothliegendes*, or Lower Permian, and the *Zechstein*, or Upper Permian, whence the period is, in that country, frequently called the Dyas. The very interesting discovery has quite lately been made in the lower Rothliegendes of Westphalia in Germany of undoubted and characteristic glacial moraines, resting upon a polished and striated pavement of Upper Carboniferous rocks. (G. Müller, 1901.) These moraines are not thick, about four feet, and suggest local rather than general glaciation, but in the Midlands of England are boulder beds which apparently show ice-action, though here the evidence is less conclusive, for no striated glacial pavements have been found, and the glacial hypothesis is not accepted by most geologists.

The Lower Permian of Europe is remarkable for the great masses of volcanic rocks, lava flows and tuffs, which it contains,

2 T

and which occur in Great Britain, France, Germany, and the Alps. This is in strong contrast to the corresponding American series, which gives no evidence of vulcanism.

Renewed disturbances at the end of the Lower Permian shifted the boundaries of the inland sea and changed its position, so that the Zechstein extends beyond the Rothliegendes and overlaps upon older rocks, and at the same time brought it into communication with the ocean, permitting the ingress of marine animals, but the conditions of life were evidently unfavourable, for the fauna is a curiously limited one, though a few species are individually abundant, and in striking contrast to the varied faunas of the truly marine facies. Later, the sea withdrew, leaving salt lakes, in which enormous bodies of rock-salt were formed in north Germany, including layers of the salts of potassium and magnesium, already referred to in a previous shapter. (See p. 225.) Smaller deposits of salt extend over central Germany to Russia. Upper Permian beds with gypsum occur in England, but not in France, which has only the Lower.

The Permian of Russia covers a very large area, but aside from typical Zechstein limestones on the Baltic coast, is quite different from that of western Europe; the principal area extends along the west side of the Ural Mountains to the Arctic Sea, and into Nova Zembla and Spitzbergen. The transition from the underlying Carboniferous is gradual, and the lower stage, the *Artinsk*, corresponds to the Wichita of Texas. Non-marine beds follow, which are again succeeded by a limestone with fossils like those of the German Zechstein, registering another invasion of the sea. The series ends with the *Tataric* stage, which is a sequence of red marls, passing upward without any apparent break into the overlying Triassic. The fossils of the Tataric stage are of peculiar interest and will be referred to again. Gypsum and salt in the non-marine beds testify to the aridity of the climate.

In *Asia* the two facies of the Permian are again met with. The marine facies occurs along a line which extends the course of the Mediterranean eastward, in Armenia, Persia, northern India,

Tibet, China, and the island of Timor. In the central Himalayas Upper Permian rests upon the upturned and eroded Lower Carboniferous, and is conformably overlaid by Triassic beds. The Salt Range of northwestern India has a very full succession of the Middle and Upper Permian, the *Productus* limestone, resting upon continental formations of the Lower Permian. At the base of the latter is a boulder clay of glacial or iceberg origin, which is an outlier of a great ground moraine that covers large areas in central India, and in places reaches a thickness of 2000 feet. The smoothed, striated and characteristically glacial pave ment upon which the boulder clay rests, has also been observed. It is certainly very remarkable to find glacial deposits formed on such a scale within the tropics and evidently at no great height above the sea-level. The boulder-clay (*Talchir*) forms the base of the *Gondwâna* system, a succession of continental deposits, with much coal, laid down by apparently unbroken sedimenta- tion, and including the Permian, Triassic, and Jurassic systems. Northern Asia has a widespread area of continental deposits, which are presumably Permian, and the Artinsk of Russia re- appears in central Asia, but the continent was mostly above sea- level and the great seas of the Carboniferous had withdrawn.

South *Africa* has a Permian development so closely parallel to that of India, that a direct land connection between the two regions may be confidently inferred. Extending almost across the continent from east to west in Cape Colony and Natal is the thick (1000 feet) glacial boulder clay of the *Dwyka*, or Lower Permian. Part of the Dwyka is shale of subaqueous origin, but most of it is a mass of boulders, striated and faceted, em- bedded in a fine unstratified matrix. (See Figs. 111, p. 231, 115, p. 233.) The formation extends northward, growing thinner on the way, into the Transvaal and perhaps into Rhodesia, and in the more northerly areas the underlying ice-worn pavement of older rocks is exposed in a state of wonderful freshness (see Figs. 70 and 275) entirely comparable to the recently abandoned beds of the shrinking glaciers of the Alps. In South Africa, therefore, the remarkable phenomenon of a continental glaciation in and near the tropics presents itself as well as in India, with the additional difficulty that in the former region the movement of the ice was *from the Equator polewards*.   Following the Dwyka boulder clay are the continental strata of the *Karroo* system, the lower part of which is Permian and in the Transvaal coal-bearing, and which corresponds to the Indian Gondwána in character, in geological date, and in the contained fossils.

FIG. 275. — Roche moutonnée, exposed by removal of Dwyka boulder clay.
— Riverton, Cape Colony. (R. B. Young)

The eastern portion of Australia and Tasmania, which had been land during the Upper Carboniferous, was largely sub-merged during the Permian, but Victoria on the south coast remained above sea-level and was glaciated, with the formation of the familiar ice-pavements and thick boulder clays, interstrati-fied with sandstones and shales. · The Upper Permian is coal-bearing, as it is also in New South Wales, where the glacial series is divided into two distinct parts by coal measures. The glacial beds occur interstratified with marine strata and some of the ice-made layers themselves contain marine fossils, which leads to the inference that the great blocks were deposited by icebergs rather than directly by glaciers. In Queensland and in northwestern Australia only the upper boulder succession is found, and the overlying coal measures are there also. The Australian ice movement was from south to north, as would naturally be expected in the southern hemisphere, but this makes the direction of movement in South Africa only the more inexplicable. In New Zealand the Permian, which is reported to be 7–10,000 feet thick, contains neither coal nor evidence of ice-action, but includes lava-flows.

In *South America* Permian beds of continental origin are found in Argentina and southern Brazil. In the latter is a glacial boulder clay, followed by a great series of strata which resemble those of the South African Karroo system.

The distribution of the Permian rocks and fossils leads to the inference that at that period the continents were so grouped as to form two great land masses, a northern including North America, most of Asia, and Europe, and a southern comprising India, equatorial and southern Africa, Australia, and South America. The existing southern continents were probably then connected by comparatively narrow land bridges across the site of the present Atlantic and Indian oceans. Between the northern and southern land-masses was the great continuous mediterranean, a sea which has been named " Thetys " (Suess) and of which we have found indications in Texas, Sicily, the eastern Alps, Asia Minor, northern India and southern China to the Pacific. That land communication was occasionally, at least, established between the northern and southern lands is evident from the distribution of the land animals and plants of both regions.

Climate. — The plain and obvious inferences from the character of the Permian rocks are so remarkable and inexplicable that they were long received with incredulity and they offer a series of fascinating problems for which no solution can yet be found. The earliest Permian in the southern hemisphere was a time of vast glaciation and of rigorous climate, as is convincingly shown by the boulder clays and ice pavements of Australia, South Africa, and South America. In the northern hemisphere the glaciation was extensive in peninsular India, apparently local in Germany and perhaps in England. In several regions, as in West Virginia, France, Germany, and Bohemia, there are workable coal measures in the Lower Permian, but arid conditions established themselves over all parts of the northern hemisphere where Upper and Middle Permian rocks are found, western North America, Texas, Kansas, England, Germany, and Russia. The extreme conditions of desert climate are registered in the great bodies of gypsum and rock-salt which characterize so much of the Permian areas.

What can have caused these climatic vicissitudes and especially the development of continental ice-sheets so near the Equator and so little above sea-level, is a problem for which many solutions have been propounded, but none of them is convincing.

Close of the Permian. — The late Palæozoic witnessed mountain-making disturbances on an almost world-wide scale, extending from the middle Carboniferous to the middle of the Lower Permian. In central Europe and Spain vigorous folding took place at the end of the Lower Carboniferous, but the most important and widespread disturbances occurred in the Permian. In North America low folds were formed in the Appalachian trough from time to time all through the Palæozoic, the evidence of which is the upheaval of the barriers described in the preceding chapters, which separated the Interior Sea from the changing bodies of water, such as the Cumberland Basin, on the east. A more energetic disturbance, with some mountain building, inaugurated the Pottsville age of the Upper Carboniferous, and this disturbance culminated at the end of the Permian in one of the greatest geographical revolutions which the history of North America has recorded. With the exception of the mountain making at the close of the Ordovician, the Palæozoic era in North America had been a time of slow, even development, with many oscillations of level, but with few violent disturbances, and with singularly few manifestations of volcanic activity. A little more land was added to the northern area during each period, but, so far as we can trace it, the geography of the Ordovician does not seem to have been very different from that of the Carboniferous. Throughout this long era the Appalachian geosyncline had been sinking, though with many shifts and oscillations, under an ever-increasing load of sediment, until the great trough contained a thickness of 25,000 feet or more of strata. Eventually the trough began to yield to lateral compression, and its contained strata were thrown into folds, or fractured by great overthrusts. Thus, in place of a sinking sea-bottom along the shore of the great Interior Sea, rose the Appalachian Mountains, which in their youth may have been a very lofty range, rivalling the Alps in height. This range extends from the Hudson River to Alabama; another range from Newfoundland to Rhode Island, and a third, the Ouachita Mountains of Arkansas, are attributed to the same set of disturbances, which thus made themselves felt for a distance of 2000 miles.

Though the entire continent felt the effects of this revolution, they are less obvious in the West. On the western side of the Rocky Mountains a great unconformity is found between the Permian and Triassic members of the Red Beds. "There are reasons to suppose that this hitherto unrecognized break is widespread, and explains many discordant features of various Red-bed sections, not only in Colorado, but in the adjacent Plateau province." (Cross.) The Great Basin region, which had been submerged through nearly the whole Palæozoic era, became land, and at the present time the surface rocks over most of this region are Carboniferous. It is, however, probable that the Permian has been stripped away by denudation, as it has been over nearly all of the northern plateau of Arizona.

Comparatively soon after the eastern part of the Great Basin had thus been converted into land, the ancient land area of its western border was depressed beneath the sea. It is probable that these two movements were connected, though they may have been separated by a considerable interval of time, In Nevada west of 117° 30′ W. long. no Palæozoic rocks have been found, and the Trias rests directly upon the Archæan.

However they may be explained, the geographical revolution which closed the Palæozoic era was accompanied by the most profound and far-reaching changes which have ever occurred in the recorded history of life, after which we find ourselves in a new world. It is probable that the change was a relatively rapid one, but there are sufficient connections between the faunas and floras of the two eras to show that the later were derived from the earlier, and that the gaps are due to the imperfections of the record.

Permian Life

We have to note, in the first place, that the animals and plants of the Permian are transitional between those of the Palæozoic and those of the Mesozoic eras. Here we find the last of many types which had persisted ever since Cambrian times, associated with forms which represent the incipient stages of characteristic Mesozoic types, together with others peculiar to the Permian.

Plants. — The flora of the Lower Permian is decidedly Palæozoic in character, and that of the Upper Permian as decidedly Mesozoic, so that if the line dividing these two great eras were drawn in accordance with the vegetation, it would pass through the Permian. Even in the Lower Permian, however, the change from the Carboniferous flora is a marked one, a change which may be largely explained by the increasing aridity of the climate. The great tree-like Lycopods, *Lepidodendron* and *Sigillaria*, which were so abundant in the Carboniferous forests, have become very rare; none of the former genus and only two of the latter have been found in the Upper Barren Measures of Pennsylvania and West Virginia. The *Calamites* continue in hardly diminished numbers and importance. The *Ferns* are exceedingly abundant and varied, and tree-ferns seem to be more common than they had been before. Especially characteristic genera of these plants are *Pecopteris, Callipteris* (Fig. 276), *Cynoglossa, Neuropteris, Sphenopteris* (Fig. 277), etc. The *Gymnosperms* mark a notable advance; in addition to the ancient *Cordaites*, are true Cycads and Conifers; of the latter are found yew-like forms, *Walchia, Saportæa* with leaves nearly four inches wide, and the *Gingkoaceæ* are probably represented by *Baiera*.

FIG. 276. — *Callipteris conferta* Brngn.
(Fontaine and White)

In the Upper Permian, Lepidodendron, Sigillaria, and Calamites are quite unknown, though probably a few stragglers still existed, and the flora is made up of *Ferns, Cycads, Gingkos,* and *Conifers,* the Angiosperms still being entirely absent.

FIG. 277.— *Sphenopteris coriacea* F. and W. (Fontaine and White)

The Permian flora of the southern land mass differs notably from that of the northern continents and is characterized especially by the broad-leaved ferns, *Glossopteris* (Fig. 278) and *Gangamopteris,* whence this is often called the " Glossopteris Flora," together with the cosmopolitan *Callipteris,* the Conifer *Voltzia* and the Calamite *Schizoneura.* In South America and South Africa, but not in India or Australia, these plants are accompanied by some survivors from the Carboniferous, such as *Lepidodendron.*

In the latter part of the Permian, or possibly not till the earliest Triassic, the Glossopteris Flora invaded the northern continents and extended its range· to northern Russia (*Tataric* stage).

Foraminifera are almost as important in the purely marine limestones as they had been in the Upper Carboniferous.

Cœlenterata. — The Corals are still mostly of Palæozoic type and belong to Carboniferous genera, but some of the modern Hexacoralla have appeared.

FIG. 278. — *Glossopteris browniana* Brngn.   Newcastle, Australia

Echinodermata. — This group has dwindled in the most remarkable way, and instead of the abundance of Crinoids which flourished in the Carboniferous seas, are found only occasional specimens.

Arthropoda. — The last few stragglers of the genus *Phillipsia* indicate the extinction of the great Palæozoic group of Crustacea, the Trilobites, which henceforth we shall meet with no more; the Eurypterids have their latest known representatives in the little coal basin of Bussaco, in Portugal, which is referred to the Permian, though with some doubt. In the Kansas Permian have been found numerous insects, which, though resembling those of the Carboniferous in a general way, all belong to species different from those of the latter. The giant insects of Commentry in France, which have been mentioned in the chapter on the Carboniferous, are now referred to the Permian by many authorities.

Bryozoa are prominent in all marine formations, sometimes forming reefs.

The **Brachiopoda** are still very abundant, especially in the Lower Permian; they are closely allied to those of the Upper Carboniferous, and, as in that period, the *Productids* play the most important rôle, though many of the species are peculiar to the Permian. The curious sessile, irregularly shaped *Richthofenia*, which began in the late Carboniferous, is especially characteristic of the Permian limestone accumulated in the Mediterranean Thetys, as in Asia, Sicily, and western Texas.

Mollusca. — In this group very striking changes are to be noted. The Bivalves increase materially in variety, and in addition to ancient genera like *Aviculopecten* (XII, 13) and *Myalina* (XII, 12) the typically marine Permian has many new forms, such as *Arca, Lucina, Lima, Bakevellia* (XII, 14), *Pleurophorus* (XII, 15), *Pseudomonotis* (XII, 16), etc. The Gastropods require no particular mention, except for the great abundance of the genus *Bellerophon* (XII, 17). It is among the Cephalopods that the great advance takes place. *Orthoceras* and *Gyroceras* continue from the older periods, and many species of the genus *Nautilus* are added, but the chief fact consists in the presence of Ammonoids with highly complex sutures, far exceeding, in this respect, the Goniatites of the Carboniferous, some of which continue to exist alongside of the more advanced forms. The more important new genera of Ammonoids are *Medlicottia, Ptychites, Popanoceras, Waagenoceras* (XII, 20), which have been found in Texas, Sicily, Russia, and India. The presence of these remarkable shells gives a strong Mesozoic cast to the Permian fauna.

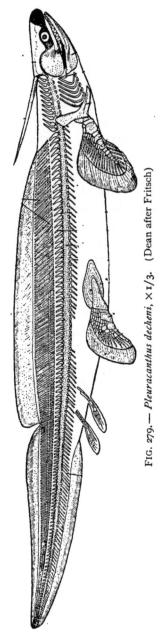

FIG. 279.—*Pleuracanthus decheni*, × 1/3. (Dean after Fritsch)

Vertebrata.— The *Fishes* are still of Carboniferous types, and many of the same genera occur, while new ones are brought in. To the Sharks are added the curious *Menaspis,* which is armed with numerous long and curved spines. Among the Dipnoi the genus *Ceratodus,* very closely allied to the modern lung-fish of Australia, makes its first appearance.

The *Amphibia* are represented, as in the Carboniferous, by the Stegocephalia, and several of the older genera persist, but many new forms appear for the first time, several of which much surpass the Carboniferous genera in size. (See Fig. 280.)

The most important character that distinguishes the life of the Permian from that of all preceding periods is the appearance in large numbers of true *Reptiles.* There is no reason to suppose that such a variegated reptilian fauna can have come into existence suddenly, and their ancestors will doubtless be discovered in the Carboniferous; but while no true reptiles are certainly known from the latter, in the Permian they are the most conspicuous elements of vertebrate life. These reptiles belong to several orders, one of which, the *Proganosauria,* is represented by *Mesosaurus* in South Africa and by *Stereosternum* in South

America. The *Proterosauria* are a very central group, from which many other reptilian orders appear to have descended:

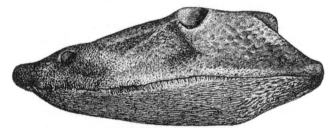

FIG. 280. — Permian Stegocephalian, *Eryops megacephalus* Cope, × 1/7. Skull Seen from Side. (Cope)

Proterosaurus and *Palæohatteria* are the most important Permian genera of this group.

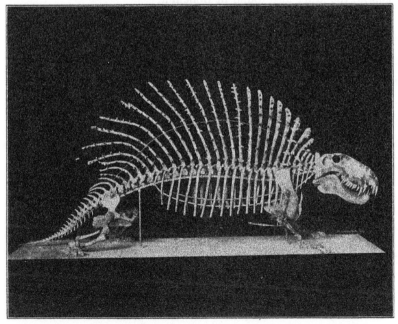

FIG. 281. — Permian Pelycosaurian, *Naosaurus claviger* Cope. (Osborn)

The *Pelycosauria* are extremely curious animals found in Texas and Bohemia. These were carnivorous land reptiles which had short tails and enormously elongated, sometimes branching, spines in the back (Fig. 281). The *Cotylosauria*, heavy massive reptiles of exceedingly primitive character, which retained several features of the stegocephalous Amphibia, are represented in Texas by two families, including several genera, of which *Diadectes* may be selected as typical. In South Africa is found the extraordinary *Pareiasaurus*, of the same order. Pareiasaurus followed the Glossopteris Flora in its northward migration and appears in the uppermost Permian of Russia (Tataric stage). In the same stage of the Russian Permian are found two other orders, which likewise seem to be migrants from South Africa, for they are abundantly represented there, the *Anomodontia*, with turtle-like beaks and either no teeth or a pair of large tusks, and the *Theriodontia*, the latter also found in Bohemia.

CHAPTER XXXII

THE MESOZOIC ERA — TRIASSIC PERIOD

THE Mesozoic era, so far as we can judge, seems to have been shorter than the Palæozoic; in North America Mesozoic rocks are very much more important and widely spread in the western half of the continent than in the eastern. The latter region was, in a measure, completed by the Appalachian revolution, and subsequent growth consisted merely in the successive addition of narrow strips to the coast-line, but in the West many great changes were required to bring the land to its present condition.

The life of the Mesozoic constitutes a very distinctly marked assemblage of types, differing both from their predecessors of the Palæozoic and their successors of the Cenozoic. In the course of the era the Plants and marine Invertebrates attained substantially their modern condition, though the Vertebrates remain throughout the era very different from later ones. Even in the Vertebrates, however, the beginnings of the newer order of things may be traced. In the earlier two periods, the Triassic and Jurassic, vegetation is almost confined to the groups of *Ferns, Cycads,* and *Conifers,* but with the Cretaceous come in the *Angiosperms,* both *Monocotyledons* and *Dicotyledons,* and since then the changes have been merely in matters of detail.

With few exceptions, the ancient *Tetracoralla* had all disappeared, and the modern *Hexacoralla* took their place. The *Echinoderms* were all markedly different from the Palæozoic types. The *Cystoids* and *Blastoids* had died out, and the *Crinoids* had been revolutionized, the *Camerata* being replaced by the *Articulata.* Likewise the modern sea-urchins, *Euechinoidea,* replaced the ancient *Palæechinoidea,* and many Mesozoic genera of the former group are still living in our modern seas. The *Starfishes* also assumed their modern condition. *Brachiopods* were far less abundant and diversified than they had been in the Palæozoic, and belonged, for the most part, to different families, while the *Bivalve* and *Gastropod Mollusca* increased to a wonderful extent. Especially characteristic are the marvellous wealth and variety of the *Ammonoid Cephalopods*, which disappear at the close of the era. The *Dibranchiate Cephalopods*, with internal shells, make their first appearance in the Mesozoic, and one group of them, the *Belemnites*, is almost exclusively confined to the era. The *Arthropods* showed the same revolutionary changes. Among the *Crustacea*, the *Trilobites* and *Eurypterids* have gone out, but all the modern groups were well represented, though many of the Mesozoic genera are no longer to be found in the seas of to-day. *Insects* reached nearly their modern condition, so far as the large groups are concerned, butterflies, bees, wasps, ants, flies, beetles, etc., being added to the older orthopters and neuropters.

Fishes became modernized before the close of the era, the *Bony Fishes* having acquired their present predominance. The *Amphibia* took a subordinate place, and after flourishing for a time, the great *Stegocephalia* died out, leaving only the pygmy salamanders and frogs of the present. *Birds* and *Mammals* made their first appearance, the former advancing rapidly to nearly their present grade of organization, though not reaching their present diversity, while the mammals remained throughout the era very small, primitive, and inconspicuous. The most significant and characteristic feature of Mesozoic life is the dominance of the *Reptiles*, which, in size, in numbers, and in diversified adaptation to various conditions of life, attained an extraordinary height of development. The Mesozoic is called the "Era of Reptiles," because these were the dominant forms of life. They filled all the rôles now taken by birds and mammals; they covered the land with gigantic herbivorous and carnivorous forms, they swarmed in the sea, and, as literal flying dragons, they dominated the air. At the present time there are only five orders of reptiles in existence, and of these only the crocodiles and a few snakes attain really large size.   In the Mesozoic era no less than twenty-five reptilian orders flourished, and many of them had gigantic members. Some were the largest land animals that ever existed, and the sea-dragons rivalled the whales in size. Nothing so clearly shows that the Mesozoic era is a great historical fact, as the dominance of its reptiles.

The Mesozoic climates offer some difficult problems.   In general, the climate was mild, as is shown by the plants found in the Mesozoic rocks of Arctic lands, for in Greenland, Alaska, and Spitzbergen was a luxuriant vegetation of warm temperate type.   On the other hand, certain geologists have maintained the existence of distinct climatic belts in the Mesozoic, indicating equatorial, northern, and southern zones, but by others this interpretation is denied.

The Mesozoic era comprises three periods, — the Triassic, Jurassic, and Cretaceous.

THE TRIASSIC PERIOD

The Triassic period is so named from the very conspicuous threefold subdivision of this system of strata in Germany, where its rocks were first studied in detail, and where they occupy a greater area than in any other European country. The German Trias is, however, not the usual facies of the system, but a very peculiar one, and cannot be taken as the standard of comparison for most other countries.

The Trias of North America is displayed under three very different facies, — that of the Pacific coast, which is marine; that of the interior, which is lacustrine; and that of the Atlantic border, which is continental. Owing to the absence of fossils common to all, it is not yet possible accurately to correlate the three facies, but the divisions of the Pacific and Atlantic borders are given in the following table together with those of Germany and the general arrangement of the Oceanic Trias, the former being chiefly of continental facies: —

2 U

TRIASSIC SYSTEM

General Oceanic		Germany	Pacific Coast	Atlantic Coast
Bajuvaric Series	Rhætic Stage		Bajuvaric Series	Newark Series
	Juvavian Stage			
Tirolic Series	Carinthian Stage	Keuper		
	Norian (or La-dinian) Stage		Tirolic Series	
Dinaric Series	Anisian Stage	Muschelkalk	Dinaric Series	
	Hydaspian Stage			
Scythic Series	Jakutian Stage	Bunter	Scythic Series	
	Brahmanian Stage	Sandstein		

Character and Distribution of Triassic Rocks

European. — The Trias of Europe has been so thoroughly studied and throws so much light upon American problems, that it will be profitable to depart from the usual order of treatment and take up first the development in that continent. As in the Permian, the Triassic rocks of Europe are found in two contrasted facies, the continental and the oceanic; the former extending, with interruptions, from Ireland, across England, France, central and southern Germany, to Poland, and consisting chiefly of red sandstones and red marls and clays, with conglomerates and some limestones. From this it follows that the rocks in the mountains which bordered the Triassic basins and plains had been profoundly decomposed, as in the southern Appalachians of to-day, where the crystalline rocks are changed into a red clay for depths of 100 feet or more, and the quartz grains are covered with a red film. As we have learned (Chapter IV) the red laterites of warm regions may be derived from very many different kinds of parent-rock, igneous rocks, crystalline schists, limestones, and dolomites. "This makes intelligible the close agreement of the continental Triassic rocks over great areas of the earth's surface. It is not at all necessary that the mountain ranges which surrounded the Triassic basins and plains should have been built up of the same rocks, which, as a matter of fact, was probably never the case. . . . It seems to be undoubted that the continental Triassic sediments were deposited in basins, or on low plains, and that their material was derived from plateaus and mountains. It may likewise be inferred from the size and rounded shape of the conglomerate pebbles that running water transported the material from the highlands to the basins." (E. Philippi.)

In *Germany*, where the plainly marked threefold division of the strata has given its name to the system, the lower series, or Bunter Sandstein, varies from 650 to 1800 feet in thickness and is chiefly made up of red sandstones and sandy shales. In northern and central Germany it is so intimately connected with the Upper Permian that any line of separation between them appears to be arbitrary, but in the west and southwest there is an overlap of the Bunter upon older rocks. Around the margins of the Triassic basins are coarse conglomerates, with finer materials toward the centres. Occasional and temporary lakes, or playas, were formed in the basins, and floods rushing from the mountains spread sheets of sand and gravel far out over the plains, while cross-bedded sands were piled up by the winds over extensive areas. Locally, the playas formed deposits of salt and gypsum, and clastic beds marked by sun-cracks, rain-prints, and tracks of animals. All of these features point unmistakably to an arid climate, though one that was probably less extremely dry than that of the Upper Permian. The mountain ranges appear to have been sufficiently high to cause abundant precipitation upon them; such a juxtaposition of rainy mountains and arid plains has nothing unusual about it.

The Middle Trias, or Muschelkalk, is marked by successive incursions of the sea, the first of which came at the end of the Bunter epoch and eventually extended over a great part of the area occupied by the Bunter Sandstein, leaving deposits with a maximum thickness but little exceeding 1100 feet. The fossils show that this was an inland sea, connected with the ocean, but having a fauna which consists of comparatively few species, though these are sometimes individually very abundant. The relation of these fossils to those of the contemporary part of the oceanic Trias is much like the relation between the modern faunas of the Black Sea and the Mediterranean. In the middle of the Muschelkalk the connection with the ocean was shut off and the German sea converted into a salt lake, as is demonstrated by the deposits of gypsum and salt, but the marine conditions were soon reëstablished.

An elevation of the land caused the withdrawal of the Middle Triassic sea and the resumption of the conditions of continental sedimentation, resulting in the formation of the *Keuper* (maximum thickness 2000 feet). The lower parts of the Keuper contain some marine beds and, locally, thin beds of coal, but most of it consists of sandy and clayey beds, which change rapidly from point to point, and are, on the whole, of finer materials than those of the Bunter. The basins had been largely filled with sediment and the mountains had been lowered by denudation, so that the coarser materials could no longer be transported. The middle Keuper was a time of extensive salt lakes, in which large bodies of gypsum and some salt were precipitated. The latest stage of the Keuper, the *Rhætic*, witnessed a renewed transgression of the sea.

Trias of continental origin occurs in other European countries. In *Great Britain* the Triassic is almost all continental, the Middle Triassic marine invasion not extending so far to the northwest, but the Rhætic transgression did, and beds of this stage form a thin, though persistent band at the top of the Keuper. The Triassie beds cover a large part of the central plains of England, extending to northeastern Ireland, and small areas occur on the east coast of Scotland. In *France* Trias of the German type, including the Muschelkalk, extended into the eastern and southern parts of the country to the Pyrenees and along eastern and southern *Spain*. Triassic rocks also occur around the margins of the central Plateau of France. In the south of *Sweden* is a considerable area of Triassic rocks: the Keuper is coal-bearing and is overlaid by marine Rhætic beds. In northeastern *Russia* is a great extent of beds belonging to the Tataric stage, which has been mentioned in connection with the Permian, but which may be in part Triassic.

In the Alps are two well-distinguished regions; in the western part conditions were not unlike those of Germany, while the eastern part displays a great development of the oceanic limestones. The western Alpine Trias is much folded and metamorphosed and becomes very thick on the Italian side of the mountains, and is conspicuous in the northern Apennines. On the east, the Palæozoic rocks of the Alps extended as a long island, or chain of islands, from the Engadine into southern Austria. "North of this old insular tract the Triassic strata are on the whole somewhat sandy. . . . On the south side the deposition of limestone and dolomite went on more continuously, though interfered with occasionally by submarine volcanic eruptions." (A. Geikie.) Almost the whole Triassic succession, except the lowest members, is represented here by great limestones and dolomites, many of the latter probably of coral-reef origin.

Asiatic. — The existence of the great Mediterranean Thetys in part of the Triassic period is indicated by the oceanic deposits which occur in Asia Minor, Central Asia, Baluchistan, Afghanistan, northern India, and Burmah, Tongking, and Southern China, but in the Lower Trias it appears that India was not in connection with Europe. In the Salt Range of northwestern India and the Himalayas is a remarkably complete succession of Triassic rocks, which overlie the Permian in a conformity that is at least apparent and may be actual, though there is a faunal break between the two systems. The *Brahmanian* stage of India is not represented in the Alps. In central India the Gondwána conditions of continental sedimentation continued apparently through the whole Triassic. Rocks of this period, which seem to belong in another faunal province, occur in Japan, the east coast of Siberia, and the Arctic islands, Spitzbergen, and Bear Island.

American. — In the early part, at least, of the period, both North and South America extended farther east than at present, and no marine Triassic rocks are known on the Atlantic slope of either

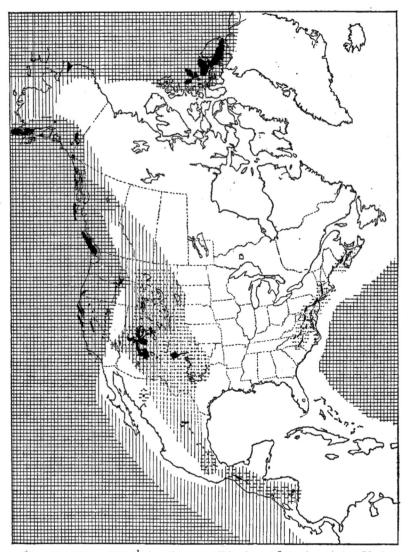

FIG. 282. — Map of North America in the Triassic and Jurassic periods. Black areas = known exposures; white areas = land; lined areas = sea; dotted areas = continental deposits. The horizontal lines indicate Triassic, and the vertical lines Jurassic, Seas.

continent, but they are extensively displayed on the Pacific side. The land barrier which during the Palæozoic era had bounded the Great Basin sea on the west was submerged and the Pacific extended over the site of the Sierras, covering western Nevada and sending a gulf into southeastern Idaho, and in British Columbia it transgressed eastward across the present mountains, and it covered part of the coast of Alaska. In California and Nevada all the series and many of the stages of the oceanic Trias may be identified and their faunal relations change in a very interesting way. The lower series (*Scythic*) " shows an intimate relationship to that of Asia and none with that of the Mediterranean region " (J. P. Smith), and this relationship is both with India and northern Asia. The difference from Europe is no doubt to be explained by the fact, above referred to, that in the Lower Trias, Thetys was interrupted somewhere between India and Europe. In the *Dinaric* series an invasion from the Mediterranean region is evident, though the track followed by this migration is not clear. In the Upper Trias (*Tirolic* series) the relations were first with India and the Mediterranean, succeeded by another migration from the north of Asia. These changes in faunal relationships have been variously explained and will again be referred to in considering the question of the Triassic climates. Little of the *Bajuvaric* series is found on the Pacific coast. In the United States the marine Triassic rocks do not exceed 4800 feet in maximum thickness, but in British Columbia this increases to 13,000, much of which is igneous material, and similar material is widely distributed in southeastern Alaska.

In central Mexico, State of Zacatecas, is an isolated area of marine Trias, belonging to the Tirolic series of the Upper Triassic and with a fauna allied to that of California. This is probably only a remnant of a formerly widespread area of such rocks, nearly all of which were eroded away during the Lower and Middle Jurassic, when most of Mexico was land.

On the Atlantic side of North America the course of events was entirely different. In the latter half of the period was formed a series of long, narrow troughs, running closely parallel to the

trend of the Appalachian Mountains, but separated from them by
the ridges of metamorphic and crystalline rocks, which follow those
mountains on the east, and which then probably had a considerable
altitude, much greater than at present.   In these troughs was laid
down the enormous thickness of non-marine rocks which constitute
the *Newark* series and are now found in several disconnected
areas from Nova Scotia to North Carolina.   The longest continuous stretch of these beds is from the Hudson River across New
Jersey, southeastern Pennsylvania and Maryland, into Virginia,
while another extensive area occupies the Connecticut valley,
through western Connecticut and Massachusetts.   The Newark
areas have been so extensively faulted that it is difficult to ascertain
their thickness, and the figures given are merely an approximation.
The series reaches its maximum in southeastern Pennsylvania,
where it is estimated at 20,000 feet, in New Jersey 12,000–15,000,
and in the Connecticut valley, 13,000.   Southward the rocks thin
quite rapidly, and about Richmond, Va., are not more than 3000
feet thick, and farther south, still less.

   The Newark rocks are prevailingly red sandstones and shales,
especially from Pennsylvania northward, but also contain some
very coarse conglomerates at the base, and higher up in the series
along the western border of the area.   Thin bands of limestone
and black, fossiliferous shales are intercalated, and in New Jersey
is a thick mass of very hard, slate-coloured shales, the Lockatong
stage.   In the northern area, Connecticut valley, New Jersey,
and Pennsylvania, many of the beds are ripple-marked, sun-
cracked, pitted with raindrops, and preserve countless footprints
of Amphibia and land Reptiles.   In Virginia and North Carolina
are workable coal-seams, and the red colour of the other rocks is
less prevalent than in the North.   Except in the black shales, fossils
are very few, and the plants show a distinct difference between the
Virginia and North Carolina area, where ferns predominate, and
the New Jersey-Connecticut region, where ferns are less abundant
and gymnosperms more so.   Whether this difference is climatic
or due to a slight difference in geological date it is difficult to say.

It is sufficiently evident that the rocks of the Newark series are not marine, but just how they were accumulated is a question as to which there is much disagreement.   It has been usual to consider the strata as of estuarine origin, but their lithological character, the surface markings of many of the beds, and the contained fossils make such an origin improbable.   It is likely that fluviatile and subaërial agencies have had more to do with the accumulation of the materials than had any body of water in communication with the sea.   At the southern end of the trough, the coal-beds point unmistakably to the existence of fresh-water swamps and bogs, while the red colour and the sun-cracks prevalent in the remainder suggest conditions resembling those of the German Bunter and Keuper, though the absence of gypsum and salt indicates that the climate was less arid than in Europe.   If we could be sure that the plant-bearing beds of Virginia were contemporaneous with those of New Jersey and the Connecticut valley, the difference in the floras would confirm the inference that the climate was sub-arid in the North, growing more moist southward, but the difference may be geological rather than geographical.   No doubt lakes of greater or less duration were formed and are now registered in the fish-bearing shales.

Igneous activity was a conspicuous feature of the Newark epoch, both in the volcanic and the intrusive form.   Lava-flows were poured out on the surface and were subsequently buried, and beds of fragmental products have been found in the Connecticut valley and in New Jersey, while dykes and sills accompany the strata throughout their extent.   Now that they are exposed by denudation, these plutonic bodies form very striking features of the topography.   One of the most remarkable of the intrusive masses is the great Palisades-Rocky Hill sill, the outcrops of which are separate, though their subterranean connection is well ascertained.   An account of the Palisades has already been given in Chapter XV.   All the known fossil-bearing horizons, with perhaps one exception, correlate the Newark rocks, with the Keuper, possibly extending up into the Rhætic.   The exception noted is in

southeastern Pennsylvania, where the beds reach their maximum
thickness and where the deeper portions are said to have yielded
Permian plants, but this awaits confirmation.

In the Mexican States of Sonora and Oaxaca are beds similar
to those of the Newark, which contain plants that show the formations to be of upper Keuper age, passing into the Rhætic. Still
farther south, similar beds occur in Honduras.

A third facies of the North American Trias is that of the western
interior, which has much the same distribution as the interior
Permian, extending from Texas, Arizona, and New Mexico, on
the south, around the Colorado island, north to the Canadian
provinces, though not continuously. In many areas the beds have
as yet yielded no fossils and are referred to the Triassic upon stratigraphical grounds, which are not always trustworthy. However,
the presence of the Trias is definitely ascertained at many points in
the region defined, though how much of the system, or what part of
it, is present at any particular locality, can rarely be determined.
As in the underlying Permian, the occurrence of gypsum and salt
is evidence of salt lakes and an arid climate, but, as was probably
true in the Newark region of the East, the climate appears to have
become less arid southward, for fresh-water Trias has been found
in Texas, northeastern New Mexico, and southwestern Colorado.

South America, like so many of the other continents, has both
the continental and marine facies of the Trias; the former is
found east of the Andes, in the Argentine Republic, and is coal-
bearing, with plants which correlate it with the Rhætic. On the
west side of the Andes the marine Trias is upturned in the mountains.

African. — In South Africa the Karroo system, as already
pointed out, is a continental formation, extending in apparently
unbroken continuity from the Permian into the Jurassic. The
Triassic portion is the upper part of the *Beaufort* series and has
yielded a remarkable array of fossil Reptiles. The land connection
with India, " Gondwána Land," was still maintained. In northern Africa Trias of the German type covers extensive areas in the

province of Constantine, Algeria, corresponding to the Muschel-kalk and Keuper, the latter containing gypsum.

Australasian. — Oceanic Trias, especially of the Upper Triassic, is found in many of the Indo-Pacific islands, Borneo, Sumatra, the Moluccas, New Caledonia, and others. In *Australia* continental Trias is found in New South Wales, where the lower portion is coal-bearing, and in Queensland, where the upper division carries coal-beds, as also in the northwest. In New Zealand, there is evidence of glaciation in this period and the glacial beds are overlaid by marine deposits.

Climate. — In the Triassic of the northern hemisphere there is evidence of very widespread aridity of climate, accompanied by general warmth; central Europe, north Africa, the western interior of North America and, in lesser degree, the northeastern part of the same continent. From the distribution of the fossils, there is reason to believe that in the Arctic Sea the water was cooler than in lower latitudes and that the remarkable changes in the faunal relations of our Pacific coast during the period are to be explained rather by the closing and reopening of Bering Straits than by the upheaval and depression of land bridges across the wider parts of the Pacific. This problem will again present itself in the subsequent periods.

THE LIFE OF THE TRIASSIC

Triassic life is entirely different from anything that had preceded it, though the way for the change was already preparing in the Permian. As we have seen, the Upper Permian, if classified by its plants alone, would be referred to the Mesozoic rather than to the Palæozoic, and we are therefore prepared to learn that the Triassic flora is very similar to that of the Upper Permian, though the Upper Trias, especially the Rhætic, marks a decided advance among the plants. Among the animals a considerable number of surviving Palæozoic types persist into the Trias which do not pass into the Jurassic.

southeastern Pennsylvania, where the beds reach their maximum thickness and where the deeper portions are said to have yielded Permian plants, but this awaits confirmation.

In the Mexican States of Sonora and Oaxaca are beds similar to those of the Newark, which contain plants that show the formations to be of upper Keuper age, passing into the Rhætic. Still farther south, similar beds occur in Honduras.

A third facies of the North American Trias is that of the western interior, which has much the same distribution as the interior Permian, extending from Texas, Arizona, and New Mexico, on the south, around the Colorado island, north to the Canadian provinces, though not continuously. In many areas the beds have as yet yielded no fossils and are referred to the Triassic upon stratigraphical grounds, which are not always trustworthy. However, the presence of the Trias is definitely ascertained at many points in the region defined, though how much of the system, or what part of it, is present at any particular locality, can rarely be determined. As in the underlying Permian, the occurrence of gypsum and salt is evidence of salt lakes and an arid climate, but, as was probably true in the Newark region of the East, the climate appears to have become less arid southward, for fresh-water Trias has been found in Texas, northeastern New Mexico, and southwestern Colorado.

South America, like so many of the other continents, has both the continental and marine facies of the Trias; the former is found east of the Andes, in the Argentine Republic, and is coal-bearing, with plants which correlate it with the Rhætic. On the west side of the Andes the marine Trias is upturned in the mountains.

African. — In South Africa the Karroo system, as already pointed out, is a continental formation, extending in apparent unbroken continuity from the Permian into the Jurassic. The Triassic portion is the upper part of the *Beaufort* series and has yielded a remarkable array of fossil Reptiles. The land connect with India, "Gondwána Land," was still maintained. In northern Africa Trias of the German type covers extensive areas in province of Constantine, Algeria, corresponding ı the Muschel-kalk and Keuper, the latter containing gypsum.

Australasian. — Oceanic Trias, especially of the ˉpper Triassic, is found in many of the Indo-Pacific islands, Boıeo, Sumatra, the Moluccas, New Caledonia, and others. In ʌstralia conti-nental Trias is found in New South Wales, where th lower portion is coal-bearing, and in Queensland, where the upper ivision carries coal-beds, as also in the northwest. In New Ze and, there is evidence of glaciation in this period and the glacia beds are over-laid by marine deposits.

Climate. — In the Triassic of the northern hemishere there is evidence of very widespread aridity of climate, accompanied by general warmth; central Europe, north Afric the western interior of North America and, in lesser degree, th northeastern part of the **same** continent. From the distribution f the fossils, there is reason

that in the Arctic Sea the wær was cooler than in lowe nd that the remarkable canges in the faur l relatio acific coast during the pend are to be

rat losing and reopening of Iring Straits

ıe d depr land bridg across the

s . Th will again resent self

ɛ or ɪc

diffe ything th had

for th ; already ppar

e sed Permian.

l be Mesozoic

are ured to

ar per P.

y

ng

es

Plants. — Triassic vegetation is composed of *Ferns, Horsetails, Cycads,* and *Conifers,* and of such plants were the Newark coal of Virginia and North Carolina, the Keuper coal of Germany and Sweden, and the Triassic coal of South Africa and Australia accumulated. The Ferns are relatively somewhat less abundant than they had been in the Carboniferous, and many of them belong to the existing tropical family of the *Marattiaceæ.* *Tæniopteris, Caulopteris, Clathropteris,* are among the most important genera. In Virginia a magnificent fern with very broad leaves, *Macrotæniopteris* (Fig. 283), is the most abundant and characteristic of the Triassic plants there found.

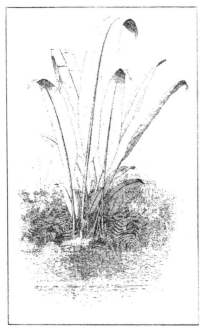

FIG. 283. — A Triassic Fern, *Macrotæniopteris magnifolia* Rogers. Restored (Russell)

The *Lycopods* have undergone a great reduction since the Carboniferous, though a few straggling specimens of plants related to *Sigillaria,* but belonging to a distinct family, have been found in the Lower Trias. The Calamites are no longer found, but on the other hand, true Horsetails of the modern genus *Equisetum* now make their first appearance, and much surpass their modern representatives in size, having stems of 4 inches in thickness. Rhizomes and stems of these plants are very common, and dense growths of them, like cane-brakes, surrounded the inland seas and salt lakes of the period. The *Cordaiteæ* have disappeared, but the *Cycadales* with their stiff leaves abounded, growing, doubtless, on the dryer lowlands above the swamps, most of them belonging to such genera as *Pterophyllum*, *Zamites*, and *Otoza-mites* (Fig. 284). This group of plants is a characteristic Mesozoic one, and the era is sometimes called the "Age of Cycads." The term *Cycadales* is employed to indicate "a group enormously wider than our recent Cycadaceæ." (D. H. Scott.) The Gingkoaceæ continue to be represented by *Baiera*. On the hills and uplands grew dense forests of Conifers, in

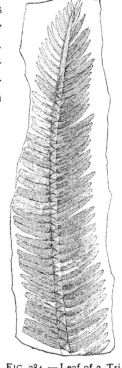

FIG. 284. — Leaf of a Triassic Cycad, *Otozamites latior* Saporta, × 1/2. (Newberry)

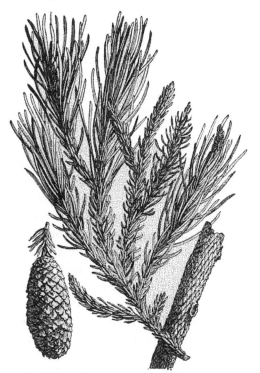

FIG. 285. — Triassic Conifer, *Voltzia heterophylla*. (Fraas)

appearance like the *Araucarians*, which are found to-day in South America, Polynesia, and Australia. *Araucarites*, and the cypress-like *Voltzia* (Fig. 285), the latter much resembling the Permian *Walchia*, are common genera.

While the Triassic flora is thus different from that of the Palæozoic, it must have given to the landscapes of the period much the same appearance of graceful and luxuriant, but somewhat gloomy and monotonous, vegetation. Probably the fern forests of New Zealand give the best modern picture of these early Mesozoic woodlands.

Of marine plants, the *Calcareous Algæ*, or Coralline Seaweeds, should be mentioned as very abundant about the coral reefs, to which they contributed largely.

Among the animals the change from Palæozoic times is much more complete than among the plants.

Cœlenterata. — Corals abounded in the seas, wherever conditions were favourable to their growth, but the Palæozoic *Tetracoralla* have nearly died out, though a few of the Tetracoralla and of the Tabulate Hexacoralla survived. Their place is taken by the modern type *Hexacoralla*, though the two groups of corals approach each other so closely that the distinction is not a sharp one.

Echinodermata. — In this type a more marked change has taken place. The *Cystoids* and *Blastoids* have disappeared, and the *Crinoids* have undergone a change of structure, the *Camerata* giving way to the *Articulata*, but the latter occur only in small numbers and in character rather transitional from the older forms than typical of the new. Of the Triassic Crinoids much the commonest is *Encrinus*, which is so characteristic of the German Muschelkalk. Similarly, the ancient type of the sea-urchins, the *Palæechinoidea*, is all but gone, only a few persisting through the Mesozoic, while the *Euechinoidea*, which began in a small way in the Carboniferous, now come to the front. The Triassic Echinoids are all of regular shape, the irregular forms not appearing till later.

Arthropoda. — The long-tailed Decapod Crustacea, *Macrura*, are found in the Trias, probably the most ancient representatives of the group. The *Ostracoda* are not uncommon. The little genus of *Phyllopoda*, *Estheria*, is very common in the German Keuper and the American Newark, and seems to be indicative of brackish-water conditions where it occurs.   Among the *Insects*, the *Coleoptera* (Beetles) are added to the two orders which are definitely known to occur in the Palæozoic.

The **Bryozoa** undergo a marked change in the disappearance of the ancient Fenestella-like genera.

Brachiopoda. — One of the most important changes from the Palæozoic to the Mesozoic consists in the great reduction of the Brachiopods in variety and numbers, and in a difference of character, the shells with long, straight hinge-line giving way to those with short, curved hinge, like *Terebratula* (Pl. XIII, Fig. 7).   Even in the Trias the reduction is very marked, though several Palæozoic genera have their latest representatives in the rocks of this system; as examples, may be mentioned *Productus*, *Athyris*, and *Cyrtina*. *Koninckina* is a new genus of the Spirifer family, which is confined to the Trias.   The still existing genera, *Terebratula* and *Rhynchonella*, are much the most abundant brachiopods of the period, and *Thecidium*, which later becomes important, has its beginning here.

Mollusca. — Almost in proportion to the decline of the brachiopods is the rise of the *Pelecypoda*, or Bivalves, which now become far more varied and abundant than they had been in the Palæozoic.   *Pecten, Pseudomonotis* (XIII, 8), *Myophoria, Halobia, Daonella* (XIII, 9), and *Cardita*, may be selected as a few examples of the commoner genera.   The higher forms of the class are, however, still rare.   The *Gastropoda* are yet in a transition stage. Several genera, such as *Murchisonia, Loxonema*, etc., here make their last appearance, and mingled with them are the forerunners and earliest representatives of modern types, such as *Cerithium* and other genera, in which the mouth of the shell is no longer a complete ring, but is drawn out into a grooved siphon.

The *Cephalopoda*, and more particularly the *Ammonoids*, have already acquired a wonderful degree of abundance and variety. The ancient Nautiloid genus *Orthoceras*, which ranges through almost the whole Palæozoic group, persists into the Triassic system, but not later, and numerous coiled Nautiloids with angulated and

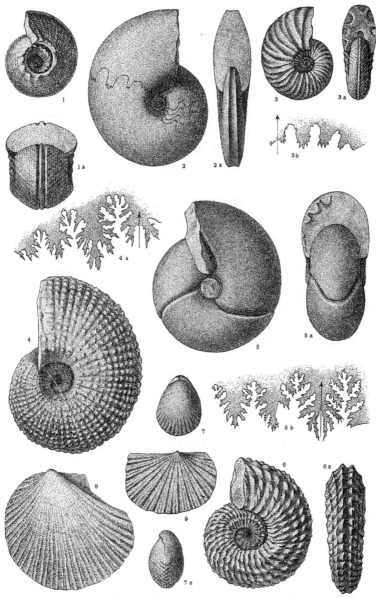

PLATE XIII. — TRIASSIC INVERTEBRATE FOSSILS

Figs. 1, 1a, *Tropites subbullatus* Hauer, side and end views, × ½, Up. Tr. 2, 2a, *Meekoceras gracilitatis* White, side & front views, Low Tr. 3, 3a, *Gymnotoceras blakei* Gabb, × ½, side and front views, Mid. Tr. 3b, *The same*, a suture line, × 1. 4, *Sagenites herbichi* Mojs. × ½, U. Tr. 4a, *The same*, a suture line, × 1. 5, 5a, *Joannites nevadanus* Hyatt & Smith, × ½, side and front views, M. Tr. 5b, *The same*, a suture line, × 1. 6, 6a, *Analcites meeki* Mojs. × ½, side and back views, M. Tr. 7, 7a, *Terebratula semisimplex* White, × 1, dorsal and side views, L. Tr. 8, *Pseudomonotis subcircularis* Gabb, × ½, M. Tr. 9, *Daonella lommeli* Wissmann, × ½, U. Tr.

ornamented shells recall those of the Carboniferous. Of the Ammonoids some still have the comparatively simple sutures of the Goniatites, others, like *Ceratites*, have slightly serrated sutures, while in the upper Triassic occur some shells in which the complexity of the sutures is carried farther than in any other known members of the group. Only a few of this great assemblage of genera can be mentioned; especially characteristic of the Trias are: *Tirolites*, *Trachyceras*, *Meekoceras* (XIII, 2), *Arcestes*, *Ceratites*, *Tropites* (XIII, 1), *Joannites* (XIII, 5), *Gymnotoceras* (XIII, 3), *Sagenites* (XIII, 4), and *Analcites* (XIII, 6). It is very interesting to observe that in the Trias occur, though but rarely, certain unusual forms of Ammonoid shells, which do not become important until the long subsequent time of the Cretaceous period. *Rhabdoceras* has a straight shell, *Cochloceras* one that is coiled in a high spiral like a gastropod, and in *Choristoceras* the coils are open. The similar Cretaceous genera were not derived from these Triassic anticipations, but are degenerate members of. many Ammonoid families. The Dibranchiate Cephalopods, and especially the characteristic Mesozoic group of the *Belemnites*, have their earliest and most primitive representatives in the Triassic genera *Atractites* and *Aulacoceras*.   ·

The **Vertebrata** are of extraordinary interest, and if the Trias had yielded only vertebrate fossils, it would still be apparent that great progress had been made since the time of the latest known Palæozoic beds. The *Fishes* display this progress least of all the Vertebrates. Shark teeth are known, but not skeletons. The Dipnoan *Ceratodus* is very characteristic, continuing up from the Permian. The Crossopterygians have greatly declined, but some very large and curious fishes of this group, like *Diplurus* (Fig. 286), still linger. The Ganoids continue to be the dominant fish-type, especially of the inland waters, and are most like the existing garpikes. *Catopterus* and *Ischypterus* are representative American genera, and *Semionotus*, *Dictyopyge*, and *Lepidotus* are nearly allied European fishes.

The *Amphibia* reach their culminating importance in the Trias,

2 X

the Stegocephalia multiplying and diversifying In a wonderful fashion, and far surpassing the genera of the Carboniferous and Permian in size. These Amphibians have been found in North America, southern Africa, and Europe; but those of the last-named continent are much the best understood, because best preserved, the Bunter sandstone of Germany being a treasure-house of such remains. *Mastodonsaurus, Cyclotosaurus*, and *Labyrinthodon* are common European genera, but there are many others. *Cheirotherium* (also European) is known only from its curious footprints, like the print of a human hand.

Reptilia. — It is in this class that we find the most remarkable changes; and although reptiles are common in the Permian, the

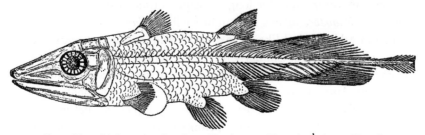

FIG. 286. — *Diplurus longicaudatus* Newberry. Newark shales. (Dean)

abundance and diversity of the Triassic reptiles are incomparably greater. Almost all the orders of Mesozoic reptiles are already represented in the Trias, though often by comparatively small and rare forms. The Triassic reptiles are much more common and better preserved in Europe than in America; but such American genera as do occur show that there was no essential difference between the reptilian faunas of the two continents.

The Gnathodontia, which are very near to the Permian Pro-ganosauria, are represented by *Telerpeton* and *Hyperodapedon*. Superficially like Crocodiles, but belonging to a different order, the *Parasuchia*, are the little *Aëtosaurus* and the formidable *Belodon* (Fig. 287), the latter found also in this country. The first of the dolphin-like *Ichthyosaurs*, which became so important in the Jurassic, are sparingly found in the Trias. Another group of sea-dragons, the *Plesiosaurs*, which attained such great development in Jurassic times, is represented in the Trias by small ancestral forms, *Nothosaurus*, etc. These are of extraordinary interest, as showing the descent of the purely marine Plesiosaurs, with their swimming paddles, from terrestrial reptiles which had feet adapted for walking. Another order of Triassic reptiles, the *Thalattosauria*, were already well adapted to a marine, predaceous life; as yet they are known only in Nevada.

One of the most characteristic of the Mesozoic groups of reptiles is the super-order *Dinosauria*, of which the Trias has many representatives; but clearly there were very many more than have yet been found, for the Newark sandstones of the eastern United

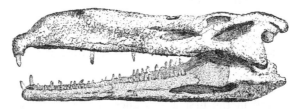

FIG. 287. — Skull of *Belodon kapffi* v. Meyer, about ⅒ natural size. (Zittel)

States have preserved a great variety of Dinosaurian footprints, but very few bones have been found in these rocks. The *Dinosauria* were much diversified, adapted for very different habits of life: some were herbivorous, others carnivorous; some walked on all fours; others were occasionally or habitually bipedal, and walked upright after the manner of birds, with which they have many structural features in common. The gigantic size attained by some of these creatures, even in the Trias, is shown by the footprints, some of which are 14 to 18 inches in length. Of the few American forms of which the bones have been found, the best known is *Anchisaurus* from the Connecticut valley, and of the European genera, the much larger *Zanclodon*.

The earliest *Turtles* are found in the Triassic of Europe, and these first-known members of the order are as typically differentiated as any of the later members. No doubt the Turtles originated in the Permian, in some region as yet unknown, and migrated to Europe in the Trias. The *Theriodontia*, which we found beginning in the Permian, culminated in the Trias, especially in southern Africa. Of this group there are two Triassic orders, the *Anomodontia* and the *Therocephalia*. The former have cutting jaws, like Turtles, and may or may not possess a pair ot great tusks in the upper jaw. *Dicynodon*, a genus of this suborder, has been found in South Africa, India, Russia, and Scotland. The Therocephalia present extraordinary approximations to the mammals, and have left a great wealth of remains — some of them very large — in the Karroo beds of South Africa, and less abundantly in India. The earliest known members of the flying reptiles, or *Pterosauria*, occur in the Rhætic of Europe.

The Trias has, as yet, yielded no Lizards or Snakes, which became very important at a later date. No birds are known from this period, nor any reptiles which can be regarded as the ancestors of birds.

Mammalia. — Still another great advance in the progress of life is registered in the first appearance of the Mammals, which occurred in the Trias. Mammals are the most highly organized forms of animals; but these, their earliest known representatives, were very small and very primitive, giving little promise of being the future conquerors of the world, as they were tiny creatures which, in a measure, represent the transition from lower vertebrates upward. Two American genera, *Dromatherium* and *Microconodon*, and one European genus, *Microlestes*, have been recovered.

CHAPTER XXXIII

THE JURASSIC PERIOD

WILLIAM SMITH, the father of historical geology, was the first to work out the divisions of the Jurassic, which he did early in the last century. The terms which he employed are local English names, and these, somewhat Latinized in form by the French geologists, are now very generally used as an international scale. These are given in the table. Smith's name for the system, " Oölitic." has been abandoned in favour of the term *Jurassic* which was first used by Brongniart and Humboldt. It was taken from the Jura Mountains of Switzerland, where the rocks of this system are admirably displayed. In Europe the Jurassic has long been a favourite subject of study, because of the marvellous wealth of beautifully preserved fossils which it contains. For this reason, the Jurassic is known with a fulness of detail, such as has been acquired regarding very few of the other systems; and no less than thirty well-defined subdivisions have been traced through many countries of the Old World. In this country the Jurassic is ill represented and its divisions are not clear.

JURASSIC SYSTEM

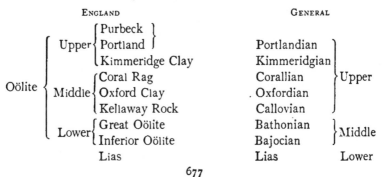

		ENGLAND		GENERAL	
Oölite	Upper	Purbeck			Upper
		Portland		Portlandian	
		Kimmeridge Clay		Kimmeridgian	
	Middle	Coral Rag		Corallian	
		Oxford Clay		. Oxfordian	
		Kellaway Rock		Callovian	
	Lower	Great Oölite		Bathonian	Middle
		Inferior Oölite		Bajocian	
		Lias		Lias	Lower

DISTRIBUTION OF JURASSIC ROCKS

American. — No undoubted Jurassic strata occur in the Atlantic border of the United States, though by some authorities the uppermost part of the Trias (*Newark* Series) is referred to this system, and by others the Potomac Series of the Cretaceous is regarded as Jurassic, at least the lowest portion of it. Whether or not these references be correct, no marine Jura is known on the Atlantic slope of North America except in Mexico. In eastern North America the Jura was a time of great denudation, when the high ranges of the Appalachian Mountains were much denuded and the newly upheaved, tilted, and faulted beds of the Trias were wasted and probably worn down to a peneplain.

At the beginning of the period Mexico was generally elevated, causing the sea to withdraw from those areas, as yet of unknown extent, which it had covered with shallow water in the Upper Triassic. At the same time the southeastern portion of the country was invaded by the sea and would seem to have remained submerged during the Lower and Middle Jurassic, but at this time most of Mexico was land, and denudation swept away most of the marine Triassic, leaving but a single known area.

In the western Interior region, what is believed to be Jurassic is, for the most part, placed there on stratigraphical grounds only, for few of the rocks have yielded fossils. The beds in question are largely sandstones which in many places rest upon Triassic strata and extend from northern Utah southward into Arizona and New Mexico, and westward from the Rocky Mountains to the Great Basin land which bordered the Pacific. Whether these doubtful beds represent the whole Jura or only a part and, if so, what part, are questions that cannot yet be answered. Some of the sandstones contain gypsum, testifying to salt-lake conditions.

On the Pacific coast the Lias occurs in California, Nevada, and Oregon, but not in British Columbia, and recurs in Alaska and some of the Arctic islands. The migration of marine animals from the north, which had been so conspicuous in the Upper

Triassic, was interrupted, and an influx of European forms, probably by way of the South American coast, took its place, and the Middle Jura appears to have retained the same connections.

The line between Middle and Upper Jurassic is not always drawn in the same place; by some writers the *Callovian* is placed in the older and by others in the younger division. In accordance with the arrangement in the table, the Upper Jurassic is here considered as beginning with the Callovian and is marked by extensive geographical changes in the southern and western regions of the continent. Mexico was very largely depressed beneath the sea, except, probably, a belt along the western coast. The small area of marine Jurassic which has quite lately been found in western Texas, is doubtless an outlier of this Mexican transgression. In the northwestern United States was formed an extensive gulf or shallow bay, which covered most of Montana, Wyoming, as far east as the Black Hills, and Utah. The beds reach their maximum thickness, 1800 feet, in the Wasatch Mountains, which is reduced to 800 in the Black Hills. The sediments laid down in this bay are limestones, shales, and marls, while the presence of gypsum shows that salt lagoons were formed by isolation of parts of the bay. The fossils are of Boreal types, like those of Alaska and Siberia, and point to an incursion from the north or northwest, though the oceanic connections of this Jurassic bay have not yet been determined, but the difference between this Boreal fauna and that of California, which is of central European character, makes any direct communication between the two areas unlikely. In the Upper Jurassic of California, of somewhat later date than the interior bay, the land connection between Alaska and Asia appears to have been again interrupted, opening the way for a current of cooler water to flow from the Arctic down the American coast. At all events, a Boreal fauna, with animals which were then abundant in northern Europe and Siberia, extended its range through California to southern Mexico, though the difference in the fossils shows that some barrier, however low and shifting, prevented the waters of the Atlantic from meeting those of the Pacific. It is remarkable that these northern animals should have extended their range so much farther south on the American than on the Asiatic coast, but the explanation is probably to be found in the cool current from the north mentioned above. Similar cases are known at the present time, in which the distribution of marine animals is controlled by warmer and colder currents in the sea.

The western interior gulf did not persist throughout the Upper Jurassic, but was drained by an elevation and was succeeded by a continental formation, partly fluviatile, partly, it may be, lacustrine, the *Morrison*.[1] The beds have quite a different distribution from those of the marine Jurassic and extend down the eastern front of the Rocky Mountains from Wyoming to Texas and New Mexico, with disconnected areas, perhaps outliers, in the Black Hills and western Colorado. The Morrison has yielded a remarkable fauna of Dinosaurs and Mammals, which have made it the object of great interest to the palæontologist. Its exact geological position has been vigorously discussed and it has been referred now to the uppermost Jurassic, again to the lowest Cretaceous. It has been suggested by Professor Williston that different areas of the Morrison are of different dates, just as we saw that the Millstone Grit (Upper Carboniferous) of the Mississippi valley is not a single uniform bed, but different beds of similar character, formed successively and corresponding to several horizons in the great mass of the Appalachian Pottsville. On this view, which is probably the solution of the problem, the Morrison includes several distinct horizons, extending from the Upper

[1] This is the formation called in the first edition of this book the *Como* stage and referred to the Cretaceous. The term Como must be abandoned in favour of Morrison, which was first proposed. "The question whether the Morrison formation is Jurassic or Cretaceous is still to be answered, and if a satisfactory answer is ever received it will doubtless be from vertebrate paleontology." (Stanton.) The stratigraphic evidence leaves it open whether these beds should be called Jurassic or Cretaceous and, on the evidence of the mammals, they are here included chiefly in the Jurassic.

Jurassic into the Lower Cretaceous, but the discrimination of these horizons is yet to be made.

In British Columbia, where the Lower and Middle Jurassic are not known, the Upper Jura occurs and shows a transgression of the sea to the eastward of the Cascade Mountains. Here and in California the rocks are prevailingly slates, with included diabase tuffs, demonstrating volcanic activity.

The close of the Jurassic in North America was a time of extensive orogenic movements on the western side of the continent, comparable to the Appalachian revolution which closed the Palæozoic era on the eastern side, though not of such far-reaching extent and consequences. The Sierra Nevada had long been a slowly subsiding geosynclinal trough, in which great thicknesses of sediment had accumulated. At the end of the Jura it yielded to the forces of lateral compression and rose into a folded mountain chain, transferring the coast-line from the eastern to the western side of the mountains. Farther north, the Klamath and Cascade ranges probably participated in the movement, as did the Coast Range, which seems to have become a chain of islands. Little is known regarding this primary condition of the Sierra Nevada, which may then have been of no great height, and which was not yet separated from the Great Basin by faults, the present mountains being due to long subsequent events. The Humboldt Range in Nevada appears to have had its beginning in the same disturbances, while a widespread movement of elevation affected the interior, though this movement began at a somewhat earlier date.

Foreign. — The greater part of *South America* was above the sea during the Jurassic period, as it had been in the Trias. Marine deposits of the former period are found only along the western border of the continent, where they extend from 5° to 35° S. lat. Throughout the Jurassic the sea which covered this western coast retained its faunal connection with the central European region, and even the minuter divisions, the substages and zones of the European Jura, are applicable to the classification of the South

American beds. The Boreal invasion, so conspicuous in California and Mexico, did not extend to South America, doubtless
because of the warmer water on that coast.

In *Europe* the Jurassic rocks are magnificently displayed, but
they differ much both in thickness and in character as they are
traced from one country to another, which results from more
frequent and more localized changes of level than had occurred
during the Palæozoic.

The Lias has a much more restricted extension than the later
Jurassic stages. At the end of the Triassic had begun a transgression of the sea (the Rhætic) which flooded many of the inland
basins, and the same transgression continued into the Lias, proceeding northward from the Mediterranean, and covering large
areas in central and southern Europe, as well as a belt across
England, but not extending to Russia. By far the greater part of
the Eurasian land mass was above the sea, and fresh-water lakes
extended across Siberia, while in China widespread deposits of
Liassic coal were accumulated.

Very early in the Upper Jura the transgression of the ocean
was renewed, and this time on a vastly larger scale, cutting the
continents by seas and straits, invading great areas that had long
been land, and covering the larger part of Europe and Asia.
This is one of the greatest transgressions of the sea in all recorded
geological history, but it did not greatly affect North America.
Central and northern Russia were submerged by an extension of the
sea from the north, and in this Russian sea was developed a highly
characteristic fauna. Strata distinguished by the Russian fauna
extend into the northeast of England and through Siberia, Spitzbergen, Nova Zembla, Alaska, the Black Hills of South Dakota,
and the uppermost Jurassic of California and Mexico, and even
penetrated to the Himalayan sea. In peninsular India the Jura
is represented by the upper division of the Gondwána system,
which, as before, was laid down as continental deposits, containing much coal. The continental mass to which India then
belonged was cut off from *Asia* by a strait or sea which covered

the site of the Himalayas, and which was connected with the Mediterranean or Thetys by way of Persia and Asia Minor. The great Jurassic transgression submerged considerable areas of northern India, as it also covered narrow areas along the eastern and western coasts of *Australia*. Much of Madagascar was under water, but the southern portion is believed to have formed part of the narrow land which extended from South *Africa* to India. Some of eastern Africa was covered by a bay of the Indian Ocean, but no marine Jurassic has been found in the southern or western portions of that continent. In South Africa, the uppermost part of the Karroo system, like the corresponding portion of the Indian Gondwána, is Jurassic.

Climate. — The very striking faunal differences which obtain between different regions have led certain observers, especially the late Professor Neumayr of Vienna, to the conclusion that climatic zones had already been established in Jurassic times, — Boreal, central European, and Alpine or Equatorial zones, with corresponding ones in the southern hemisphere. This conclusion as to climatic belts is, however, a very doubtful one, and is in conflict with the distribution of the several geographical faunas, for the central European fauna is found in equatorial Africa, and the supposed equatorial fauna occurs in the Andes 20° of latitude south of its proper boundary. It is much more likely that some of the marked faunal differences are due to varying facies, depth of water, character of bottom, etc., and even more to the partly isolated sea-basins and the changing connections which were established between them. On the other hand, the Boreal fauna, both in its restrictions and in its migrations, seems definitely to lead to the conclusion that the Arctic Sea was distinctly colder than the other oceans, though there is no likelihood that the difference of temperature was nearly so great as at present, or that the Arctic contained ice, even in winter. "The suggestion of climatic influence on the dispersion of marine animals in the Upper Jurassic is very strong." (J. P. Smith.)

Jurassic Life

The life of the Jurassic has been preserved in wonderful fulness and variety; but with comparatively few exceptions, our knowledge of it has been principally derived from Europe, where a host of eminent geologists have long studied the great wealth of material. The contrast between North America and Europe in regard to the relative abundance of Jurassic marine fossils is seen from the fact that while in Great Britain alone more than 4000 species have been described, in America hardly one-tenth of that number has so far been found.

Plants. — The flora of the Jurassic differs little; on the whole, from that of the Trias, and is made up of *Ferns, Horsetails, Cycads, Conifers,* and *Gingkos.* Tree ferns flourished in northern Europe in great variety. The Cycads attain their culmination of abundance and diversity in this period, no less than forty species occurring in a single horizon of the English Upper Jura and an extinct order of the Cycadales, the *Bennettiteœ,* flourished remarkably. The Conifers are of somewhat more modern aspect than those of the Trias, and, from their resemblance to genera which are still extant, have received such names as *Thujites, Taxites, Cupressites, Pinites,* etc. The Araucarian pines abounded in Europe. The Gingkos, or Maidenhair Trees, continued to be represented by *Baiera.* Monocotyledons have been reported from the Jurassic, but the evidence for their existence is very doubtful.

Foraminifera are found in great numbers and variety in the soft Jurassic clays, many of them belonging to genera which still abound in the modern seas. It must not be supposed that these organisms first became so abundant in Jurassic times; it is merely that the conditions for the preservation of these microscopic and exquisite shells had not been so favourable before.

Radiolaria. — The beautiful siliceous tests of the Radiolarians are also found in multitudes. In the Alps occur whole strata of red flints and jaspery slates, which are composed almost entirely of these tests.

PLATE XIV. — JURASSIC INVERTEBRATE FOSSILS

Fig. 1, *Pentacrinus asteriscus* M. and H., section of stem, × 2, U. J. 2, *Cidaris coronata* Goldf., × ½, Kimmeridgean, Europe. 3, *Antinomia catulli* Pictet, × ½, Tithon., Alps. 4, *Eumicrotis curta* Hall, × ³/₂, U. J. 5, *Tancredia corbuliformis* Whitf., × 1, U. J. 6, *Gervillia montanaensis* Meek, × ½, U. J. 7, *Volsella subimbricata* Meek, × ½, U. J. 8, *Trigonia americana* Meek, × ½. U. J. 9, *Pholadomya kingi* Meek, × ½. U. J. 10, *Camptonectes bellistriata* Meek, × ½, U. J. 11, *Pleuromya inconstans* Aguilera and Costello, × ½, U. J. 12, *Myacites subcompressus* Meek, × ½. U. J. 13, *Gryphæa arcuata* Lam., × ²/₃, Lias, France. 14, *Ostrea marshi* Sowerby, × ³/₄, Bajocian, Germany. 15, *Lyosoma powelli* White, × ½, U. J. 16, *Pleurotomaria conoidea* Deshayes, × ½, France. 17, *Nerinea dilatata* d'Orb, × ½, France. 18, *Peltoceras* cf. *athleta* Phillips, × ¼, Callovian, Europe. 19, *Lytoceras fimbriatum* Sowerby, × ¼, Lias, England. 20, *Crioceras bifurcatum* Quenstedt, × ½, Bathonian, Germany.

Spongida. — Sponges are found in wonderful profusion and diversity and in such perfect preservation that every detail of their beautiful structure may be made out with the microscope. In some localities these sponges are heaped up in such masses that they fill the strata, while other localities of the same horizon are entirely free from them.

Cœlenterata. — Corals abound, especially in the Upper Jurassic of central Europe. The *Anthozoan Corals* all belong to the modern *Hexacoralla*, in decided contrast to the Tetracoralla of the Palæozoic seas. *Isastræa*, *Montlivaultia*, and *Thecosmilia* are the dominant genera.

The **Echinodermata**, especially the *Crinoids* and *Sea-urchins*, are of great importance. The Crinoids are vastly more abundant than they had been in the Trias, and although the number of genera and species is not at all comparable to the great assemblage of Carboniferous times, yet for profusion and size of individuals the Jurassic has never been surpassed. Especially characteristic are the superb species of *Pentacrinus*, a close relative of which still exists in the West Indian seas. Other common genera are *Apiocrinus* and *Eugeniacrinus*. These genera all belong to the *Articulata*, which have a very different type of structure from the Palæozoic forms, but, like nearly all the latter, they were attached to the sea-bottom by their long stems. In the Jurassic appear the first of the Articulate free-swimming Crinoids, like *Comatula*, the commonest of modern genera. These animals possess a stem only in their early stages of development; subsequently they become detached and free. *Star-fishes* and *Brittle Stars* are not very common, but have attained a completely modern structure.

The *Echinoids* have undergone a wonderful expansion and diversification by the time of the Middle Jurassic. In the Lias, as in the Trias, we find only the regular, radially symmetrical sea-urchins, with mouth and anus at the opposite poles of the shell, such as *Cidaris* (Pl. XIV, Fig. 2), but in the Middle and Upper Jura appear the irregular *Spatangoids* and *Clypeastroids*. In these the shell is bilaterally symmetrical, rather than radially so, the anus, and even the mouth, losing their polar positions, and the shape of the ambulacral areas being greatly changed. This is another instance of the attainment of modern structure which so many of the Mesozoic Invertebrates display.

Arthropoda. — The *Crustacea* are not found in very many localities, but places like the famous lithographic limestones of Solenhofen in Bavaria, where the conditions of preservation were favourable, show that this group was very abundant and far advanced in the Jurassic seas. The long-tailed (macrurous) Decapods (of which the lobster is a familiar example) are in the ascendant and are represented by many genera, several of which still exist. The Crabs, or short-tailed Decapods, which are now so very common, make their first known appearance in the Jurassic, but they were still rare, and connecting links between the long-tailed and short-tailed series were more abundant. *Isopods* and *Stomatopods* also abounded.

The *Xiphosura* are reduced to the single genus *Limulus*, which then occurred in the European seas, while the living horseshoe crabs of that genus are found only on the east coast of the United States and in the Molucca Islands.

Insects are found in multitudes in certain localities, and display a great advance in the number of types over any of the Palæozoic periods. The Orthopters and Neuropters which we found in the Palæozoic are enriched by many new forms, such as grasshoppers, while beetles (*Coleoptera*) become very abundant. The Hymenopters (ants, bees, wasps, etc.) and the Dipters (flies) date from the Jurassic, and Lepidopters (butterflies) have also been reported, though doubtfully. As the latter insects are dependent upon a flowering vegetation, definite proof of their presence in the Jura will establish the existence of the Angiosperms at that time.

Brachiopoda. — These shells are still common in the Jura, but they are simply a profusion of individuals belonging to a few genera, most of which persist in our recent seas; compared even with the Trias, Jurassic brachiopods are much diminished. Terebratuloids like *Antinomia* (XIV, 3), *Waldheimia*, and *Rhyn-*

chonella are much the most important genera, and the last stragglers of the long-lived Palæozoic *Spiriferina* are here found.

Mollusca. — The Bivalves, which had already become such important elements of the Triassic fauna, greatly increase in the Jura,
their shells forming great banks and strata. Many of the genera
are still living, and only a few of the more abundant ones can be
mentioned here. Oysters like *Gryphæa* (XIV, 13), *Exogyra*, and
Ostrea itself (XIV, 14) are common, and the Scallop shell, *Camptonectes* (XIV, 10), is important. *Trigonia* (Fig. 288 and Pl. XIV,

FIG. 288. — Slab of *Trigonia clavellata*, from the English Oxfordian

Fig. 8) is especially characteristic of the Jura, but a few representatives of that genus have persisted to the present time and
are found in the Australian seas. *Diceras* and *Pholadomya*
(XIV, 9) are likewise common genera, and there are very many
others. Among the *Gastropoda* the most significant change lies
in the importance which the siphon-mouthed shells now for the
first time assume; examples of this group are *Nerinea* (XIV, 17),
Alaria, *Purpurina*, etc. Of the shells with entire mouths the

ancient Palæozoic genus *Pleurotomaria* (XIV, 16) is as abundant as ever, not beginning to decline until the Cretaceous period.

The *Cephalopods* are at the very height of their culmination, and are present in an astonishing profusion and diversity, filling whole strata with their heaped-up shells. The Nautiloids differ from those of the Trias in their smoother and more involute shells. The Ammonoids do not display so many types of shell structure as we have found in the Trias, and the genera are mostly different

FIG. 289. — Slab of *Belemnites compressus* Blainv, from the Lias of England

from those of the latter period; but in number of distinct species the Jura much surpasses the other Mesozoic periods. *Phylloceras* and *Lytoceras* (XIV, 19) continue on from the Trias, but the most abundant, characteristic, and widely spread genera are new. Of these may be mentioned: *Peltoceras* (XIV, 18) *Arietites*, *Ægoceras*, *Harpoceras*, *Stephanoceras*, *Perisphinctites*, and many others, each with large numbers of species. *Crioceras* (XIV, 20) is an uncoiled ammonoid. The *Belemnites* (Fig. 289), which were in-

2 V

troduced in a small way in the Trias, in the Jurassic blossom out
into an incredible number of forms, exceeding even the Ammonites
in abundance of individuals, if not of species. These extinct Cephalopods belonged to the Dibranchiata, as do all the living forms
except the Pearly·Nautilus; they in some measure serve to connect
the extinct genera having external shells with the existing naked
squids and cuttle-fishes, which have only rudimentary internal
shells, the pen or cuttle-bone. The Belemnites have a straight,

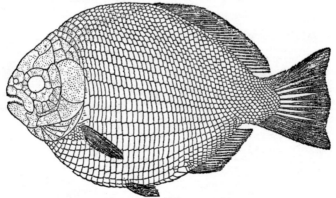

FIG. 290. — Dapedius politus.   (Smith Woodward)

conical, chambered shell, called the phragmocone, which ends above
in a broad, thin plate. The phragmocone was partly external
to the animal, and its lower, pointed end was inserted into a dart-
or club-shaped body called the guard or rostrum which is composed
of dense, fibrous, crystalline calcite.   Usually only the guard
is preserved in the fossil state, and specimens are so common that
they have attracted popular interest and bear the folk-name of
" thunderbolts."   In a few instances the animal has been preserved almost entire, so that the structure is well understood.

Vertebrata. — The Fishes have· advanced much beyond those
of the Trias. The Sharks have attained practically their modern
condition, and the broad, flattened Rays are a new type of the
order. The Chimæroids were much more numerous and rela-

tively important than they are at present, when only a few are left. The Dipnoans had become very scarce and are hardly represented in the northern hemisphere, save for the persistence of *Ceratodus*. The Crossopterygians were greatly reduced, though a few exceedingly curious forms, like *Undina*, still linger. Of the

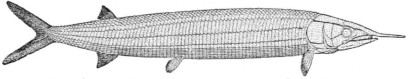

FIG. 291. — *Aspidorhynchus acutirostris* Ag. (Smith Woodward)

Teleostome fishes the Ganoids were still the dominant type, as they had been since the Devonian. Some of these Jurassic forms are evidently the forerunners of the Sturgeons, but most of them resemble the Gar-pike of our Western rivers (*Lepidosteus*), and are covered with a heavy armour of thick, shining, rhomboidal scales.

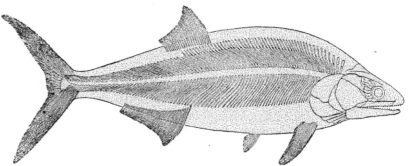

FIG. 292. — *Hypsocormus insignis* Wagner. (Smith Woodward)

Many of these Ganoids are of small or moderate size, such as *Dapedius* (Fig. 290) and *Aspidorhynchus* (Fig. 291), while others, like the superb *Lepidotus*, were very large, evidently the kings of their race. Some of the Jurassic fishes approximate the Teleosts so closely that it seems arbitrary to call them Ganoids. *Caturus*, *Leptolepis*, *Hypsocormus* (Fig. 292), and *Megalurus* are much like what the ancestral Teleosts must have been.

No *Amphibia* are certainly known from the Jurassic.

The *Reptiles* have attained a higher and more diversified plane of existence than in the Trias. Most of the Triassic genera and several entire orders have become extinct, but new and more advanced forms come in to take their places. The *Rhynchocephalians* abound and give rise to many diversified types of terrestrial, semi-aquatic and fully aquatic reptiles, and the first of the true *Lizards* (*Lacertilia*) appear. *Turtles* have grown much more numerous than in the Trias and have distributed themselves over the world. The *Ichthyosauria* are a highly characteristic Jurassic group; for though they are found in both the Trias and the Cretaceous, the Jura, and especially the Lias, is the time of their principal expansion. Certain localities in the Lias of England and Germany have yielded an incredible number of skeletons, and some of the specimens have preserved the impressions of the outline of the body and limbs, showing recognizably the nature of the skin. These reptiles were entirely marine in their habits and preyed upon fishes, and their limbs were converted into swimming paddles; there is a dorsal fin and a large tail-fin, the principal organ of propulsion (see Fig. 293). The muzzle is drawn out into an elongate slender snout, armed with numerous sharp teeth, which were set in a continuous groove, not in separate sockets. The eye is very large and protected by a number of bony plates, which are often preserved in the fossil state. The neck is very short and hardly distinguished from the porpoise-like body. The skin was smooth, having neither horny scales nor bony scutes, which was of advantage in lessening the friction of the water. In length, these reptiles sometimes exceeded 25 feet, and in appearance they must have been very like the modern porpoises and dolphins, but the resemblance is entirely superficial, for porpoises and dolphins are warm-blooded Mammals. *Baptanodon*, found in Wyoming, is an Ichthyosaur without teeth and must have fed upon small and soft marine invertebrates, as do the toothless whales.

Another group of carnivorous marine reptiles is that of the

FIG. 293. — *Ichthyosaurus quadriscissus* Quenst., LiaS. Restoration by C. R. Knight under the direction of Prof. H. F. Osborn. (Copyright, American Museum of Natural History, N.Y.)

Plesiosauria, which began in the Trias and culminated in the Jura, and which forms a curious contrast to the Ichthyosaurs. In the typical genus *Plesiosaurus* (Fig. 294) the head is relatively very small, and the jaws are provided with large, sharp teeth, set in distinct sockets. The neck is exceedingly long, slender, and serpent-like, and marked off distinctly from the small body. The swimming paddles are much larger than in the Ichthyosaurs and probably had more to do with locomotion; the skeleton of the paddle departs much less widely from the structure of a terrestrial reptile's foot than does that of an Ichthyosaur. With their long necks, the Plesiosaurs could lie motionless far below the surface, occasionally raising their heads above the water to breathe, or darting them to the bottom after their prey, which consisted chiefly of fish. The Jurassic species of *Plesiosaurus* do not much exceed a length of 20 feet, but *Pliosaurus* of the same group was gigantic, a single paddle sometimes measuring 6 feet in length; the reptiles of the latter genus had, however, proportionately larger heads and shorter necks.

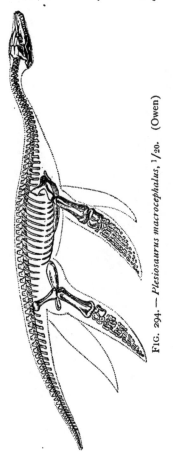

FIG. 294.— *Plesiosaurus macrocephalus*, 1/20. (Owen)

The seas and rivers of Jurassic times were swarming with *Crocodiles*, the most ancient yet known, *Teleosaurus* being the commonest genus of the period. In appearance these reptiles much resembled the modern Gavial of India and had a similar elongate and slender snout. The fore legs were much smaller than the hind, and these animals were doubtless of more exclusively aquatic habits than the crocodiles and alligators of the present day. One suborder of the Crocodiles, the *Thalattosuchia*, was almost entirely marine in habits, the skin being smooth and without scales and the fore limbs converted into paddles, while the very long tail ended in a large fin. In the Jurassic of South Africa has been found the other extreme of crocodilian development, a little reptile which was terrestrial and had long, running legs.

FIG. 295. — *Allosaurus agilis* Marsh, a carnivorous Dinosaur from the Morrison. Restoration by C. R. Knight under the direction of Prof. H. F. Osborn. (Copyright, American Museum of Natural History, N.Y.)

The *Dinosauria* became much larger, more numerous and diversified than they had been in the Trias, though, as the footprints in the Newark sandstones teach us, only a small fraction of the Triassic Dinosaurs has yet been recovered. Making all due allowance for this, it seems, nevertheless, to be true that the group had made notable progress in the Jurassic. The known American Jurassic Dinosaurs are from the Morrison, and so some of those mentioned below may be Cretaceous. The group of Dinosauria is a greatly varied one, comprising reptiles of very different

size, appearance, structure, and habits of life.   Some were heavy,
slow-moving quadrupeds, having fore and hind legs of not very
unequal length, with hoof-like toes, and usually with very small
heads.   Dinosaurs of this type were mostly plant-feeders and had
rows of grinding teeth adapted for such a diet.   *Brontosaurus*,
from the Morrison, is an example of this kind of Dinosaur, which
attained a length of 60 feet, and *Diplodocus* was a not very dissimilar and even larger reptile.   *Stegosaurus* was another herbivorous

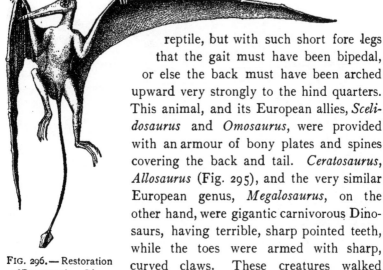

reptile, but with such short fore legs
that the gait must have been bipedal,
or else the back must have been arched
upward very strongly to the hind quarters.
This animal, and its European allies, *Sceli-
dosaurus* and *Omosaurus*, were provided
with an armour of bony plates and spines
covering the back and tail.   *Ceratosaurus*,
Allosaurus (Fig. 295), and the very similar
European genus, *Megalosaurus*, on the
other hand, were gigantic carnivorous Dinosaurs, having terrible, sharp pointed teeth,
while the toes were armed with sharp,
curved claws.   These creatures walked
upon their elongated hind legs and were the
most   formidable   beasts   of   prey   that

FIG. 296. — Restoration
of Pterosaurian, *Rham-
phorhynchus.* (Zittel)

scourged the Jurassic lands.   Not all of the Jurassic Dinosaurs
were gigantic; very small ones also ranged through the forests,
or may even have been arboreal in their habits.   *Compsogna-
thus*, for example, was a bipedal, carnivorous Dinosaur hardly
larger than a house cat.

Another very remarkable order of reptiles, the *Pterosauria*, the
earliest known appearance of which is in the Rhætic, became
important and characteristic in the Jurassic (Fig. 296).   These

animals were provided with wings, and were true fliers; thus realizing the old myth of flying dragons. The head is relatively large, but very lightly constructed, and set at right angles with the neck, as in birds. In the Jurassic species, the jaws are more or less completely armed with teeth, which by their form show the carnivorous propensities of the animal. The joints of the external or little finger of the hand are much thickened and elongated, this finger being longer than the body and legs together. A membrane, or *patagium*, was stretched between the elongate finger on one side, and the body and leg on the other, thus forming the wing, which rather resembled the wing of the bat than that of a bird, though differing from the former in being supported by one finger instead of four. A few exceptionally well-preserved specimens found in the Solenhofen limestones have retained the clearly-marked impressions of these wing membranes. The legs, like those of bats, were small and weak, and the tail was very short in some species, very long in others. Some, at least, of the latter had a membranous, oar-like expansion at the tip of the tail. That the Pterosaurs had the power of true flight, and did not merely take great leaps like the flying squirrels, is shown by the hollow, pneumatic bones (like those of birds), and by the keel on the breast-bone for the attachment of the great muscles of flight. This keel is found in both birds and bats. The skin was naked, having neither scales nor feathers. The Jurassic Pterosaurs were small, the spread of wings not exceeding three feet.

Birds. — One of the most remarkable advances which Jurassic life has to show consists in the first appearance of the birds. As yet, only a single kind of Jurassic bird has been found, and that in the Solenhofen limestones. This, the most ancient known bird, is called *Archæopteryx* (Fig. 297), and has many points of resemblance to the reptiles, and many characters which recur only in the embryos of modern birds. The peculiarities which strike one at the first glance are the head and tail; there was no horny beak, but the jaws are set with a row of small teeth, while the tail is very long, composed of separate vertebræ, and with a pair of quill-feathers attached to each joint. The wing is constructed on the same plan as that of a modern bird, but is decidedly more primitive. The four fingers are all free (in recent birds two of the three fingers are fused together); they have the same number of joints as in the lizards, and are all provided with claws. The plumage is thoroughly bird-like in character, but is peculiar in the presence of quill-feathers on the legs, and apparently also in the absence of contour feathers from the head, neck, and much of the body, leaving those parts naked. This very extraordinary creature was, then, a true bird, but had retained many features of its reptilian ancestry, and shows us that those ancestors have still to be sought in the Trias or even the Permian.

FIG. 297. — Restoration of *Archæopteryx lithographica* v. Meyer. (Andreae)

Mammalia. — The mammals of the Jurassic are still very rare and imperfectly known, and have been found in only a few places. How many mammalian genera should be referred to the Jurassic will depend upon where the somewhat arbitrary line is drawn, which separates that system from the Cretaceous.

The *Multituberculata* are regarded as belonging to the most primitive kind of Mammals, the *Monotremata*, at present represented only by the duck-billed Mole (*Ornithorhynchus*) and Spiny Ant-Eater (*Echidna*) of Australia, animals which, though warm-blooded and suckling their young, reproduce by laying an egg. Of the Multituberculates the most prominent Jurassic representatives are the English genus *Plagiaulax*, from the Purbeck, and the American genera *Ctenacodon* and *Allodon*, from the Morrison of Wyo-

ming. In another group, which may be related to the Marsupials, the teeth are simpler and more numerous; examples of this
group are the Purbeckian genera, *Stylodon*, and *Triconodon*, and
the Morrison genera, *Dryolestes* and *Dicrocynodon*. Thus, at the
very end of the Jurassic, the mammals are still tiny, insignificant
creatures, which play but a very subordinate rôle in the luxuriant
terrestrial life of the period.

CHAPTER XXXIV

THE CRETACEOUS PERIOD

THE name *Cretaceous* is derived from the Latin word for chalk (*Creta*), because in England, where the name was early used, the thick masses of chalk are the most conspicuous members of the system. Though first made known in England, the main subdivisions of the Cretaceous, as employed in geological literature, bear French names, which have proved themselves better adapted to general use.

In very marked contrast to the scanty development of the Jura, the Cretaceous strata of North America are displayed on a vast scale, and cover enormous areas of the continent, eloquent witnesses of the great geographical changes in that long period. Continental, estuarine, and marine rocks are all well represented, and, in consequence, our information regarding the life of North America and its seas during Cretaceous times is incomparably more complete than it is for the Triassic and Jurassic.

The Cretaceous rocks of North America are of very different character in the different parts of the continent, and require separate classification.

DISTRIBUTION OF CRETACEOUS ROCKS

American. — At the opening of the Cretaceous, the Atlantic coast of North America appears to have been farther to the eastward than it is at present; but just as had happened in the Triassic period, a long, narrow depression was formed, running roughly parallel with the coast, and in this depression for a long period of time, sediments in the form of gravels, sands, and clays were de-

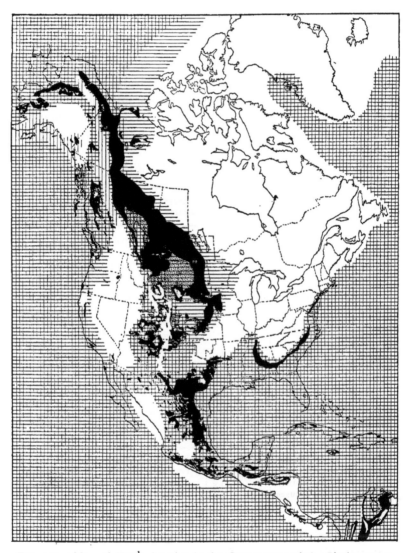

FIG. 298.—Map of North America in the Cretaceous period. Black areas =
known exposures; white = land; dotted areas = continental formations;
lined areas = sea. Vertical lines indicate Lower, and horizontal lines Upper
Cretaceous Seas

posited. This is the *Potomac* series, which is divisible into several stages. There are unconformities within the series, which contains driftwood, some lignite, and iron ore. The beds are of continental origin and probably differed locally in their circumstances of deposition, flood-plain, delta, and marsh, being apparently all represented. The Potomac has been traced through the islands of Martha's Vineyard, Nantucket, Block Island, Long Island, Staten Island, across New Jersey, and thence southward to Georgia,

CRETACEOUS FORMATIONS OF THE UNITED STATES

	European	Atlantic Border	Gulf Border	Southern Interior	Northern Interior	Pacific Border
Upper Cretaceous	Danian	Manasquan	Wanting	Wanting	Livingstone and Denver	Wanting
		Rancocas		Laramie	Laramie	
	Senonian	Matawan	Ripley Selma	Montana 2. Glauconitic 1. Ponderosa Marls	Montana 2. Fox Hills 1. Fort Pierre (Belly River)	Chico
		Monmouth	Eutaw			
	Turonian			Colorado 2. Austin 1. Eagle Ford	Colorado 2. Niobrara 1. Benton	
Lower Cretaceous	Cenomanian Albian	Wanting	Wanting	Dakota	Dakota	
	Aptian Neocomian and Wealden	Potomac {4. Raritan 3. Patapsco 2. Arundel 1. Patuxent}	Tuscaloosa	Comanche {3. Washita 2. Fredericks- burg 1. Trinity}	Washita Fuson Lakota (Kootanie)	Shastan {Horsetown Knoxville}

where it turns northwestward, following the Mississippi embayment into Tennessee, and from there turning southwestward through Arkansas. In the northern part of this region, from Nantucket to the Delaware River, only the upper part of the Potomac has been found, and the same appears to be true of the *Tuscaloosa,* as the extension around the Mississippi embayment is called. The Potomac is nowhere marine, and everywhere rests unconformably upon the underlying Triassic and older rocks. As the thickness of sediment is not great (not exceeding 600 feet), the process of deposition must have been very slow, or broken by long interruptions with intervals of erosion.

While along the Atlantic border the land was more extended than at present, in the southern part of the continent a different order of events was brought about. Nearly the whole of Mexico had been submerged by the great Upper Jurassic transgression, and in the Lower Cretaceous the sea extended over Texas and New Mexico into Arizona, and gradually expanded northward in the successive stages of the Comanche epoch, or Lower Cretaceous. At the base of the Lower Cretaceous strata in Texas is found a deposit of continental sands, the *Trinity* stage, which is the recognized equivalent of the basal Potomac. The advancing sea covered these sands, and the continued depression soon established a clear and quite deep sea, in which were formed the great masses of the *Comanche* limestones, that are the surface rocks of large areas in Mexico, and cover much of Texas. The Ouachita Mountains of Arkansas stood out as a promontory in the Lower Cretaceous sea, and the ancient shore-line has been traced around their foot. Over a great part of Texas the Comanche limestones are soft, and beds of chalk occur among them; while in Mexico, where they have been folded into mountain ranges, they have become much harder and more compact. The thickness of the limestones increases southward; from 1000 feet in northern central Texas, it rises to 5000 feet on the Rio Grande, and on the Mexican plateau to an even greater amount. No less than six distinct, successive marine faunas are found in the Comanche limestones of Texas, and the faunal relationships of this region are closest with the Mediterranean province of Europe, and especially with the Lower Cretaceous of Portugal. The greatest expansion of the Comanche sea northward took place in the last of its stages, the *Washita*, when Oklahoma, southern Kansas, and eastern Colorado were covered by it.

Just how far north this sea reached has not yet been determined, but there is reason to think that it extended into central Wyoming. This Lower Cretaceous marine invasion of the northern interior lasted but a relatively short time, and, until quite recently, its thin deposits had escaped detection.

In the northern interior region the Lower Cretaceous beds, except those of the Washita, are of continental origin, and it is not practicable to correlate those of different areas where the stratigraphic connections cannot be traced. Part of the Morrison is probably Lower Cretaceous, though we cannot yet say how much of it, nor what particular areas. Another non-marine formation found east of the Gold Range of British Columbia, extending southward into Montana, is the *Kootanie* stage, the plant remains of which correlate it with the lower Potomac, and it certainly is not the oldest Cretaceous, for in British Columbia it has been found lying unconformably upon marine Lower Cretaceous. In part, the Kootanie was formed in tracts of low-lying, swampy lands; on which a luxuriant vegetation produced valuable deposits of coal. Lower Cretaceous beds have been found surrounding the Black Hills where they have been divided into the Lakota and Fuson stages, of continental origin, with abundant remains of land plants.

Along the Pacific coast Lower Cretaceous rocks are displayed on a great scale. The Great Basin land then extended from southern Nevada to 54° N. lat. in British Columbia, with the Sierra Nevada rising along part of its western border, to which the Pacific extended. North of the Gold Range in British Columbia, the ocean spread eastward, though no doubt broken by many islands, to the eastern base of the Rocky Mountains. The Coast Range of California formed a chain of islands and reefs. In the Sierra Nevada occurs an unconformity between the Lower Cretaceous and the uppermost Jurassic, but it does not imply the lapse of a very long period of time.

. The older division of the Californian Lower Cretaceous is called the *Knoxville*, and has an estimated maximum thickness of 20,000 feet, laid down upon a slowly subsiding sea-bottom. This enormous thickness is no doubt due to an extremely rapid deposition of the débris abundantly supplied from the waste of the newly upheaved Sierras. At the end of the Knoxville age, the subsidence became more rapid and the sea began to encroach upon the land, for the *Horsetown* beds, which have a thickness of 6000 feet, overlap the Knoxville shoreward and extend over upon the underlying Jurassic and other pre-Cretaceous systems. Although the two stages of the Californian Lower Cretaceous are entirely conformable throughout, and appear to have been formed by a continuous process of sedimentation, yet there is a very marked faunal change between them. The Knoxville beds have a northern fauna, allied to that of Russia, showing that the connection with Russian seas, which had been established in late Jurassic times, was still kept up. With the beginning of the Horsetown age, however, this northern communication was interrupted doubtless by the closing of Bering Sea, and a connection was formed with the waters of southern Asia, and in that way with central Europe. The decided contrast which we find between the Lower Cretaceous faunas of California and those of Texas points to the existence of a land barrier between the seas of the two regions.

In the southern region the Lower Cretaceous was terminated by an upheaval, which caused the Comanchean Sea to withdraw from Texas and the area to the west and north of it. This mid-Cretaceous land epoch must have continued for a considerable time, permitting extensive denudation and a complete change in the fauna. Wherever the marine Upper Cretaceous is in contact with the Comanche limestones north of Mexico, the two are unconformable, and no species of animal is known to pass from one to the other. In Mexico the Lower Cretaceous passes into the Upper without a break, the disturbances there taking place at a later date.

The Upper Cretaceous rocks have a far wider distribution over North America than have those of the lower division, which is due to an enormous transgression of the sea over the land, one of the greatest in all recorded geological history. Over the region of the

2 z

Great Plains the Upper Cretaceous was inaugurated by the formation of a non-marine stage, the *Dakota*. These strata cover much
of Texas, lying unconformably upon the Comanche series, and extend northward into Canada. In Kansas, however, the connection of the Dakota with the Washita is very close, bands of sandstone carrying Dakota species of plants being interstratified with
the marine beds. On the western side of the Colorado uplift, the
Dakota is less distinctly a sandstone formation, and is characterized
by beds of shale and even coal seams of workable thickness. In
most parts of the Rocky Mountain region the Dakota rests in
apparent conformity upon the lowest continental Cretaceous, and
even upon the Jurassic. In the Uinta and Wasatch ranges there
is no apparent break in sedimentation from the Palæozoic to the
end of the Mesozoic, though the whole Lower Cretaceous is there
wanting. From this we may infer that during the long Lower
Cretaceous time all these regions had been low-lying lands, nearly
or quite at base-level, and therefore not subject to profound denudation.

It was at the end of the Dakota age that the great subsidence
took place which affected nearly all parts of the continent, and
brought the sea in over vast areas where for ages had been dry
land. South of New England the Atlantic coastal plain was submerged, and in New Jersey, at least, the waters covered even the
nearly base-levelled Triassic belt, bringing the sea up to the foot of
the crystalline highlands. The lowlands of Maryland, Virginia,
and the Carolinas, and all of Florida were under the ocean, and
the Gulf of Mexico was extended northward in a great bay (the
Mississippi embayment), covering western Tennessee and Kentucky and extending into southern Illinois. In Mexico important
changes occurred during the Upper Cretaceous. Early in this
epoch a general elevation restricted the sea to the northeastern
part of the country, but in the south was an opposite movement,
the sea transgressing upon the ancient land which lay to the west
of the Isthmus of Tehuantepec, finally covering it late in the period.
In this region the Upper Cretaceous overlaps upon the Archæan.

Elsewhere, the disturbance referred to was orogenic, resulting in the formation of most of the Mexican mountain ranges and continuing nearly or quite to the end of the period. Texas was again extensively submerged, and a wide sea connected the Gulf of Mexico with the Arctic Ocean. The eastern coast of this interior sea began in northwestern Texas, running through Kansas and Iowa nearly to the present line of the Mississippi River. Westward the coast-line was the uplift which ran from the west coast of Mexico into British Columbia. The Colorado region was again converted into islands. North of the Great Basin land the interior sea was connected with the Pacific and Arctic Oceans, which united over the northwestern part of the continent.

On the Pacific side, the Sierras, which had suffered greatly from denudation, were again folded. A moderate transgression of the sea caused the Upper Cretaceous to extend farther east than the Lower. Volcanic activity continued and immense bathyliths were formed deep within the mountains. The sea extended from Lower California northward along the Sierra into eastern Oregon at the foot of the Blue Mountains.

The North American continent was thus divided into two principal land masses, the larger one to the east and comprising the pre-Cambrian and Palæozoic areas. In the limits of the United States this land lay almost entirely east of the Mississippi, except for a southwestern peninsula, including Missouri, Arkansas, Oklahoma, and part of Texas. The western area was much smaller, extending from Mexico into British Columbia, and having its greatest width between the fortieth and forty-fifth parallels of latitude. Between the two lands lay the Colorado Islands, and doubtless many smaller ones as well.

The character of sedimentation differed so much in the various regions of the continent that the subdivisions of the Upper Cretaceous have received different names in the separate provinces, and only approximately correspond in time. The greater number of these subdivisions, which are founded chiefly upon physical changes, gives to the Upper Cretaceous the appearance of being longer and more important than the Lower, but this is only an appearance. In Europe the Lower Cretaceous has, in recent times, been divided into six series, an arrangement which proves to be of general validity.

Along the Atlantic border the Upper Cretaceous strata are a series of marine sands and clays, which are still almost horizontal in position and of loose, incoherent texture. In New Jersey there are extensive developments of green sands locally called marl. The Appalachian Mountains, which had been subjected to the long-continued denudation of Triassic, Jurassic, and Lower Cretaceous times, were now reduced nearly to base-level, the Kittatinny plain of geographers (see p. 512). This peneplain was low and flat, covering the whole Appalachian region, and the only high hills upon it were the mountains of western North Carolina, then much lower than now. Across this low plain the Delaware, Susquehanna, and Potomac must have held very much their present courses, meandering through alluvial flats.

On the Gulf border the Upper Cretaceous beds of Alabama and Mississippi, which were laid down in the Mississippi embayment, are in 3 stages. Below are the sands and clays of the *Eutaw* (300 feet) which is correlated with the Matawan of New Jersey; next follows the soft limestone, or chalk, of the *Selma* (500–1200 feet), and at the top are 200 to 500 feet of the *Ripley* sands. Eastward the water shallowed, and in Georgia we find about 1400 feet of clays and sands. Northward along the Mississippi embayment the beds thin greatly and are mostly clays and sands.

In the interior region lying upon the Dakota are the marine beds of the *Colorado*, of which the lower division is the *Benton*, a mass of shales and limestones with a maximum thickness of 1000 feet, though varying much from point to point. The depression still continuing, the sea became quite deep, making favourable conditions for the formation of the chalk and harder limestones of the *Niobrara*. This chalk is best seen in Kansas, but extends into South Dakota; elsewhere are sandstones and limestones with a maximum thickness of 2000 feet. A movement of reëlevation of the sea-bottom began even in Colorado times, and in the northern part of the interior region oscillations of level produced alternating fresh-water, or estuarine, and marine conditions. In Montana and the Canadian province of Alberta is a thick body of estuarine or fresh-water strata with seams of coal (the *Belly River* formation) interposed between the marine deposits of the Colorado below and the Montana above. In Utah is another fresh-water deposit of coal-bearing rocks of Colorado age.

In the *Montana* epoch marine conditions still prevailed, but the waters of the northern sea had generally become much shallower, and a marked change of fauna was produced. In Alberta are coal measures of this date. Two divisions of the Montana are distinguished, although not everywhere separable; the *Fort Pierre*, which is composed of shales and sandstones with a maximum thickness of 8000 feet, and the *Fox Hills*, sandstones and some shales, which do not exceed 1000 feet. This movement of upheaval in the interior was accompanied or followed by an uplift on the Atlantic and Gulf coasts, for along these borders the uppermost Cretaceous beds are either wanting or represented by exceedingly thin deposits. In the interior the continued upheaval caused fresh-water and swampy conditions to prevail over very wide areas, though not so widely extended as had been the Upper Cretaceous sea. This great continental formation is the *Laramie*, which has no such eastward extension as the marine Cretaceous, but is restricted to the western side of the basin and is, in part, probably equivalent to the latest marine Cretaceous, the difference being in facies rather than in time. However, this applies only to the older part of the Laramie, which as a formation continues much later than any of the marine stages and may even have persisted into the Eocene. The northwestern part of the continent had been converted into dry land, but a broad area of non-marine deposition extended up the course of the present Mackenzie River to 62° N. lat. Another and vastly larger area began about 57° N. lat., and reached, though perhaps with interruptions, to northeastern Mexico, surrounding the Colorado island. This area was 2000 miles long and 500 miles wide, and reproduced the conditions which obtained around the Mississippi valley in the Upper Carboniferous, immense swamps and peat-bogs in which gathered the quantities of vegetable matter now converted into coal seams. The clastic rocks interstratified with the coal are probably fluviatile and lacustrine deposits, and occasional brackish-water conditions are reported from some areas. Workable coal is found in all the stages of the western Cretaceous, but none of these stages is comparable to the Laramie for the extent and thickness of its coal measures.

The Laramie was a time of tranquillity, with only slow and gentle changes of level, but towards its close some important disturbances took place, especially along the Rocky Mountains. The first of these movements affected only the Colorado island, and its effects are especially well shown in the Denver basin, where some 800 feet of conglomerates (the *Arapahoe*) rest upon the Laramie unconformably. The second series of movements was much more extensively felt, producing marked unconformities both in Colorado and Montana. In Colorado there was a great volcanic outburst, and the *Denver* stage, which overlies the Arapahoe unconformably, is principally composed of andesitic tuffs. In Montana the equivalent stage (*Livingstone*), which also contains considerable volcanic material, is 7000 feet thick and unconformable with the Laramie. It is probable that the Arapahoe, Denver, and Livingstone, all of which occur along the Rocky Mountains, correspond to beds which elsewhere are a part of the Laramie. The latter in eastern Wyoming passes into undoubted Eocene above, by what appears to be an unbroken continuity of sedimentation.

The Upper Cretaceous of the Pacific coast comprises the *Chico* series, with a maximum thickness of 4000 feet. In Vancouver's Island the Chico is coal-bearing. The faunal connections of the Chico are with southern Asia, that series having very little in common with the fossils of the interior region. The uppermost Cretaceous is wanting along the Pacific coast, except for certain coal-bearing beds in Washington, which appear to represent the Laramie.

The resemblance of the Chico fossils to those of southern Asia indicates the closing of Bering Sea and thus the possibility of a migration of shoal-water animals all around the shores of the North Pacific, at the same time providing a way for the interchange of land animals and plants between North America and the Old World. The Upper Cretaceous faunas of Mexico are surprisingly different from those of the United States, and so like those of the Mediterranean region of Europe as to suggest an east and west shore-line across the Atlantic in the latitude of Brazil, while the northern connection of America with Europe probably continued also, for the shallow-water fossils of New Jersey resemble those of central Europe.

The Mesozoic era was closed in the West, as the Palæozoic had been in the East, by a time of great mountain making, and to this movement is attributed the formation of most of the great Western mountain chains. From the Arctic Ocean to Mexico the effects of the disturbance were apparent. The Rocky Mountains, the Wasatch and Uinta ranges, the high plateaus of Utah and Arizona, and the mountains of western Texas date from this time, though subsequent movements have greatly modified them. Vast volcanic outbreaks accompanied the upheaval, which was on a far grander scale than the Appalachian revolution had been.

Foreign. — In *South America* the Cretaceous history is much like that of the northern continent. The subsidence which inaugurated the Lower Cretaceous extended the sea over the northern part of South America and covered northeastern Brazil, with fresh-water deposits in central Brazil. All along the Cordillera, from Venezuela to Patagonia, marine Cretaceous is found, but east of the mountains, with the exceptions already noted, the system is represented chiefly by non-marine sandstones. In Patagonia, however, is an area of marine Lower Cretaceous east of the Andes, though its extent is not known. Thick continental sandstones represent most of the period, but toward the close, the entire Patagonian plain appears to have been submerged for a short time by a transgression of the Upper Cretaceous sea. The faunal relations of the South American Lower Cretaceous are very intimate with northern and western Africa. Gigantic volcanic activity went on along the Cordillera in Mesozoic times; in Chili and Peru the marine Cretaceous is principally made up of stratified igneous material, and the Andes contain the largest known area of Mesozoic eruptives. The mountain-making upheaval probably came at the close of the Cretaceous.

In *Europe*, toward the end of the Jura, the sea retired from nearly all of the central region, which in part became dry land and in part was covered with lakes and inland seas. One of the largest of these covered much of southern England, extending far into Germany, and in it was deposited a great thickness of sand and clay, with some shell limestone, the *Wealden*. The Alpine region remained submerged under a clear and deep sea, and the transition from the Jurassic is very gradual. In the oldest Cretaceous epoch (*Neocomian*) a renewed transgression submerged large parts of central Europe, though the sea was far less extensive than that of the Middle and Upper Jurassic. In consequence, a great gulf was established over southern England, northern France, and north Germany to Poland, a gulf bounded on the north by the highlands of Britain, Scandinavia, and northwestern Russia, and on the south by a land stretching from Ireland to Bohemia; Belgium was mostly an island. The expanded Mediterranean covered southeastern Asia Minor and northern Africa. In the Upper Cretaceous the northern gulf was greatly extended, covering many areas that had been land since Palæozoic or pre-Cambrian times. Parts of this basin became very deep, and its most characteristic deposit, especially over southern England and northern France, was chalk, which the microscope shows to be made up of the shells of Foraminifera and to resemble the modern foraminiferal oozes. Over the Alpine region upheavals in the Upper Cretaceous had established land areas, indicated by extensive fresh-water deposits recurring at intervals from Spain to Hungary, in the latter country containing coal. The Cretaceous was closed in Europe by a gradual upheaval which excluded the sea from wide areas that it had occupied.

In *Africa* the only extensive Cretaceous areas are those of the north, where the Atlas Mountains, and much of the surface of the Libyan desert are made up of these rocks. A limited transgression of the sea also took place along the western coast. In South Africa are traces of two Cretaceous invasions of the sea, both of which merely occupied old valleys for quite a short distance from the coast of Cape Colony and Natal. The first invasion is of Neocomian date (*Uitenhage* beds) and its fossils have a distinct likeness to those of Patagonia. The second incursion took place later in the Cretaceous, at a horizon not yet determined.

Southern and eastern *Asia* display many areas of Cretaceous rocks, as, for example, in southern India and Japan. *Australia* also has extensive areas of this system, which are best known in Queensland, where they are chiefly Lower Cretaceous and contain coal. The New Zealand Cretaceous is also coal-bearing.

Climate. — The evidence for the existence of climatic zones is more distinct in the Cretaceous than in the Jurassic, though the difference between the zones must have been slight, for the Upper Cretaceous flora extends to Greenland with hardly any change. On the other hand, the marine animals show a decided difference according to latitude. In the Mediterranean region of Europe and Asia, the West Indies, Mexico, and the north coast of South America the seas abounded in reef-building corals, in the extraordinary groups of bivalve molluscs, or Pelecypoda, called the *Rudistes* and *Caprotinæ*, and in certain genera of Ammonites, such as *Lytoceras*, *Haploceras*, and *Phylloceras*. In northern and central Europe and on the Atlantic coast of the United States these forms are rare or absent and other groups take their place. The probable explanation of the seeming contradiction in the testimony of land plants and marine animals is in the existence of a cool polar sea and southward currents from it.

CRETACEOUS LIFE

The life of the Cretaceous displays so great an advance over that
of the Jurassic that the change may fairly be called a revolution.

Plants. — If the separation between the Mesozoic and Cenozoic
eras were made entirely with reference to the plants, it would pass
between the Lower and the Upper Cretaceous, just as a similar
criterion would remove the Upper Permian to the Mesozoic. The
vegetation of the Lower Cretaceous, especially of the lowest, is still
much like that of the Jura. Ferns, Horsetails, Cycads, and Conifers

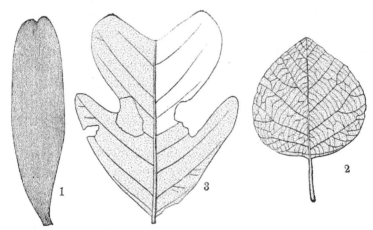

FIG. 299. — Cretaceous leaves. Dakota stage.   1. *Dammarites emarginatus* Lesq.,
1/2.   2. *Betulites Westi* Lesq., 3/4.   3. *Liriodendron giganteum* Lesq., 1/2.
(Lesquereux)

continue to make up most of the flora. The Cycadales, in particular,
abound in the Lower Cretaceous, many of them belonging to the
Bennettiteæ. On the other hand, the impending revolution is an-
nounced by the appearance of *Dicotyledons* of archaic and primitive type. In the higher parts of the Potomac the Cycads become
less abundant and the Dicotyledons very much more so. Here
we find many leaves which belong to genera that cannot be distinguished from those of modern forest trees, such as *Sassafras*, *Popu-*

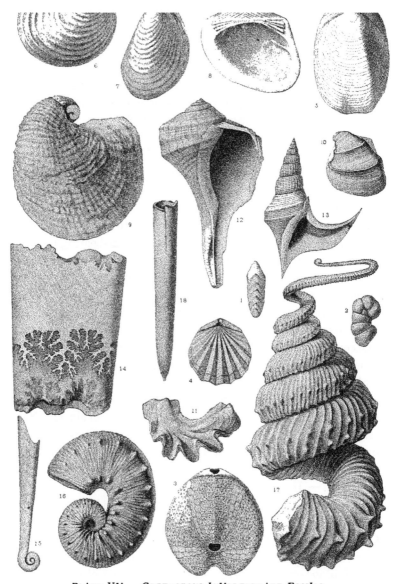

PLATE XV. — CRETACEOUS INVERTEBRATE FOSSILS

Fig. 1, *Frondicularia major* Bornemann, × 3, Up Cret., N.J. 2, *Haplophragmium concavum* Bagg, × 12, U. C. 3, *Cardiaster cintus* Morton, × ½, U. C. 4, *Terebratella plicata* Say, × 1, U. C. 5, *Terebratula harlani* Whitf. × ½, U. C. 6, *Inoceramus vanuxemi* M. and H., × ½, Ft. Pierre. 7, *Aucella pioche* Gabb, × 1, Knoxville. 8, *Idonearca nebrascensis* Owen, × ½, Fox Hills. 9, *Exogyra costata* Say, × ⅜, U. C. 10, *Veniella conradi* Morton, × ⅝, U. C. 11, *Ostrea larva* Lam., × ⅝, U. C. 12, *Pyropsis bairdi* M. and H , × ½, Fox Hills. 13, *Anchura americana* × 1, Evans and Shumard, Fox Hills. 14, *Baculites compressus* Say, × ½, fragment of adult with suture-lines, Ft. Pierre. 15, *The same*, a very young shell, × 5, showing apical coil. 16, *Scaphites nodosus* Owen, × ⅜, Ft. Pierre. 17, *Heteroceras stevensoni* Whitf., × ²/₅, Ft. Pierre. 18, *Belemnitella americana* Morton, × ½, U. C.

lus, Liriodendron, etc. No Dicotyledons have been found in the Kootanie of the Northwest, or in the Wealden of northern Europe, but they occur in the Lower Cretaceous of Portugal, Greenland, and Spitzbergen. In the latter part of the Lower and in all the Upper Cretaceous of North America the flora assumes an almost completely modern character, and nearly all of our common kinds of forest trees are represented: Sassafras, Poplars, Willows, Oaks, Maples, Elms, Beeches, Chestnuts, and very many others. A new element is the Monocotyledonous group of Palms, which speedily assumes great importance. Each successive plant-bearing horizon of the Cretaceous is characterized by its own special assemblage of plants, but in its general features the Upper Cretaceous flora is essentially modern, and this is true of the world at large, while in the Lower division it was only in North America and a few other scattered areas that the Angiosperms had gained a foothold. Cretaceous *animals* are sufficiently different from those of the Jura, but the change is not so revolutionary as we have found among the plants.

Foraminifera play an important part in the construction of Cretaceous rocks, especially of the great masses of chalk (Pl. XV, Figs. 1, 2), while the green sands are casts of foraminiferal shells in glauconite. The most abundant genus, as in the recent Atlantic oozes, is *Globigerina.*

Spongida. — In the Cretaceous of Europe Sponges are more numerous and varied than at any other time, but in North America they are far less common.

Cœlenterata. — The Corals were very much as they are to-day and require no special description.

The **Echinodermata** undergo some very marked changes. The Crinoids are much reduced since the Jurassic, and were never again to assume their ancient importance; characteristic Cretaceous genera are the stemless, free-swimming *Uintacrinus* and *Marsupites.* The Sea-urchins are incomparably more numerous in Europe than in North America; of the Regular forms may be mentioned *Pseudodiadema, Cidaris,* and *Salenia,* and of the

Irregular forms, *Toxaster*, *Holaster*, *Cassidulus*, *Cordiaster* (XV, 3), etc.

Arthropoda. — Among the Crustacea we need only note the great increase in the Brachyuran Decapods, or Crabs, in the beds of the Gulf border and of Europe.

Brachiopoda are very much as in the Jurassic; the common genera are *Terebratula* (XV, 5), *Terebratella* (XV, 4), and *Rhynchonella*.

Mollusca. — This group is very richly developed, and many genera are peculiar to the period. The large, curious oysters belonging to the genera *Ostrea* (XV, 11), *Gryphæa*, and especially *Exogyra* (XV, 9), are common, and the many species of *Inoceramus* (XV, 6) are very characteristic, especially of the northern facies. More modern types are *Idonearca* (XV, 8) and *Veniella* (XV, 10). Confined to the Cretaceous are the extraordinary shells classed as *Rudistes*, in which one valve is long and horn-shaped, and the other a mere cover for it. These shells of the genera *Hippurites*, *Radiolites*, and *Coralliochama* are much commoner in Europe than in America and are preëminently southern in distribution. Other peculiar Cretaceous Bivalves are *Requienia* and *Caprotina*. *Aucella* (XV, 7) is of interest both in the Upper Jurassic and the Cretaceous as a typically Boreal group of shells. The *Gastropods* (XV, 12, 13), are very much as in the Jura, but in the latter part of the period come in many genera which reach their fullest development in Tertiary and recent times, such as *Fusus*, *Murex*, *Voluta*, *Cypræa*, and many others.

The *Cephalopods* are very peculiar; in addition to numerous Ammonoid genera with closely coiled shells of normal type, such as *Hoplites*, *Schlœnbachia*, *Placenticeras*, we find very many shells entirely or partially uncoiled, or rolled up in peculiar ways, which give to the Cretaceous Cephalopod fauna a character all its own. In *Crioceras* (XIV, 20) the shell is coiled in an open, flat spiral, the whorls of which are not in contact. *Ancyloceras* has a similar open coil, followed by a long, straight portion, and recurved terminal chamber. *Scaphites* (XV, 16) is like a shortened *Ancylo-*

ceras. In *Ptychoceras* the shell consists of two parallel parts, connected by a single sharp bend. *Turrilites* is coiled into a high spiral, like a Gastropod, and *Baculites* (XV, 14, 15) has a perfectly straight shell except for a minute coil at one end. *Heteroceras* (XV, 17) displays the extreme of irregularity of growth. *Nautilus* is represented by many species, some of them very large. Belemnites are very abundant, but in the Upper Cretaceous the genus *Belemnitella* (XV, 18) replaces the true *Belemnites*.

The **Vertebrata** form the most characteristic element of the Cretaceous fauna. Among the *Fishes* a revolution has occurred. Sharks of modern type abound, and their teeth are found in countless numbers; but the principal change consists in the immense expansion of the *Teleosts*, or Bony Fishes, which now take the dominant place, while Ganoids become rare. Most of the Cretaceous Teleosts belong to modern families and even genera, such as the Herrings, Cod, Salmon, Mullets, Catfishes, etc.; but a characteristic Cretaceous type, now extinct, is that of the *Saurodonts*, fierce, carnivorous fishes of great size and power. The genus *Portheus*, common in the Kansas chalk, was 12 to 15 feet long, and was provided with great reptile-like teeth.

The *Reptiles* continued to be the dominant types of the land, the sea, and the air, and it may fairly be questioned whether the Jura or the Cretaceous should be regarded as the culminating period of Reptilian history. *Ichthyosaurs* and *Plesiosaurs* are perhaps less abundant than in the Jura, but are of greatly increased size. *Elasmosaurus*, a Plesiosaur from the Kansas chalk, had a length of 40 to 50 feet, of which 22 feet belonged to the slender neck. Confined to the Cretaceous are the remarkable marine reptiles of the group *Mosasauria*, which swarmed on the Atlantic and Gulf coasts, and especially in the interior sea. These were gigantic, carnivorous marine lizards, with the limbs converted into swimming paddles (see Fig. 300). *Turtles*, both fresh-water and marine, abound, and some were very large. *Lizards* and *Snakes* are but scantily represented, not displaying the manifold variety of structure which they afterwards acquired. An order of aquatic reptiles, the *Choristodera*, nearly allied to the Rhyncho-cephalia, appeared in the latter part of the period. *Crocodiles*, like those of modern days, were ubiquitous in both fresh and salt waters, and in North America, at least, some of the long-snouted Jurassic type of crocodiles, *Teleosaurus*, continued into the Upper Cretaceous.

FIG. 300. — *Tylosaurus dyspelor* Marsh. A Mosasaurian in pursuit of Saurodont Fishes (*Portheus*). Restoration by C. R. Knight, under the direction of Professor H. F. Osborn. (Copyright by Amer. Mus. Nat. Hist., N.Y.)

The *Pterosaurs* of the Cretaceous are remarkable for their great size, far exceeding that of the Jurassic species. The closely allied genera, *Ornithostoma* of Europe and *Pteranodon* of Kansas, had a head of nearly 3 feet in length, with a long, pointed, tooth-less bill, like that of a bird; the spread of wings exceeded 20 feet.

The *Dinosaurs* continue in even greater profusion than in the Jurassic; they are, of course, much commoner and better pre-served in continental deposits than in marine, and hence are best known from the base and the summit of the system. Many of the genera were the largest land animals that ever lived, and the size of the bones is astonishing. The Wealden of Europe has yielded some magnificent Dinosaurs; especially the genus *Iguanodon*, of which many complete skeletons have been found in Belgium. Dinosaurs are much less common in the marine Upper Cretaceous, but the green sands of New Jersey have yielded *Hadrosaurus*, a herbivorous Dinosaur much like *Iguanodon*, and some car-

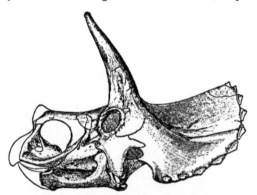

FIG. 301. — Skull of *Triceratops flabellatus* Marsh, from the Side, 1/30. (Marsh.)

nivorous types also. The Laramie and Denver beds have preserved many fine specimens, which show that the Dinosaurs flourished in almost undiminished variety till the end of the Cretaceous. The erect, herbivorous type is represented in these beds by *Monoclonius* and *Diclonius* (Fig. 302), which are nearly related to *Hadrosaurus*. *Triceratops* (Fig. 301) and *Torosaurus* are huge, quadrupedal reptiles, with three large horns on the head and an extraordinary frill-like

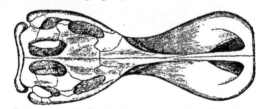

FIG. 302. — Skull of *Diclonius mirabilis* Cope, from above, 1/19. (Cope)

extension of the skull over the neck. Carnivorous Dinosaurs likewise continued, such as *Lælaps*, *Tyrannosaurus* and *Ornithomimus*, the latter with hind limbs which are especially birdlike in structure.

The *Birds* of the Cretaceous are much more abundant and

advanced than the known Jurassic birds. In the Upper Cretaceous of Kansas, and probably of England also, are found two
remarkable birds, *Hesperornis* and *Ichthyornis*. In the former,
which was nearly 6 feet high, the wings were rudimentary, while
Ichthyornis, a much smaller bird, had powerful wings. Both of
these genera possessed teeth, like *Archæopteryx*, but except in
that feature and in certain minor details of structure, they are
entirely like modern birds. Bird bones like the corresponding
parts of the Cormorants and Waders have been found in the green
sands of New Jersey, but it is not known whether they had teeth.

Mammalia. — Cretaceous Mammals are more numerous and
varied than those of the Jurassic, but they continue to play a very
modest rôle, and are nearly all of minute size. In America they
have been found only in the uppermost Cretaceous, and in Europe
they are not known as yet, though doubtless they existed in that
part of the world. The mammals of the Laramie already begin
to show affinities with the forms which are to succeed them
in the Tertiary. The *Multituberculata* are represented by two
genera, *Meniscoëssus* and *Ptilodus*, while other mammals of
doubtful affinities are *Didelphops*, *Pediomys*, and *Cimolestes*.
Many others are known, but they are too imperfect for reference.
With one exception, *Thlæodon*, which is of moderate size, all these
mammals are exceedingly small.

In brief, Cretaceous life is still typically Mesozoic, but a change
toward Cenozoic conditions is already manifest, especially in the
Plants, the Gastropods, the Teleostean Fishes, and the Birds.
There is still a gap between the life systems of the two eras, but it
is not so wide as it was once believed to be, and it may be hoped
that future discoveries will bridge it entirely.

CHAPTER XXXV

CENOZOIC ERA — TERTIARY PERIOD

THE history of the Cenozoic era brings us by gradual steps to the present order of things. Of no part of geological history have such full and diversified records been preserved as of the Cenozoic, and yet this very fulness is a source of difficulty and embarrassment when we attempt to arrange the various phenomena in their chronological order.

The sedimentary rocks of the Cenozoic era are, for the most part, quite loose and uncompacted; it is relatively rare to find hard rocks, such as so generally characterize the older formations. They are also most frequently undisturbed, retaining nearly their original horizontal positions, except when they have been upturned in the formation of great mountain chains. Another characteristic feature of Cenozoic strata is their locally restricted range; only in the oldest parts of the group do we find such widely extended formations as are common in the Palæozoic and Mesozoic groups, and the later Cenozoic strata become more and more local in their character. This implies the restriction of the changes of level, the great transgressions and withdrawals of the sea no longer taking place as they had in the preceding eras. On the other hand, mountain making was effected on a very grand scale in the Cenozoic, and vulcanism was prevalent to, an extent that seems never to have been reached before.

The *climate* of the era underwent some very remarkable and inexplicable changes. At the beginning it resembled that of the Cretaceous in its generally mild and equable character, a luxuriant vegetation flourishing far within the Arctic Circle; but by very slow degrees and with many fluctuations, the climate grew colder, culminating in the Glacial Age, when much of the land in the northern hemisphere was covered with sheets of ice and snow and reduced to the condition of modern Greenland.

The *life* of the Cenozoic era is very clearly demarcated from that of the Mesozoic, though many modern characteristics began in the Cretaceous or even earlier. The peculiar Mesozoic Ammonoids, Belemnites, and many curious Bivalves disappeared almost entirely at the end of the Cretaceous, leaving only a few stragglers here and there to persist into the older Tertiary. Even more striking is the dwindling of the Reptiles; the Ichthyosaurs, Plesiosaurs, Mosasaurs, Dinosaurs, and Pterosaurs, which had given such a marked individuality to the Mesozoic fauna, have become totally extinct, leaving only Lizards and Snakes, Turtles and Crocodiles, and a few *Choristodera* to represent the class. But Cenozoic life is not distinguished from Mesozoic merely by negative characters; it has its positive features as well. The plants and invertebrate animals nearly all belong to *genera* which are still living, and the proportion of modern *species* steadily increases as we approximate the present time. The Fishes, Amphibia, and Reptiles differ but little from those of modern times, and the Birds take on the diversity and relative importance which characterize them now. Above all, the Mammals undergo a wonderful expansion and take the place of the vanished reptiles, giving to Cenozoic time an altogether different character from all that went before it. The great geographical and climatic changes produced migrations of land animals and plants upon a grand scale, from continent to continent and from zone to zone, the result of which is the distribution of living beings over the earth's surface as we find it to-day.

There is some difference of usage regarding the subdivisions of the Cenozoic group, though the difference is principally with reference to the rank of those subdivisions. We shall follow the usual American practice of dividing the group into two systems, the *Tertiary* and *Quaternary*.

THE TERTIARY PERIOD

The names *Tertiary* and *Quaternary* are remnants of an old geological nomenclature which has lost its significance, and were proposed when the whole succession of strata was believed to be divisible into four groups, called the *Primary, Secondary, Tertiary,* and *Quaternary*, respectively. When it was learned that there were groups and systems much older than the so-called *Primary*, the name *Palæozoic* was substituted for *Primary*, as was *Mesozoic* for *Secondary*, though the latter term is still used, especially in England. The name *Tertiary* has thus lost its meaning, but is nevertheless retained as a division of the Cenozoic group or era.

TERTIARY FORMATIONS OF THE UNITED STATES

	European	Western Interior	Gulf Border	Pacific Border
Pliocene	Sicilian Astian Plaisancian Messinian	? Blanco Republican River	Florida or Caloosahatchie	MERCED
Miocene	Tortonian Helvetian Langhian	Loup Fork Deep River Arikaree	Chesapeake	MONTEREY (Cal.) Empire (Oregon)
Oligocene	Aquitanian Tongrian Ligurian	John Day White River Uinta	Chipola Chattahoochee Vicksburg	Astoria (Oregon) Kenai (Alaska)
Eocene	Bartonian Lutetian Suessonian	Bridger Wind River (Green River) Wasatch	Jackson Claiborne Chickasaw	TEJON
Paleocene	Thanetian Montian	Torrejon ⎰ Fort Puerco ⎱ Union	Midway	?

The great revolution which closed the Cretaceous and inaugurated the Tertiary has left its effects visible in all the continents, but the gap between the two periods is not everywhere the same. This revolution gave to North America nearly its present outlines, except for the land connections with Europe and Asia, which were

FIG. 303. — Map of North America in the Tertiary period. Black areas = known exposures of marine Tertiary; white = land; lined areas = sea; dotted areas = continental formations

from time to time interrupted and renewed. In consequence of this, marine Tertiary beds occur only along the borders of the continent, while the Tertiary of the interior is all of continental origin. In other continents, and especially in Europe, the distribution of land and sea was very different in the Tertiary from what it is now, and the topography of the land was profoundly altered in the course of the period. Some of the highest mountain ranges of the earth were upheaved in Tertiary times, such as the Atlas, the Alps, the Caucasus, and the Himalayas, and many ranges of earlier date were subjected to renewed compression and upheaval. That Tertiary mountains are high is not due to any extreme degree of compression as compared with that which produced older ranges, but merely to the youth of the former; denudation. has not yet had time to sweep them away.

The Tertiary system or period is divisible into five well-distinguished series or epochs, which may usually be identified in both the marine and continental formations; but for lack of common fossils it is not yet possible to correlate the stages and substages of the interior region with those of the coast. In the preceding table, therefore, no exact comparison of these minor subdivisions is intended.

The name Tertiary was given by Cuvier and Brongniart, early in the last century, to the succession of marine, brackish-water, and fresh-water beds in the Paris basin. Sir Charles Lyell many years later proposed the division of the Tertiary into three parts, Eocene (from the Greek *eos*, the dawn, and *kainos*, recent), Miocene (*meion*, less, and kainos), and Pliocene (*pleion*, more, and kainos), a scheme which is still used, modified by Beyrich through the insertion of a fourth epoch, the Oligocene (*oligos*, little or in small degree, and kainos). Last of all, the lower Eocene has been separated under the name Paleocene (*palaios*, ancient, as in Palæozoic) a change proposed thirty years ago by the botanist Schimper, but only lately coming into wider favour. It has become customary to distinguish between the older and newer parts of the Tertiary by grouping together the Eocene and Oligocene into the *Palæogene*, and the Miocene and Pliocene into the *Neogene*. *Eocene* and *Neocene* are employed in the same way, but this is objectionable because it is using *Eocene* in two different senses.

The Paleocene Epoch

The term *Paleocene* has not been used by American geological writers, who have, however, frequently employed the more non-committal name of post-Cretaceous. It will be an advantage to follow the European usage wherever this can be done to express facts of correspondence between the two continents.

American. — Marine formations of this epoch have not yet been distinctly identified in North America, though in the table the Midway of the Gulf region has been provisionally placed in that series. On the other hand, extensive areas in the western interior are referable to it. In the region of the Rocky Mountains and northern plains, the Denver and Livingstone beds may eventually prove to be a part of the Paleocene series, a correlation which is favoured by the plants which·they contain. They also contain remains of Dinosaurs, and though it is not at all impossible that some of these great reptiles should have survived in the earliest Tertiary, they are not yet known to have done so. The oldest known beds which are definitely assignable to the Paleocene are those of the *Fort Union*, a formation with a maximum thickness of 2000 feet, which covers very large areas in Canada, Montana, North Dakota, and eastern Wyoming, and is composed of sandstones and clay rocks. In Montana it lies conformably on the Livingstone, and in Wyoming there is an apparently unbroken succession from the Laramie into the Fort Union, barren sandstones between the two· probably representing the Livingstone-Denver series. Originally referred to the Tertiary on account of its plants, the position of the Fort Union has been confirmed by the finding of a considerable number of mammals in it. The conditions under which these beds were formed have not been clearly determined. That they may have been partly lacustrine is indicated by the presence of fresh-water shells in some localities. Other parts are probably flood-plain deposits and others again entirely subaërial. Beds with similar plants have been found in Greenland and Alaska.

A somewhat different facies of the Paleocene occurs in north-western New Mexico and southwestern Colorado in a formation which also lies in apparent conformity upon the Laramie. In these beds, which are about 800 feet thick and mostly barren of fossils, are two separate horizons, which have yielded numerous fossil mammals, and each of which has its own characteristic fauna. The lower and older of these horizons is the *Puerco*, and the higher the *Torrejon*, the Fort Union corresponding to both together.

Foreign. — In northern Patagonia is a continental formation, called, from one of its most characteristic fossils, the *Notostylops beds*, the mammals of which suggest correlation with the Puerco of North America.

In *Europe* a rapid elevation of the continent had occurred in the latest phases of the Cretaceous, followed at the beginning of the Tertiary by numerous minor oscillations of level, which occasioned a continual struggle between land and sea and, as a result, " the Paleocene does not consist of purely marine deposits, but of a repeated alternation of marine, brackish, and fresh-water formations." (Kayser.) The sands, marls, and limestones thus produced cover a large area in the north of France, Belgium, and the south of England. The upper portion of the series near Rheims, in France, contains a mammalian fauna so like that of the American Torrejon as to indicate not only a correlation with that horizon, but also the existence of a land bridge between the two continents, which permitted the migration of terrestrial animals from one to the other. Paleocene beds occur also in Denmark and in central Russia and in southern Europe; in the south of France and the Pyrenees, it is represented by fresh-water beds, while in Egypt it is marine.

Climate. — The earliest Tertiary floras of Greenland and Alaska show that the equable conditions of the late Cretaceous continued, but those of England indicate merely a temperate climate in that latitude, where in the true Eocene it became tropical.

Paleocene Life

The vegetation of the Paleocene is of essentially the same kind as that of the latest Cretaceous, the difference between the two being principally a matter of species. Among the marine invertebrates, which are still very incompletely known, the most important feature to be noted is a negative one; namely, the almost universal and complete absence of the characteristic Mesozoic types, Ammonites, Belemnites, and the like. Of the land animals, the more or less aquatic reptiles allied to the Rhynchocephalia, the *Choristodera*, survived, but all the other Mesozoic orders of reptiles vanished in what appears to be a startlingly sudden way; no trace of the Dinosaurs, Pterosaurs, Ichthyosaurs, Plesiosaurs, Mosasaurs, etc., has been found in the Paleocene. On the contrary, the *Mammals* have greatly increased in numbers and importance, but are still extremely primitive and very like those of the late Cretaceous; with one, or perhaps two, exceptions, they all belong to extinct orders and are of small or moderate size, not a single large one having been found. The close relation with the Mesozoic is seen in the presence of several of the *Multituberculata;* the last and largest of the order, *Polymastodon* and *Ptilodus*, the latter a survival from the Laramie, are the common genera. The primitive flesh-eaters, *Creodonta*, and primitive hoofed animals, *Condylarthra* and *Amblypoda*, are abundant, and the very curious extinct orders of the *Tillodonta* and *Tæniodonta*, and primæval *Lemuroids*, and perhaps *Insectivora*, complete the list. The assemblage is an extremely archaic one and notable for the absence of the characteristically Tertiary groups; there are no Monkeys, Rodents, Bats, true Carnivores, Artiodactyls, Perissodactyls, or any of the other higher orders of mammals.

THE EOCENE EPOCH

American. — Along the Atlantic and Gulf borders the coast-line of the Eocene closely follows that of the Cretaceous, of which only a narrow strip separates the Eocene from the Triassic and crystalline rocks of the Piedmont plain. The unconformity between the Cretaceous and Eocene indicates that along this coast the latter period had been inaugurated by an encroachment of the sea upon the land. The Mississippi embayment had nearly the same size and form as before, extending up to the mouth of the Ohio. Florida was entirely submerged, as was most of Central America, cutting off the northern from the southern continent. On the Atlantic coast the Eocene rocks are unconsolidated sands and clays, with some glauconitic greensand, particularly in New Jersey. They form a narrow belt through New Jersey, Maryland, and Virginia, widening into a quite broad band through the Carolinas and the Gulf States, and extending around the borders of the Mississippi embayment into Texas. In the Gulf region the rocks are more consolidated, and are quite hard limestones, sandstones, and shales, with extensive deposits of lignite, formed in ancient peat bogs which followed the low-lying Gulf shores; some of these lignitic formations may be more properly referable to the Paleocene.

On the Pacific coast a long, narrow arm of the sea occupied the great valley of California, extending northward into Oregon and Washington; its deposits are at present principally displayed along the eastern flank of the Coast Range. These deposits form a single series, the *Tejon*, which lies upon the Chico in apparent conformity; but the lowest Eocene is not represented in the Tejon, and in Oregon an unconformity between the two series has been detected. The presence of Atlantic species in the Eocene of California indicates that the barrier between the oceans which had existed in the Cretaceous, was submerged for a time.

In the western interior the Eocene formations, with the exception of a few small areas in Colorado, are all confined to the plateau region west of the Rocky Mountains. Formerly, they were considered to be lacustrine deposits, but this explanation is applicable to only a small part of them; they are much more generally due to the subaërial agents, rivers, rain, and the wind, in the work of filling basins of slow depression. In some instances, at least, the material of these beds is principally fine volcanic ash and dust, so extensively distributed as to prove great volcanic activity in the region. It has not yet been determined how generally the formations are built up of these volcanic materials. Whatever the mode of their accumulation, the Eocene stages are remarkably complete and have preserved a marvellous record of the successive faunas of the epoch, which registers not only the changes in animal life, but also the shifting land connections with the other continents.

The oldest and most widely spread of these Eocene stages is the *Wasatch*, the principal area of which extends from New Mexico, over eastern Utah and western Colorado, to the Uinta Mountains, around the eastern end of which it passes in a narrow band, expanding again north of the mountains and covering southwestern Wyoming. As the Wasatch is in many places buried under newer rocks, it is not certain that the beds are continuous over this great area of 450 miles long by 250 miles wide in the broadest part, but there is no reason to suppose that they were not originally so. Another great body of Wasatch strata occupies the Big Horn Basin of northwestern Wyoming, extending into southern Montana, and two small areas in southern Colorado are referred to the same date. The thickness of the Wasatch varies from 1500 to 2500 feet.

The *Wind River*, the second of the Eocene stages, appears to be present in two distinct facies, though, as the two contain no common fossils, the correlation is open to some doubt. The beds of the typical facies occupy the Wind River Basin, north of the mountains of that name, in central Wyoming, and contain many remains of mammals, and also occur in the Huerfano Cañon, Colorado. The second facies (*Green River Shales*) occupies the same position stratigraphically that the Wind River does palæontologically, following upon the Wasatch and overlaid by the Bridger. This facies is typically displayed in southern Wyoming in the valley of the Green River and is a thick body of very finely laminated " paper shales," which seem to have been deposited in a very shallow lake, and have preserved an extraordinary number of plants, insects, and fishes, but no mammals have been found except in the form of footprints. Occasional crystals of gypsum indicate that the water became salt, at least locally, while the presence of such fishes as the Rays points to communication with the sea. The third of the Eocene stages is the *Bridger*, of southwestern Wyoming and northeastern Utah, which lies upon the Green River Shales, overlapping these east and west and extending over upon the Wasatch. The Bridger beds are very largely made up of volcanic ash and dust deposited partly on land and partly in shallow or periodical lakes. The remains of fishes, crocodiles, and non-marine shells scattered through the beds is a proof of subaqueous deposition, and the large selenite crystals (see p. 20), which frequently occur, indicate that the water was occasionally strongly saline.

These Eocene continental deposits are principally sands and clays, with occasional banks of conglomerate. They are more or less indurated; but usually quite soft and weather rapidly, giving rise to the characteristic bad-land scenery so frequently mentioned.

While the mammals of the Wasatch age are so largely identical with those of Europe that an easy way of land migration must have been open between the continents, the similarity becomes less and less and the two faunas follow such divergent lines of development through the Wind River and Bridger ages, that the connection between America and Europe must evidently have been interrupted. A brief connection with South America is suggested by the very unexpected discovery of an Armadillo in the Bridger, though it is possible that this animal was a survival from the connection which existed between North and South America in late Cretaceous or Paleocene times. If so, the ancestors of the Bridger armadillo yet remain to be discovered in the Wind River and Wasatch.

The Eocene epoch was brought to a close by a series of movements which added a narrow belt of land to the Atlantic and Gulf coasts. In the interior, the plateau region was elevated and drained and did not again become an area of extensive deposition. A great mountain-making disturbance followed the Bridger age, elevating the mountain ranges to which the post-Cretaceous disturbances had given birth and, in the neighbourhood of these mountains, tilting and upturning the Eocene strata.

Foreign. — The Old World Eocene has a very different development from that of North America, the eastern continents not assuming their present outlines till much later. At the close of the Cretaceous period extensive geographical changes had taken place in *Europe*, consisting chiefly in the retreat of the sea from wide areas which it had occupied in the Cretaceous. This was especially the case in Russia, northern Germany, and France, and southern England, and in place of the great gulf which had occupied these regions (see p. 712) were found only scattered bodies of fresh and brackish water in which the Paleocene deposits were laid down. At a later time the sea again advanced over part of these areas, which explains the general unconformity between the Cretaceous and Tertiary strata. In southern Europe the Mediterranean regained the great expansion which it had partly lost in the latter part of the Cretaceous, extending far over northern Africa, where nearly the whole continent north of the equator was submerged in the early Eocene sea, and transgressing over southern Europe. A long, narrow arm of this sea extended from southern France, past the north side of the future Alps and Carpathians, into western Asia. Another narrow sea, or strait, extended down the east side of the Ural Mountains, from the Arctic Ocean to the expanded Mediterranean, completely cutting off Europe from Asia. This complete severance of Europe from Asia necessitates an independent land connection of the former with North America to explain the community of terrestrial animals and plants between these continents. From Asia Minor the Mediterranean extended across Persia and Turkistan, northern India, Borneo, and Java, to the Pacific, separating the southern peninsulas from the Asiatic mainland. There was thus a continuous sea around the earth, everywhere separating the southern continents from the northern, though transient connections between them may have been established.

In the Alpine and north *African* regions were accumulated thick masses of limestone, largely composed of the gigantic foraminiferal shells called *Nummulites*, a hard massive limestone which reaches a thickness of several thousands of feet. Closely associated with the Nummulitic facies of the Eocene is the *Flysch*, an extremely thick mass of sandstones and shales, which occurs in the Alps and Apennines, the Carpathians and Balkan Peninsula, the Caucasus, Asia Minor, and southern Asia generally. In the Alpine region the Flysch contains enormous erratic blocks of granite, gneiss, etc., which appear to have come from southern Bohemia, and which have been interpreted as due to transportation by glaciers. This interpretation has not, however, been established. In northern Europe no such widely spread formations occur. After the Eocene had continued for some time, a marine basin, the Anglo-Gallic, was formed over southern England, northern France, and Belgium, which contains a succession of alternating marine, brackish, and fresh-water strata. This basin is classic ground, for in it were made the studies of Cuvier and Brongniart, which led to the recognition of the Tertiary as a distinct system and founded the science of Palæontology.

On the west coast of Africa the sea encroached in a narrow belt. The correlations of the early Tertiary rocks of *Australia* and New Zealand are still the subject of debate, but there seems to be little doubt that the Eocene is present. In *South America* the Eocene of Patagonia consists of a series of continental deposits containing a highly interesting succession of mammals. These grow more and more divergent from the mammals of the northern continents.

The Eocene thus had broad seas where now is land, and continents now connected were then separated by straits and sounds. On the other hand, there were then land bridges joining land areas which are now far apart. Some of these land bridges may be reconstructed with much confidence, while others are more or less probable. America was connected with Asia across what is now Bering's Sea, and also with Europe, probably by an extension of Greenland and Iceland. The Antarctic continent apparently had a much greater extension than it has now, and seems to have been joined with both Australia and South America. It is quite possible that Africa was more or less directly connected with the same land mass. If this be true, then in Eocene times the northern continents, Europe and Asia, were joined in the Arctic latitudes by way of North America, while South America, Africa, and Australia radiated in three great lines from the South Pole. Between the two series of continents, northern and southern, swept the transverse seas, of which the Mediterranean and Caribbean are remnants.

Climate. — The Eocene climate, especially as inferred from the plants, was warmer than that of the Paleocene. In England, for example, the temperate flora of the latter epoch was followed by one of subtropical character, and in North America the subtropical zone extended much farther north than in recent times.

Eocene Life

Except for the Vertebrates, Eocene life is chiefly instructive from the manner of its distribution over the globe. Invertebrates and plants are nearly the same as modern forms, the genera, for the most part, still existing, though the species are nearly all extinct.

Plants. — The Eocene flora of North America, which is very rich and varied, is found preserved in widely separated localities, — Canada, Montana, Wyoming, and Idaho. Besides Ferns and Horsetails, this flora includes some Grasses, Bananas, and many noble Palms (Fig. 304), Myrtles, Beeches, Oaks, Willows (XVI, 1), Poplars, Elms, Sycamores, Laurels, Magnolias, Maples, Walnuts, Pines, Spruces, Arbor Vitæ, and the like. Even in Greenland and Alaska was a luxuriant growth of forests of a temperate character, such as could not exist there now.

The European flora has a more decidedly tropical character than that of North America, and contains plants whose nearest living allies are now widely scattered, occurring in the warmer parts of America, Africa, Asia, and Australia. Cypresses,

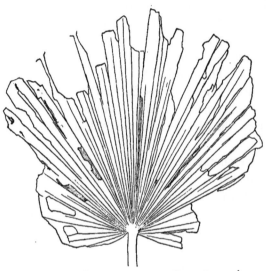

Yews, and Pines are abundant, including the *Sequoia*, now confined to California, and the *Gingko* of China and Japan. Aloes, Palms, and Screwpines occur, mingled with the ordinary temperate forest trees, Elms, Poplars, Willows, Oaks, etc. The distribution of plants in the

FIG. 304.—*Flabellaria* sp., × 1/12. Green River Shales

Eocene was thus very different from what it is at present.

Animals. — *Foraminifera* of relatively enormous size abounded, and their shells make up great rock masses. *Orbitolites* is a conspicuous genus along our Gulf coasts, *Nummulites* in the Old World. Corals are completely modern in character. The Sea-urchins and especially the *Irregulares* are much the most important representatives of the *Echinoderms*. Of the *Mollusca* both Bivalves (Pl. XVI, Figs. 2, 3) and Gastropods (XVI, 4, 5) increase greatly and are very rich in species. Nautiloid Cephalopods are more varied and widely distributed than now (XVI, 8), and in a few localities, particularly in India, *Ammonites* and *Belemnites* have been found, but these are mere belated stragglers from the Cretaceous and are much too rare to be at all characteristic. Among the Crustacea should be noted the increase of the *Crabs*, which are more numerous and varied than in the Cretaceous.

The *Fishes*, both fresh-water and marine, differ only in minor

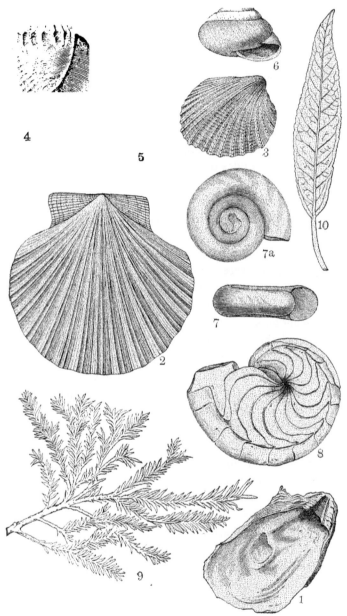

PLATE XVI. — AMERICAN TERTIARY FOSSILS

Fig. 1, *Ostrea virginiana*, × ½, Miocene. 2, *Pecten madisonicus*, × ½, Miocene. 3, *Cardita perantiqua*, Eocene. Whitfield. 4, *Volutolithes sayana*, × ¾, Eocene. 5, *Oliva carolinensis*, × ¾, Miocene. 6, *Helix dalli*, Miocene. White. 7, *Planorbis convoluta*, ? Fort Union. Meek. 8, *Aturia vanuxemi*, × ¼, Eocene. 9, *Glyptostrobus ungeri*, × ½, Eocene. Lesquereux. 10, *Salix* sp., × ¾, Miocene. (Figs. 1–5, and 8, after Whitfield.)

details from modern fishes. The *Reptiles* are likewise essentially modern in character, the Choristodera having died out with the Paleocene, and only two groups, the Lizards and Snakes, are more numerous than they had been in Mesozoic times, though the venomous snakes had not yet appeared. The Eocene beds of the West contained multitudes of large Crocodiles and a great variety of Turtles.

Eocene *Birds* are very much more numerous, advanced, and diversified than those of the Cretaceous; one characteristic feature of the times was the presence in Europe and America of extremely large, flightless birds, more or less like the ostriches in appearance. Of flying birds there were many kinds; Owls, Eagles, Buzzards, Vultures, Gulls, Waders, Woodcock, Quail, Ibis, and Pelicans are represented by ancestral forms, somewhat different from their modern descendants.

The *Mammals* have developed in a marvellous way since the Cretaceous, assuming in terrestrial life that dominant place which they have ever since held. Compared with the evolution of other animal groups, that of the mammals has been so rapid that each stage of the Eocene has its own mammalian fauna, differing from those of the preceding and succeeding stages. Besides these geological differences between the successive mammalian assemblages, there are often marked geographical differences between the faunas which are of approximately contemporaneous age, but widely separated in space. Comparing Europe and North America in this respect, we find that in the Eocene each continent had its own peculiarities, but that the land connection between them allowed intermigration and thus kept up a close general similarity in their mammals in the Lower Eocene, but this connection was interrupted and the faunas of the middle and later portions of the epoch diverge more and more in the two continents. The southern continents, on the other hand, had altogether different mammalian faunas, due to their long separation from the northern lands.

The change from the Paleocene mammals to those of the

WASATCH was very abrupt, though no great time interval is involved, and in Europe the change was equally sudden, most of the archaic Mesozoic types going out and those of more modern character replacing them. Evidently this was due to an influx of mammals from some region still unknown, and hardly at all to a development of the Paleocene mammals. *Rodents* come in for the first time in North America. *Perissodactyls* make their first appearance with ancestral members of the Horse family (*Hyracotherium*), the tapirs (*Systemodon*), and other families now extinct. The Wasatch horse was a curious little creature, not larger than a domestic cat, with four toes on the fore foot and three on the hind, instead of having only a single functional toe, like the modern horses. The curious extinct group of hoofed animals called the *Amblypoda* greatly increases in numbers and in stature, and both in Europe and America the predominant genus is *Coryphodon*. *Artiodactyls* also appear for the first time in ancestral members of the Pigs (*Eohyus*), and the Ruminants (*Trigonolestes*). The *Creodonts* increase in numbers and in the size and strength of the individuals, *Pachyæna* being as large as a bear. Numerous *Lemuroids* and primitive types of *Monkeys* (*Anaptomorphus*) swarmed in the trees. The correspondence between the mammals of Europe and North America was never closer than in Wasatch times.

The WIND RIVER fauna is a development of the Wasatch, apparently without the admixture of foreign elements by immigration, and there is no such complete change as at the end of the Paleocene. Noteworthy is the appearance of the first known member of the Perissodactyl family of the *Titanotheres*, a family which was destined to great expansion in the Upper Eocene and Lower Oligocene; also of the earliest true Carnivora.

The BRIDGER mammals represent a steady advance upon those of the Wind River. The Perissodactyls may be said to culminate in the Bridger; for though they afterwards reached much higher stages of development, they never again had the same relative importance. Horses, still of minute size, but more highly developed

FIG. 305. — *Eobasileus cornutus* Cope. One of the horned Amblypoda of the Bridger. Restoration by C. R. Knight, under the direction of Prof. H. F. Osborn. (Copyright by American Museum Natural History, N.Y.)

than those of the Wasatch, Tapirs, Rhinoceroses, and Titanotheres (*Palæosyops*) are extraordinarily abundant. *Coryphodon* has vanished, but the wonderful elephantine, six-horned
Uintatherium and *Eobasileus* (Fig. 305) take its place in North
America, though not in Europe. Artiodactyls, Creodonts, Rodents,
Tillodonts, and Lemurs were more diversified than ever, and
Bats are found here for the first time. The remarkable discovery in the Bridger of such a distinctively South American type as
an armadillo has already been mentioned.

In the Upper Eocene seas great whale-like forms (*Zeuglodon*)
of extraordinary appearance and structure had grown abundant.

The recently discovered Middle and Upper Eocene fauna of
Egypt is of very great interest. The mammals differ much from
those of Europe, but there are some forms common to both regions.
The long-sought ancestors of the Elephants have been found in the
Egyptian beds, and a very curious animal (*Arsinoëtherium*), which
might be described as a small elephant with a pair of huge horns
upon its narrow head, accompanies them.

Volcanic eruptions continued in the Rocky Mountain region
during the Eocene. The Yellowstone Park was a centre of great
activity, with numerous cones ejecting acid lavas and tuffs.

The Oligocene Epoch

American. — The marine Oligocene is better developed and
better understood on the Gulf coast than elsewhere, and therefore forms the standard of comparison. " It was a period of
profuse invertebrate life and steady sedimentation, especially of
oceanic deposits in water of not always great depth. Some 2000
feet of strata, formed almost wholly of organic débris, were deposited in the peninsular region of Florida." (Dall.) In the Gulf
region there is no decided stratigraphic break between the Eocene
and Oligocene, but a change in the marine fauna. The Oligocene
is but scantily shown on the Atlantic coast; some beds in the
Carolinas are referred to it and traces of it occur in New Jersey,

but it is generally concealed beneath the overlying Miocene. In the western Gulf region certain fluviatile beds are placed in this epoch, but upon insufficient reasons. Oligocene limestones are found in the Greater Antilles and very extensively in Central America, which seems to have been nearly or quite submerged. At the end of the lower division of the series there was some disturbance, raising northern Florida into an island, and shoaling the water where deposition continued. The marine fauna of the Oligocene is an assemblage of warm-water animals, very much like those which now live on the coasts of the West Indies and Central America, some of the West Indian forms extending as far north as New Jersey.

On the Pacific coast the Oligocene is found in western Oregon and British Columbia and very extensively in Alaska, where the *Kenai* formation, 10,000 feet thick and containing beds of lignite, is exposed along the coast and at many places in the interior down to British Columbia. The overlying Miocene follows in apparent conformity.

In the interior, Oligocene formations are among the most important of all the continental Tertiaries. The lower division, the *Uinta*, is found in a relatively small area of northeastern Utah and northwestern Colorado, where it lies unconformably upon the Bridger, overlapping the latter upon the southern flanks of the Uinta Mountains. The Uinta, which is the last of the Tertiary horizons in the plateau region, is usually regarded as uppermost Eocene, but its fauna allies it more closely with the Oligocene. Its mammals show that the isolation from Europe, which had begun after the Wasatch, still continued.

The Middle Oligocene, or White River, covers a vast area, northeastern Colorado, western Nebraska, eastern Wyoming, and southwestern South Dakota, with outliers in the Black Hills and North Dakota, and a separate area in the Northwest Territory of Canada. The mode of formation of the White River beds has long been a subject of discussion; originally they were considered to be lacustrine, a view which is supported by their very regular stratification, but it is now very generally believed that they are chiefly fluviatile, and several of the old stream-channels, filled with cross-bedded sandstone and banks of conglomerate, have been'observed (see Fig. 36, p. 110). The fine clays which make up most of the beds are chiefly flood-plain deposits, but there are also beds of pure white volcanic ash, showing that the volcanic activity which was so marked in the Eocene still continued. A system of streams meandering over a nearly base-levelled plain, with very low divides between them, would in times of flood unite into a vast, but shallow and temporary lake, and such would appear to have been the conditions under which the White River beds were laid down.

The mammals of the White River prove that a way of intermigration for terrestrial animals had again been established with Europe. While many families did not join in this migration and each continent had several groups peculiar to itself, the number of identical and closely allied genera common to both, and the appearance in America of types which Europe had had in earlier times, is sufficient proof of the renewed connection. The contrast between the Uinta and White River in this respect is very marked.

The Upper part of the interior Oligocene is the *John Day*, which covers a large part of eastern Oregon, and a small area in central Montana. The Oregon beds are a very thick mass (3000–4000 feet) of stratified volcanic ash and tuff, with some fresh-water beds at the top. Evidently, gigantic eruptions were in progress and the vents were at no great distance, though too far away for the formation of a coarse agglomerate.

Foreign. — During the Eocene nearly all Germany had been land, but in the Oligocene it was invaded by the sea, which broke in from the north and covered all the northern plain, extending into Belgium, and sending long bays to the south. One of these reached to the strait on the north of the Alps, expanding into a large basin near Mayence and Frankfort. Over Germany the sea was shallow, permitting the formation of extensive peat-bogs, where were accumulated masses of lignite or brown coal. The

Oligocene is very extensively displayed in southern Russia, marine below and lignitic above. In the basin of Paris the sea had a greater extent than in Eocene times, though with lacustrine beds intercalated. The Lower Oligocene of the Parisian area contains thick bodies of gypsum, which were formed in very strongly saline lagoons. In England the beds are more of brackish- and fresh-water origin. In southern Europe the sea retreated from wide areas, and in its place were extensive bodies of fresh and brackish water, in many of which peat-bogs accumulated masses of lignite. Such lignitic deposits occur at intervals in the south of France, Switzerland, and Bavaria. In the Alps, Apennines, Carpathians, Caucasus, Asia Minor, and southern Asia, the Oligocene is represented by the upper part of the Flysch, the formation of which began in the Eocene.

The Oligocene is found in north *Africa*, but in the other continents, beside those enumerated, it has not yet been separated from the Eocene below or the Miocene above.

Oligocene Life

The marine invertebrates so resemble those of the Eocene that any general statement of the differences is difficult; these differences are, for the most part, of species only.

The UINTA contains large and numerous crocodiles, their last appearance in the northern interior, and a highly interesting mammalian fauna, which, however, is only partially known and demands further exploration. The great Uintatheres, which dominated the Upper Eocene, have disappeared, and the Perissodactyls have begun to decline in relative importance, though not absolutely; the small three-toed Horses continue to develop steadily; Rhinoceroses and Tapirs are abundant, and the Titanotheres increase notably in stature and in the prominence of their horns. The most characteristic feature of the Uinta fauna, however, is the great increase in the Artiodactyls, which then began to assume the place they have ever since held as the most numerous and important of the hoofed animals. In the Uinta the Artiodactyls mostly belong to a great and typically American group (*Tylopoda*), of which the camel and llama are among the few modern survivors. The most primitive known ancestor of the camels and llamas is found in these beds (*Protylopus*) associated with a curious extinct family, the *Oreodonts*, which were extremely abundant and varied throughout the American Oligocene and Miocene, and with other families of small, graceful animals, which throve also in the White River and John Day. This large assemblage of the Artiodactyls distinguishes the Uinta fauna very sharply from that of the Bridger.

The Creodonts are still common, though distinctly less so than they had been in the Eocene, and the true Carnivora are beginning to replace them.

In the WHITE RIVER, or Middle Oligocene, the Crocodiles have become extremely rare, and only a dwarf species is known, but Lizards are much more numerous. The Mammals, which are preserved in astonishing numbers, resembled those of the Uinta, but had made great progress since that time. The Creodonts had almost disappeared, leaving only one or two curious genera (*e.g. Hyænodon*), while the Carnivora became abundant, Dogs, Sabre-tooth Cats, Weasels, and primitive Raccoons, being represented. The Lemurs and Monkeys have vanished from North America. The Perissodactyls continue to be abundant; the Horses are represented by the little three-toed *Mesohippus*, about as large as a sheep, the Tapirs by *Protapirus*, and Rhinoceroses by three very distinct series: thus, *Metamynodon* was a heavy, short-legged, aquatic animal, not unlike a hippopotamus in appearance; *Cænopus* a more slender, terrestrial animal with the proportions of a tapir, and *Hyracodon* was a long-necked, long-limbed, lightly built, running type, yet still a rhinoceros. The Titanotheres culminate in the massive, elephantine *Titanotherium* and its allied genera, which developed huge nasal horns (see Fig. 306) and died out early in the White River.

The Artiodactyls continue to increase; the native stock which came over from the Uinta age shows a distinct advance in develop-

FIG. 306. — *Brontotherium dolichoceras* Scott and Osborn.   Restoration by C. R. Knight, under the direction of Prof. H. F. Osborn.
(Copyright by the American Museum Natural History, N.Y.)

ment; the Camels (*Poëbrotherium*) and allied families are very common, among them *Protoceras*, a very curious animal, the male of which had four horns, and a pair of tusks in the upper jaw, while the Oreodonts must have covered the plains in great herds, so abundant are their remains. The Peccaries, or American repr sentatives of the pigs, are not yet known from the Uinta, but occur in the White River (*Perchærus*), and the extraordinary, long-limbed, two-toed, pig-like *Elotherium* may have descended from Uinta ancestors, or may have been a migrant from the Old World, as certainly were the members of the European family of Anthracotheres (*Anthracotherium* and *Hyopotamus*) which appear in the White River beds; there is nothing like them in the Uinta. The Rodents of the White River are much more numerous and varied than they had been before; Marmots, Squirrels, Beavers, Mice, Pocket-gophers, and Rabbits were already well established.

The Mammals of the JOHN DAY are much like those of the White River, but are more advanced and modernized, and some ancient groups have vanished, among them the Creodonts, the aquatic and cursorial Rhinoceroses, the immigrant Anthracotheres, and the huge Titanotheres. On the other hand, the Carnivora, especially the Dogs and Sabre-tooth Cats, greatly increase in numbers and diversity, and the same is true of the Rodents. The Horses and true Camels are larger than those of the White River, as are also the Oreodonts, but the Rhinoceroses are reduced to the two-horned *Diceratherium*.

The Oligocene Mammals of Europe have much in common with those of North America, but there are many local differences. In Europe, the Weasels were much more varied and common than in America, and the Civet-cats, a family which never reached this continent at all, were well represented. Some families of Perissodactyls, such as the Palæotheres, and a host of Artiodactyl families, some extinct, like the Anoplotheres, Cainotheres, and Xiphodonts, and the true Ruminants, were then peculiar to the Old World.

Climate. — The disappearance of the crocodiles from the northern interior seems to show that the climate had grown rather cooler, though on the Atlantic coast warm-water conditions still continued, and the vegetation shows that Europe still had a subtropical climate, palms growing up to the north of Germany. The *Kenai beds* of Alaska contain a temperate vegetation, and probably the leaf-bearing beds which are distributed so generally around the Arctic Sea and have yielded similar plants, should be referred to the Oligocene, though they are usually called Miocene.

THE MIOCENE EPOCH

American. — The marine Miocene rocks, which have an enormous development on the Pacific coast, are rather scantily displayed along the Atlantic and Gulf borders. The eastern coast, which had emerged during the Oligocene, was slightly depressed, and the Miocene beds were deposited unconformably upon the Eocene, in some places overlapping the latter landward, and it may be that the narrow belt of coastal Eocene has all been exposed by the denudation of the overlying Miocene. In any event, the Miocene coast-line was nearly the same as that of the Eocene had been, save for the reduction of the Mississippi embayment and the presence of the Florida island. Miocene beds occur in the island of Martha's Vineyard, apparently are concealed beneath the sea along the New England coast, and, from New Jersey southward, are almost continuous. In New Jersey their thickness is only 700 feet, thinning to 400 feet in Maryland, but reaching 1500 feet in eastern Texas, where they are concealed under later deposits, but their presence is revealed by deep borings. In the North the strata are unconsolidated sands and clays with local accumulations of diatom ooze, as at Richmond, Va., (see p. 314), but in Florida they are compact limestones, and in Georgia, limestones and conglomerates. Owing to the nearly complete closing of the Mississippi embayment, Miocene strata do not extend into Tennessee and Arkansas.

The Oligocene of the Atlantic coast had been a time of warm

waters, but in the Miocene a cool current flowed southward along the shore and through the straits between the Florida island and the mainland into the Gulf of Mexico. " The change by which the Oligocene was brought to a close and the typical Miocene inaugurated, caused . . . the most remarkable faunal break in the geological history of the United States after the Cretaceous." (Dall.)

On the Pacific coast the Miocene rocks, though reaching the enormous thickness of 5000 to 7000 feet, form only a narrow belt, and lie unconformably upon the Eocene. The Coast Range formed a chain of reefs and islands in the Miocene sea. Volcanoes were very active and showered great quantities of ash into the sea, where it was extensively mingled with diatoms, which largely compose the *Monterey* series, though sandstones and bituminous shales also occur. The sea did not extend into the northern part of the Sacramento valley, which is filled with continental sediments, fluviatile and subaërial and perhaps partly lacustrine. Orogenic disturbances took place in California, for the older part of the series in the Santa Cruz Mountains near San Francisco is folded and metamorphosed and the newer part there rests unconformably upon it.

The foothills of the Sierras had been worn down to a peneplain, which was elevated, perhaps early in the Miocene, and carved into valleys and ridges, and in the lower stream courses the " deep *Auriferous Gravels* " were laid down. In the Upper Miocene came a depression and very thick masses of the " bench *Auriferous Gravels* " accumulated in the valleys. Then followed a time of great volcanic activity in the Sierras, at first forming lavaflows and tuffs of rhyolite, then, after an interval, andesite tuffs and breccias, which poured down the valleys as great torrents of mud.

The coast of Washington and Oregon was covered by the sea, which extended up the valley of the Columbia and its tributary the Willamette, but the beds are far thinner than in California and, in places, lie upon folded and eroded Eocene. The sea also extended over parts of British Columbia. Early in the Miocene, Alaska was depressed, especially to the north, and the valley of the Yukon invaded by the sea and much of western Alaska was submerged, yet in the Middle and Upper Miocene, at least, some land connection with the Old World must have existed.

The Miocene fauna of California was largely indigenous and shows no evidence of communication with Asia, which would indicate that Bering Strait was open; if so, the undoubted connection of America with Eurasia must have been by some other route, perhaps by way of Greenland and the north Atlantic. Atlantic Miocene species are not known in the Pacific fauna, whence it may be inferred that the upheaval of Central America and the Isthmus of Panama, joining South and North America and separating the two oceans, took place at the close of the Oligocene, though there are some difficulties in accepting this view.

In the interior region Miocene continental deposits, mostly fluviatile, cover a vast area, though to no very great depth. The *Arikaree*, or *Rosebud*, stage, which is in part transitional from the uppermost Oligocene, is found overlying the White River beds in South Dakota, western Nebraska, and eastern Wyoming, with small areas in northeastern Colorado and Montana. The middle Miocene *Deep River* stage, occurs in widely scattered areas of restricted extent, in central Montana, central Wyoming, northeastern Colorado, northwestern Texas, and eastern Oregon. In this stage the migration of land mammals from the Old World, which ceased at the close of the White River Oligocene, was resumed, bringing in several new types, particularly the primitive elephants, which migrated from Africa to Asia and reached Europe and North America at nearly the same time. In this stage also appear the first forerunners of the migration from South America for which the junction of the two Americas opened the way. The *Loup Fork* stage covers much of the Great Plains region with a thin sheet of fine sands and marls, in successive disconnected areas from South Dakota far into Mexico, with outlying areas in Montana and New Mexico.

In addition to these comparatively well-known and well-defined stages of the Miocene, there are several others which are referred

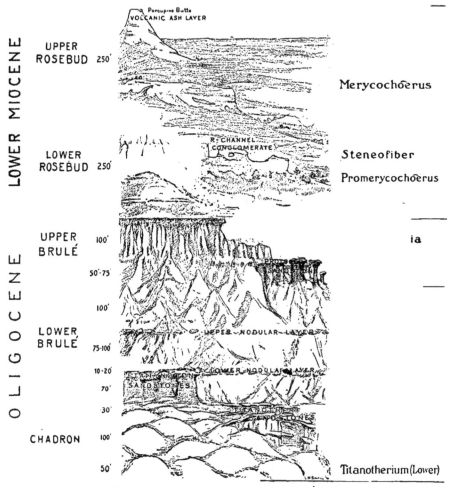

FIG. 307. — Idealized section of the great Bad Lands of South Dakota. (Osborn)

to the Miocene, although for no very convincing reasons. In southwestern Nevada is an immense thickness (14,000 feet) of supposably Miocene beds, described as being lacustrine, but con-

taining some coal and sulphur.   Several other areas are found in Nevada, Washington, and British Columbia.   A small area of probably Miocene rocks occurs in the South Park of Colorado, the *Florissant* beds, which have usually been called Oligocene, but which recent and more extended studies have shown to be probably Miocene.   The deposits are thin, papery shales, composed of fine volcanic ash showered into a small body of water, and have preserved countless insects and plants, many fish and a few birds, but no mammals.

The Miocene was a time of great volcanic activity in the Pacific mountain ranges and along the principal range of the Rocky Mountains; the great volcanoes of the Cascades and of Mexico are believed to date from this epoch, and in the Yellowstone Park were immense eruptions of andesites and basalts, both lavas and tuffs.   The great Columbia River basaltic flows are of early Miocene date, for they lie upon the slightly disturbed and eroded John Day, while Middle Miocene beds were deposited upon them. Vulcanism was also displayed in the West Indies, the Andes, and Patagonia.

As we have seen, orogenic movements went on in California between the Lower and Upper Miocene.   Later in the epoch and at its close, these movements grew very important and widely extended, affecting the mountains of all the Pacific States and causing the principal upheaval of the Coast Range in California and Oregon. The great fault bounding the Sierras on the east was made and the block mountains of the Great Basin raised by an extensive system of faults.   The high plateaus of southern Utah and northern Arizona were raised, beginning the great erosion-cycle of the Colorado River.   In the East the West Indian islands were raised and the Florida island was joined to the mainland.

Foreign. — In the north of *Europe* the sea retreated from large areas; northern Germany was now dry land, with only a relatively small bay invading it, while England was entirely above water, and has no marine Miocene beds.   On the west coast of Europe, the Atlantic encroached largely, as in France, Spain, Portugal, and also the northwest of Africa. Spain was joined to Africa, but straits across northern Spain and southern France connected the Atlantic with the Mediterranean. Another change of great importance was the shutting off of the long-standing connection of the Mediterranean with the Indian seas. The former covered much of eastern Spain, and flooded the lower Rhone valley, sending an arm along the northern border of the Alps to the neighbourhood of Vienna. Here it expanded into a broad basin, connected with another great basin covering Hungary. Most of Italy, Sicily, and a large part of northern Asia Minor were under water, but the Adriatic and Ægean Seas were mostly land, and the Alps formed a chain of islands, mountainous, but not nearly so high as at present.

At the end of the Lower Miocene came a great upheaval of the Alps, by which the sea was again excluded from that region, and, just as in the Oligocene, inland seas and lakes took the place of the marine straits. The basins of Vienna and Hungary had a very complex history, with repeated changes of size and position, resulting in the formation of an immense inland sea (the Sarmatian Sea), which reached from Vienna to the Black, Aral, and Ægean Seas, and was nearly as large as the present Mediterranean. This vast basin had but a limited connection with the ocean, and represented conditions much like those of the Black Sea at present. Europe had also a number of fresh-water lakes, particularly in France, Switzerland, and Germany, which have preserved a very interesting record of Miocene land life. A comparison with that of North America shows that a way of migration was still open between the two continents. In the basin of the Ebro, in Spain, the Miocene consists of red sandstones and marls, with beds of gypsum and salt, demonstrating arid conditions which were no doubt only local.

In the Old World the Miocene was a time of mountain making. The Pyrenees had been elevated in the later Eocene; the Alps received nearly their present altitude in the Miocene. The Apennines had two distinct phases of upheaval, one in the Eocene and

3 C

one in the Miocene, the latter coinciding with that of the Alps. The Caucasus dates from the close of the Miocene, while the date of the Himalayas is yet uncertain, but was either Eocene, or Miocene.

Marine Miocene beds occur in north *Africa*, on the coast of the Soudan and in Asia Minor. In *Asia* marine Miocene is known to be present in northwestern India, in Burmah and Japan, also in the island of Java.

The whole of Patagonia was submerged in a great transgression of a shallow, epicontinental sea, the *Patagonian* stage, and after some oscillation the sea withdrew and the terrestrial *Santa Cruz* beds were deposited. These are very largely composed of volcanic tuffs, but also contain cross-bedded sandstones and other fluviatile deposits. Marine deposits which are correlated with the Patagonian stage are on the west coast of Chili.

The Tertiary rocks of *Australia* and *New Zealand* which cover extensive regions, especially in Victoria, have not yet been definitely classified. Certain of these, usually referred to the Oligocene but more probably Lower Miocene, show so close a resemblance in their fossils to those of Patagonia, as to require the assumption of a continuous coast-line with South America, probably by way of Antarctica. The probability of this assumption is much strengthened by the occurrence of the marine Patagonian with its characteristic fossils in the South Shetland Islands, an Antarctic group.

Miocene Life

The life of the Miocene is in all respects a great advance upon that of the Eocene and Oligocene. The *Grasses* greatly multiply and take possession of the open spaces, producing a revolution in the conditions of food for the herbivorous animals. The vegetation of North America, as far north as Montana, perhaps even to northern British Columbia, still bore a southern character. In the Upper Miocene tuffs of the Yellowstone Park and contemporary strata of Oregon are found such trees as Poplars, Walnuts, Hicko-

FIG. 308. — *Rhus* Sp. A sumach from the Florissant Shales.

ries, Oaks, Elms, Maples, Beeches, noble forms of Magnolias and Sycamores. One species of *Aralia* had leaves 2 feet long by 3 inches wide. Curiously enough, the Breadfruit (*Artocarpus*) flourished in Oregon, and probably on the Yellowstone also. Conifers were numerous and varied. At Florissant the plants are of a similar warm-temperate character, with very few palms, but with *Sequoia*, the California Redwood, abundant.

In Europe the Lower Miocene flora was quite like that of modern India; over the central and western regions Palms continue to flourish, together with Live Oaks, Myrtles, Magnolias, Figs, etc.

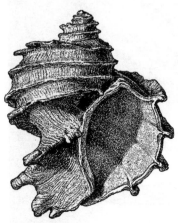

In the latter part of the epoch a change is noted, and such trees as Beeches, Poplars, Elms, Maples, Laurels, and the like become dominant.

Marine *Invertebrates* belong almost entirely to genera which still live in the seas, and many of the species persist to our own day. In Europe the older Miocene has numbers of shells such as now live only in warm seas, like *Cyprœa, Mitra, Purpura, Strombus,* etc. (See

FIG. 309.—*Ecphora quadricostata* Say, × 2/3, Yorktown, Va.

Pl. XVII, p. 765.) The Miocene of our Atlantic coast was evidently a time of cooler waters, and a similar change took place in Europe in the Upper Miocene. A very characteristic shell of the Atlantic coast Miocene is *Ecphora quadricostata* (Fig. 309).

The terrestrial *Vertebrates* of the interior are of much interest. Little is known of Miocene *Birds* in this country, but in Europe they are abundantly preserved and are of distinctly African character. Parrots, Indian Swallows, Secretary Birds, Adjutants, Cranes, Flamingoes, Ibises, Pelicans, Sand-grouse, and numerous Gallinaceous birds, were mingled with birds of European type, such as Eagles, Owls, Woodpeckers, Gulls, Ducks, etc.

The Lower Miocene (*Arikaree*) Mammals of the interior have only lately been discovered and are not yet fully described. In general, these animals are a continuation of the John Day fauna, in a higher stage of advancement, without admixture of exotic elements. The *Ancylopoda*, a very curious group of hoofed animals in which the hoofs had been converted into huge claws, and of which a few traces have been found in the White River and John Day, assume great importance in these beds.

In the Middle Miocene (*Deep River*) came a renewed migration from the Old World, bringing in the first of the elephant group (*Proboscidea*) which had simpler teeth than the modern elephants and a pair of tusks in the lower jaw, as well as in the upper. The genus is *Tetrabelodon*. The first of the true *Ruminants* to appear in North America came in with this migration and were, in a measure, intermediate between deer and antelopes, while European Rhinoceroses accompany them. Of the native stock, the Horses and Camels deserve particular mention as having increased in size and in variety and having made great advances toward the modern standard. The Oreodonts in a variety of bizarre genera, some of them aquatic, are very common.

The Upper Miocene (*Loup Fork*) Mammals resemble closely those of the Deep River stage, rather more advanced and modernized. The true Cats and a number of weasel- and otter-like Carnivores came in from the Old World, while the *Wolves*, *Panthers*, and *Sabre-tooth Tigers* were very numerous. Besides the true Ruminants, the American type of *Camels* and *Llamas* continued to flourish in such genera as *Procamelus*, *Pliauchenia*, and others. One very remarkable camel, *Alticamelus*, had nearly the same proportions as the giraffe. The extraordinary four-horned genus, *Protoceras*, of the White River, is represented by *Syndyoceras*, in which the four horns are much increased in length, but the tusks are reduced. The Loup Fork Horses (*Protohippus* and *Hipparion*) are much more modern in character and larger in size than their predecessors, but still have three toes on each foot. The Rhinoceroses are very abundant, and form a peculiar American genus (*Aphelops*) of massive, hornless animals. The Peccaries, or American swine, were commoner in the Loup Fork than in the earlier Miocene stages. The Atlantic coast Miocene has yielded numbers of Dolphins, Sperm and Whalebone Whales.

In Europe the Upper Miocene mammals were, in general, like those of North America, but a salient difference is in the much greater number of early types of Deer and Antelopes which are found there, together with various forms of Swine and ancestral Bears. Besides the Mastodons, which were common to both continents, Europe had in *Dinotherium* a remarkable kind of elephant; this animal had a much flattened head and a pair of massive, backwardly curved tusks in the lower jaw. The weasel and otter tribe of Carnivora was much more abundant and varied in Europe, and the Civet-cats, which were also common there, did not migrate to America.

Little is known of the Miocene Mammals of other continents except South America, where a magnificent assemblage has been preserved in the *Santa Cruz* tuffs of Patagonia. This fauna is so entirely different from that of the northern hemisphere that it seems to belong to another world. It contains no Carnivora, Proboscidea, Artiodactyla, or Perissodactyla. The flesh-eaters were carnivorous Marsupials, like those of Australia, and another family of Marsupials like the Australian Phalangers, was also present in addition to the American Opòssums. The Rodents, of which there were very many, all belong to the great porcupine group (Hystricomorpha) and closely resemble modern South American types, but among them are no rats or mice, squirrels, marmots, beavers, hares, or rabbits. Edentates are extraordinarily numerous and varied, Armadillos, Glyptodonts, and Ground Sloths forming one of the most conspicuous elements of the fauna. Hoofed animals were present in multitudes, but though having a certain likeness to those of the northern continents, they are but remotely related to them. The Toxodontia (*Nesodon*) were slow and massive animals, and the little Typotheria had a superficial resemblance to Rodents. The *Litopterna* had one group which imitated the horses in a surprising manner and another which had some likeness to the llamas (see Fig. 310). The Homalodotheria were a parallel to the northern Ancylopoda and the Astrapotheria, largest of Santa Cruz mammals, were not altogether unlike Rhinoceroses.

The *climate* of the early Miocene was much like that of the Oligocene and decidedly warmer in Europe than in North America, though it was mild even in the latter. The difference seems to have been largely due to the manner in which Europe was intersected by arms and gulfs of the warm southern sea. In the Upper Miocene the climate became distinctly cooler on both sides of the ocean.

THE PLIOCENE EPOCH

American. — The Pliocene is not a conspicuous formation in this country, and only of comparatively late years has it been recognized at all on the Atlantic coast. The movements which closed the Miocene gave to the Atlantic and Gulf shores nearly their present outlines, but some differences may be noted. Much of southern Florida was still under water, and a gulf invaded northern Florida, covering a narrow strip of Georgia and South Carolina. Isolated patches of Pliocene rocks in North Carolina and Virginia may be remnants of a continuous band. The *Gay Head Sands* on the island of Martha's Vineyard have marine fossils and lie unconformably on the Miocene, forming the most northerly known exposure of marine Pliocene on the Atlantic coast. All of these marine formations in the eastern United States are very thin and in notable contrast to the Pacific coast. Florida also has some fresh-water Pliocene. A small part of eastern Mexico, much of Yucatan, and some of Central America were still submerged.

On the Pacific coast the post-Miocene upheaval had laid bare the western foothills of the Sierra and greatly disturbed the Miocene strata of the Coast Range. The latter range sank again early

FIG. 310. — *Theosodon lydekkeri* Amegh. One of the Litopterna from the Santa Cruz beds of Patagonia. On the right are two examples of *Borhyæna*, a carnivorous Marsupial. Restoration by C. R. Knight, under the direction of the author

in the Pliocene, whose strata lie unconformably upon the Miocene, and extend over upon older beds. The transgression of the sea was limited, and Pliocene rocks form only a narrow band along the coast in California, Oregon, and Washington. The San Francisco peninsula was an area of subsidence and maximum deposition, for here no less than 5800 feet of sandstone (the *Merced* series) were formed, and quite lately Professor Lawson has described a series of beds, containing much volcanic material, 7000 feet thick, lying below the Merced and above the Monterey Miocene. This would make the Pliocene near San Francisco have a thickness of nearly 13,000 feet, by far the thickest mass of Pliocene in North America. On the other hand, a deduction, perhaps a very considerable one, should probably be made from this thickness, for the upper part of the Merced appears to be. Quaternary. (Dall, Arnold.) The mountains of British Columbia are believed to have been at a higher level than now, an elevation which probably connected Vancouver's and the Queen Charlotte Islands with the mainland. Marine Pliocene also occurs in southern Alaska. The marine Pliocene faunas of California and Japan became closely similar, due to a migration along the shore around the North Pacific, where the climate was temperate, no Indian species joining in the migration of the Japanese forms. In the Upper Pliocene the waters of the California coast appear to have been somewhat colder than they are now.

In the interior region a few areas of Pliocene, resembling the Upper Miocene in physical character and in mode of formation, have been described. The oldest of these, the *Republican River* stage, overlies and is intimately associated with the upper Loup Fork, in northwestern Kansas, northern Nebraska, and eastern Oregon. The *Blanco* stage is typically displayed in the Staked Plains of northwestern Texas, where it contains South American mammals, and is also found in Nebraska and Oregon. The Upper Pliocene is not definitely known to be represented in the interior, but its presence is suspected in Texas and elsewhere.

Some isolated areas of Pliocene which cannot yet be correlated

with the stages mentioned, are found in southern Idaho, eastern
Washington, etc., and no doubt much of the surface deposits of
the Great Basin and other regions is Pliocene, but lack of fossils
prevents their determination.

The volcanic activity in the Rocky Mountain and Pacific coast
regions, which had begun in the Cretaceous, continued through
the Pliocene. The great outflow of rhyolite which built up the
Yellowstone Park plateau is referred to the Pliocene. Some of
the enormous fissure eruptions, which flooded northern California and Nevada, southern Idaho, eastern Oregon, and Washington, with thick sheets of basalt, obliterating the valleys and
revolutionizing the system of drainage, are probably Pliocene,
as some are demonstrably Miocene.

A problematical formation is the *Lafayette*, whose geological
position and mode of origin are still debated. The Lafayette is
a belt of sands and gravels which runs through Maryland, Virginia, the Carolinas, and the Gulf States, around the southern end
of the Appalachians, up to southern Illinois, whence it turns southwestward to Texas. As in the typical exposures the Lafayette
rests unconformably upon the Miocene and is uncomformably
overlaid by the Pleistocene, many authorities refer it to the Pliocene. The mode of formation is somewhat obscured by the absence of fossils, but this very lack and the physical characters of
the beds make a marine origin improbable. It is more likely
that the deposition was subaërial, "resulting from a comparatively rapid Pliocene uplift in the Appalachian region." (Dall.)
According to another view, the Lafayette is due to a depression
of the coastal plain while the Piedmont region was elevated,
"and streams gorged with detritus from the decayed, uplifted
Piedmont above rushed down to the sea and poured their contents into the ocean. Either the waves were weak or the sea
advanced rapidly or this decayed material was discharged in
enormous quantities, for the sea was unable to cope with it and
deposited it unsorted on the bottom." (Shattuck.)

At or near the close of the Pliocene, extensive upheavals took place in several different parts of the continent, especially on the Pacific slope. The rise of the Rocky Mountains continued, raising the western part of the Miocene beds 3000 feet higher than the eastern. The height of the Sierra was greatly increased by the rise of the mountains along the eastern fault-plane and the tilting of the whole block westward. The new valleys cut through the late basalt sheets of the Sierras are much deeper than the older valleys excavated in Cretaceous and Tertiary times, which is due to the greater height of the mountains and consequent greater fall of the streams. The fault-blocks which form the Basin Ranges were still further displaced, increasing their height. The Wasatch Mountains and the High Plateaus of Utah and Arizona were again upraised. The great mountain range of the St. Elias Alps, in southeastern Alaska, was upheaved at this time, or even later, and the mountains of British Columbia were probably raised still higher. In Washington and Oregon the uplift was small, but became much greater in southern California, reaching 2500 feet in the Monte Diablo range. On the eastern side of the continent the uplift was on a much more restricted scale, not generally exceeding 100 feet. The Florida anticline underwent renewed compression, which increased its height; in Georgia, the continuation of this fold rose to 400 feet. The same movement extended the coast of Mexico and Central America and brought the continent to nearly its present outlines.

It is not necessary to suppose that all these movements were contemporaneous; merely that they occurred, now in one place, now in another, at or near the end of the Pliocene epoch.

Foreign. — In *Europe* the sea generally retreated at the end of the Miocene, leaving in the north only Belgium and a small part of northern France under water. In England the sea advanced upon the land; while in the Mediterranean region large areas remained submerged, as in Spain, Algeria, nearly all of central and southern Italy and Sicily, and Greece. In this region volcanic activity was intense, and Ætna, Vesuvius, and the volcanoes of central Italy had begun their operations. Germany has no marine

Pliocene, but extensive areas of fluviatile and other continental deposits belong to this epoch; especially famous are the stratified sands of the Eppelsheim basin, called *Dinotherium sands*, which also occur in several other parts of South Germany. In the lower Main valley are Pliocene lignites, the plants of which are nearly one-half Conifers, but also include many American trees, such as Walnuts and Hickories. Continental Pliocene, containing the same land mammals, occurs in many separate areas of southern Europe and Asia Minor, Mt. Léberon in the south of France, Pikermi near Athens, the island of Samos, which was then part of the continent, and Maragha in Persia, are all celebrated localities of the older Pliocene, while the newer Pliocene fauna is found in great abundance in the river deposits of the lower Arno valley in Italy (*Val d'Arno* stage). Over the region of the great Sarmatian Sea of the Upper Miocene were numerous bodies of brackish water, in which lived shells much like those which now inhabit the Caspian.

On the south side of the Himalayas, in northern India, are several thousand feet of sandstones and conglomerate, with some clay and lignite, formed principally from the piedmont accumulations transported from the Himalayas during the Pliocene, though probably the process of accumulation began in the Upper Miocene. These deposits now make the Siwalik Hills, famous for their fossil bones; and similar deposits with the same fossils occur in Borneo, and probably Java, which then were connected with Asia. In *South America* a Pliocene transgression of the sea took place, submerging the entire eastern coast of Argentina and Patagonia (*Paraná*, or *Cape Fairweather* stage) and along the line of 47° S. lat., at least, extending to the foothills of the Andes. The marine Pliocene beds were involved in the last upheaval of the southern Andes.

Pliocene Life

The life of the Pliocene is very modern in character. Little is known of the vegetation in North America, but in Europe it is marked by the continued disappearance of the characteristically tropical plants and by an approximation to the modern European flora. Many trees persisted, however, which are no longer native to that continent, but are still found in eastern Asia or in North America, such as Tulip Trees, Walnuts, Hickories, Magnolias, Sequoias, etc.

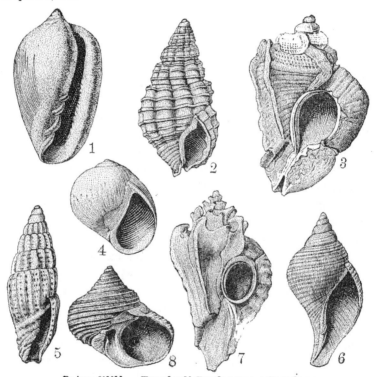

PLATE XVII. — TERTIARY FOSSILS FROM FLORIDA

Fig. 1, *Marginella aurora*, × 3/4, Miocene. 2, *Nassa bidentata*, × 3/4, Miocene and Pliocene. 3, *Purpura conradi*, × 2/3, Miocene. 4, *Natica floridana*, × 1/2, Miocene. 5, *Mitra wilcoxi*, × 1/2, Miocene. 6, *Fasciolaria tulipa*, × 1/2, Pliocene. 7, *Typhis floridana*, Pliocene. 8, *Turbo rectogrammicus*, × 1/2, Pliocene. (After Dall)

Marine *Invertebrates* are nearly identical with modern forms, and the great majority of Pliocene species of shells are still living.

The *Mammals* are still somewhat behind their modern successors, though much more advanced than their predecessors.

Those of North America are still incompletely known, and the list is a short one. Mastodons, Horses, Rhinoceroses, Peccaries, and very large Llamas represent the hoofed animals, beside the Dogs, Cats, and Mustelines. The effects of the connection with South America are seen in the appearance of the gigantic Ground Sloths and Armadillos, and of southern families of Rodents. If this connection was actually established at the close of the Oligocene, it is difficult to see why the South American mammals should so long have delayed their migration to the north.

The early Pliocene mammals of southern Europe closely resemble those of modern Africa, — Wolves, Cats, Civets, Hyænas, Monkeys, Rhinoceroses, three-toed Horses, Deer (of which Africa has none), a great variety of Antelopes and of Giraffe-like forms, and Swine. Mastodon and Dinotherium persisted, the latter attaining great size. India had a similar fauna, with certain geographical differences. Especially to be noted are the great variety of Oxen, the presence of Bears, true Elephants, and the Hippopotamus, of the first Old World Camels, and of the extraordinary *Sivatherium* and *Brahmatherium*, great four-horned creatures allied to the Giraffes. In the Upper Pliocene the true Elephants, Oxen, Hippopotamus, and Bears had extended their range to Europe, but not, so far as we know, to North America. In volcanic tuffs of probably Pliocene age on the island of Java was discovered a fossil, *Pithecanthropus erectus*, which has attracted great interest and has been the subject of much discussion. According to one view, these bones, part of the skull and a thigh bone, represent one of the " missing links " in the ancestry of Man; according to another, they are human, but abnormal, while a third opinion regards them as belonging to a large ape. The material is insufficient for a definite solution of the problem.

On the coast of Zululand, southeastern Africa, a small area of continental Upper Pliocene has yielded a few fossil mammals, which suffice to show that Africa had already acquired the fauna which characterizes it now. The list includes an elephant, rhinoceroses, hippopotamuses, and several antelopes.

The **Climate** of the Pliocene was, on the whole, evidently cooler than that of the Miocene, as is shown by the changes in the character of the vegetation and of the marine shells. On the American Atlantic coast this is not true, for here the Miocene waters were exceptionally cold and the Pliocene was warmer, but, on the other hand, the thin beds of Florida and the Carolinas represent but a small part of the Pliocene. In the English Pliocene the proportion of Arctic shells rises from 5 % in the oldest to over 60 % in the newest beds. The refrigeration was greater in the sea than on the land, for the vegetation shows that the air had not yet grown cold. That was to come later.

CHAPTER XXXVI

THE QUATERNARY PERIOD

THE Quaternary is the last of the great divisions of geological time and may be said to be still in progress, for its events led by gradual steps to the present climatic and geographical order of things, and to the present geographical distribution of animals and plants over the surface of the earth. Quaternary deposits are to a very large extent continental in their origin, marine sediments in most regions being of very subordinate extent, and consist generally of loose, uncompacted sands, gravels, boulder clays, clays, and the like. These deposits never reach any very great thickness, but their horizontal extent is at least equal to that attained by any preceding system, for in one form or other they cover almost the entire surface of the globe. In an even greater degree than in the Tertiary are the Quaternary formations of different areas difficult to correlate, because of the locally restricted character of many of them, the frequent and radical changes of facies from point to point, and the scantiness of fossils or their absence over wide regions.

The line of division between the Tertiary and Quaternary is not easy to draw, especially in those regions which were not reached by the great glaciers and ice-sheets of the Pleistocene. The seas at the end of the Pliocene had much the same extent as later, and on the same coasts the same kinds of material continued to accumulate, and some of the Pliocene coral reefs continued to grow uninterruptedly into the Pleistocene. Even in the regions of glaciation, the end of the Tertiary is fixed differently by different authorities.

According to general practice, the Quaternary is divided into two epochs, (1) the *Pleistocene*, and (2) the *Recent*, though various names are used for them.

The Pleistocene or Glacial Epoch

The conception of immense ice-sheets, like those of Greenland and Antarctica, covering large parts of North America and Europe at comparatively recent geological dates, is one that at first seems to be incredible, and made its way very slowly in the face of determined opposition.   Originally suggested by Agassiz about 1846, it required nearly thirty years to gain the general acceptance of geologists. This change of view was brought about by evidence too strong to be resisted; not that all difficulties have been removed and all problems solved, but no other hypothesis can rival the glacial in satisfactorily explaining so many and such varied classes of phenomena.   It will be well to summarize this evidence briefly.   The work of erosion, transportation, and deposition accomplished by existing glaciers was described in Chapters VI and VIII, but many of the illustrations and descriptions of those chapters were drawn from the areas of Pleistocene glaciation, partly for the convenience of dealing with things near home, and partly in order to bring the past and present into immediate juxtaposition.   The characteristics of glacial erosion, the rounded and flowing outlines, the smoothed and polished rocks, the parallel striæ cut by the stones and boulders held in the bottom of the ice, are all to be found abundantly in the glaciated area wherever the rocks are hard enough to receive and retain the marks and have been protected from the weather since the withdrawal of the ice; in many instances, even prolonged exposure to weathering has not sufficed to destroy the markings.   The striæ, which are parallel in small areas, when examined on a large scale, are found to be arranged in definite systems, which show the outward movement of the ice from the centres of dispersion.   The *roches moutonnées* and hummocks of rock, gently sloping and smoothed on the side against

3 D

which the ice impinged (stoss side) abrupt and often rough on the sheltered side (lee side) characterize the areas of Pleistocene glaciation, just as they do the rocky beds recently abandoned by retreating glaciers. In short, all the characteristic features of glacial erosion, which can be reproduced by no other known agency, occur where, by the theory, they should be found. When the contact of the drift with the smoothed and striated ice-floor can be observed (see Fig. 69, p. 160), the change is sudden and abrupt, the drift resting upon the hard, clean, unaltered rock, not at all like the gradual transition of soil, subsoil, and rotten rock (cf. Fig. 34, p. 106) which occurs when the soil arises from the decomposition of rock in place. There are some exceptions to this rule, where the erosive action of the ice was feeble and insufficient to remove all the old soil and rotten rock, but such exceptions offer no difficulty of explanation.

The drift itself is as convincing in its testimony of glacial deposition as the ice-floors are in their evidence of glacial erosion. The unstratified drift, made by the ice alone, in the form of moraines, chiefly terminal, but also lateral around projecting lobes of the ice, are highly characteristic of glaciers, as are the huge erratic and perched blocks, which often have travelled hundreds of miles. The ground moraine, or till, made up of finely ground rock-flour, in which are embedded boulders large and small, many of them faceted, smoothed, and striated, as only ice-worn boulders can be, and spread out in sheets of very variable thickness, its large boulders often deposited high above the points whence they were taken, testifies eloquently that it was accumulated by moving ice, which alone can deposit the finest and the coarsest materials together and can move to a large extent independently of the topographical features. The materials of the till are mostly of local origin and have travelled but a few miles, but there is almost always a greater or less admixture of stones from a long distance, and generally these are smaller in proportion to the distance travelled, because of the wear to which they have been subjected.

The stratified drift is no less indicative of glacial action. Whether advancing or retreating, the ice margin was melting, and the drift left by retreating ice-sheets was more or less worked over by water. Subglacial streams discharging in valleys made valley trains (see p. 234), and water descending in broad, shallow sheets, where the ice ended on a plain, made overwash plains, both connected with morainic deposits at their head. Eskers are gravel-filled subglacial tunnels, but drumlins and kames offer much difficulty of interpretation and have not yet been explained in a way that commands general assent. The ice barriers frequently formed lakes large and small, and in these lakes water-made and ice-made deposits were intimately associated. The glacial theory "distinctly affirms that rivers, lakes, the sea, icebergs, and pan-ice must have coöperated with glacier ice in the production of the drift, each in its appropriate way and measure." (Chamberlin and Salisbury.)

Finally, the evidence of the fossils, marine and terrestrial, animal and plant, strongly supports the glacial theory, by demonstrating a general refrigeration of the climate, when Arctic molluscs lived on the coasts of New England and northern Europe, and Arctic vegetation covered the lands in low latitudes, and Arctic mammals, like the reindeer and the musk-ox, descended to the south of France and to Arkansas. The testimony is thus all harmonious as to the great expansion of the Pleistocene glaciers and ice-sheets.

No one would pretend that there are no difficulties, still unexplained, in the way of accepting the theory. Some of these have been alluded to above; another is the enormous thickness of the ice-sheets required by the evidence, several thousand feet, for the glacial marks sweep over the tops of the highest mountains in New England and New York. On the other hand, it is held by physicists that 1600 feet is the maximum possible thickness of ice, as a greater amount would cause the bottom to melt from pressure, and in confirmation of this it is pointed out that the Antarctic ice-cap does not exceed the theoretical maximum, and that " at the

present day no ice more than 1600 feet (thick) has been recorded."
(Ferrar.)· This conflict of evidence it must be left to future investigations to reconcile, but the probable solution is to be found
in the temperature relations of the ice. The theoretical maximum
depends upon the assumption that the bottom of the ice is at or
near 32° F., but if it were considerably below this, the thickness
might be greatly increased.

Distribution of Pleistocene Glaciers. — The ice-sheets were localized, not universal, though it is probable that the entire world felt
the effects of the lowered temperature. At the time of maximum
extension of the ice, it covered nearly 4,000,000 square miles in
North America, especially toward the northeast of the continent;
in Europe the ice-cap which covered the north, Great Britain,
Scandinavia, North Germany, etc., is computed at 770,000 square
miles (A. Geikie), and the Alps were deeply buried in ice, which
flowed far out over the surrounding lowlands. Glaciation in Asia
was principally confined to the mountain ranges, as in Asia Minor;
on the south side of the Himalayas the ice descended to within
3000 feet of the present sea-level. On Mt. Kenya, which is
almost on the equator in eastern Africa and still has glaciers, the
presumably Pleistocene ice covered the whole mountain like a cap,
descending 5400 feet below the present glacier limit. In New
Zealand the ice also descended below the present sea-level and
some of the old moraines stand in the sea. The Australian Alps
and the western highlands of Tasmania bore extensive glaciers,
which, however, ended 1000 to 2000 feet above the sea. The
glaciers of the Patagonian Andes extended to the foot of the
mountains and out upon the plains, which were then probably
submerged. Thus, the northern hemisphere, above all, North
America, was the region of the most extensive Pleistocene glaciation, but in the southern hemisphere, and even in the tropics, its
effects are visible.

Glacial and Interglacial Stages. — It is still a debated question,
whether there was a single Glacial age, or epoch, during which the
ice-sheet, though having many episodes of advance, never entirely

FIG. 311. — North America in the time of maximum glaciation. The letters indicate the centres of dispersal of the ice. C = Cordilleran Glacier; K = Kewa-
tin Glacier; L = Laurentide Glacier; N.F. = Newfoundland Glacier

disappeared, or whether there were several distinct Glacial ages, when the snow accumulated to form an ice-cap which spread out widely from its centres of dispersal, separated by Interglacial ages, when the ice-cap completely melted away. Among students of these problems the present tendency is to accept the multiple character of the Glacial and Interglacial stages, one of the strongest arguments for which is the evidence of fossils showing the return of mild and even warm conditions in some of the Interglacial ages. At the same time, there is much difference of opinion regarding the number of these disappearances and reappearances of the ice.

Obviously, the problem is one of much difficulty, because each advance of the ice would tend to remove the older drift, or to bury it out of sight under new accumulations, when erosion was insufficient to remove it. Only on the margins of the successive ice-sheets, where they but partially coincided, should we expect to find the evidence preserved. A series of such advances and retreats of the ice must have produced an exceedingly complex succession of stratified and unstratified drift, and it is not surprising that the interpretations of such obscure phenomena should differ. If the superposed sheets of drift, one over the other (an arrangement which is not questioned), were separated by long, truly interglacial times, then each drift-sheet in turn must have been exposed to the denuding agencies for corresponding lengths of time and should exhibit the various stages of chemical and mechanical disintegration proportioned to the length of exposure. Between the earlier and later drifts there should be manifest differences in this respect. Further, to complete the evidence, interglacial deposits, with testimony of climatic amelioration from the fossils, should be observed.

The following comparative table gives the views on this subject of Professors Chamberlin and Salisbury (I) for the Mississippi valley, of Professor James Geikie (II) for Europe, and of the Prussian Geological Survey (III) for North Germany.

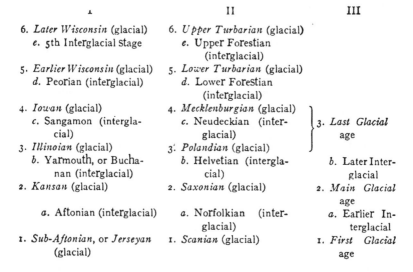

I	II	III
6. *Later Wisconsin* (glacial)	6. *Upper Turbarian* (glacial)	
e. 5th Interglacial stage	e. Upper Forestian (interglacial)	
5. *Earlier Wisconsin* (glacial)	5. *Lower Turbarian* (glacial)	
d. Peorian (interglacial)	d. Lower Forestian (interglacial)	
4. *Iowan* (glacial)	4. *Mecklenburgian* (glacial)	
c. Sangamon (interglacial)	c. Neudeckian (interglacial)	3. *Last Glacial* age
3. *Illinoian* (glacial)	3. *Polandian* (glacial)	
b. Yarmouth, or Buchanan (interglacial)	b. Helvetian (interglacial)	b. Later Interglacial
2. *Kansan* (glacial)	2. *Saxonian* (glacial)	2. *Main Glacial* age
a. Aftonian (interglacial)	a. Norfolkian (interglacial)	a. Earlier Interglacial
1. *Sub-Aftonian*, or *Jerseyan* (glacial)	1. *Scanian* (glacial)	1. *First Glacial* age

In the Alps, Professor ·Penck has determined three Glacial stages, and Huntington has found evidence of five in the mountains of Turkestan.

The table must not be understood as attempting to correlate the events in Europe and America, as that would be premature. At any rate, the events in the two continents did not correspond in the way which the table seems to imply. For example, in the last of Geikie's Glacial stages, the Upper and Lower Turbarian, the glaciers are described as being restricted to the high lands and mountains, not forming a general ice-sheet. The Mecklenburgian and Polandian more nearly correspond to the Wisconsin.

American. — At the time of greatest expansion the ice-sheets covered nearly all of North America down to lat. 40° N., anticipating the conditions of modern Greenland, though on a vastly larger scale. Three distinct centres or areas of maximum accumulation of the ice have been identified in northern Canada, from which the great ice-sheets flowed outward in all directions, though each one of the sheets had its own episodes of advance and retreat, so that the same region of country was overflowed, now by extensions from one sheet, and again by those from another. One of these centres of accumulation and distribution lay to the north of the St. Lawrence River, and on the highlands of Labrador, sending its ice-mantle southward over the Maritime Provinces, New England, and the Middle States, as far west as the Mississippi, River. This is called the *Laurentide*, or *Labradorean Ice-sheet* or *Glacier*. A second centre was near the west coast of Hudson's Bay, and from this area the ice streamed outward in all directions westward toward the Rocky Mountains, northward to the Arctic Ocean, eastward into Hudson's Bay, southward through Manitoba into the Dakotas, Minnesota, and Iowa. This great ice-sheet has been named the *Keewatin Glacier*, from the Canadian district of that name. A third centre was formed by the Cordillera of British Columbia, which for a distance of 1200 miles was buried under a great ice-mantle that flowed both to the northwestward and southeastward. To these large and well-defined centres should probably be added a fourth, Newfoundland, from which, there is reason to think, came the ice which crossed Cape Cod and extended over Nantucket Sound to the island.

In addition to the great northern ice-cap, large local glaciers accumulated in all the western mountains ranges: the Rocky Mountains, as far south as New Mexico, the Uinta, Wasatch, and Bighorn ranges, and the Sierras and the Cascades, even the San Francisco Mountains of northern Arizona, and the other ranges of the western Cordillera, all bore thousands of glaciers. In these mountains almost every valley shows the evidences of former glaciation, in cirques at the head, in the smoothed and striated walls and bed, and in the moraines at the foot. The mountains of Alaska were heavily glaciated, but not the lowlands.

In the Mississippi valley the Pleistocéne sequence is best displayed. The first known advance of the ice (sub-Aftonian) is registered in much disintegrated drift, which is exposed by denudation in Iowa. A similar sheet of very old and much worn drift which extends from beneath much later drift in New Jersey and Pennsylvania may be of the same date.

A great retreat of the ice, if not its entire disappearance, brought about interglacial conditions at least in the Mississippi valley (*Aftonian* stage). The surface exposed by the retiring ice was occupied by vegetation, which in many places in Iowa formed accumulations of peat, sometimes to the depth of 25 feet. The *Kansan* stage represents the greatest extension southwestward of the ice-sheet, when the glacier descended from the north (perhaps the Keewatin glacier) nearly to the mouth of the Ohio River, and spread across Iowa and Missouri far into Kansas and Nebraska. East of the Mississippi the Kansan drift has not been recognized. Again came a time of retreat, when soil was formed, and the Kansan drift was eroded and deeply decomposed, and peat deposited upon it (*Yarmouth*, or *Buchanan* stage). A renewed extension of the ice laid down the *Illinois* till-sheet, which is found not only in that State, but in Iowa also, overlapping the Kansan drift, and it extends to Wisconsin, eastward into Ohio and Indiana, and passes under later till-sheets to the northeast. This Illinoian drift appears to be derived from the Laurentide glacier.

The *Sangamon* interglacial deposits are of peat, old soil, etc. A fourth recrudescence of the glacier (*Iowan* stage) occasioned the deposit of another till-sheet, of an extent not yet determined, which is best displayed in northeastern Iowa, where it is intimately associated with the largest accumulations of loess in the Mississippi valley. The Iowan till-sheet is followed (*Peorian* stage) by interglacial deposits which are perhaps contemporaneous with those so well shown near Toronto on Lake Ontario. The latter beds form a succession of fine shales and sand that lie between two sheets of glacial drift and are divisible into two parts; the lower (*Don* formation) contains many fossils of which the plants, such as the Pawpaw and the Mock-orange, indicate a climate distinctly warmer than that of the region at present and about like that of the middle United States. The fossils of the upper part (*Scarboro* formation) indicate a cold temperate climate and herald the approach of a renewed glaciation, which in turn is recorded in the overlying till.   Such facts are difficult to explain, except as the result of truly interglacial conditions.   The *Wisconsin* stages are much the most conspicuous and best known of all, and the sheets of till and drift are far thicker than those of the other Glacial stages.   Especially conspicuous is the great terminal moraine, or rather morainic belt, which itself records many episodes in the history of the ice and which has been traced across the continent.   Beginning at Nantucket, the moraine runs through Long Island and Staten Island to New Jersey, which it crosses into Pennsylvania; here it bends sharply to the northwest to the boundary of New York, but turns southwest almost at a right angle, reaching nearly to the Ohio River at Cincinnati.   It crosses in an irregular, sinuous line the states of Indiana, Illinois, Iowa, and thence northwestward through the Dakotas into Montana, where it nearly follows the international boundary to the Pacific coast mountains (see Fig. 311).

In the Eastern States there is no such clear indication of several successive ice-invasions as in the Mississippi valley, the Wisconsin erosion and its thick mantle of drift removing or obscuring the records of earlier events.   The remnants of very ancient till-sheets in New Jersey and Pennsylvania have been mentioned, and in New York and New England not more than two or three invasions can be identified.   In part, this may be due to the later development of the Laurentide glacier.   The geologists of the Canadian Survey believe that, " beginning at the west and going eastward, these three great glaciers [*i.e.* the Cordilleran, Keewatin, and Laurentide] reached their widest extent and retired in succession."   (Tyrrell.)

The final retreat of the ice was by slow stages with many halts. In the central West are preserved many lines of moraine, with kettle-holes, kames, and drumlins, which mark readvances and pauses in the retreat.

Probably every retreat of the ice was accompanied by the formation of barrier lakes held in by the ice-front, but only those of the final recession have left intelligible records of themselves. A comparatively simple case is that of Lake Agassiz, which covered Manitoba and Minnesota with a great sheet of water, 760 miles from north to south. The lake was formed when the Keewatin glacier in its retreat had freed nearly all of Manitoba from the ice and was joined by the Laurentide glacier from the east, making a great wall of ice which shut off the drainage toward Hudson's Bay, while to the south high land held back the lake in that direction. The water of the lake rose until it overflowed the lowest point in the southern barrier and formed a river (*Warren River*, now abandoned) which joined the Mississippi. The level of the lake was gradually lowered as Warren River deepened its bed, and was finally discharged when the retreat of the ice opened the course to Hudson's Bay.

The history of the great Laurentian lakes is extremely complex and is slowly being deciphered by the combined efforts of many workers. The changing positions of the lobes which projected from the ice-front, the numerous basins, now connected and now severed, as the water rose and fell, combined with slow diastrophic movements, make up a very intricate succession of temporary lakes and shifting outlets. Considerations of space forbid more than a brief and simplified outline of this interesting story. When the ice had retreated so far as to uncover land to the north of the divide between the basins of the St. Lawrence and the Mississippi and eastward the Hudson, the waters produced by the melting of the ice were held in between those divides and the ice-front, forming a great number of small lakes from New York to Minnesota, three of which require mention as the earliest recorded stages of the Great Lakes at a time when most of their present basins was filled by the ice. Of these three lakes which embraced three prominent lobes of the ice-front, one was in the axis of Lake Superior, one at the southern end of Lake Michigan, and the third in a line with Lake Erie, but west of it, and each discharged by a separate outlet to the Mississippi.

Omitting several intermediate stages, and coming to a time when the basins of Lake Michigan and Lake Erie and part of that of Lake Ontario had been freed by the retreat of the ice, we find *Lake*

Whittlesey, which filled the basin of Lake Erie, but was very much larger; it was connected along the ice-front to the north with the crescentic *Lake Saginaw,* that discharged westward into *Lake Chicago,* a larger Lake Michigan, which retained its original outlet to the Mississippi. Lake Whittlesey was succeeded by *Lake Warren,* which was formed by a junction with Lake Saginaw on the northwest, and by an extension along the ice-front, eastward into New York and northeastward into Ontario, but still discharging westward to Lake Chicago and the Mississippi. Later, however, Lake Warren extended into central New York and emptied by way of the Mohawk into the Hudson. The condition of the Erie basin now becomes obscure, for when, after several changes, *Lake Iroquois* was established, it was merely an enlarged Lake Ontario, and the three upper lakes, now clear of ice, had coalesced into the great, irregular *Lake Algonquin,* which had lost connection with the Mississippi and discharged into Lake Iroquois, at first probably by the line of the St. Clair and the Erie basin and later probably along the course of the present Trent across the province of Ontario. As the Mohawk outlet persisted, the entire discharge of the lakes was into the Hudson, but whether the Erie basin was filled with water as it certainly was somewhat later, has not been clearly made out. Lake Algonquin was eventually cut off from its connection with Lake Iroquois and found an outlet by way of Lake Nipissing and the Ottawa River, thus severing the series of lakes into two independent systems, while the lowering of the water level in Lake Algonquin had divided it into three lakes which had very nearly the present outlines of Superior, Michigan, and Huron. The final stage in the history is connected with the Champlain subsidence and reëlevation presently to be described.

The Champlain Subsidence. — In the Glacial epoch a suhsidence had begun which continued until it became a very marked feature of the times. The depression was greatest toward the north and especially in the valley of the St. Lawrence; at the mouth of the Hudson, for example, the land stood about 70 feet below its present level, on the coast of Maine 150 to 300 feet, and in the St. Lawrence valley 500 to 600 feet below. The consequence of the depression was that an arm of the sea extended up the St. Lawrence to Lake Ontario, which was little, if at all, above sea-level. Two long and narrow gulfs reached out from this sea, one up the valley of the Ottawa River and the other over Lake Champlain, while the Hudson River appears to have been converted into a narrow strait connecting the marine waters of the Champlain basin with those of the Atlantic where New York Bay now is. The lines of raised beaches, the sands and gravels filled with marine shells, and the bones of whales and walruses, are the present evidences of this submergence.

The Champlain subsidence and the reëlevation which expelled the sea from the Hudson and Ottawa rivers and from the basins of Lakes Ontario and Champlain also affected the Great Lakes. Lake Iroquois had found a lower outlet than the Mohawk, when the ice withdrew from the Adirondacks, into Lake Champlain, which then discharged into the Hudson, because of the ice barrier to the north. Subsequently this outlet of Lake Iroquois was into the Champlain sea, when the subsidence had drowned the St. Lawrence valley. Just when the Niagara began to flow is not certain, nor when the basin of Lake Erie was refilled, if it were ever emptied, but so long as the upper lakes had their outlet through the Ottawa, the Niagara carried only the discharge of Lake Erie. The elevation which followed the Champlain subsidence was accompanied by a slight tilt of the lake region to the southwest, cutting off the Ottawa outlet and causing the three upper lakes to discharge into Lake Erie. The consequent change in the volume of the Niagara is registered in the sudden increase in the width of its gorge.

THE NON-GLACIATED AREAS

In the non-glaciated parts of the continent occur stratified Pleistocene deposits, which it is very difficult to associate with the events taking place in the glaciated area, for lack of any means of direct comparison. On the Atlantic slope from New Jersey southward a succession of Pleistocene gravels and sands constitutes the *Columbian* formation, so called because of its typical development in the District of Columbia. These deposits are differently interpreted by those who have examined them, but they appear to be largely fluviatile and subaërial much like the Pliocene Lafayette. Three parts of the Columbia formation have been recognized which by some authorities are regarded as three successive depressions and emergences from the sea, but the difficulties in this interpretation are such that a non-marine origin is more probable. On the other hand, marine fossils in the uppermost of the three divisions in the Chesapeake Bay region indicate some depression in that area. All the divisions contain large boulders transported by floating ice.

Over the Great Plains from South Dakota to Texas the surface formation is a fine, calcareous, sandy clay, which lies unconformably upon the eroded surfaces of older strata, from the Blanco to the Cretaceous. This formation has been called the *Sheridan* stage ("*Equus Beds*"), from Sheridan County, Nebraska, where it is typically displayed. It is, to a large extent, of æolian origin and in places contains great numbers of fossil bones. In South Dakota the Sheridan passes under a drift sheet, and probably it corresponds to one of the earlier interglacial stages. ·

In the Great Basin, the later Pleistocene had, temporarily at least, a much less arid climate than at present, as is indicated by the many lakes which it contained, and two of these, Lakes Bonneville and Lahòntan, were very large (see p. 219). The former, which was in the eastern part of the Great Basin, had an outlet northward to the Snake River, and had two periods of expansion, separated by one of almost complete desiccation. Lake Lahontan, which had no outlet, had similar episodes of rise and fall.

On the Pacific coast, marine Pleistocene in two unconformable stages occurs in southern California; the fauna of the lower stage has still a somewhat northern character, but in the upper stage the water became warmer than it is now, and tropical species which no longer live on the California coast were present. Pleistocene movements affected the Pacific coast to the amount of 3000 feet in the Inyo Mountains of California, 200 feet or more on the coast of Oregon, and 4000 feet in southeastern Alaska, and increased the height of the Sierra, Wasatch, and Basin ranges and of the high plateaus of Utah and Arizona.

The volcanic activity which had been so very striking during the Tertiary period in the western region continued into the Pleistocene, as is to be seen in the lava flows of the Great Basin, Arizona, New Mexico, Idaho, all the Pacific States, and Alaska.

The late Pleistocene was a time of ameliorated climate and heavy rainfall, when the flooded rivers moved sluggishly, owing to the diminished slope, and spread sheets of sands, gravels, and clays over their flood plains and in their estuaries, through which they have subsequently cut terraces, when elevation had given them renewed power.

The events cf the Glacial epoch, and the diastrophic movements which accompanied and followed it, have had the most important and widespread effects upon the topography of the glaciated regions. The sheets of drift, stratified and unstratified, have completely changed the surface of the country, and by filling up the pre-Glacial valleys, have revolutionized the drainage, only the largest streams being able to regain their old courses. Innumerable lakes, large and small, were formed in depressions, rock basins, and behind morainic dams, the contrast between the glaciated and non-glaciated regions in regard to the number of lakes in each being very striking.

The Pleistocene was closed and the Recent epoch inaugurated by a movement of upheaval which raised the continent to its present height. These Pleistocene movements have been correlated with the accumulation and removal of the ice, and it is at least a curious coincidence that the continent should have slowly sunk under the maximum load of ice and have risen again after the ice had melted. These movements were greatest where the ice was thickest.

FOREIGN PLEISTOCENE

The Glacial epoch in Europe ran a course remarkably parallel with its history in North America. After the first Glacial and Interglacial stages (perhaps representing the sub-Aftonian and Aftonian), came the time of the greatest expansion of the ice, the *Saxonian* stage of Geikie, which is believed to correspond to the Kansan of America. The great centre of dispersion was the Scandinavian peninsula, where the ice was probably 6000 to 7000 feet thick, and whence it flowed outward, filling the Baltic and North seas, and covering Finland, northwestern Russia, the lowlands of Germany, and extending to England. The Highlands of Scotland were a secondary centre, its ice-sheets flowing into the North Sea and uniting with those from Scandinavia, and westward to the ocean. The Irish Channel was also filled up. From the southwest of Ireland to the North Cape of Norway, a distance of 2000 miles, was probably a continuous wall of ice fronting the sea, like that which now surrounds the Antarctic continent. At the same time the Alps were the seat of enormous glaciers, only the highest peaks rising above the sheets of ice, and these great glaciers extended far out from the foot of the mountains, covering all the lowlands of Switzerland and extending from Austria and Bavaria, on the east, to the Rhone valley near Lyons, on the west. The high plateau of Asia, from the Himalaya to Bering's Sea, shows evidences of glaciation, and great valley glaciers were formed on the southern slopes of the Himalayas, extending in some places to within 2000 feet of the sea-level.

A second great Glacial stage (the fourth Glacial or *Mecklenburgian* of Geikie) is generally recognized in Europe and correlated with the Wisconsin stage of America. This ice-sheet was much less extensive than the former one, being confined principally to Finland, Scandinavia, the Baltic Sea, which it filled, Denmark, and a little of north Germany. The prevailing motion of this sheet was from east to west. The Alpine glaciers were also extended far beyond their present limits, but not so widely as before.

Following the Mecklenburgian stage came alternating periods of milder and colder climates, the fourth and fifth Interglacial, and fifth and sixth Glacial stages of Geikie, the Glacial stages marked, not by the formation of great continental ice-sheets, but by the extension or recrudescence of local snow-fields and valley glaciers. Oscillations of level also occurred along the coasts, allowing limited transgressions of the sea.

The Pleistocene of the other continents has been considered in the general introductory statements.

Causes of the Glacial Climates. —This is but a special case of the general problem of climatic changes in the history of the earth. We now know that the Pleistocene glaciation is not what it was once supposed to be, a unique phenomenon in geological history. On the contrary, in at least three other periods, Algonkian, Cambrian, and Permian, we have found evidence of glaciation rivalling or equalling that of the Pleistocene. At other times mild and equable climates have extended far into the Arctic regions.

In attempting to explain these remarkable changes, three kinds of agencies have been called upon: (1) Astronomical, or change in the position of the earth's axis, in the eccentricity of its orbit, in the heat radiated from the sun, etc., but these are now very generally discarded. (2) Geographical, or changes in the arrangement of land and sea, in the height of the land, direction of ocean currents, etc., but none of these seems to afford any real help in solving the problem. (3) The most promising and widely favoured agency of climatic changes is now sought in variations in the composition of the atmosphere, especially in the quantity of carbon dioxide present. Whatever may eventually come of this, it has not yet advanced beyond the stage of an interesting speculation.

Pleistocene Life

The frequent and extreme climatic changes, of which we find such abundant evidence in the Pleistocene, caused extensive migrations and dispersions of animals and plants, and the rapid

3 E

succession of Arctic and temperate forms in the same region. Many land bridges between different continents, or between continents and what are now islands, were not severed until the end of the Pleistocene, permitting migrations which are no longer possible. The extension of the ice-sheets brought with them Arctic floras and faunas, which retreated in the Interglacial times, while temperate animals and plants spread northward to replace them. These conditions produced a very severe struggle for existence and were fatal to a great many large mammals, causing numerous extinctions over the larger part of the world.

Pleistocene *Plants* are almost all of the same species as those now living, but they are often very differently distributed. The Glacial cold greatly impoverished the European forests, which in the Pliocene had many kinds of trees now found only in North America or in eastern Asia. Owing to the east and west trend of the European mountains, the forests could not retire before the ice, and return, as they did in the United States, where no mountain barriers shut them off from the warm latitudes of the south. When the ice-sheets melted and the climate was ameliorated, the Arctic flora and fauna were forced to retreat in their turn; they did so not only by following the retiring ice-front, but also by ascending the mountains as the latter were cleared of ice. Thus, high mountains in the northern hemisphere have on their upper slopes an Arctic flora and fauna, separated, perhaps, by hundreds of miles from the nearest similar colony. For example, the higher parts of the White Mountains have plants which do not occur on the lowlands until Labrador is reached, and the snowy Alps have truly Arctic plants and animals. In Europe the disappearance of the ice-sheets was followed by a dry climate, when a fauna like that of the Russian Steppes extended to western Europe.

Of Pleistocene *animals* it is only the mammals that require mention. Here also we find the same mingling of northern and southern forms, and association of types now widely separated. North America had Mastodons (*i.e.* an extinct type of elephant which had smaller and simpler grinding teeth than the true elephants), Elephants, Horses, Tapirs, the first Bisons (which had migrated from the Old World, as had several kinds of Deer and the Musk-ox), Peccaries and huge Llamas, Wolves, great Cats as large as lions, Sabre-tooth Tigers, and the first Bears, also immigrants. A great variety of Rodents is found, most of them kinds which still inhabit the country, but mingled with these are South American forms like the Cavies and Water Hog (*Hydrochærus*), and the giant Beaver (*Castoroides*) is an altogether peculiar form. Enormous Ground Sloths (*Megatherium, Mylodon,* and *Megalonyx*) and Glyptodonts show that the way of migration from the south was still open.

In South America were an astonishing number of huge Edentates: Sloths nearly as large as elephants, Ant-eaters, and a marvellous variety of giant Armadillos. Some of the immigrants from the north, which are now extinct, still lingered in the Pleistocene, such as the Mastodons and Horses, as did also some of the peculiarly South American hoofed animals, *Typotheria, Toxodontia,* and *Litopterna,* the ancestors of which may be traced back almost continuously to the *Notostylops* beds of the Paleocene.

Europe was the meeting-ground of mammalian types now widely scattered. Together with Arctic forms like the Reindeer, Musk-ox, Mammoth (Hairy Elephant), Hairy Rhinoceros, and the Lemming (*Myodes*) were found southern animals, such as the Hippopotamus, several kinds of Elephants and Rhinoceroses, Lions, and Hyænas, and likewise species allied to those still living in Europe, such as the huge Cave Bear, the gigantic Irish Deer (*Megaceros*), and great Oxen. Northern Africa was joined to Europe by way of Malta and Sicily, and probably at Gibraltar also, permitting frequent intermigrations. The junction of Ireland with Great Britain and of both with the continent continued until after the ice-sheets had disappeared, so that these islands, and especially Great Britain, were stocked by the continental animals and plants.

In the Pleistocene of India are found many animals which now live only in Africa, such as the Baboon, Spotted Hyæna, etc.

Australia had a Pleistocene mammalian fauna composed, with the exception of the Wild Dog (*Canis dingo*), of Marsupials, allied to those which still inhabit that region, but many of them were of vastly greater size than the living forms.

The Pleistocene Mammals are remarkable for the great size which distinguishes many of them, and it is just these which have passed away, leaving a world that is " zoölogically impoverished," but is nevertheless a much more agreeable place of residence without them.   Further we note, (1) that the Pleistocene mammals are in general like the smaller forms which have succeeded them in the same regions, but (2) that in Europe and North America there was a commingling of types now found only in widely separated regions.

In Europe *Man* first appears in the early Pleistocene.   It is altogether probable that the human race originated in Asia, quite aside from the doubtful testimony of the Pliocene *Pithecanthropus*, and reached Europe by migration.   The most ancient European men, such as the "Man of Spy " in Belgium, are of a much lower type physically than those of the later Pleistocene; only in the *Recent* age do human remains and implements become at all common and so the Recent or Postglacial time is frequently called the Human Period, the description of which is rather in the province of prehistoric Archæology than in that of Geology.   Whether Man reached North America in the Pleistocene is still an open question, though there is no reason why he should not have accompanied the Old World mammals in their frequent migrations, and there is some evidence that he did and that a race older than the American Indian occupied this continent.   This evidence, however, is not altogether conclusive and has been subjected to a vigorous destructive criticism, so that many authorities are altogether sceptical.   On the other hand, the undoubted presence of human bones in the Pleistocene of South America, associated with a mammalian fauna which is almost entirely extinct, lends additional strength to the position of those who contend that Man had reached North America before the ice finally disappeared.

APPENDIX

FOR convenience of reference, the system of classification of the animals and plants which has been used in the book is here given in tabular form, omitting those groups which possess no importance as fossils. Groups marked with an asterisk (*) are extinct.

ANIMAL KINGDOM

Sub-Kingdom I. PROTOZOA.
  Class 1. **Rhizopoda.**
      Order 1. *Foraminfera.*
         2. *Radiolaria.*
Sub-Kingdom II. CŒLENTERATA.
  Sub-Branch A. PORIFERA.
    Class 1. **Spongia,** Sponges.
      Sub-Class 3. **Silicispongia,** Siliceous Sponges.
          4. **Calcispongia,** Calcareous Sponges.
  Sub-Branch B. CNIDARIA.
    Class 1. **Anthozoa.**
      Sub-Class 1. ***Tetracorálla,** Palæozoic Corals.
         2. **Hexacoralla,** Modern Corals.
         3. **Octocoralla,** Modern Corals.
    Class 2. **Hydrozoa.**
      Sub-Class 1. **Hydromedusæ.**
        Order 2. *Hydrocorallinæ,* Hydroid Corals.
           3. *Tubulariæ,* Organ Corals.
           4. **Graptolitoidea,* Graptolites.
      Sub-Class 2. **Acalephæ,** Jellyfishes.
Sub-Kingdom III. ECHINODERMATA.
  Sub-Branch A. PELMATOZOA.
    Class 1. **Crinoidea,** Sea Lilies.
      Order 1. **Inadunata.*
         2. **Camerata.*
         3. *Articulata.*
    Class 2. ***Cystoidea.**
      3. ***Blastoidea.**
  Sub-Branch B. ASTEROZOA.
    Class 1. **Ophiuroidea,** Brittle Stars.
      2. **Asteroidea,** Starfishes.
  Sub-Branch C. ECHINOZOA.
    Class 1. **Echinoidea,** Sea-urchins.
      Sub-Class 1. ***Palæechinoidea.**
         2. **Euechinoidea.**
        Order 1. *Regulares,* Regular Sea-urchins.
          2. *Irregulares,* Spantangoids, Sand-dollars.
    Class 2. **Holothuroidea,** Sea Cucumbers.
Sub-Kingdom IV. VERMES, Worms.

Sub-Kingdom V. BRYOZOA, Sea Mosses.
Sub-Kingdom VI. BRACHIOPODA, Lamp Shells.
    Order 1. *Inarticulata.*
       2. *Articulata.*
Sub-Kingdom VII. MOLLUSCA.
    Class 1. **Pelecypoda**, Bivalves.
      2. **Scaphoda**, Tusk-shells.
      3. **Amphineura**.
      4. **Gastropoda**, Conchs, Whelks, Cowries, etc.
      5. **Pteropoda**.
      6. **Cephalopoda**.
    Sub-Class 1. **Tetrabranchiata**.
      Order 1. *Nautiloidea*, Pearly Nautilus.
        2. **Ammonoidea*, Ammonites.
    Sub-Class 2. **Dibranchiata**.
      Order 1. **Belemnoidea*, Belemnites.
        2. *Sepioidea*, Cuttlefishes, Squids.
        3. *Octopoda*, Octopuses.
Sub-Kingdom VIII. ARTHROPODA.
    Class 1. **Crustacea**.
    Sub-Class 1. **Merostomata**.
      Order 1. **Eurypterida*.
        2. *Xiphosura*, Horse-shoe Crabs.
        3. **Synxiphosura*.
    Sub-Class 2. ***Trilobita**.
    Sub-Class 3. **Eucrustacea**, Typical Crustacea.
    Super-Order 1. *Phyllopoda.*
       3. *Ostracoda*.
       4. *Cirripedia*, Barnacles.
       5. *Malacostraca*.
    Order 1. *Phyllocarida*.
      2. *Schizopoda*.
      3. *Stomatopoda*.
      4. *Decapoda*.
    Sub-Order a. *Macrura*, Lobsters, etc.
        b. *Brachyura*, Crabs.
  Class 2. **Arachnida**, Spiders and Scorpions.
    3. **Myriapoda**, Centipedes.
    4. **Insecta**, Insects.
    Order 1. *Orthoptera*, Cockroaches, Grasshoppers, etc.
      2. *Neuroptera*, Caddis-flies, Ant-lions, etc.
      . *Hemiptera*, Cicadas, etc.
      4. *Diptera*, Flies.
      5. *Lepidoptera*, Butterflies and Moths.
      6. *Coleoptera*, Beetles.
      7. *Hymenoptera*, Bees, Wasps, Ants, etc.
Sub-Kingdom IX. VERTEBRATA.
  Class 2. ***Ostracodermata**.
    3. **Pisces**, Fishes.
    Sub-Class 1. **Selachii**, Sharks, Rays, etc.
      2. **Holocephali**, Chimæras, or Spook-fishes.
      3. **Dipnoi**, Lung-fishes.

4. *Arthrodira.
5. Teleostomi.
    Order 1. *Crossopterygii.*
        2. *Actinopteri.*
        Sub-Order 1. *Chondrostei* or *Ganoids*, Sturgeon, Gar-pike, etc
                 2. *Teleocephali* or *Teleosts*, Herring, Salmon, etc.
Class 3. **Amphibia.**
    Order 1. **Stegocephali.*
        3. *Urodela*, Mud-puppies, Salamanders.
        4. *Anura*, Frogs and Toads.
Class 4. **Reptilia.**
    Super-Order *THERIODONTIA.
      Order 1. *Cotylosauria.*
          2. * *Therocephalia.*
          3. *Cynodontia.*
          4. *Anomodontia.*
      Order 5. *Placodontia.*
          6. *Plesiosauria.*
          7. *Testudinata*, Tortoises and Turtles.
    Super-Order DIAPTOSAURIA.
      Order 8. *Procolophonia.*
          9. *Proterosauria.*
         10. *Proganosauria.*
         11. *Gnathodontia.*
         12. *Pelycosauria.*
         13. *Choristodera.*
         14. *Rhynchocephalia*, New Zealand Lizard.
      Order 15. *Parasuchia.*
         16. *Ichthyosauria.*
         17. *Thalattosauria.*
         18. *Crocodilia.*
        Sub-Order a. *Mesosuchia*, Crocodiles, Alligators.
               b. *Thalattosuchia.*
    Super-Order *DINOSAURIA.
      Order 19. *Theropoda.*
         20. *Opisthocœlia.*
         21. *Orthopoda.*
    Super-Order SQUAMATA.
      Order 22. *Lacertilia*, Lizards.
         23. *Mosasauria.*
         24. *Ophidia*, Snakes.
      Order 25. *Pterosauria.*
Class 5. **Aves**, Birds.
    Order 1. *Saururæ* (Archæopteryx).
        2. *Ratitæ*, Wingless Birds, Ostriches, etc.
        3. *Carinatæ*, Flying Birds.
Class 6. **Mammalia.**
    Sub-Class 1. PROTOTHERIA.
      Order 1. *Monotremata*, Spiny Ant-eater, Duck-billed Mole.
        2. *Multituberculata.*
    Sub-Class 2. EUTHERIA.
    Super-Order 1. MARSUPIALIA, Opossum, Kangaroo, etc.

2. PLACENTALIA, Placentals.

Order   1. *Edentata*, Sloths, Armadillos, etc.
          2. *Cetacea*, Whales, etc.
          3. *Sirenia*, Sea-cows and Dugongs.
          4. *Insectivora*, Moles, Hedgehogs, etc.
          5. *Cheiroptera*, Bats.
          6. **Creodonta*, Primitive Flesh-eaters.
          7. *Carnivora*, Dogs, Cats, Bears, Seals, etc.
          8. **Tillodonta*.
          9. **Tæniodonta*.
         10. *Rodentia*, Rats, Porcupines, Squirrels, Hares, etc.
         11. **Condylarthra*.
         12. **Amblypoda*.
         13. **Typotheria*.
         14. **Toxodontia*.
         15. **Homalodotheria*.
         16. **Astrapotheria*.
         17. **Litopterna*.
         18. *Hyracoidea*, Dassies.
         19. *Proboscidea*, Elephants.
         20. *Artiodactyla*, Pigs, Camels, Deer, Antelopes, etc.
         21. *Perissodactyla*, Horses, Tapirs, Rhinoceroses.
         22. **Ancylopoda*.
         23. *Lemuroidea*, Lemurs.
         24. *Primates*, Monkeys, Apes, Man.

VEGETABLE KINGDOM

Sub-Kingdom A. **CRYPTOGAMÆ, Flowerless Plants.**
  I. THALLOPHYTA.
    Class 1. **Algæ**, Seaweeds, etc.
         2. **Fungi**, Mushrooms, etc.
  II. BRYOPHYTA, Mosses.
  III. PTERIDOPHYTA.
    Class 1. **Filicales**, Ferns.
         2. **Equisetales**, Horsetails.
         3. **Lycopodiales**, Club Mosses.
         4. **Sphenophyllales.**
         5. **Cycadofilices.**
Sub-Kingdom B. **PHANEROGAMÆ, Flowering Plants.**
  IV. GYMNOSPERMÆ.
    Order 1. *Cycadales*.
      Sub-Order 1. *Cycadaceæ*.
           2. *Zamieæ*.
           3. **Bennettiteæ*
    Order 2. **Cordaiteæ*.
         3. *Gingkoaceæ*, Maiden-hair Tree.
         4. *Coniferæ*, Pines, Spruces, etc.
  V. ANGIOSPERMÆ.
    Class 1. **Monocotyledones**, Palms, Grasses, Lilies, etc.
         2. **Dicotyledones**, Oaks, Roses, Crucifers, etc.

INDEX

Page numbers marked with an asterisk (*) indicate figures. Where several references are given under one heading, the most important is in heavy-face type. Names of the *genera* of animals and plants are in italics; species are not listed.

Abysmal deposits, 245, **269**
Abyssal rocks, 284
Acadian epoch, 549; province, 613, 620; Range, 504
Acanthodes, 635
Accidents to rivers, 491
Accordance of mountain summits, 513; of river valleys, 140, 443
Accumulations, organic, 309, residual, 186
Acervularia, *602
Acidaspis, 573, *576, 603
Acid rocks, 292, 293, 294
Actinocrinus, 602, 631
Actinolite, 16
Actinopteri, 608, 635
Adirondack Mts, 534; faults in, 465
Adjutant birds, 756
Ægoceras, 689
Æolian rocks, 191, **316**, 317
Æsiocrinus, 631
Ætna, 50, 141, 763
Aetosaurus, 674
Africa, 117, 187, 472; Archæan, 538; Carboniferous, 623; continental formations, 567; Cretaceous, 713; Devonian, 599; Eocene, 734; ice-ages, 543; Jurassic, 683; Miocene, 754; Oligocene, 744; Ordovician, 567; Permian, 643; Pleistocene, 772; Rift Valley, 469; Silurian, 584; Triassic, 666; volcanoes, 54
Aftershocks, 44
Aftonian stage, 775, **777**, 784

Agassiz, A, 260, 270, 271
Agassiz, L, 769
Age, geological, 530; topographical, 439
Agencies, diastrophic, 436; dynamical, **25**, 281; igneous, 26, **28**; subterranean, 26, **28**; surface, 26, **97**
Agglomerate, volcanic, 81, 286, **300**, 389, 459
Aggradation of land, 436
Agnostus, *557, 558
Aktian deposits, 245, **266**
Alabaster, 20
Alaria, 688
Alashuk River, 141
Alaska, earthquakes, 41, 46; fjords, 496; glaciers, *154, *157
Albany stage, 640
Albian series, 702
Albite, 13, 14
Aleutian Ids, 68; earthquakes, 41; frost action, 115
Algæ, 192, 194, 309, 314, 570; calcareous, 670
Algonkian, 531, **540**
"Alkali" lakes, *226
Alkalies, 294, 297, 298; precipitates of, 307
Alkaline, carbonates, 192; earths, precipitates of, 307; sulphides, 194; waters, 194
Allegheny stage, 610
Allodon, 698
Allosaurus, *695, 696
Allotriomorphic grains, 286
Alluvial cones, **202**, *203, 205, *479; fans, 202, *203
Alluvium, river, 279
Almandine, 17

Aloes, 736
Alpine ranges, denudation of, 510
Alps, 506, 510, 514; compression, 506; elevation, 726, **753**; glaciers, 156, 165; Pleistocene glaciation, 772, 775, **784**
Altai Mts, 538
Alteration of minerals, **18**, 290, 406, 431
Alticamelus, 757
Alumina, 294, 407
Aluminium, 6, 427
Aluminous silicates, 104, 186
Amazon, 205, 269; delta, 213; material carried by, 147
Amblypoda, 729, *740, 741
Ambonychia, *574
Ammonium carbonate, 266; chloride, 82
Ammonoidea, **604**, 634, 651, 656, 671, 689, 717, 723, 729, 736
Amorphous substances, 8
Amphibia, 548, 608, **636**, 652, 656, 673, 692, 723
Amphibole-trachyte, 297
Amphiboles, 15, **16**, 290, 297, 409
Analcites, *672, 673
Anaptomorphus, 739
Anchisaurus, 675
Anchura, *715
Ancyloceras, 717
Ancylopoda, **757**, 759
Andes, 203, 298, 712; elevation, 712, 764
Andesine, 13, 14
Andesite, 293, **298**, 299, 393; -breccia, 301; metamorphism of, 420; -ob-

793

3 F

Lightning Source UK Ltd.
Milton Keynes UK
UKHW021833230119
336088UK00012B/272/P

9 781332 586455